Latin American Mathematics Series

Latin American Mathematics Series – UFSCar subseries

Published under the Latin American Mathematics Series, which was created to showcase the new, vibrant mathematical output that is emerging from this region, this subseries aims to gather high-quality monographs, graduate textbooks, and contributing volumes based on mathematical research conducted at/with the Federal University of São Carlos, a technological pole located in the State of São Paulo, Brazil. Submissions are evaluated by an international editorial board and undergo a rigorous peer review before acceptance.

Shiferaw Berhanu • Nordine Mir •
Gustavo Hoepfner
Editors

Geometric Analysis of PDEs and Several Complex Variables

In Honor of Jorge Hounie's 75th Birthday, Serra Negra, Brazil, July 31–August 4, 2023

Springer

Editors
Shiferaw Berhanu
Department of Mathematics
University of Maryland
College Park, MD, USA

Nordine Mir
Department of Arts and Sciences
Texas A&M University at Qatar
Doha, Qatar

Gustavo Hoepfner
Department of Mathematics
Federal University of São Carlos
São Carlos, Brazil

ISSN 2524-6755 ISSN 2524-6763 (electronic)
Latin American Mathematics Series
ISSN 2524-6755 ISSN 2524-6763 (electronic)
Latin American Mathematics Series – UFSCar subseries
ISBN 978-3-031-69701-2 ISBN 978-3-031-69702-9 (eBook)
https://doi.org/10.1007/978-3-031-69702-9

Mathematics Subject Classification: 35H10, 35A27, 32A07, 35Q55, 35P20

This Springer imprint is published by the registered company Springer Nature Switzerland AG
The registered company address is: Gewerbestrasse 11, 6330 Cham, Switzerland

Preface

This issue in Springer is dedicated to Jorge Hounie on the occasion of his 75th birthday and in recognition of his outstanding achievements in several areas of Analysis. Jorge has made many deep contributions to Partial Differential Equations, Complex Variables, Harmonic Analysis, and Involutive Structures. He has also been a very kind and great mentor of numerous graduate students and postdocs who went on to pursue successful academic careers. Jorge Hounie has been a member of the Brazilian Academy of Sciences since 1999.

All the papers in the issue were solicited by the editors and all the submissions were reviewed by experts. The contents of these papers reflect the breadth of Jorge's research interests. They range over areas that include Several Complex Variables, CR Geometry, and Partial Differential Equations. We take this opportunity to thank the authors, the anonymous referees, and Springer for making this issue possible. On behalf of the contributors of this issue, we express our best wishes to Jorge for many more years of good health, happiness, and continued success.

College Park, MD, USA — Shiferaw Berhanu
Doha, Qatar — Nordine Mir
São Carlos, Brazil — Gustavo Hoepfner

Contents

On Some of the Works of Jorge Hounie

Shiferaw Berhanu

1 Introduction

Jorge Hounie has made many remarkable and lasting contributions to partial differential equations and complex analysis. He co-authored the book [BCH] which has been useful in introducing the subject of involutive structures to junior and senior researchers. This chapter summarizes some of his achievements in joint works with S. Berhanu and J. Tavares. Many of these works generalize well-known properties of holomorphic functions to solutions of a single or systems of complex vector fields. The collaboration with Berhanu led to twenty publications. Some of the results were further extended in different directions by his former doctoral students and their collaborators.

We begin with his results with J. Tavares:

2 The Hartogs Property

A celebrated theorem of Hartogs states that if $\Omega \subseteq \mathbb{C}^n$ is a domain ($n \geq 2$) and $K \subseteq \Omega$ is compact such that $\Omega \setminus K$ is connected, then any holomorphic function f on $\Omega \setminus K$ extends as a holomorphic function to Ω.

This result was extended by Hounie and Tavares to the solutions of tube structures.

S. Berhanu (✉)
Department of Mathematics, University of Maryland, College Park, MD, USA
e-mail: sberhanu@umd.edu

S. Berhanu et al. (eds.), *Geometric Analysis of PDEs and Several Complex Variables*, Latin American Mathematics Series – UFSCar subseries,
https://doi.org/10.1007/978-3-031-69702-9_1

Let m and n be positive integers. We will denote by $x = (x_1, \ldots, x_m)$ and $t = (t_1, \ldots, t_n)$ variable points in $\mathbb{R}^m$ and $\mathbb{R}^n$, respectively. Let

$$\phi(t) = (\phi_1(t), \ldots, \phi_m(t))$$

be a smooth mapping, $\phi : V \to \mathbb{R}^m$, where $V \subseteq \mathbb{R}^n$ is an open set.

Let

$$Z_i(x, t) = x_i + \sqrt{-1}\phi_i(t), \ 1 \le i \le m,$$

and consider the associated n complex vector fields on $\mathbb{R}^m \times \mathbb{R}^n$ given by

$$L_j = \frac{\partial}{\partial t_j} - \sqrt{-1}\sum_{k=1}^{m} \frac{\partial \phi_k}{\partial t_j}(t)\frac{\partial}{\partial x_k}, \ 1 \le j \le n.$$

The vector fields L_j define a system of linearly independent complex vector fields and are said to define a tube structure.

Clearly,

$$L_j Z_i = 0, \ 1 \le i \le m, 1 \le j \le n.$$

Ehrenpreis showed that the classical theorem of Hartogs follows from the following fact: If f is a compactly supported smooth $1-$form in $\mathbb{C}^n, n \ge 2$ such that $\overline{\partial} f = 0$, there exists $u \in C^\infty(\mathbb{C}^n)$ of compact support such that $\overline{\partial} u = f$. In the work [HT1], Hounie and Tavares characterized the tube structures for which the analogous solvability property holds. They showed that this solvability property is equivalent to the property that for any nonzero vector $\xi = (\xi_1, \ldots, \xi_m)$, the function $\xi_1\phi_1(t) + \cdots \xi_m\phi_m(t)$ does not attain a local minimum over a compact set. They then showed that these equivalent properties imply the Hartogs property for the tube structure defined by the L_j.

As shown in their paper, their result has an easy application to the Lewy hyperquadrics $\{(z, s+\sqrt{-1}(|z_1|^2+\cdots |z_k|^2-|z_{k+1}|^2-\cdots |z_n|^2)) : z \in \mathbb{C}^n, s \in \mathbb{R}\}$: The Hartogs property is valid iff $1 < k < n$.

3 Removable Singularities for Differential Operators

The classical Riemann's Removable Singularity Theorem states that if f is a bounded and holomorphic function on a punctured disc $D \setminus \{p\}$ in $\mathbb{C}$, then it extends as a holomorphic function to D. Ahlfors generalized this theorem as follows: Let $K \subseteq \mathbb{C}$ be compact. The analytic capacity of K, $\gamma(K)$, is defined by $\gamma(K) = \sup\{|f'(\infty)| : f \in H^\infty(\mathbb{C} \setminus K), |f| \leq 1\}$. Ahlfors proved that the set K is removable for bounded analytic functions if and only if $\gamma(K) = 0$.

In [HP] Harvey and Polking introduced a notion of capacity for a class of constant coefficient differential operators that includes elliptic and parabolic operators and characterized removable sets for solutions: For each compact set $K \subseteq \Omega$, the capacity with respect to a differential operator $P(x, D)$ is $\sup\{|\langle P(x, D)f, 1\rangle| : |f| \leq 1, P(x, D)f = 0 \text{ on } \Omega \setminus K\}$. Their notion is not localizable. In the work [HT2] Hounie and Tavares introduced a localized notion of capacity for differential operators and characterized removable sets for solutions of locally solvable differential operators. In addition, for differential operators of order one, they gave a geometric characterization of removable sets in terms of orbits in the sense of Sussmann.

4 Rado's Property

A classical theorem of Rado states that a continuous function u defined on an open set $D \subseteq \mathbb{C}$ that is holomorphic on $D \setminus u^{-1}(0)$ is holomorphic on D. The result also holds in $\mathbb{C}^n$. Generalizations to solutions of complex vector fields that are not elliptic have been explored in [RS], [HT3], [H], and [T].

In [RS], the authors proved the analogue of Rado's theorem for continuous CR functions on strictly pseudoconvex hypersurfaces in $\mathbb{C}^n$. In [HT3], Hounie and Tavares showed that if L is a smooth complex vector field on an open set $D \subseteq \mathbb{R}^n$ that satisfies the local solvability condition of Nirenberg and Treves, u is continuous on D and $Lu = 0$ on $D \setminus u^{-1}(0)$, then $Lu = 0$ on D. A further generalization was proved by Hounie [H] for the solutions of locally integrable structures of co-rank one that are locally solvable in degree one. These are an overdetermined system of equations

$$L_1 u = 0, \ldots, L_n u = 0,$$

where $L_1, \ldots, L_n$ are pairwise commuting smooth complex vector fields defined on an open subset of $\mathbb{R}^{n+1}$, and it is assumed that this system of vector fields has local first integrals at every point and it is solvable in the sense that the equations

$$L_1 u = f_1, \ldots, L_n u = f_n$$

can be solved locally for all smooth right-hand sides that satisfy the compatibility conditions $L_j f_k = L_k f_j$, $1 \leq j, k \leq n$.

5 Generalized FBI Transforms

The classical FBI transform is a nonlinear transform which has the form

$$\mathcal{F}u(x,\xi) = \int_{\mathbb{R}^m} e^{i.\xi.(x-y)-|\xi||x-y|^2} u(y)\, dy,\ x,\xi \in \mathbb{R}^m, \tag{1}$$

where u is a continuous function of compact support in $\mathbb{R}^m$ or a distribution of compact support in which case the integral is understood in a duality sense. This transform characterizes microlocal analyticity [S], microlocal smoothness [Tr], and microlocal Gevrey regularity. It has been used extensively to study the local and microlocal regularity of solutions of linear and nonlinear partial differential equations.

In [S] (see also [D]) a more general FBI transform was considered where the phase function behaved much like the quadratic phase $i\xi.(x-y) - |\xi||x-y|^2$ in that the real part of the Hessian was required to be negative definite.

The work [BH7] introduced a class of FBI transforms where the real part of the Hessian of the phase function was allowed to degenerate. Examples of such transforms include, for each $k = 2, 3 \ldots$,

$$\mathcal{F}_k u(x,\xi) = \int_{\mathbb{R}^m} e^{i\xi.(x-y)-|\xi||x-y|^{2k}} u(y)dy,\ x,\xi \in \mathbb{R}^m. \tag{2}$$

Note that when $k > 1$, these transforms have a Hessian that degenerates at the origin. The more general FBI transforms of [BH7] were shown to characterize microlocal analyticity and microlocal smoothness. They also characterize microlocal regularity of Gevrey functions (see [BHa]). In the work [B], these transforms were used in an essential way to prove the following microlocal hypoellipticity result:

Consider the system of smooth tube vector fields

$$L_j = \frac{\partial}{\partial t_j} - \sqrt{-1}\sum_{k=1}^{m} \frac{\partial \phi_k}{\partial t_j}(t)\frac{\partial}{\partial x_k},\ 1 \le j \le n,$$

which have the complete set of first integrals given by $Z_j(x,t) = x_j + \sqrt{-1}\phi_j(t),\ 1 \le j \le m, \phi_j(0) = 0$.

Theorem 4.1 *Let $\xi^0 \in \mathbb{R}^m \setminus 0$. Assume that there is a sequence $t_j^* \neq 0$ converging to 0 such that for some fixed positive integer N,*

$$|\phi(t_j^*)|^{2N} < -\phi(t_j^*) \cdot \xi^0. \tag{3}$$

Then if h is any Lipschitz continuous solution, $(0, \xi^0)$ is not in the analytic wavefront set of $h_0(x) = h(x, 0)$.

6 The F. and M. Riesz Theorem

Consider a Borel measure μ defined on the boundary T of the unit disc D of the complex plane. A celebrated theorem proved in 1916 by F. and M. Riesz states that if the Fourier coefficients of μ vanish for all negative integral values, i.e.,

$$\hat{\mu}(k) = \int_0^{2\pi} \exp(-2\pi i k\theta)\, d\mu(\theta) = 0, \quad k = -1, -2, \ldots,$$

then μ is absolutely continuous with respect to the Lebesgue measure $d\theta$. The hypothesis is equivalent to the existence of a holomorphic function $f(z)$ defined on D whose weak boundary value is μ.

The F. and M. Riesz theorem has undergone an extensive generalization mainly in two different directions: (i) generalized analytic function algebras, which have as a starting point the fact that the vanishing condition on the Fourier coefficients means that μ is orthogonal to the algebra of continuous functions on T that extend holomorphically to D; (ii) ordered groups, which emphasize instead the role of the group structure of T in the classical result. Thus, although absolute continuity with respect to Lebesgue measure is a local property (i.e., if each point has a neighborhood where it holds, then it holds everywhere), both directions focus on global objects. It turns out that the theorem can be localized.

A local version of the theorem can be stated as: If h is holomorphic on $Q = (-a, a) \times (0, b)$ with a weak boundary value bh which is a measure on $(-a, a)$, then $bh \in L^1_{\text{loc}}(-a, a)$.

In the works [BH1] and [BH2], Berhanu and Hounie generalized this classical result to classes of complex vector fields. In [BH2], this theorem was generalized to smooth, locally integrable vector fields in the plane. In [BH1] a generalization was proved for CR functions on embedded CR manifolds M where the trace is taken on a strongly noncharacteristic submanifold such as a noncharacteristic hypersurface, or a maximally real submanifold $N \subseteq M$.

7 Bochner's Extension Theorem for Rough Tubes

A classical theorem of Bochner states that a holomorphic function defined on a tube $M = \Omega + i\mathbb{R}^n$ in $\mathbb{C}^n$ for some domain $\Omega \subseteq \mathbb{R}^n$ extends as a holomorphic function to the convex hull $\text{ch}(M) = \text{ch}(\Omega) + i\mathbb{R}^n$. A CR version of this extension theorem was proved by Kazlow in [K] and Boivin and Dwilewicz in [BD] (see also [BO1] and [Ko]). In these papers, the tube above is replaced by a tube CR manifold $M = X + i\mathbb{R}^n$ where X is a connected submanifold of $\mathbb{R}^n$ of class at least $\mathbb{C}^2$. It was then proved that for any h a CR function, there exists a function H holomorphic on the interior of the hull of M and which extends h. The study of extension of CR functions on tube manifolds with tempered growth in the tube direction was also considered in [BO2].

In [BH5] we investigated the validity of such extension phenomenon for tubes M when X is a more general set without assuming any growth restrictions, in particular, X and hence M may not be a manifold. Berhanu and Hounie [BH5] considered sets $X \subseteq \mathbb{R}^n$ that are arcwise connected by a rectifiable curve (X not necessarily a manifold), and defined a continuous function f to be a generalized CR function on $X + i\mathbb{R}^n$ if locally, it is the uniform limit of CR functions. By the Baouendi–Treves approximation theorem, this latter property holds for continuous CR functions when X is a submanifold. We showed that a generalized CR function f can be extended to F on the tube over $X \cup \text{int ch}(X)$ that is a continuous generalized CR function on the tube over int ch(X). Here ch denotes the convex hull. Thus, when int ch(X) is open in $\mathbb{R}^n$, $F(z)$ will be holomorphic on the tube over int ch(X). The tools used in [BD] when X is a manifold include Tumanov's theorem (on holomorphic extendability to wedges of CR functions at minimal points), and the edge-of-the-wedge theorem of Airapetjan and Henkin. These methods require that X be a manifold and that it be of regularity at least of class $\mathbb{C}^2$. For example, it is known (see [R] and [S]) that even the classical edge-of-the-wedge theorem may fail when the edge is only Lipschitz.

8 The Rudin–Carleson Theorem

A classical result known as the Rudin–Carleson theorem can be stated as follows: Let $E \subseteq \mathbb{T} = \{z \in \mathbb{C} : |z| = 1\}$ be a closed set of Lebesgue measure 0. If $g \in C(\mathbb{T})$, $\exists h \in Hol(D) \cap C(\overline{D}) : h|_E = g|_E$. ($D = \{z \in \mathbb{C} : |z| < 1\}$.) This theorem has applications to the study of interpolation submanifolds in several complex variables. In [BH3] and [BH4], this theorem was generalized to planar complex vector fields as follows:

Suppose L is a smooth complex vector field in the plane:

(i) If the Rudin–Carleson result holds for L, then L is locally solvable.
(ii) Suppose L is a real analytic, locally solvable vector field on a neighborhood U of $\overline{D}$.

Assume that L does not have a relatively compact Sussmann orbit in U. Let $E \subseteq \partial D$, $|E| = 0$, E closed. Suppose $g \in C(\partial D)$ that is constant on $\gamma \cap \partial D$ whenever γ is a one-dimensional orbit of L. Then

$$\exists h \in C(\overline{D}) : Lh = 0 \text{ in } D \quad \text{and} \quad h|_E = g|_E.$$

9 Generalized Analytic Functions

The work [BH6] investigated several boundary properties of solutions of the equation

$$Lu = a(z)\partial_{\bar{z}}u + b(z)\partial_z u + A(z)u + B(z)\bar{u} = 0, \tag{4}$$

where $a(z)$ and $b(z)$ are Hölder continuous and A and B are L^p functions for some $p > 2$ and $a(z)\partial_{\bar{z}} + b(z)\partial_z$ is an elliptic vector field. Necessary and sufficient conditions for the existence of a weak boundary value for solutions were established. Other properties that were established include the F. and M. Riez theorem and the Rudin–Carleson property.

Special cases of this equation arise in several geometrical and physical problems. In geometry, it is intimately connected with the infinitesimal bending of surfaces in $\mathbb{R}^3$ of positive Gaussian curvature K. The equation also arises in the study of the elasticity of thin shells. An important example is the classical and widely studied Beltrami vector field $\partial_{\bar{z}} - q(z)\partial_z$, $|q(z)| \le c < 1$, that is important in the theory of quasiconformal mappings. Let $A(x, y)$ and $B(x, y)$ be complex-valued functions defined on a domain $\Omega \subseteq \mathbb{C}$ of the complex plane. A function u that satisfies, in the sense of distributions, Vekua's equation

$$\overline{\partial} u + Au + B\overline{u} = 0$$

where $A, B \in L^p(\Omega)$ for some $p > 2$ is said to be a generalized analytic function on Ω. In particular, if A and B vanish identically, u is a holomorphic (or analytic) function. It is well known that holomorphic functions and generalized analytic functions share many local qualitative properties. Indeed, many of these properties follow from the Similarity Principle which states that for any solution u, there exist a holomorphic function h and a Holder continuous function g such that

$$u = e^g h.$$

In particular, this principle implies that if u is not constant, its zeros are isolated and of finite order. The Rado property is another consequence of the Similarity Principle. On the other hand, there are important boundary properties of holomorphic functions and solutions of linear and nonlinear equations that are not shared by the boundary values of solutions of Vekua's equation even when A and B are constants. For example, it is well known that if h is holomorphic on a smoothly bounded domain D with a weak boundary value bh and bD is smooth, then bh is microlocally smooth (real analytic if bD is real analytic) in a direction. However, [BH6] showed that there is a solution of $\overline{\partial} u - \overline{u} = 0$ with the property that the C^∞ wavefront set of the boundary values contains both directions. The delimitation of the wavefront set of the boundary values for holomorphic functions and more generally solutions of locally integrable vector fields had been used to establish the localized version of the F. and M. Riesz property for the homogeneous solutions of some classes of (not necessarily elliptic) vector fields. Berhanu and Hounie [BH6] employed a different method to establish the F. and M. Riesz property for the solutions of

$$Lu = a(z)\partial_{\bar{z}} u + b(z)\partial_z u + A(z)u + B(z)\bar{u} = 0. \tag{5}$$

10 The Similarity Principle

In [BHS] a similarity principle was established for solutions of equations of the form

$$Lu = Au + B\overline{u}$$

for a large class of locally solvable planar vector fields in the plane with smooth coefficients. The result was applied to establish uniqueness in the Cauchy problem for semilinear equations $Lw = f(t, x, w)$, where f is assumed to have bounded first derivatives and $w \in L^p_{\text{loc}}$, $p > 1$. This work was generalized to the class of all locally solvable vector fields under a minimal smoothness assumption on the coefficients of L, and the functions A and B in [HS] where an application to uniqueness in the Cauchy problem was also derived.

11 The H^p Property

Suppose $h(z)$ is a holomorphic function on the rectangle

$$Q = (-a, a) \times (0, b)$$

with a weak boundary value bh at $y = 0$. If $bh \in L^p(-a, a)$ for some $1 \leq p \leq \infty$, then it is well known that:

(1) For any $0 < c < a$, the traces $h(\cdot, y)$ are uniformly bounded in L^p:

$$\int_{-c}^{c} |h(x + iy)|^p \, dx \leq C, \qquad y \downarrow 0.$$

(2) $h(x + iy) \equiv 0$ if bh vanishes on a set of positive measures.

The work [BH8] generalized the preceding results to the class of locally solvable vector fields in the plane. This latter work was further extended by G. Hoepfner and Hounie in [HH] where they characterized the boundary values of homogeneous solutions of planar one-sided locally solvable vector fields (in the sense of [BH12]) with analytic coefficients with the property that the L^p norm of their traces is locally uniformly bounded, for the range $0 < p \leq 1$. For $p \neq \frac{1}{n}$, $n = 1, 2, \ldots$ the boundary value must locally belong to the local Hardy space $h^p(\mathbb{R})$ of Goldberg, while for $p = \frac{1}{n}$, $n = 1, 2, \ldots$ they needed to introduce a new class of atomic Hardy spaces when the vector field is of infinite type at some boundary point.

12 Results on Unique Continuation

In [HKMP] and [BR] the authors proved the following boundary unique continuation result:

Theorem 11.1 *Let u be harmonic in $B_r^+ = \{x \in \mathbb{R}^n : |x| < r,\ x_n > 0\}$, continuous on $\overline{B_r^+}$. Assume that:*

(1) $u(x', 0) \geq 0$ for $|x'| \leq r$.
(2) The function u is flat at 0, that is, $u(x) = O(|x|^N)$ for all N.

Then $u \equiv 0$.

In [BH9], this theorem was generalized to the solutions of the Helmholtz equation $\Delta u + cu = 0$ in B_r^+, continuous on $\overline{B_r^+}$. The preceding theorem was applied in [BH10] to derive unique continuation results for CR mappings between classes of embedded CR manifolds. The work [BH11] contains other results on unique continuation for solutions of systems of complex vector fields that vanish on a set of positive measures.

References

[BR] M.S. Baouendi and L. Rothschild, *A local Hopf lemma and unique continuation for harmonic functions*, International Mathematics Research Notices **8** (1993), 245–251.

[B] S. Berhanu, *A generalization of a microlocal version of Bochner's theorem*, Transactions of the AMS, **374** (2021), 5269–5285.

[BCH] S. Berhanu, P.D. Cordaro, and J. Hounie, *An Introduction to Involutive Structures*, Cambridge University Press (2008).

[BHa] S. Berhanu and Abraham Hailu, *Characterization of Gevrey regularity by a class of FBI transforms*, Novel methods in Harmonic Analysis, **2**(2017) Springer -Verlag (Birkhauser) 451–482.

[BH1] S. Berhanu and J. Hounie, *An F. and M. Riesz theorem for a system of vector fields*, Inventiones Mathematicae, **162**(2005), 357–380

[BH2] S. Berhanu and J. Hounie, *An F. and M. Riesz theorem for planar vector fields*, Mathemtaische Annalen, **320**(2001), 463–485

[BH3] S. Berhanu and J. Hounie, *A Rudin-Carleson theorem for planar vector fields*, Mathemtaische Annalen, **347**(2009), 95–110

[BH4] S. Berhanu and J. Hounie, *A generalization of the Rudin-Carleson theorem*, Advances in phase space analysis of partial differential equations, Birkhauser **78**(2009), 37–57.

[BH5] S. Berhanu and J. Hounie, *A Generalization of Bochner's Extension Theorem to Rough Tubes*, Journal of Geometric Analysis, **21** (2011), 455–475

[BH6] S. Berhanu and J. Hounie, *Boundary behavior of generalized analytic functions*, Journal of Functional Analysis, **266** (2014), 4121–4149

[BH7] S. Berhanu and J. Hounie, *A class of FBI transforms*, Comm.Part. Diff. Equations, **37** (2012), 38–57

[BH8] S. Berhanu and J. Hounie, *On boundary properties of solutions of complex vector fields*, Journal of Functional Analysis, **192** (2002), 446–490.

[BH9] S. Berhanu and J. Hounie,*A local Hopf lemma and unique continuation for the Helmholtz equation*, Communications in PDEs, **43** (2018), 448–466.

[BH10] S. Berhanu and J. Hounie, *A Hopf lemma for holomorphic functions in Hardy spaces and applications to CR mappings*, Journal d'Analyse Mathematique, **138** (2019), 835–855.

[BH11] S. Berhanu and J. Hounie, *Uniqueness for locally integrable solutions of overdetermined systems*, Duke Math. Journal, **105** (2000), 387–410.

[BH12] S. Berhanu and J. Hounie, *On boundary regularity for one-sided locally solvable vector fields*, Indiana University Mathematics Journal, **52** (2003), 1447–1477.

[BHS] S. Berhanu, J. Hounie, P. Santiago, *A similarity principle for complex vector fields and applications*, Transactions of the AMS,**353** (2000),1661–1675.

[BO1] A. Boggess, *The holomorphic extension of C^k CR Functions on Tube Manifolds*, Annales Polonici Mathmatici **LXX** (1998), 35–42.

[BO2] A. Boggess, *The holomorphic extension of H^p -CR functions on tube submanifolds*, Proc. Amer. Math. Soc. **127** (1999), 1427–1435. [BD]

[BD] A. Boivin and R. Dwilewicz, *Extension and Approximation of CR Functions on Tube Manifolds*, Trans. Amer. Math. Soc. **350** (1988), 1945–1956.

[D] Jean-Marc Delort, *F.B.I. Transformation, Second Microlocalization and Semilinear Caustics*, Lecture Notes in Mathematics **1522**, Springer-Verlag (1992).

[HP] Harvey, R., Polking, J.*A notion of Capacity which characterizes removable singularities*, Trans. Amer. Math. Soc.,**169** (1972),183–195.

[HH] G Hoepfner, J Hounie, *Locally solvable vector fields and Hardy spaces*, Journal of Functional Analysis **247** (2) (2007), 378–416

[H] J. Hounie, *A Rado Theorem for Locally Solvable Structures of Co-rank One*, Trends in Mathematics series, Birkhauser.

[HS] J. Hounie, E.R. da Silva, *A similarity principle for locally solvable vector fields*, J. Math. Pures Appl. **81** (2002), 715–746.

[HT1] Hounie, J. and Tavares J., *The Hartogs property for Tube structures*, Indagationes Mathematicae, **1** (1990), 51–61.

[HT2] Jorge Hounie and Joaquim Tavares, *On removable singularities of locally solvable differential operators*, Inventiones Math.,**126** (1996), 589–623.

[HT3] Jorge Hounie and Joaquim Tavares, *Rado's theorem for locally solvable vector fields*, Proceedings of the AMS, **119**, No. 3 (1993), 829–836.

[HKMP] X.J. Huang, S.G. Krantz, D. Ma, Y. Pan, *A Hopf lemma for holomorphic functions and applications*, Complex Variables Theory Appl. **26** (1995), 273–276.

[K] M. Kazlow, *CR functions and tube manifolds*, Trans. Amer. Math. Soc. **255** (1979), 153–171.

[Ko] H. Komatsu, *A local version of Bochner's tube theorem*, J. Fac. Sci. Univ. Tokyo Sect. 1A **19** (1972), 201–204

[R] J. P. Rosay, *A propos de wedges et d'edges et de prolongements holomorphes*, Trans. Amer. Math. Soc. **297** (1986), 63–72.

[RS] Jean Pierre Rosay and Edgar Lee Stout, *Rado's theorem for CR functions*, Proceedings of the AMS, **106**, Number 4 (1989),1017–1026.

[S] J. Sjöstrand, *Singularités analytiques microlocales, astérisque* **95**, Société Mathématique De France (1982), 1–166.

[T] Joaquim Tavares, *On the Rado's theorem for locally integrable structures*, Journal of Mathematical Analysis and Applications, **191** (1995), 490–496.

[Tr] F. Treves, *Hypo-analytic structures*, Princeton University Press (1992).

On the Research of J. Hounie: A Personal Statement

Paulo D. Cordaro

To the lovely memory of Emilie

It is an honor for me to contribute to this volume in homage to my dear friend Jorge Hounie. This (quite short) note will have a highly personal character. It does not pretend to be a survey of the entire Jorge's work, not even close. I just want to relate a few instances in our long friendship and point out some of his outstanding contributions to our research field. I first met Jorge at Rutgers, in the fall of 1980. At that time I was starting my PhD program and Jorge was beginning a term as a visiting professor. During the academic year our families had a very cordial relationship, but Jorge and I did not have any mathematical interaction. It was during his stay at Rutgers that Jorge wrote a beautiful work on the global hypoellipticity for vector fields on a surface [4]. This work has had an important impact in that area of analysis; indeed it was the starting point for much of the research on this question.

One of the aspects that I admire most in Jorge's research work is his deep understanding of Harmonic Analysis, a consequence of his awareness of the work carried out in Buenos Aires by the research group gathered around Alberto P. Calderón. He worked on the problem of continuity of pseudodifferential operators either alone (see, e.g., [5]) or in collaboration with his Argentinian colleague Josefina Álvarez (see, e.g., [1]). Meanwhile, he introduced many of his PhD students to this field (e.g., Paulo Antonio Liboni Filho, Rafael Kapp, Evandro Raimundo da Silva, Gustavo Hoepfner, Tiago Picon); they are now members of our research group in Brazil, with a solid background in Harmonic Analysis. In this connection I must mention one of my favorite Jorge's works, the solvability under the Nirenberg–

Partially supported by grants from CNPq and FAPESP (2018/14316-3), Brazil.

P. D. Cordaro (✉)
University of São Paulo, São Paulo, Brazil
e-mail: cordaro@ime.usp.br

S. Berhanu et al. (eds.), *Geometric Analysis of PDEs and Several Complex Variables*, Latin American Mathematics Series – UFSCar subseries,
https://doi.org/10.1007/978-3-031-69702-9_2

Treves condition (P) of first-order linear partial differential operators (PDOs) with coefficients of low regularity [6]. This work highlights, again, his mastery of Harmonic Analysis techniques. During the Brazilian summer of 1990, I met him again at Rutgers, and I recall how he was happy for this accomplishment.

I returned to Brazil in 1985, and almost immediately Jorge and I started a collaborative work, mainly focusing on the question of solvability for systems of vector fields associated to locally integrable structures. We based all our work on the seminal work by Treves [8], in which he introduces the problem of local exactness for the differential complex associated to a corank one locally integrable structure. In [8] Treves presents the proof in degree one and when the structure is real analytic. Simultaneously, Treves also stated a conjecture which would be equivalent to local exactness in all degrees of the complex and in the general smooth case. Much work has been done on this conjecture in the following 18 years, with contributions by Gerardo Mendoza and later Sagun Chanillo, both in collaboration with François Treves. Finally, Jorge and I completed the proof of the Treves conjecture in the general $\mathcal{C}^\infty$ case (cf. [2]). It is significant that one of the main steps in the proof is the use of a variation of an integral formula due to Bochner, which Jorge had the idea to bring into the picture and which he recalled from his previous work on Harmonic Analysis.

I want to stress that, during that time, Jorge also developed a highly productive collaboration with Shif Berhanu, starting ca. 2000 and extended over the years. A description of this collaborative research can be appreciated in the paper by Shif in this same volume.

In 2011, during our usual Workshop in Serra Negra, Nicholas Lerner kindly explained to us a question regarding Hörmander's theorem on the semiglobal solvability for pseudodifferential operators satisfying Condition (P). For instance, for linear PDO, it would be expected, as in the local case, to obtain solutions with a loss of $m - 1$ derivatives (here m is the order of the operator), but Hörmander's theorem only gives, for any $\varepsilon > 0$, a solution with loss of $m - 1 + \varepsilon$ derivatives. In [3], Jorge and I prove that semiglobal solvability with $\varepsilon = 0$ occurs for linear PDO in two dimensions.

More recently, Jorge has supervised the post-doc programs of two outstanding young researchers, Camilo Campana and Giuliano Zugliani. Both were PhD students of Adalberto P. Bergamasco and now hold permanent positions in two major Brazilian universities: Camilo at the Federal University of Santa Catarina and Giuliano at the Campinas University in São Paulo state. I call attention to a very beautiful result obtained with Giuliano in [7] on global solvability for a nonvanishing complex vector field on a compact manifold, satisfying Condition (P).

The collaboration with Jorge was very important to my own career. I learnt a lot from him. Our meetings were always very cordial, with lots of jokes and not a hint whatsoever of competitiveness. I really enjoyed doing mathematics with him, and these remembrances will stay with me forever.

References

1. J. Álvarez and J. Hounie. Estimates for the kernel and continuity properties of pseudo-differential operators. *Arkiv för Matematik*, 28, 1–22, 1990.
2. P. D. Cordaro and J. Hounie. Local solvability for a class of differential complexes. *Acta Mathematica*, 187(2), 191–212, 2001.
3. P. D. Cordaro and J. Hounie. Semi-global solvability with loss of one derivative of partial differential operators on surfaces. *Advances in Mathematics* 301, 458–485, 2016.
4. J. Hounie. Globally hypoelliptic vector fields on compact surfaces. *Communications in PDE*, 7, 343–370, 1982.
5. J. Hounie. On the L2 continuity of pseudo- differential operators. *Communications in PDE*, 11, 765–778, 1986.
6. J. Hounie. Local solvability of first order linear operators with Lipschitz coefficients. *Duke Mathematical Journal* 62, 467–477, 1991.
7. J. Hounie and G. Zugliani. Global solvability and global hypoellipticity of complex vector fields on surfaces. *Journal of Differential Equations*, 340, 616–641, 2022.
8. F. Treves. On the local solvability and the local integrability of systems of vector fields. *Acta Mathematica*, 151(1), 1–48, 1983.

The Taylor Expansion of CR Germs on a Class of Levi Degenerate Models

Mariam Alakhdar and Giuseppe Della Sala

1 Introduction

The general context of this chapter is the following basic question: Which formal power series are realized as the Taylor series of germs of CR functions on a real submanifold $M \subset \mathbb{C}^n$? We call the map $\mathfrak{b}_p$ associating to a germ $f \in \mathrm{CR}^\infty(M, p)$ at $p \in M$ its Taylor series the *Borel map* at p. Since CR functions are the solution of a particular system of PDEs, it is easy to see that their Taylor series can be expressed in terms of a set of *basic solutions* of this system—for instance, the restriction to M of the complex coordinate functions. Aside of that, not many general results are known about the Borel map.

This question has been the subject of several works in recent years [1, 6, 4, 3]. In most of the known results, the main interest is to understand in which conditions the Borel map is surjective, or sometimes injective. There are however also examples in which the map is neither, as can be directly computed in the case of a real hyperplane (or a Levi-flat hypersurface) of $\mathbb{C}^n$. This leads to the general question (cf. [4, Question 5]) of characterizing the possible subalgebras of $\mathbb{C}[[z_1, \ldots, z_n]]$ which correspond to the image of the Borel map for a suitably chosen M. Are they

The second author was supported by the Center for Advanced Mathematical Sciences at the American University of Beirut.

M. Alakhdar
Department of Mathematics, American University of Beirut, Beirut, Lebanon

G. D. Sala (✉)
American University of Beirut, Beirut, Lebanon
e-mail: gd16@aub.edu.lb

S. Berhanu et al. (eds.), *Geometric Analysis of PDEs and Several Complex Variables*, Latin American Mathematics Series – UFSCar subseries,
https://doi.org/10.1007/978-3-031-69702-9_3

all realized (up to taking a quotient) as tensor products of a ring of convergent power series and a ring of formal power series, as in the case of the hyperplane?

The main result of this chapter is the construction of examples of manifolds M for which the image has a more complicated structure. Indeed, the most natural candidate was suggested in [4, Remark 6.3] as the hypersurface $\{\operatorname{Im} w = |z_1^2 - z_2^3|^2\} \subset \mathbb{C}^3$, which also features prominently (as archetypical counterexample) in the discussion of hypersurfaces of finite D'Angelo type, see [5]. We will show that the image of the Borel map can be computed explicitly for models of the kind $M = \{\operatorname{Im} w = |z_1^p - z_2^q|^2\} \subset \mathbb{C}^3$, $p, q \geq 2$.

The interest of such models derives from the presence of a singular complex variety $\Sigma \subset M$, namely $\Sigma = \{w = z_1^p - z_2^q = 0\}$. Clearly, any germ of CR function at 0 must be holomorphic on Σ, and at the formal level, this implies that its Taylor series must belong to the subring $\mathrm{CR}^{\mathrm{for}}(M)$ of power series which are "convergent along Σ" (cf. Sect. 2.2 for a more accurate definition). As it turns out, the obvious necessary condition is also sufficient:

Theorem 1 *Let M, $\mathrm{CR}^{\mathrm{for}}(M)$ be defined as above. Then $\mathfrak{b}_0(\mathrm{CR}^\infty(M, 0)) = \mathrm{CR}^{\mathrm{for}}(M)$.*

The behavior of the Borel map is in turn linked to the study of the algebraic properties of the ring of CR functions. Indeed, while the algebra of holomorphic germs has been investigated deeply since at least the nineteenth century, many surprisingly elementary questions involving germs of CR functions appear to be still open. Among them, the most fundamental is maybe the characterization of the maximal ideal $\mathfrak{m}$ of $\mathrm{CR}^\infty(M, p)$: The straightforward conjecture that $\mathfrak{m}$ is generated by any set of basic solutions has been so far verified in a number of important special cases (see [4]), but the default expectation is that it should always hold. The corresponding question is much easier to study in the setting of algebras of formal power series, and the Borel map offers the natural approach to link the two pictures. Thus, as a consequence of Theorem 1, we can verify the conjecture for the model hypersurfaces $M = \{\operatorname{Im} w = |z_1^p - z_2^q|^2\}$:

Theorem 2 *Let M be as above. The maximal ideal $\mathfrak{m}$ of $\mathrm{CR}^\infty(M, 0)$ is generated by the basic solutions $z_1|_M, z_2|_M, w|_M$.*

The scope of the results above can be considered somewhat restrictive, since they apply to a specific family of polynomial models. On the other hand, our main aim in the present work is to produce examples of hypersurfaces for which the behavior of the Borel map is more complicated than in the manifolds studied so far. The first natural generalization of the models considered here would be hypersurfaces of the form $\{\operatorname{Im} w = |f(z_1, z_2)|^2\}$, for an arbitrary germ of holomorphic function $f(z_1, z_2)$ such that $f(0, 0) = 0$. The nature of the singularity of the variety $\{f = 0\}$ at the origin should dictate the nature of image of the Borel map. The case of nonvanishing gradient of f is equivalent to $\{\operatorname{Im} w = |z_1|^2\}$ (the image consists of series holomorphic in z_2); the other extreme case $f = 0$ corresponds to series holomorphic in z_1, z_2. The examples studied in this chapter show that some kinds of intermediate behaviors are also possible.

We believe that the techniques used in this chapter can be extended to the case of a general $f(z_1, z_2)$, but a refinement of the approach will be needed. The main technical issue lies in the generalization of Lemma 6 (see Sect. 3.1) to situations in which the explicit parametrizations used in the proof are not available, starting with the case when $\{f = 0\}$ is an irreducible variety.

The proof of Theorems 1 and 2 builds on techniques introduced in the recent previous work on the Borel map, in particular [6, 4]. However, an essential ingredient of these techniques which is missing in our setting is the existence of a *peak function* ψ of finite order. For instance, the proof of the surjectivity of the Borel map in [6] is based on a rather explicit construction, producing a CR germ realizing a given arbitrary formal power series S: In the construction, each term of S is multiplied by an "approximate cut-off" CR function obtained by suitable manipulations of ψ.

The basic idea behind the proof of Theorem 1 is that the procedure of [6] can be still applied to a CR function ψ whose image is non-open, even if it is not a peak of finite order: For the models M, one such function ψ is for instance given the restriction of w to M. However, in this case the cut-off series associated to S will fail to converge to a smooth CR function for arbitrary choices of S (indeed this would imply that the Borel map is surjective). In order to show that the construction is convergent for series $S \in \mathrm{CR}^{\mathrm{for}}(M)$ we first need to provide a more concrete description of such series: This is achieved in Proposition 4. As it turns out, the terms that are allowed in the series belonging to $\mathrm{CR}^{\mathrm{for}}(M)$ are precisely the ones that are required to offset the divergence of the derivatives of the approximate cut-off functions.

This chapter is organized as follows. In Sect. 2, after a brief discussion of the geometry of M, the ring $\mathrm{CR}^{\mathrm{for}}(M)$ and the Borel map from the space of CR germs to the space of formal power series are introduced. Section 3 contains the proof of the main result $\mathfrak{b}_0(\mathrm{CR}^{\infty}(M, 0)) = \mathrm{CR}^{\mathrm{for}}(M)$, while in Sect. 4 this result is applied to characterize the maximal ideal of $\mathrm{CR}^{\infty}(M, 0)$.

2 Preliminaries

Here and in the rest of the chapter, we will consider complex coordinates (z_1, z_2, w) for $\mathbb{C}^3$, with $z_j = x_j + iy_j$ and $w = u + iv$. We will denote as usual by $\mathbb{C}[[z_1, z_2, w]]$ the ring of formal power series in z_1, z_2, w, and by $\mathbb{C}\{z_1, z_2, w\}$ the ring of convergent power series. Given $R > 0$, we will also employ the notation $\mathbb{C}_R\{t_1, \dots, t_n\}$ for the ring of power series in the variables $t_1, \dots t_n$ which converge on a polydisc of radius R.

2.1 Model Hypersurfaces

Fix positive integers $p, q \in \mathbb{Z}^+$, $2 \le p \le q$. We will be interested in the model hypersurfaces $M_{p,q} \subset \mathbb{C}^3$ defined as follows:

$$M_{p,q} = \left\{(z_1, z_2, w) \in \mathbb{C}^3 : \operatorname{Im} w = |z_1^p - z_2^q|^2\right\}. \tag{1}$$

Each $M_{p,q}$ is a Levi degenerate real hypersurface of $\mathbb{C}^3$, which contains the singular complex 1-dimensional variety

$$\Sigma_{p,q} = \left\{(z_1, z_2, w) \in \mathbb{C}^3 : w = 0,\ \theta_{p,q}(z_1, z_2) = 0\right\},$$

where we define $\theta_{p,q}(z_1, z_2) = z_1^p - z_2^q$. To simplify the notation we will omit the indices p, q from this point on.

Let $d = \gcd(p, q)$ be the greatest common divisor of p and q, so that $p = da$, $q = db$ for some relatively prime $a, b \in \mathbb{Z}^+$. As it is well known, Σ admits exactly d irreducible components: Indeed

$$\theta = (z_1^a - z_2^b)(z_1^a - \omega_a z_2^b)(z_1^a - \omega_{2a} z_2^b) \dots (z_1^a - \omega_{(d-1)a} z_2^b),$$

where we denote by $\omega_j = \cos(2\pi j/p) + i \sin(2\pi j/p)$ $(0 \le j \le p-1)$ the p-roots of unity, and hence the irreducible components are given by $\{z_1^a - \omega_{ja} z_2^b = 0, w = 0\}$. These can be parametrized by the following complex curves $\gamma_j : \mathbb{C} \to \mathbb{C}^3$:

$$\gamma_j(\tau) = (\omega_j \tau^b, \tau^a, 0), \quad j = 0, \dots, d-1.$$

We denote the ideal induced by the variety Σ in $\mathcal{O}_3 = \mathbb{C}\{z_1, z_2, w\}$ by $\mathcal{I}(\Sigma)$, and the coordinate ring of Σ by $\mathcal{O}_\Sigma = \mathcal{O}_3/\mathcal{I}(\Sigma)$. It is not hard to check that $\mathcal{I}(\Sigma) = \langle \theta, w \rangle$ and $\mathcal{O}_\Sigma$ is isomorphic to the space

$$\left\{b_0(z_2) + b_1(z_2)z_1 + \dots + b_{p-1}(z_2)z_1^{p-1} \in \mathcal{O}_3 : b_j(z_2) \in \mathbb{C}\{z_2\}\right\} \cong \mathcal{O}_\Sigma \tag{2}$$

as can be verified by applying the Weierstrass division theorem. For a fixed $R > 0$, we will further denote by $\mathcal{O}_\Sigma(R)$ the subset given by the elements of $\mathcal{O}_\Sigma$ for which $b_0, \dots, b_{p-1} \in \mathbb{C}_R\{z_2\}$.

2.2 Algebras of Formal Power Series

Later it will also be convenient to consider the formal version of $\mathcal{I}(\Sigma)$ (i.e., the ideal generated by θ, w in $\mathbb{C}[[z_1, z_2, w]]$), which we will denote by $\mathcal{I}^{\text{for}}(\Sigma) \subset$

$\mathbb{C}[[z_1, z_2, w]]$. We then have $\mathcal{I}(\Sigma) = \mathcal{I}^{\text{for}}(\Sigma) \cap \mathbb{C}\{z_1, z_2, w\}$. Note that $\mathcal{I}^{\text{for}}(\Sigma)$ can also be interpreted as the ideal induced by the variety Σ in $\mathbb{C}[[z_1, z_2, w]]$, in the following sense: A formal power series S belongs to $\mathcal{I}^{\text{for}}(\Sigma)$ if and only if $S \circ \gamma \equiv 0$ for all holomorphic curves $\gamma : \mathbb{C} \to \mathbb{C}^3$ satisfying $\gamma(\mathbb{C}) \subset \Sigma$.

Let $\mathbb{C}[[z_1, z_2, w]]/\mathcal{I}^{\text{for}}(\Sigma)$ be the "formal coordinate ring" of Σ, and denote by $\pi : \mathbb{C}[[z_1, z_2, w]] \to \mathbb{C}[[z_1, z_2, w]]/\mathcal{I}^{\text{for}}(\Sigma)$ the projection onto the quotient. Then $\mathcal{O}_\Sigma, \mathcal{O}_\Sigma(R)$ can be naturally seen as subrings of $\mathbb{C}[[z_1, z_2, w]]/\mathcal{I}^{\text{for}}(\Sigma)$ because any two germs $f_1, f_2 \in \mathcal{O}_3$ are equivalent modulo $\mathcal{I}^{\text{for}}(\Sigma)$ if and only if they are equivalent modulo $\mathcal{I}(\Sigma)$. The set $\pi^{-1}(\mathcal{O}_\Sigma) \subset \mathbb{C}[[z_1, z_2, w]]$ is then the subalgebra of formal power series whose restriction to Σ is holomorphic (cf. Lemma 6).

We will be interested in the following algebra of formal power series:

$$\mathrm{CR}^{\text{for}}(M) = \left\{ S \in \mathbb{C}[[z_1, z_2, w]] : \exists R > 0 \text{ s.t. } \frac{\partial^{k+m+\ell} S}{\partial z_1^k \partial z_2^m \partial w^\ell} \circ \gamma_j \right.$$
$$\left. \in \mathbb{C}_R\{\tau\} \, \forall k, m, \ell \geq 0, 0 \leq j \leq d-1 \right\};$$

an equivalent definition (as we will check later) is

$$\mathrm{CR}^{\text{for}}(M) = \left\{ S \in \mathbb{C}[[z_1, z_2, w]] : \exists R > 0 \text{ s.t. } \pi \left(\frac{\partial^{j+k+\ell} S}{\partial z_1^j \partial z_2^k \partial w^\ell} \right) \right.$$
$$\left. \in \mathcal{O}_\Sigma(R) \text{ for all } j, k, \ell \geq 0 \right\}.$$

We think of $\mathrm{CR}^{\text{for}}(M)$ as the algebra of "formal CR functions" on M at 0. It might be worth noting that $\mathrm{CR}^{\text{for}}(M)$ does not coincide with the largest subalgebra of $\pi^{-1}(\mathcal{O}_\Sigma)$ which is closed under derivations: Additional control on the radius of convergence of the restriction of derivatives to Σ is necessary, since the Taylor series of CR germs defined on M will satisfy such a condition for small enough $R > 0$.

2.3 *Germs of CR Functions and the Borel Map*

We will denote by $\mathrm{CR}^\infty(M)$ the ring of smooth CR functions defined on M, that is, the space of the elements of $C^\infty(M, \mathbb{C})$ annihilated by the CR vector fields

$$L_1 = \frac{\partial}{\partial \bar{z}_1} - 2ip\bar{z}_1^{p-1}\theta \frac{\partial}{\partial \overline{w}}, \quad L_2 = \frac{\partial}{\partial \bar{z}_2} + 2iq\bar{z}_2^{q-1}\theta \frac{\partial}{\partial \overline{w}},$$

and by $\mathrm{CR}^\infty(M, 0)$ the ring of germs of smooth CR functions defined on a neighborhood of 0 in M. Furthermore, we will denote by $\mathfrak{b}_0$ the Borel map at 0

$$\mathfrak{b}_0 : \mathrm{CR}^\infty(M, 0) \to \mathbb{C}[[z_1, z_2, w]]$$

that assigns to a germ f its formal Taylor series $\mathfrak{b}_0(f)$ at 0, expressed in terms of the fundamental family of solutions z_1, z_2, w (see, e.g., [4] for a more precise formal definition of the Borel map).

Consider now the following complex vector fields M_1, M_2, M_3 in $\mathbb{C}^3$:

$$M_1 = \frac{\partial}{\partial z_1} - 2ipz_1^{p-1}\overline{\theta}\frac{\partial}{\partial \overline{w}}, \quad M_2 = \frac{\partial}{\partial z_2} + 2iqz_2^{q-1}\overline{\theta}\frac{\partial}{\partial \overline{w}}, \quad M_3 = \frac{\partial}{\partial w} + \frac{\partial}{\partial \overline{w}}.$$

A straightforward computation shows that $M_j|_M$ is a section of $\mathbb{C}T(M)$ for all $1 \leq j \leq 3$. The vector fields $M_j|_M$ satisfy the following properties (cf. [4]):

- The vector fields L_1, L_2, M_1, M_2, M_3 commute with each other and generate the space of derivations of $C^\infty(M, \mathbb{C})$ at every point.
- If $f \in \mathrm{CR}^\infty(M)$ is the restriction of a holomorphic function $f \in \mathcal{O}(\mathbb{C}^3)$, then $M_j f$ is the restriction of $\partial f/\partial z_j$ $(j = 1, 2)$ or $\partial f/\partial w$ $(j = 3)$.
- The vector fields M_j are derivations of the algebras $\mathrm{CR}^\infty(M)$, $\mathrm{CR}^\infty(M, 0)$, and furthermore, the following diagram commutes:

$$\begin{array}{ccc} \mathrm{CR}^\infty(M,0) & \xrightarrow{\mathfrak{b}_0} & \mathbb{C}[[z_1, z_2, w]] \\ \downarrow{\scriptstyle M_j} & & \downarrow{\scriptstyle \partial_j} \\ \mathrm{CR}^\infty(M,0) & \xrightarrow{\mathfrak{b}_0} & \mathbb{C}[[z_1, z_2, w]], \end{array} \tag{3}$$

  where ∂_j denotes either the operator $\partial/\partial z_j$ $(j = 1, 2)$ or $\partial/\partial w$ $(j = 3)$.
- The Borel map $\mathfrak{b}_0$ at 0 in terms of the fundamental set of CR functions $\{z_1|_M, z_2|_M, w|_M\}$ can be written explicitly by using the vector fields M_j, in the following way:

$$\mathrm{CR}^\infty(M, 0) \ni f \to \mathfrak{b}_0(f) = \sum_{j,k,l=0}^{\infty} \frac{M_1^j M_2^k M_3^l f}{j!k!l!}(0) z_1^j z_2^k w^l \in \mathbb{C}[[z_1, z_2, w]].$$

3 Image of the Borel Map

Using the fact that the restriction of any CR function to Σ is holomorphic, and applying the vector fields M_1, M_2, M_3, it is easy to see that $\mathfrak{b}_0(\mathrm{CR}^\infty(M, 0)) \subset \mathrm{CR}^{\mathrm{for}}(M)$. Our main result is that the opposite inclusion also holds:

Theorem 3 *Fix positive integers $p, q \in \mathbb{Z}^+$, $2 \le p \le q$, let $M \subset \mathbb{C}^3$ be the real hypersurface defined by (1), and let $\mathfrak{b}_0 : \mathrm{CR}^\infty(M, 0) \to \mathbb{C}[[z_1, z_2, w]]$ be the Borel map at 0. Then*

$$\mathfrak{b}_0(\mathrm{CR}^\infty(M, 0)) = \mathrm{CR}^{\mathrm{for}}(M).$$

In order to prove Theorem 3, our first step will be to achieve a concrete description of all power series in $\mathrm{CR}^{\mathrm{for}}(M)$, as provided by Proposition 4. Afterward, we will show that the series in $\mathrm{CR}^{\mathrm{for}}(M)$ can be realized by applying a rather explicit construction based on a sequence of approximate cut-off functions (see Lemmas 8, 9, and Corollary 10). The final point is to prove that the series obtained gives rise to a smooth function. This is done in Proposition 14, by working around similar lines as in [6]; the crucial point is to show that the approximate cut-off functions obtained by manipulating the peak function available for M are enough to achieve the convergence of the derivatives of the series given by Proposition 4.

Proposition 4 *A formal power series $S \in \mathbb{C}[[z_1, z_2, w]]$ belongs to $\mathrm{CR}^{\mathrm{for}}(M)$ if and only if*

$$S = \sum_{j,k=0}^{\infty} a_{j,k}(z_1, z_2)\theta^j w^k, \tag{4}$$

where $a_{j,k} \in \mathcal{O}_\Sigma$ is of the form given in (2) for all $j, k \ge 0$ and, furthermore, all the germs $a_{j,k}$ converge on a common neighborhood of 0.

Remark 1 A more abstract formulation, equivalent to the statement in Proposition 4, can be given as follows:

$$\mathrm{CR}^{\mathrm{for}}(M) = \bigcup_{R>0} \left(\mathcal{O}_\Sigma(R) \mathbin{\widehat{\otimes}_{\mathbb{C}}} \mathbb{C}[[\theta, w]]\right), \tag{5}$$

where $\mathcal{O}_\Sigma(R)$ denotes the space of elements of $\mathcal{O}_\Sigma$ which converge on a polydisc of (poly)radius R, while $\widehat{\otimes}_{\mathbb{C}}$ denotes the completion of the $\mathbb{C}$-tensor product according to the Fréchet space topologies of $\mathcal{O}_\Sigma(R)$ and $\mathbb{C}[[\theta, w]]$.

A somewhat interesting consequence of Theorem 3 is that Σ is the largest complex structure contained in the (local) polynomial hull of M, in the sense specified by the following:

Corollary 5 *For any ball B centered at 0, any complex curve through 0 contained in the polynomial hull $\widehat{M}_B$ of $M \cap \overline{B}$ is contained in Σ.*

Proof Any CR function f defined on a neighborhood of 0 in M can be locally extended holomorphically to a function $\widetilde{f}$ defined on $\widehat{M}_B$. If C is the local parametrization of a complex curve contained in $\widehat{M}_B$, it follows that $\widetilde{f} \circ C$ is holomorphic around 0 in $\mathbb{C}$; hence, $\mathfrak{b}_0(f) \circ C$ is holomorphic. This would contradict the fact that all series of the form (4) can be realized as $\mathfrak{b}_0(f)$ for some f, unless the image of $\mathcal{C}$ is contained in $\Sigma = \{w = \theta = 0\}$ (otherwise one can find a series as in (4), which is not holomorphic along $\mathcal{C}$, by choosing an appropriate sequence of constant functions $a_{j,k}(z_1, z_2) = a_{j,k}$). □

While the statement in the Corollary above might be well known (and can likely be obtained by more direct means), we were not able to find a precise reference in the literature.

To prove Theorem 3, we will first focus on the identification provided by Proposition 4. Then we will show that any series of the form (4) can be realized as the Taylor series of a CR germ on M.

3.1 *Proof of Proposition 4*

We will start with the following lemma characterizing the formal power series in $\pi^{-1}(\mathcal{O}_\Sigma)$:

Lemma 6 *Let $S \in \mathbb{C}[[z_1, z_2, w]]$ be a formal power series. Then $S \in \pi^{-1}(\mathcal{O}_\Sigma)$ if and only if for every germ $\gamma(\tau)$ of holomorphic curve $\gamma : \mathbb{C} \to \mathbb{C}^3$ with image contained in Σ one has $S \circ \gamma \in \mathbb{C}\{\tau\}$. More precisely, for any $R > 0$, we have that $S \in \pi^{-1}(\mathcal{O}_\Sigma(R))$ if and only if $S \circ \gamma_j \in \mathbb{C}_R\{\tau\}$ for all $0 \leq j \leq d-1$, where the curves γ_j are the parametrizations defined in Sect. 2.1.*

Proof It is clear that if $S \in \pi^{-1}(\mathcal{O}_\Sigma)$ (resp., $S \in \pi^{-1}(\mathcal{O}_\Sigma(R))$), then $S \circ \gamma \in \mathbb{C}\{\tau\}$ (resp., $S \circ \gamma_j \in \mathbb{C}_R\{\tau\}$).

To prove the opposite implication, let $R > 0$ and let $S \in \mathbb{C}[[z_1, z_2, w]]$ be a power series satisfying $S \circ \gamma_j \in \mathbb{C}_R\{\tau\}$ for all $0 \leq j \leq d-1$. In order to show that $S \in \pi^{-1}(\mathcal{O}_\Sigma(R))$, we will first write

$$S = b_0(z_2) + b_1(z_2)z_1 + \ldots + b_{p-1}(z_2)z_1^{p-1} + \theta R_1 + w R_2 \tag{6}$$

for certain $b_j \in \mathbb{C}[[z_2]]$, $R_1 \in \mathbb{C}[[z_1, z_2]]$, and $R_2 \in \mathbb{C}[[z_1, z_2, w]]$, where $\theta = z_1^p - z_2^q$. This can be achieved by applying the formal version of the Weierstrass division theorem. The claim $\pi(S) \in \mathcal{O}_\Sigma(R)$ is thus equivalent to $b_j \in \mathbb{C}_R\{z_2\}$, $0 \leq j \leq p-1$.

Since by construction $\theta \circ \gamma_j \equiv 0$, $w \circ \gamma_j \equiv 0$, substituting into (6), we get

$$\begin{aligned} g_j(\tau) := S(\gamma_j(\tau)) &= b_0(\tau^a) + \omega_j \tau^b b_1(\tau^a) + \ldots + \omega_j^{p-1} \tau^{b(p-1)} b_{p-1}(\tau^a) \\ &= \sum_{n=0}^{p-1} \omega_j^n \tau^{bn} b_n(\tau^a). \end{aligned} \tag{7}$$

By assumption, then $g_j \in \mathbb{C}_R\{\tau\}$ for $j = 0, \dots, d-1$. Now, for $0 \le c \le a-1$, consider the following linear operators

$$T_c : \mathbb{C}[[\tau]] \to \mathbb{C}[[\tau]], \;\; h = \sum_{m=0}^{\infty} h_m \tau^m \to T_c(h) = \sum_{\ell=0}^{\infty} h_{a\ell+c} \tau^{a\ell+c},$$

that is, $T_c(h)$ keeps only the terms $h_m \tau^m$ such that $m \equiv c \mod a$. It is clear that the radius of convergence of $T_c(h)$ is larger than or equal to the one of h, in particular $T_c(\mathbb{C}_R\{\tau\}) \subset \mathbb{C}_R\{\tau\}$.

Furthermore, since the powers appearing in $b_n(\tau^a)$ have exponents which are multiples of a, we have $T_c(\tau^m b_n(\tau^a)) = \tau^m b_n(\tau^a)$ if $m \equiv c \mod a$, and $T_c(\tau^m b_n(\tau^a)) = 0$ otherwise. Using this fact and applying the operator T_c (for a fixed $0 \le c \le a-1$) to both sides of (7), we obtain

$$\sum_{\ell=0}^{d-1} \omega_j^{n_c+a\ell} \tau^{b(n_c+a\ell)} b_{n_c+a\ell}(\tau^a) = T_c(g_j(\tau)), \;\; j = 0, \dots, d-1, \tag{8}$$

where $0 \le n_c \le a-1$ is the (uniquely determined) solution of the congruence $bn \equiv c \mod a$. Defining for convenience $\kappa_{c,\ell}(\tau) = \tau^{b(n_c+a\ell)} b_{n_c+a\ell}(\tau^a)$, $0 \le \ell \le d-1$, we can reformulate the system in (8) as

$$\sum_{\ell=0}^{d-1} \omega_{n_c+a\ell}^{j} \kappa_{c,\ell}(\tau) = T_c(g_j(\tau)), \;\; j = 0, \dots, d-1,$$

where we have used the properties of the complex roots of unity (namely, $(\omega_r)^s = \omega_1^{rs} = (\omega_s)^r$). Putting $\kappa_c = (\kappa_{c,0}, \dots, \kappa_{c,d-1})^t \in (\mathbb{C}[[\tau]])^d$, $T_c(g) = (T_c(g_0), \dots, T_c(g_{d-1}))^t \in (\mathbb{C}_R\{\tau\})^d$ (both thought of as column vectors of power series), we can then rewrite (8) in matrix form as $A_c \kappa_c = T_c(g)$, where A_c is the $d \times d$ Vandermonde matrix with constant coefficients

$$A_c = \begin{pmatrix} 1 & 1 & 1 & \dots & 1 \\ \omega_{n_c}^1 & \omega_{n_c+a}^1 & \omega_{n_c+2a}^1 & \dots & \omega_{n_c+(d-1)a}^1 \\ \omega_{n_c}^2 & \omega_{n_c+a}^2 & \omega_{n_c+2a}^2 & \dots & \omega_{n_c+(d-1)a}^2 \\ \omega_{n_c}^3 & \omega_{n_c+a}^3 & \omega_{n_c+2a}^3 & \dots & \omega_{n_c+(d-1)a}^3 \\ \vdots & \vdots & \vdots & \ddots & \vdots \\ \omega_{n_c}^{(d-1)} & \omega_{n_c+a}^{(d-1)} & \omega_{n_c+2a}^{(d-1)} & \dots & \omega_{n_c+(d-1)a}^{(d-1)} \end{pmatrix}.$$

It follows that the components of the vector $\kappa_c = A_c^{-1} T_c(g)$ are all convergent power series. Since it is clear that

$$\kappa_{c,\ell}(\tau) = \tau^{b(n_c+a\ell)} b_{n_c+a\ell}(\tau^a) \in \mathbb{C}_R\{\tau\} \Leftrightarrow b_{n_c+a\ell}(z_2) \in \mathbb{C}_R\{z_2\},$$

we deduce that $b_{n_c+a\ell} \in \mathbb{C}_R\{z_2\}$ for all $0 \le c \le a-1, 0 \le \ell \le d-1$, that is, $b_j \in \mathbb{C}_R\{z_2\}$ for all $0 \le j \le p-1$ (indeed we have $j = n_c + a\ell$ for uniquely determined $0 \le c \le a-1, 0 \le \ell \le d-1$). □

The argument in the preceding lemma can be extended to higher-order derivatives:

Lemma 7 *Let $S \in \mathbb{C}[[z_1, z_2, w]]$ be a formal power series, and fix $R > 0$. The following conditions are equivalent:*

- $\dfrac{\partial^{j+k+l} S}{\partial z_1^j \partial z_2^k \partial w^l} \in \pi^{-1}(\mathcal{O}_\Sigma(R))$ *for all $j, k, l \ge 0$.*
- *S is of the form given in (4).*

Proof An application of the formal Weierstrass division theorem (using (6) iteratively on the quotient series R_1, R_2) allows to write S as in (4),

$$S = \sum_{j,k=0}^{\infty} a_{j,k}(z_1, z_2)\theta^j w^k,$$

$$a_{j,k}(z_1, z_2) = b_0^{j,k}(z_2) + b_1^{j,k}(z_2)z_1 + \ldots + b_{p-1}^{j,k}(z_2)z_1^{p-1}$$

with $b_0^{j,k}, \ldots, b_{p-1}^{j,k}$ just formal (rather than convergent) series. We need to show that $b_m^{j,k} \in \mathbb{C}_R\{z_2\}$ for all indices j, k, m.

Suppose by contradiction that this is not the case, and let k_0 be the smallest index for which the series a_{j,k_0} fail to be convergent on the R-polydisc P_R for all j. Fixed k_0, denote then by j_0 the smallest index for which a_{j_0,k_0} is not convergent on P_R. Subtracting from S the part convergent on P_R, we can replace it by its (j_0, k_0)-tail

$$\sum_{k>k_0 \text{ or } k=k_0, j \ge j_0} a_{j,k}(z_1, z_2)\theta^j w^k,$$

while still satisfying the conditions of the lemma. We denote this new series again by S. By the product rule, we can then write

$$\frac{\partial^{j_0+k_0} S}{\partial z_1^{j_0} \partial w^{k_0}} = a_{j_0,k_0}(z_1, z_2)\, j_0! k_0! \left(\frac{\partial \theta}{\partial z_1}\right)^{j_0} + \theta Q_1 + w Q_2$$

for suitable formal power series Q_1, Q_2. Following the same procedure as in Lemma 6 and observing that $(\frac{\partial \theta}{\partial z_1})^{j_0} = p^{j_0} z_1^{j_0(p-1)}$, we have that

$$a_{j_0,k_0}(\omega_j \tau^b, \tau^a)\tau^{bj_0(p-r)} \in \mathbb{C}_R\{z_1, z_2\}, \text{ and thus } a_{j_0,k_0}(\omega_j \tau^b, \tau^a) \in \mathbb{C}_R\{z_1, z_2\}$$

for all $j = 0, \ldots, d-1$. Exactly the same argument of Lemma 6 shows that a_{j_0,k_0} is convergent on P_R, giving a contradiction. □

3.2 Proof of Theorem 3

Let $f \in \mathrm{CR}^\infty(M,0)$ be a germ of CR function defined on a neighborhood U of 0 in M. For a small enough $R > 0$ we can assume that U contains $P_R \cap M$, where P_R denotes the polydisc of radius R centered at the origin. The first inclusion in Theorem 3 is, in essence, a consequence of the fact that the restriction of f (and of all its M_j-derivatives) to the complex curve Σ must be holomorphic on $\Sigma \cap P_R$; however, because of the singularity of Σ, to make the argument precise we use the holomorphic parametrizations of the irreducible components of Σ introduced in Sect. 2.1. Let, then, $\mathfrak{b}_0(f) \in \mathbb{C}[[z_1, z_2, w]]$ be the formal Taylor expansion of f at 0. By (3), we have that $M_1^k M_2^m M_3^l f \in \mathrm{CR}^\infty(M,0)$ and furthermore

$$\mathfrak{b}_0(M_1^k M_2^m M_3^l f) = \frac{\partial^{j+k+l}(\mathfrak{b}_0(f))}{\partial z_1^k \partial z_2^m \partial w^l}$$

for all $k, m, l \geq 0$. For all $0 \leq j \leq d-1$ the Taylor series expansion of $M_1^k M_2^m M_3^l f \circ \gamma_j$ is $\mathfrak{b}_0(M_1^k M_2^m M_3^l f) \circ \gamma_j = \frac{\partial^{k+m+l}(\mathfrak{b}_0 f)}{\partial z_1^k \partial z_2^m \partial w^l} \circ \gamma_j$. Furthermore, $M_1^k M_2^m M_3^l f \circ \gamma_j : \mathbb{C} \to \mathbb{C}$ is holomorphic on a disc of radius R (since it is the composition of a CR function with a holomorphic curve with image contained in M), and thus $\frac{\partial^{k+m+l}(\mathfrak{b}_0(f))}{\partial z_1^k \partial z_2^m \partial w^l} \circ \gamma_j$ is convergent for all $k, m, l \geq 0, 0 \leq j \leq d-1$. By definition this means that $\mathfrak{b}_0(f) \in \mathrm{CR}^{\mathrm{for}}(M)$.

To prove the opposite inclusion in Theorem 3, let $S \in \mathbb{C}[[z_1, z_2, w]]$ be a power series in $\mathrm{CR}^{\mathrm{for}}(M)$. By Proposition 4 we have that $S \in \bigcup_{R>0} \left(\mathcal{O}_\Sigma(R) \,\widehat{\otimes}_{\mathbb{C}}\, \mathbb{C}[[\theta, w]]\right)$, i.e., we can assume that S is of the form given in (4) for a certain $R > 0$. Our aim is to construct a function $f \in \mathrm{CR}^\infty(P_R \cap M)$ such that $\mathfrak{b}_0(f) = S$.

We will follow closely the procedure applied in [6], taking as starting point a CR function with non-open image, in our present case

$$\psi = -iw|_M = -iu + |z_1^p - z_2^q|^2,$$

and manipulating it in order to obtain a suitable sequence $\varphi_{j,k}$ of approximate cut-off functions. The crucial difference with the case of [6] is that ψ is no longer a peak function of finite order in the sense defined there (indeed no such function can exist as that would imply that $\mathfrak{b}_0$ is surjective). We will show that, nevertheless, the approximate cut-offs that can be constructed from ψ are enough to ensure convergence when applied to a series of the kind (4).

Let then

$$S = \sum_{j,k=0}^{\infty} a_{j,k}(z_1, z_2)\theta^j w^k$$

with $a_{j,k}$ as in (2) for some fixed $R > 0$. We will assume that R is small enough that $|w| \leq 1$, $|\theta| \leq 1$ on $M \cap P_R$, and furthermore that each $a_{j,k}$ is bounded there.

For $\epsilon > 0$ we define

$$D_\epsilon = \{p \in M : (|w(p)|^2 + |\theta(p)|^2)^{1/2} \le \epsilon\},$$

which is a neighborhood of 0 in M. Furthermore we put $D_{j,k} = D_{\epsilon_{j,k}}$, where

$$\epsilon_{j,k} = \min\{1/(|j| + |k|), 2^{-jk}/\|a_{j,k}\|_\infty\}, \quad \|a_{j,k}\|_\infty = \sup_{p \in P_R \cap M} |a_{j,k}(p)|,$$

and $A_{j,k} = \sup_{p \in P_R \setminus D_{j,k}} |1/\psi(p)|$.

We will consider, for a certain $y = y(j,k) \in \mathbb{R}^+$ (to be determined later), the functions

$$\psi_y = -\frac{\lambda(y)}{\psi^y} + B(y), \quad \varphi_y = e^{-\exp(\psi_y)},$$

where $B(y) = \frac{1}{\sin(y)}$ and $\lambda(y) = B(y)^{1-y}$. Note that the image of ψ is contained in $\{z \in \mathbb{C} : \operatorname{Re} z \ge 0\}$: To compute ψ^y, we choose the standard branch of the logarithm defined on $\mathbb{C} \setminus \mathbb{R}^-$. While the functions ψ_y, φ_y are a priori defined on $\psi \neq 0$, it can be checked that φ_y extends (by 1) across $\{\psi = 0\}$ as a flat function of class C^∞ (see Corollary 12). Moreover, φ_y is a CR function, being the composition of ψ with holomorphic functions.

The following lemmas correspond to [6, Lemmas 4.1 and 4.2] and subsequent remarks:

Lemma 8 *Let* $V_y = \psi_y^{-1}(\{w \in C : \operatorname{Re} w \ge 0\})$*, and fix any* $L > 0$*. For all small enough* $y \in R^+$*, we have that* $\operatorname{Re} \psi_y(p) > L$ *for* $p \in \{|\psi| > 1/A_{j,k}\}$ *and* $\operatorname{Im} \psi_y(p)| \le \cos(y)^{-1}$ *for any* $p \in V_y$*. In particular, it follows that* $U \setminus D_{j,k} \subset \{|\psi| > 1/A_{j,k}\} \subset V_y$.

Lemma 9 *There exists* $y \in \mathbb{R}^+$ *such that* $\varphi_y(p) \le \epsilon_{j,k}$ *for all* $p \in \{|\psi| > 1/A_{j,k}\}$ *(hence for all* $p \in U \setminus D_{j,k}$*), and* $\varphi_y(p) \le e$ *for* $p \in D_{j,k}$.

Remark 2 The proof of Lemma 9 uses Lemma 8, by showing that the claimed estimate is achieved when $p \in W_{j,k} = \{\operatorname{Re} \psi_{y(j,k)} > L(j,k)\}$ for suitably chosen $L(j,k)$, $y(j,k)$.

For ease of notation we will write $\varphi_{j,k} = \varphi_{y(j,k)}$, where $y(j,k)$ is as in the previous lemma. The functions $\varphi_{j,k}$ can be applied to the (possibly divergent) series S in order to obtain a convergent one:

Corollary 10 *The series*

$$\sum_{j,k \geqslant 0} a_{j,k} \theta^j w^k \varphi_{j,k}$$

is uniformly convergent on $P_R \cap M$.

Proof Indeed, by construction and Lemma 9, we have $|a_{j,k}||w^j||\theta^k||\varphi_{j,k}| \leq C\|a_{j,k}\|_\infty \epsilon_{j,k}^{jk} \leq C2^{-jk}$ for $p \in D_{j,k}$, and $|a_{j,k}||w^j||\theta^k||\varphi_{j,k}| \leq \|a_{j,k}\|_\infty \epsilon_{j,k}| \leq 2^{-jk}$ for $p \in P_R \setminus D_{j,k}$. □

The needed function will then be given by

$$f = \sum_{j,k \geqslant 0} a_{j,k}\theta^j w^k \varphi_{j,k}. \tag{9}$$

The main task is now to show that f is of class C^∞. To this end, consider real variables x_1, y_1, x_2, y_2, u on M, and for any multi-index $J \in \mathbb{N}^5$ denote by D^J the differential operator associated to J in the standard way. The proof of the following results is identical to the one given in [6, Lemma 4.3, Corollary 4.2-4.3].

Lemma 11 *Let $J \in \mathbb{N}^5$. Then there exist $A > 0$ and $\alpha_1 < \alpha_2$, $\gamma \in N$ (all depending only on J) such that*

$$|D^J \varphi_{j,k}| \leq A|\varphi_{j,k}|(\max\{|\log \varphi_{j,k}|^{\alpha_1}, |\log \varphi_{j,k}|^{\alpha_2}\})\frac{1}{|\psi|^\gamma}.$$

Furthermore, $\max\{|\log \varphi_{j,k}|^{\alpha_1}, |\log \varphi_{j,k}|^{\alpha_2}\} = |\log \varphi_{j,k}|^{\alpha_1}$ *near* 0.

Corollary 12 *$\varphi_{j,k}$ is smooth in a neighborhood of* 0*, and flat at* 0*, for any $j, k \in \mathbb{N}$. In particular it is CR around* 0.

In order to study the derivatives of the terms of the series in (9), we also need to consider the derivatives of the functions $a_{j,k}$:

Lemma 13 *Fix $0 < \varrho < R$, $J \in \mathbb{N}^5$. Then there exists $B > 0$ such that*

$$\|D^J a_{j,k}\|_\infty^\varrho \leq B|J|!\frac{\|a_{j,k}\|_\infty}{\varrho^{|J|}},$$

where $\|\cdot\|_\infty$ denotes the sup-norm on $M \cap P_R$, while $\|\cdot\|_\infty^\varrho$ denotes the sup-norm on $M \cap P_{R-\varrho}$.

Proof Any differential operator D^J can be expressed as a suitable composition of the vector fields M_1, M_2, M_3 defined in Sect. 2.3 and the CR vector fields L_1, L_2. Since $M_3 a_{j,k} = L_1 a_{j,k} = L_2 a_{j,k} \equiv 0$, while $M_1 a_{j,k} = \partial a_{j,k}/\partial z_1$, $M_2 a_{j,k} = \partial a_{j,k}/\partial z_2$, and $a_{j,k}$ is holomorphic on P_R, and the claim in the lemma is a direct consequence of the Cauchy estimates (applied on polydiscs with center in $P_{R-\varrho}$). □

Finally, we can show that the derivatives of the series in (9) converge uniformly by adapting the argument used in [6, Lemmas 4.4 and 5.1]:

Proposition 14 *The function f is of class C^∞ on $M \cap P_R$.*

Proof We want to estimate $|D^J(a_{j,k}\theta^j w^k\varphi_{j,k})|$. Fix j,k with $\max\{j+k\} > |J|$. In particular we assume $j > |J|$; the estimate for $k > |J|$ is analogous (but easier). Moreover, we fix $0 < \varrho < \min\{1, R\}$ and carry out the following computations at points of $P_{R-\varrho}$. By using repeatedly the product rule, we obtain

$$|D^J(a_{j,k}\theta^j w^k\varphi_{j,k})| \leq \sum_{H+L+I_1+I_2=J} E(H,L,I_1,I_2)|D^H(a_{j,k})||D^L(\varphi_{j,k})||D^{I_1}\theta^j||D^{I_2}w^k| \leq$$

(with purely combinatorial constants $E(H,L,I_1,I_2)$)

$$\leq \sum_{H+L+I_1+I_2=J} E'(H,L,I_1,I_2)|D^H(a_{j,k})||D^L(\varphi_{j,k})|(jk)^{|J|}|\theta|^{|j|-|I_1|} \leq$$

(where the constants $E'(H,L,I_1,I_2)$ now depend also on the modulus of the derivatives of w,θ of order at most $|J|$)

$$\leq \sum_{H+L+I_1+I_2=J} E'(H,L,I_1,I_2)B|H|!\frac{\|a_{j,k}\|_\infty}{\varrho^{|H|}}|D^L(\varphi_{j,k})|(jk)^{|J|}|\theta|^{j-|I_1|}, \tag{10}$$

where Lemma 13 was used in the last step. Now, Lemma 11 allows to estimate

$$|D^L(\varphi_{j,k})| \leq K|\varphi_{j,k}|\log\varphi_{j,k}|^\alpha\frac{1}{|\psi|^\gamma} \quad \text{for all } |L| \leq |J|,$$

where the constants K, α, and γ depend on J but are uniform in L. Moreover, since by construction

$$|\psi| = (u^2 + |\theta|^4)^{1/2} \geq |\theta|^2,$$

we can write further

$$|D^L(\varphi_{j,k})| \leq K|\varphi_{j,k}||\log\varphi_{j,k}|^\alpha\frac{1}{|\theta|^{2\gamma}} \quad \text{for all } |L| \leq |J|.$$

(In the case $k > |J|$, we can instead simply use that $|\psi| = |w|$). Then, taking into account that the sum in (10) is finite, we obtain

$$|D^J(a_{j,k}\theta^j w^k\varphi_{j,k})| \leq \frac{K'(J)}{\varrho^{|J|}}\|a_{j,k}\|_\infty|\varphi_{j,k}||\log\varphi_{j,k}|^\alpha|(jk)^{|J|}|\theta|^{j-|J|-2\gamma} \tag{11}$$

for another constant $K'(J) > 0$ depending on J but not on j, k. We can assume further that $j - |J| - 2\gamma \geq 1$.

Given (11), the smoothness of f on $M \cap P_R$ follows by applying the exact same argument as in [6, Lemmas 5.1], estimating separately on $P_{R-\varrho} \cap W_{j,k}$ and on $P_{R-\varrho} \setminus W_{j,k}$ (where $W_{j,k}$ is defined in Remark 2). □

4 The Maximal Ideal

The study of the algebraic properties of the ring of CR germs provides one of the motivations for considering the Borel map (cf. [4, Sections 9–10]). As it turns out, results about its image can be applied to study the structure of the maximal ideal of $\mathrm{CR}^\infty(M, 0)$,

$$\mathfrak{m} = \{f \in \mathrm{CR}^\infty(M, 0) : f(0) = 0\}.$$

In the rings of germs of smooth (resp., holomorphic) functions, the maximal ideals are generated by the basic (real or complex) coordinate functions. It is natural to ask whether the same holds for CR functions:

Theorem 15 *The maximal ideal* $\mathfrak{m}$ *is generated by the basic solutions* $z_1|_M, z_2|_M, w|_M$.

In order to prove Theorem 15, we follow a similar strategy as in [4]. We will first show that each germ in $\mathfrak{m}$ belongs to $\langle z_1, z_2, w\rangle$ modulo elements of the ideal $\mathcal{F} = \ker(\mathfrak{b}_0) \subset \mathrm{CR}^\infty(M, 0)$ of *flat* germs, that is, germs with all derivatives vanishing at 0. In [4], the surjectivity of the Borel map was exploited to achieve this step; while in our context the Borel map is no longer surjective, we will still make use of the isomorphism

$$\mathfrak{b}_0 : \mathrm{CR}^\infty(M, 0)/\mathcal{F} \to \mathfrak{b}_0(\mathrm{CR}^\infty(M, 0)) = \mathrm{CR}^{\mathrm{for}}(M).$$

Accordingly, our first step is to check that the formal version of Theorem 15 is verified, i.e., that the corresponding property is satisfied in $\mathrm{CR}^{\mathrm{for}}(M)$:

Lemma 16 *The maximal ideal of* $\mathrm{CR}^{\mathrm{for}}(M)$ *is generated by* z_1, z_2, w.

Proof Let S be a series as in (4), with $S(0) = 0$. Putting $a_{j,k}(z_1, z_2) = \sum_{\ell=0}^{p-1} b_{\ell,j,k}(z_2) z_1^\ell$, we can first of all write

$$S = wQ + z_1 T + \sum_{j=0}^{\infty} b_{0,j,0}(z_2)\theta^j,$$

where

$$Q = \sum_{j\geq 0, k\geq 1} a_{j,k}(z_1, z_2)\theta^j w^{k-1}, \quad T = \sum_{j=0}^{\infty}\sum_{\ell=1}^{p-1} b_{\ell,j,0}(z_2) z_1^{\ell-1}\theta^j,$$

both belong to $\mathrm{CR}^{\mathrm{for}}(M)$. It is then enough to show that $\sum_{j=0}^{\infty} b_{0,j,0}(z_2)\theta^j$ is an element of $\langle z_1, z_2\rangle$. Note that, since $S(0) = 0$, we must have $b_{0,0,0}(0) = 0$, and hence $b_{0,0,0}(z_2) = z_2 c(z_2)$ with $c \in \mathcal{O}_\Sigma(R)$. Using that $\theta^j = z_1^p\theta^{j-1} - z_2^q\theta^{j-1}$ for $j \geq 1$, we have

$$\begin{aligned}\sum_{j=0}^{\infty} b_{0,j,0}(z_2)\theta^j &= z_2 c(z_2) + z_1 \sum_{j=1}^{\infty}(b_{0,j,0}(z_2) z_1^{p-1})\theta^{j-1} - z_2 \\ &\quad \times \sum_{j=1}^{\infty}(z_2^{q-1} b_{0,j,0}(z_2))\theta^{j-1},\end{aligned}$$

which concludes the proof. □

Corollary 17 *In* $\mathrm{CR}^\infty(M, 0)$, *we have* $\mathfrak{m} = \mathcal{F} + \langle z_1, z_2, w\rangle$.

Proof Let $f \in \mathrm{CR}^\infty(M, 0)$ be an element of $\mathfrak{m}$; then $\mathfrak{b}_0(f)$ belongs to the maximal ideal of $\mathrm{CR}^{\mathrm{for}}(M)$. By Lemma 16, we can write

$$\mathfrak{b}_0(f) = z_1 R_1 + z_2 R_2 + wQ$$

for suitable $R_1, R_2, Q \in \mathrm{CR}^{\mathrm{for}}(M)$. By Theorem 3, there exist $g_1, g_2, h \in \mathrm{CR}^\infty(M, 0)$ such that $\mathfrak{b}_0(g_1) = R_1, \mathfrak{b}_0(g_2) = R_2, \mathfrak{b}_0(h) = Q$. Putting $\widetilde{f} = z_1 g_1 + z_2 g_2 + wh$, we have $\widetilde{f} \in \langle z_1, z_2, w\rangle$ and

$$\begin{aligned}\mathfrak{b}_0(f - \widetilde{f}) &= \mathfrak{b}_0(f) - z_1\mathfrak{b}_0(g_1) - z_2\mathfrak{b}_0(g_2) - w\mathfrak{b}_0(h) \\ &= \mathfrak{b}_0(f) - z_1 R_1 - z_2 R_2 - wQ = 0,\end{aligned}$$

which shows that $f - \widetilde{f} \in \mathcal{F}$ is a flat function. □

In view of Corollary 17, to prove Theorem 15, it is enough to show that $\mathcal{F} \subset \langle z_1, z_2, w\rangle$. One can show more—in fact, $\mathcal{F} \subset \langle w\rangle$:

Lemma 18 *Let* $f \in \mathcal{F}$. *Then* $f = wg$ *with* $g \in \mathcal{F}$.

Proof Let $g = f/w$; the function g is a priori defined on the set $\{w \neq 0\}$, but we will show that it extends smoothly up to $\{w = 0\} = \Sigma \subset M$. This will also imply that g is a CR function, since it is CR on the open dense subset $M \setminus \Sigma$.

Since $f \in \mathcal{F}$, we have that $f \circ \gamma_j$ is holomorphic and flat at $\tau = 0$ for all $0 \le j \le d-1$ (where the γ_j are as in Sect. 2.1). Thus $f \circ \gamma_j \equiv 0$ for all j, which implies that $f|_\Sigma \equiv 0$. Furthermore, $M_1^i M_2^j M_3^\ell f$ is also CR and flat at 0 for all $i, j, \ell \in \mathbb{N}$—by definition, the ring $\mathcal{F}$ is closed under the derivations M_1, M_2, M_3. Using the same argument we obtain that $M_1^i M_2^j M_3^\ell f|_\Sigma \equiv 0$ for all $i, j, \ell \in \mathbb{N}$, which implies that f is flat at all points of Σ and not only at 0.

We now follow an argument analogous to the one in [4, Lemma 9.2]. Note that, in the coordinates (z_1, z_2, u) of M, Σ can be expressed as the zero set of the analytic function $|\psi|^2 = |w|^2 = u^2 + |\theta|^4$ (cf. Sect. 3.2). Combining the Lojasiewicz inequality (cf. [2]) with Taylor's theorem applied to f to any finite order along Σ (exploiting the flatness of f at all points of Σ), we obtain that

$$|f| \le C_k |w|^k \quad \text{for some constant } C_k > 0$$

for any fixed $k \in \mathbb{N}$. This means that $|f| = O(|w|^k)$ for all k, which immediately implies that f/w extends continuously (by 0) up to $\Sigma = \{w = 0\}$.

In order to conclude that g is C^∞, it is enough to show that $M_1^i M_2^j M_3^\ell g$ extend continuously across Σ for any fixed $i, j, k \in \mathbb{N}$, because $L_1 g = L_2 g \equiv 0$. Since M_1 and M_2 annihilate w and the fields M_1, M_2, M_3 commute, we can write

$$M_1^i M_2^j M_3^\ell g = M_3^\ell \frac{\widetilde{f}}{w} = \sum_{m=0}^{\ell} (-1)^m m! \frac{M_3^{\ell-m} \widetilde{f}}{w^{m+1}},$$

where $\widetilde{f} = M_1^i M_2^j f$. As before, by the closeness of $\mathcal{F}$ under M_1, M_2, M_3, we have $\widetilde{f}, M_3 \widetilde{f}, \ldots, M_3^\ell \widetilde{f} \in \mathcal{F}$. The first part of the proof can now be applied to show that each one of the functions $\frac{M_3^{\ell-m} \widetilde{f}}{w^{m+1}}$ and thus $M_1^i M_2^j M_3^\ell g$ extend continuously across Σ. □

Proof (Theorem 15) It is clear that $\langle z_1, z_2, w \rangle \subset \mathfrak{m}$. On the other hand, applying Corollary 17 and Lemma 18, we have $\mathfrak{m} = \mathcal{F} + \langle z_1, z_2, w \rangle \subset \langle w \rangle + \langle z_1, z_2, w \rangle = \langle z_1, z_2, w \rangle$. □

References

1. Rafael F. Barostichi, Paulo D. Cordaro, and Gerson Petronilho, *On the Borel property for solutions to systems of complex vector fields*, Math. Nachr. **286** (2013), no. 14–15, 1439–1451. MR3119692
2. Edward Bierstone and Pierre D. Milman, *Semianalytic and subanalytic sets*, Inst. Hautes Études Sci. Publ. Math. **67** (1988), 5–42. MR972342
3. Paulo D. Cordaro, Giuseppe Della Sala, and Bernhard Lamel, *The Borel map for compact sets in the plane*, J. Funct. Anal. **278** (2020), no. 6, 108402, 17. MR4054104

4. ——, *The Borel map in locally integrable structures*, Math. Ann. **377** (2020), no. 3–4, 1155–1192. MR4126892
5. John P. D'Angelo, *Several complex variables and the geometry of real hypersurfaces*, Studies in Advanced Mathematics, CRC Press, Boca Raton, FL, 1993.
6. Giuseppe Della Sala and Bernhard Lamel, *Asymptotic approximations and a Borel-type result for CR functions*, Internat. J. Math. **24** (2013), no. 11, 1350091, 16. MR3143607

On a Model "Sum of Squares" Operator

Antonio Bove and Marco Mughetti

Dedicated to Jorge Hounie, as a token of our appreciation and friendship.

1 Introduction

The purpose of this chapter is the study of a model operator being a "sum of squares" of vector fields in two variables. More precisely we are interested in the Gevrey or real analytic regularity of the solutions when the coefficients are real analytic.

The reason why we restrict ourselves to a two-dimensional case deserves some elucidation.

First of all we always assume that Hörmander's bracket condition [20, 22] is satisfied, i.e., the iterated commutators of the vector fields generate a two-dimensional vector space. As a consequence all the operators we consider are C^∞ hypoelliptic. Our main concern is thus their real analytic (C^ω) or Gevrey s (G^s) hypoellipticity.

In 1972 and then in 1973 Baouendi and Goulaouic, [2], and Oleĭnik and Radkevič, [27], showed that C^∞ hypoellipticity does not imply C^ω hypoellipticity by studying in detail two model operators.

On the other hand in 1973 Derridj, [18], showed that a sum of squares operator with analytic coefficients is always G^r hypoelliptic, where r denotes the maximum length (the number of vector fields) of the iterated brackets needed to satisfy Hörmander condition.

In 1978 Treves, [30], and in 1980 Tartakoff, [29], showed with different proofs that for a sum of squares with real analytic coefficients, satisfying Hörmander condition, we have C^ω hypoellipticity if the characteristic variety is a symplectic manifold and if the principal symbol vanishes exactly to the second order on

A. Bove (✉) · M. Mughetti
Dipartimento di Matematica, Università di Bologna, Bologna, Italy
e-mail: antonio.bove@unibo.it; marco.mughetti@unibo.it

S. Berhanu et al. (eds.), *Geometric Analysis of PDEs and Several Complex Variables*, Latin American Mathematics Series – UFSCar subseries,
https://doi.org/10.1007/978-3-031-69702-9_4

the characteristic manifold. We recall that a submanifold, M, of $T^*\mathbb{R}^n \setminus \{0\}$, the cotangent bundle minus the zero section, is symplectic if the restriction to M of the symplectic form $d\sigma = d\xi \wedge dx = \sum_{j=1}^n d\xi_j \wedge dx_j$ has rank $\dim M$. Moreover we say that the principal symbol, p, vanishes exactly to the second order on $\text{Char}(P)$ if $\ker F_p(\rho) = T_\rho \text{Char}(P)$, for $\rho \in \text{Char}(P)$. Here $F_p(\rho)$ denotes the fundamental matrix of p, which we may define in coordinates as $F_p(\rho) = dH_p(\rho)$, and $H_p(\rho) = (\partial_\xi p(\rho), -\partial_x p(\rho))$ denotes the Hamilton vector field of p at ρ.

In 1996 Treves formulated a conjecture for the real analytic hypoellipticity of a sum of squares operator, based on a real analytic stratification of the characteristic variety of p in [12, 31]. According to the conjecture we have C^ω hypoellipticity if and only if the strata of the stratification defined by Treves are real analytic symplectic submanifolds.

In 2017 and 2018, it has been proved, [7, 1], that Treves conjecture is false, in dimension $n \geq 4$, in its sufficient part.

We believe that the same is true in dimension $n = 3$, even though the Hamilton leaves of the possible strata—no definition for such strata yet—project injectively onto the fibers of the cotangent bundle.

Finally, we believe that the conjecture is true in the two-dimensional case.

In this chapter we study a model operator in two variables and show that its behavior is in agreement with the conjecture, even though we do not provide a proof of the optimality, which, to our knowledge, is a difficult open problem.

More precisely we consider, for $(x, y) \in \mathbb{R}^2$, the operator

$$P(x, y, D_x, D_y) = D_x^2 + \big(f(x, y)D_y\big)^2 = X_1^2 + X_2^2, \tag{1}$$

where $D_x = i^{-1}\partial_x$ and analogously for D_y. Moreover

$$f(x, y) = y^{2a}x^m + x^\ell, \quad a, m, \ell \in \mathbb{N} = \{1, 2, \dots\}, \tag{2}$$

with the condition that

$$|\ell - m| \text{ is an even integer.} \tag{3}$$

We immediately see that the Hörmander condition is satisfied with brackets of length $\ell + 1$.

Note that if $a = 0$ the operator in (1) is a Grušin operator whose analytic hypoellipticity has been proved in [19], Theorem 5.1.

The characteristic variety of P is $\text{Char}(P) = \{(x, y) \mid \xi = 0, f(x, y) = 0, \eta \neq 0\}$. In other words

$$\text{Char}(P) = \{(0, y; 0, \eta) \mid \eta \neq 0\}$$

is a symplectic real analytic submanifold of codimension 2.

Assume that $\ell \leq m$. Then we may write $f(x, y) = x^\ell(1 + y^{2a}x^{m-\ell})$, with $m - \ell \geq 0$ and even. In this case, according to Treves conjecture, there is only one type of stratum coinciding with (the connected components of) the characteristic manifold.

If, on the other hand, $\ell > m$, then $f(x, y) = x^m(y^{2a} + x^{\ell-m})$, and we see that the characteristic manifold is symplectic and is made up by a nontrivial Treves stratification. In fact taking the Poisson bracket $\{X_1, X_2\}$, where X_i denotes the symbol of the vector field X_i, and $\{\cdot, \cdot\}$ the Poisson bracket, we see that $\{X_1, X_2\} = 0$ on the characteristic set. Hence taking the iterated Poisson bracket,

$$\underbrace{\{X_1, \{X_1, \{\cdots \{X_1}_{m \text{ times}}, X_2\} \cdots\}\}\}_{|\mathrm{Char}(P)} = m! y^{2a}\eta.$$

This vanishes when $y = 0$ so that the stratification is given by

$$\Sigma_{1,\pm} = \{(0, y; 0, \eta) \mid y \neq 0, \pm\eta > 0\}; \tag{4}$$

and

$$\Sigma_{2,\pm} = \{(0, 0; 0, \eta) \mid \pm\eta > 0\}, \tag{5}$$

since we have

$$\underbrace{\{X_1, \{X_1, \{\cdots \{X_1}_{\ell \text{ times}}, X_2\} \cdots\}\}\}_{|\Sigma_{2,\pm}} = \ell!\eta \neq 0,$$

which is the generating bracket satisfying Hörmander condition. Hence we see that Hörmander condition is satisfied at length $m + 1$ on $\Sigma_{1,\pm}$ and at length $\ell + 1$ on $\Sigma_{2,\pm}$.

In particular we see that the $\Sigma_{1,\pm}$ are symplectic submanifolds of dimension two and the $\Sigma_{2,\pm}$ are one- dimensional, and hence nonsymplectic. Moreover, since the η half line is the integral curve of H_y, the strata $\Sigma_{2,\pm}$ can be thought of as the Hamilton leaves of $\Sigma_{2,\pm}$.

In this chapter we prove the following theorem:

Theorem 1.1 *Let P be given as in* (1)–(3). *Then:*

(i) If $\ell \leq m$, P is analytic hypoelliptic.
(ii) If $\ell > m$, P is Gevrey s hypoelliptic for any $s \geq s_0$, where

$$s_0 = \left(1 - \frac{1}{2a}\frac{\ell - m}{\ell + 1}\right)^{-1} > 1. \tag{6}$$

A few remarks are in order. First we observe that the above statement is in agreement with the Treves conjecture: when the Poisson strata are symplectic, we get analytic hypoellipticity.

When the Poisson strata are not symplectic, we prove Gevrey s_0 hypoellipticity. We believe that this number is actually optimal, meaning that for any s, $1 \leq s < s_0$, there are solutions to the equation $Pu = f$, with $f \in C^\omega$, such that $WF_s(u) \neq \emptyset$. However, since a proof of this fact is long and highly technical, we do not attempt to provide one in this chapter. We refer to [11, 14, 15, 24] for results in similar, although particular, cases.

Assumption (3) is not necessary for our conclusion, meaning that we could drop it. It implies that in the case $\ell > m$ the vector field X_2 vanishes only on the line $x = 0$ outside of the origin. In general we could have cases where X_2 vanishes on a certain number of analytic real curves outside of the origin—see, e.g., [9] for a description of the geometry of the characteristic variety in two variables. In any case the points on those curves are points of analytic regularity for the solution, so that their analytic wave front set projects onto the origin $x = 0$, $y = 0$.

Unfortunately including such a case would have forced us to produce a much longer and technical proof, and we decided to sacrifice generality to attain a greater level of clarity. We refer to [3, 5] for a result in this direction.

The proof of Theorem 1.1 is contained in the next two sections.

2 Proof of Theorem 1.1 in Case *(i)*

The basic tool for the proof is the so-called a priori subelliptic estimate for P—see [20, 28] for a general proof.

Let Ω denote an open set in $\mathbb{R}^2$, $(0, 0) \in \Omega$, then the subelliptic estimate for P in case *(i)* is

$$\|u\|^2_{\frac{1}{\ell+1}} + \sum_{j=1}^{2} \|X_j u\|_0^2 \leq C \left(\langle Pu, u\rangle + \|u\|_0^2 \right), \tag{7}$$

where $u \in C_0^\infty(\Omega)$, C is a positive constant, and $\|\cdot\|_s$ denotes the Sobolev norm of order s in $\mathbb{R}^2$ (when $s = 0$, we get the L^2 norm).

To prove the real analyticity of the solutions to $Pu = g$, where g is a real analytic function in Ω, it is enough to show that

$$\|\varphi D_y^p u\|_{\frac{1}{\ell+1}} + \sum_{j=1}^{2} \|X_j \varphi D_y^p u\|_0 \leq C_0^{p+1} p!, \tag{8}$$

for any $p \in \mathbb{N}$, where φ is an Ehrenpreis cutoff function supported in Ω (see the definition below). We point out that φD_y^p is microlocally elliptic on the characteristic variety near the origin, so that (8) implies the real analyticity of u.

Definition 2.1 ([17, 21]) For any natural number p, denote by φ_p a function in $C_0^\infty(\mathbb{R}^2)$, $\varphi_p(x, y)$, such that:

(S) φ_p is equal to 1 in a small neighborhood, W, of the origin and supported in W', with $W \subset W' \Subset \Omega$.

We say that φ_p is an Ehrenpreis sequence of cutoff functions if there is a positive constant R such that for $|\alpha| \leq R(p + 1)$ we have, for every p

$$|\partial_x^{\alpha_1} \partial_y^{\alpha_2} \varphi_p(x, y)| \leq C_\varphi^{|\alpha|+1} p^{|\alpha|}, \tag{9}$$

where $C_\varphi > 0$ and independent of p.

We now recall how to localize the function u in order to take advantage of the a priori estimate (7) and of the analytic regularity of Pu.

We observe that $\varphi_p(x, y)$ can be constructed as a product $\varphi_{1,p}(x)\varphi_{2,p}(y)$ of two Ehrenpreis type cutoff functions, each having support in a neighborhood of the origin of the line. Since, in case *(i)*, $\mathrm{Char}(P) = \{(0, y; 0, \eta) \mid \eta \neq 0\}$, we see that any x-derivative produces a function $(\partial_x^{\alpha_1} \varphi_{1,p}(x))\varphi_{2,p}(y)$ supported far from $x = 0$, where P is an elliptic operator and hence an analytic hypoelliptic operator. From the equation

$$Pu = g \in C^\omega(\Omega),$$

we have that $u \in C^\infty(\Omega)$ and, setting $\tilde{\Omega} = \Omega \cap \{(x, y) \in \Omega \mid x \neq 0\}$, $u \in C^\omega(\tilde{\Omega})$. As a consequence, defining $v = \varphi_{1,p}(x)u$, we have

$$Pv = \varphi_{1,p}(x)Pu + [P, \varphi_{1,p}(x)]u = \varphi_{1,p}(x)g + [D_x^2, \varphi_{1,p}(x)]u.$$

The r.h.s. of the above equation satisfies estimates of type (9), with a different constant, independent of p, for $\alpha_1 \leq R(p + 1)$.

We may thus assume that

$$Pu = g,$$

where the derivatives of g have analytic growth rate for $|\alpha| \leq R(p + 1)$, and u is compactly supported with respect to x.

Denoting by $\varphi_p(y)$ the function $\varphi_{2,p}(y)$, by (7) we have

$$\|\varphi_p D_y^p u\|_{\frac{1}{\ell+1}} + \sum_{j=1}^{2} \|X_j \varphi_p D_y^p u\|_0 \le C\left(\langle P\varphi_p D_y^p u, \varphi_p D_y^p u\rangle + \|\varphi_p D_y^p u\|_0^2\right). \tag{10}$$

Let us treat first the "error term" $\|\varphi_p D_y^p u\|_0$ in the r.h.s. above.

Proposition 2.2 ([6]) *Using the above notation we have the inequality*

$$\|\varphi_p D_y^p u\| \le C p^{-\frac{1}{m+1}} \|\varphi_p D_y^p u\|_{\frac{1}{m+1}} + C^{p+1} p^p, \tag{11}$$

where C is positive and independent of p.

Proof Let χ be a smooth function such that $\chi(t) = 1$ if $|t| \ge 2$ and $\chi(t) = 0$ if $|t| \le 1$. Consider the pseudodifferential operator $\chi(p^{-1}D_y)$. We have that $\chi(p^{-1}D_y) \in OPS^0_{0,0}$, the Hörmander (ρ,δ)-class of order 0 with $\rho = \delta = 0$. Then

$$\|\varphi_p D_y^p u\| \le \|(1-\chi(p^{-1}D_y))\varphi_p D_y^p u\| + \|\chi(p^{-1}D_y)\varphi_p D_y^p u\|. \tag{12}$$

Consider the first summand on the r.h.s. above. Even though $1-\chi$ has a compact support, we have no composition formula for χ. Observe that

$$\varphi_p D_y^p u = \sum_{s=0}^{p} (-1)^s \binom{p}{s} D_y^{p-s}\left(\varphi_p^{(s)} u\right), \tag{13}$$

so that

$$\|(1-\chi(p^{-1}D_y))\varphi_p D_y^p u\| \le \sum_{s=0}^{p} \binom{p}{s} \left\|(1-\chi(p^{-1}D_y))D_y^{p-s}\left(\varphi_p^{(s)} u\right)\right\|. \tag{14}$$

We immediately verify that

$$\sigma\left((1-\chi(p^{-1}D_y))D_y^{p-s}\right) = (1-\chi(p^{-1}\eta))\eta^{p-s} \in S^0_{0,0},$$

because $p^{-1}|\eta| \le 2$ in the support of $1-\chi$. Here $\sigma(A)$ denotes the symbol of the pseudodifferential operator A.

We also verify that

$$\max_{0\le\alpha\le\ell} \sup_{\eta} \left|\partial_\eta^\alpha \left(1-\chi(p^{-1}\eta)\eta^{p-s}\right)\right| \le C^{p-s+1} p^{p-s}.$$

Here $C > 0$ is a suitable constant independent of p, and ℓ is given in Theorem A.2. This bounds the $S^0_{0,0}$-seminorms of $(1 - \chi(p^{-1}D_y))D_y^{p-s}$ needed to apply the Calderón–Vaillancourt theorem (see Theorem A.2 in the appendix) so that we obtain that

$$\|(1 - \chi(p^{-1}D_y))D_y^{p-s}\|_{\mathscr{L}(L^2,L^2)} \leq C^{p-s+1}p^{p-s},$$

for a new positive constant C. Hence, using the definition of the cutoff function φ_p, we deduce

$$\begin{aligned}\|(1 - \chi(p^{-1}D_y))\varphi_p D_y^p u\| &\leq \sum_{s=0}^{p} \binom{p}{s} C^{p-s+1} p^{p-s} \|\varphi_p^{(s)} u\| \\ &\leq C^{p+1} C_\varphi p^p \|u\| \sum_{s=0}^{p} \binom{p}{s} \left(\frac{C_\varphi}{C}\right)^s \\ &\leq C_1^{p+1} p^p,\end{aligned}$$

where C_1 is independent of p if we choose $C \geq C_\varphi$, which is always possible.

Consider now the second summand in (12). We have

$$\|\chi(p^{-1}D_y)\varphi_p D_y^p u\| \;=\; p^{-\frac{1}{m+1}} \|p^{\frac{1}{m+1}} \chi(p^{-1}D_y) D^{-\frac{1}{m+1}} \circ D^{\frac{1}{m+1}} \varphi_p D_y^p u\|,$$

where $D^s = \mathrm{Op}\left((1 + |\xi|^2 + \eta^2)^{s/2}\right)$, for any $s \in \mathbb{R}$. Again using the support of χ, we see that

$$\sigma\left(p^{\frac{1}{m+1}} \chi(p^{-1}D_y) D^{-\frac{1}{m+1}}\right) = p^{\frac{1}{m+1}} \chi(p^{-1}\eta)(1 + |\xi|^2 + \eta^2)^{-\frac{1}{2(m+1)}} \in S^0_{0,0},$$

with the $S^0_{0,0}$-seminorms uniformly bounded w.r.t. p. Thus the Calderón–Vaillancourt theorem yields

$$\|p^{\frac{1}{m+1}} \chi(p^{-1}D_y) D^{-\frac{1}{m+1}}\|_{\mathscr{L}(L^2,L^2)} \leq C,$$

where C is a positive constant independent of p. So we have

$$\begin{aligned}\|\chi(p^{-1}D_y)\varphi_p D_y^p u\| &\leq C p^{-\frac{1}{m+1}} \|D^{\frac{1}{m+1}} \varphi_p D_y^p u\| \\ &\leq C p^{-\frac{1}{m+1}} \|\varphi_p D_y^p u\|_{\frac{1}{m+1}}.\end{aligned}$$

This completes the proof of the proposition. □

Consider next the scalar product on the r.h.s. of (10). Since $[X^2, A] = 2X[X, A] - [X, [X, A]]$, we have

$$\begin{aligned}\langle P\varphi_p D_y^p u, \varphi_p D_y^p u\rangle &= \langle \varphi_p D_y^p P u, \varphi_p D_y^p u\rangle + \langle [P, \varphi_p D_y^p] u, \varphi_p D_y^p u\rangle \\ &= \langle \varphi_p D_y^p P u, \varphi_p D_y^p u\rangle + 2\sum_{j=1}^{2} \langle [X_j, \varphi_p D_y^p] u, X_j^* \varphi_p D_y^p u\rangle \\ &\quad - \sum_{j=1}^{2} \langle [X_j, [X_j, \varphi_p D_y^p]] u, \varphi_p D_y^p u\rangle \\ &= \langle \varphi_p D_y^p g, \varphi_p D_y^p u\rangle + 2\langle [X_2, \varphi_p D_y^p] u, X_2^* \varphi_p D_y^p u\rangle \\ &\quad - \langle [X_2, [X_2, \varphi_p D_y^p]] u, \varphi_p D_y^p u\rangle, \end{aligned} \tag{15}$$

since X_1 commutes with $\varphi_p D_y^p$, due to our choice of the cutoff function φ_p.

Since the derivatives of g up to the order $R(p+1)$ have analytic growth rate on the support of φ_p, we see that

$$\left|\langle \varphi_p D_y^p g, \varphi_p D_y^p u\rangle\right| \le \|\varphi_p D_y^p u\|_0^2 + \left(C_g^{1+p} p!\right)^2,$$

where $C_g > 0$ and the first term is treated using Proposition 2.2. Let us consider the second term in the expression of the scalar product in (15). Writing $X_2^* = X_2 + g_2$ we have

$$\begin{aligned}&\langle [X_2, \varphi_p D_y^p] u, X_2^* \varphi_p D_y^p u\rangle \\ &\quad = \langle [X_2, \varphi_p D_y^p] u, X_2 \varphi_p D_y^p u\rangle + \langle [X_2, \varphi_p D_y^p] u, g_2 \varphi_p D_y^p u\rangle. \end{aligned} \tag{16}$$

Consider the first term on the r.h.s. of the above relation:

$$\left|\langle [X_2, \varphi_p D_y^p] u, X_2 \varphi_p D_y^p u\rangle\right| \le \delta \|X_2 \varphi_p D_y^p u\|_0^2 + C_\delta \|[X_2, \varphi_p D_y^p] u\|_0^2,$$

where δ is small in such a way that its coefficient can be absorbed on the l.h.s. of (10). Consider then the second term above.

$$[X_2, \varphi_p D_y^p] = (x^m y^{2a} + x^\ell)\varphi'_p D_y^p + x^m \varphi_p [y^{2a}, D_y^p] D_y,$$

where $\varphi'_p = D_y \varphi_p$. Since

$$\varphi'_p D_y^p = \sum_{r=0}^{p-1} (-1)^r D_y \varphi_p^{(r+1)} D_y^{p-1-r} + (-1)^p \varphi_p^{(p+1)}, \tag{17}$$

and

$$\left[y^{2a}, D_y^p\right] = -\sum_{k=1}^{2a} \binom{p}{k} \frac{(2a)!}{(2a-k)!} (-i)^k y^{2a-k} D_y^{p-k}, \tag{18}$$

we have

$$\begin{aligned}
[X_2, \varphi_p D_y^p] &= \sum_{r=0}^{p-1} (-1)^r X_2 \varphi_p^{(r+1)} D_y^{p-1-r} + (-1)^p (x^m y^{2a} + x^\ell) \varphi_p^{(p+1)} \\
&\quad - \sum_{k=1}^{2a} \binom{p}{k} \frac{(2a)!}{(2a-k)!} (-i)^k y^{2a-k} x^m \varphi_p D_y^{p-k+1} \\
&= \sum_{r=0}^{p-1} (-1)^r X_2 \varphi_p^{(r+1)} D_y^{p-1-r} + (-1)^p (x^m y^{2a} + x^\ell) \varphi_p^{(p+1)} \\
&\quad - \sum_{k=1}^{2a} \sum_{r=0}^{p-k} \binom{p}{k} \frac{(2a)!}{(2a-k)!} (-i)^k (-1)^r y^{2a-k} x^{m-\ell} \\
&\qquad \cdot (1 + y^{2a} x^{m-\ell})^{-1} X_2 \varphi_p^{(r)} D_y^{p-k-r} \\
&\quad - \sum_{k=1}^{2a} \binom{p}{k} \frac{(2a)!}{(2a-k)!} (-i)^k y^{2a-k} x^m (-1)^{p-k+1} \varphi_p^{(p-k+1)}.
\end{aligned} \tag{19}$$

Let us now get back to (16).

$$\begin{aligned}
\left| \langle [X_2, \varphi_p D_y^p] u, X_2^* \varphi_p D_y^p u \rangle \right| &\le \delta \| X_2 \varphi_p D_y^p u \|_0^2 \\
&\quad + \gamma \| \varphi_p D_y^p u \|_0^2 + C'_\delta \| [X_2, \varphi_p D_y^p] u \|_0^2,
\end{aligned}$$

where $\gamma > 0$ and C'_δ denotes a new positive constant depending on δ. The first term—as already said—can be absorbed on the l.h.s. of (10) provided δ is small. The second is dealt with using Proposition 2.2. Hence let us consider the third.

By (19) we have

$$\begin{aligned}
\| [X_2, \varphi_p D_y^p] u \|_0 &\le \sum_{r=0}^{p-1} \| X_2 \varphi_p^{(r+1)} D_y^{p-1-r} u \|_0 + C_X \Bigg(\| \varphi_p^{(p+1)} u \|_0 \\
&\quad + \sum_{k=1}^{2a} \sum_{r=0}^{p-k} p^k \| X_2 \varphi_p^{(r)} D_y^{p-k-r} u \|_0 + \sum_{k=1}^{2a} p^k \| \varphi_p^{(p-k+1)} u \|_0 \Bigg),
\end{aligned}$$

where C_X denotes a positive constant depending on the problem data only, that is small if we suitably shrink the open set Ω.

As we can see the second and fourth terms above are bounded by $C_0^{p+1} p!$ due to the properties of the Ehrenpreis sequence φ_p, while the first and third terms contain norms that can be bounded once more using the subelliptic estimate (10).

Let us now examine the double commutator term in (15). To keep the notation simple we write $f(x, y) = y^{2a}x^m + x^\ell$, so that $X_2 = f D_y$. First we need an expression for the double commutator. To keep it short, we write $f^{(k)}$ for $D_y^k f$, since only y-derivatives are involved at this stage.

$$
\begin{aligned}
[X_2, [X_2, \varphi_p D_y^p]] &= [f D_y, f\varphi_p' D_y^p - \sum_{k=1}^{p} \binom{p}{k} \varphi_p f^{(k)} D_y^{p-k+1}] \\
&= f[D_y, f\varphi_p'] D_y^p + f\varphi_p' [f, D_y^p] D_y - \sum_{k=1}^{p} \binom{p}{k} [f D_y, \varphi_p f^{(k)} D_y^{p-k+1}] \\
&= f[D_y, f\varphi_p'] D_y^p - f\varphi_p' [D_y^p, f] D_y - \sum_{k=1}^{p} \binom{p}{k} f[D_y, \varphi_p f^{(k)}] D_y^{p-k+1} \\
&\qquad - \sum_{k=1}^{p} \binom{p}{k} \varphi_p f^{(k)} [f, D_y^{p-k+1}] D_y \\
&= f(f\varphi_p')' D_y^p - \sum_{k=1}^{p} \binom{p}{k} f\varphi_p' f^{(k)} D_y^{p-k+1} - \sum_{k=1}^{p} \binom{p}{k} f(\varphi_p f^{(k)})' D_y^{p-k+1} \\
&\qquad + \sum_{k=1}^{p} \sum_{k_1=1}^{p-k+1} \binom{p}{k} \binom{p-k+1}{k_1} \varphi_p f^{(k)} f^{(k_1)} D_y^{p-k-k_1+2} \\
&= \sum_{j=1}^{4} T_j. \qquad (20)
\end{aligned}
$$

Thus we have to examine the terms

$$
\langle [X_2, [X_2, \varphi_p D_y^p]] u, \varphi_p D_y^p u \rangle = \sum_{j=1}^{4} \langle T_j u, \varphi_p D_y^p u \rangle.
$$

Let us start with the summand involving T_1. Applying (17), we have

$$
\begin{aligned}
\langle T_1 u, \varphi_p D_y^p u\rangle &= \langle (f\varphi_p')' D_y^p u, f\varphi_p D_y^p u\rangle \\
&= \sum_{r=0}^{p-1} (-1)^r \langle (f\varphi_p')^{(r+1)} D_y^{p-r-1} u, D_y f\varphi_p D_y^p u\rangle \\
&\qquad + (-1)^{p+1} \langle (f\varphi_p')^{(p+1)} u, f\varphi_p D_y^p u\rangle \\
&= \sum_{r=0}^{p-1}\sum_{\alpha=0}^{r+1} (-1)^r \binom{r+1}{\alpha} \Big(\langle f^{(\alpha)} \varphi_p^{(r+2-\alpha)} D_y^{p-r-1} u, X_2 \varphi_p D_y^p u\rangle \\
&+ \langle f^{(\alpha)} \varphi_p^{(r+2-\alpha)} D_y^{p-r-1} u, f'\varphi_p D_y^p u\rangle \Big) + (-1)^{p+1} \langle (f\varphi_p')^{(p+1)} u, f\varphi_p D_y^p u\rangle .
\end{aligned}
$$

We are going to treat each term above by essentially pulling back a y-derivative by means of (17). We do this in detail here and will then consider such procedure as routine for the remaining terms.

Consider the scalar product $\langle f^{(\alpha)} \varphi_p^{(r+2-\alpha)} D_y^{p-r-1} u, X_2 \varphi_p D_y^p u\rangle$. The second factor can be absorbed on the l.h.s. of (10), so that we need to estimate the norm of the (square of the) first factor.

As we saw above, only a fixed number of derivatives of f are nonzero, and moreover we have that $|f^{(\alpha)}| \leq C_f |f|$, since $\ell \leq m$. We note explicitly that this fact has already been used for the estimate of the simple commutator term.

Hence for the first set of summands we have

$$
\begin{aligned}
&\binom{r+1}{\alpha} \| f^{(\alpha)} \varphi_p^{(r+2-\alpha)} D_y^{p-r-1} u \|_0 \\
&\qquad \leq (r+1)^\alpha \Big(\sum_{r_1=0}^{p-2-r} \| f^{(\alpha)} D_y \varphi_p^{(r+2-\alpha+r_1)} D_y^{p-r-r_1-2} u \|_0 \\
&\qquad\qquad + \| f^{(\alpha)} \varphi_p^{(r+2-\alpha+p-1-r)} u \|_0 \Big) \\
&\qquad \leq C_f (r+1)^\alpha \Big(\sum_{r_1=0}^{p-2-r} \| X_2 \varphi_p^{(r+2-\alpha+r_1)} D_y^{p-r-r_1-2} u \|_0 \\
&\qquad\qquad + \| \varphi_p^{(r+2-\alpha+p-1-r)} u \|_0 \Big). \qquad (21)
\end{aligned}
$$

The last summand is estimated by

$$
C_f C_\varphi^{p-\alpha+1} \|u\|_0 p^{p+1} \leq C_f (2C_\varphi)^{p+1} \|u\|_0 p^p .
$$

As for the other summands we see that the order of D_y drops by $r+r_1+2$, but there is a gain of p^{r+r_1+2} in the cutoff functions as well as in the term $(r+1)^\alpha$. Thus the above quantity exhibits an analytic growth rate.

Next we deal with the term involving T_4. The other terms can then be treated along the same lines. Thus

$$\begin{aligned}
&\langle T_4 u, \varphi_p D_y^p u\rangle \\
&\qquad = \sum_{k=1}^{p} \sum_{k_1=1}^{p-k+1} \binom{p}{k}\binom{p-k+1}{k_1} \langle f^{(k_1)} \varphi_p D_y^{p-k-k_1+2} u, f^{(k)} \varphi_p D_y^p u\rangle .
\end{aligned}$$

We point out explicitly that since $f^{(k)} \not\equiv 0$ for $k \leq 2a$, it is enough to estimate a single summand.

For the sake of brevity, we may bound the two binomials by p^{k+k_1} and work on a single summand, k and k_1 being understood to run on their intervals.

Therefore, using (17), we have

$$\begin{aligned}
&p^{k+k_1} \left| \langle f^{(k_1)} \varphi_p D_y^{p-k-k_1+2} u, f^{(k)} \varphi_p D_y^p u\rangle \right| \\
&\quad \leq p^{k+k_1} \Bigg(\sum_{r=0}^{p-k-k_1+1} \left| \langle f^{(k_1)} D_y \varphi_p^{(r)} D_y^{p-k-k_1+1-r} u, f^{(k)} \varphi_p D_y^p u\rangle \right| \\
&\qquad\qquad + \left| \langle f^{(k_1)} \varphi_p^{(p-k-k_1+2)} u, f^{(k)} \varphi_p D_y^p u\rangle \right| \Bigg) \\
&\leq p^{k+k_1} \Bigg(\sum_{r=0}^{p-k-k_1+1} \sum_{r_1=0}^{p-1} \left| \langle f^{(k_1)} D_y \varphi_p^{(r)} D_y^{p-k-k_1+1-r} u, f^{(k)} D_y \varphi_p^{(r_1)} D_y^{p-1-r_1} u\rangle \right| \\
&\qquad + \sum_{r=0}^{p-k-k_1+1} \left| \langle f^{(k_1)} D_y \varphi_p^{(r)} D_y^{p-k-k_1+1-r} u, f^{(k)} \varphi_p^{(p)} u\rangle \right| \\
&\qquad + \sum_{r_1=0}^{p-1} \left| \langle f^{(k_1)} \varphi_p^{(p-k-k_1+2)} u, f^{(k)} D_y \varphi_p^{(r_1)} D_y^{p-1-r_1} u\rangle \right| \\
&\qquad\qquad + \left| \langle f^{(k_1)} \varphi_p^{(p-k-k_1+2)} u, f^{(k)} \varphi_p^{(p)} u\rangle \right| \Bigg).
\end{aligned}$$

We may now apply the Cauchy–Schwartz inequality to each scalar product and keep in mind that, if it's not zero, we have $|f^{(k)}| \leq C_f |f|$, where C_f denotes a suitable positive constant.

Eventually we obtain

$$
\begin{aligned}
& p^{k+k_1}\left|\langle f^{(k_1)}\varphi_p D_y^{p-k-k_1+2}u, f^{(k)}\varphi_p D_y^p u\rangle\right| \\
& \quad \le C_f^2\Bigg(\sum_{r=0}^{p-k-k_1+1}\sum_{r_1=0}^{p-1} p^{k+k_1-1}\|X_2\varphi_p^{(r)}D_y^{p-k-k_1+1-r}u\|_0 \\
& \qquad \cdot p\|X_2\varphi_p^{(r_1)}D_y^{p-1-r_1}u\|_0 \\
& \quad + \sum_{r=0}^{p-k-k_1+1} p^{k+k_1-1}\|X_2\varphi_p^{(r)}D_y^{p-k-k_1+1-r}u\|_0\cdot p\|\varphi_p^{(p)}u\|_0 \\
& \quad + \sum_{r_1=0}^{p-1} p^{k+k_1-1}\|\varphi_p^{(p-k-k_1+2)}u\|_0\cdot p\|X_2\varphi_p^{(r_1)}D_y^{p-1-r_1}u\|_0 \\
& \qquad + p^{k+k_1-1}\|\varphi_p^{(p-k-k_1+2)}u\|_0\cdot p\|\varphi_p^{(p)}u\|_0\Bigg).
\end{aligned}
$$

The above norms exhibit an analytic growth rate, and choosing C_0 in (8) large enough compared to C_φ in (9), we can safely sum the various contributions. Hence, as in the case of the simple commutator, we can conclude.

We leave the details to the reader.

3 Proof of Theorem 1.1 in Case *(ii)*

Assume now $\ell > m$. Again we are going to use the subelliptic estimate to prove the second statement of Theorem 1.1.

As a preliminary observation we point out that if $(x_0, y_0) \neq (0, 0)$ and Ω is a neighborhood of (x_0, y_0), with $(0, 0) \notin \Omega$, then, by well-known facts, see [16] and [26], we have that the solutions of $Pu = g \in C^\omega(\Omega)$ are also in $C^\omega(\Omega)$.

In fact, if (x_0, y_0) does not belong to the space projection of the characteristic manifold, so does Ω and P is microlocally elliptic, which implies C^ω hypoelliptic.

On the other hand if (x_0, y_0) belongs to the space projection of the first strata, $\Sigma_{1,\pm}$, see (4), which means that $y_0 \neq 0$, then it is well known that P is analytic, and hence G^s, for every $s \ge 1$, microhypoelliptic on Ω.

This allows us to simplify a bit the proof by eliminating the cutoff function φ_p. Let us consider the equation

$$Pu = g \in G^{s_0}(\Omega).$$

From the above remarks we know that $u \in G^{s_0}(\Omega \setminus \{(0,0)\})$. Let $\varphi \in G_0^{s_0}(\Omega)$, $\varphi \equiv 1$ in a neighborhood, U, of the origin, $U \Subset \Omega$. Let $v = \varphi u$. Then

$$Pv = \varphi P u + [P, \varphi] u.$$

The r.h.s. of the above equation is in $G_0^{s_0}(\Omega)$ since the first-order differential operator $[P, \varphi] = 0$ in U and u is Gevrey s_0 outside $(0,0)$, as mentioned above. Hence writing u instead of v, we may assume that u solves

$$Pu = g \in G_0^{s_0}(\Omega), \tag{22}$$

and $u \in C_0^\infty(\Omega)$, supported in a neighborhood of the origin.

As above we are going to show that

$$\|D_y^p u\|_{\frac{1}{\ell+1}} + \sum_{j=1}^{2} \|X_j D_y^p u\|_0 \le C_0^{p+1} p!^{s_0}, \tag{23}$$

where s_0 is defined in (6).

By (7) we have

$$\|D_y^p u\|^2_{\frac{1}{\ell+1}} + \sum_{j=1}^{2} \|X_j D_y^p u\|_0^2 \le C \left(\langle P D_y^p u, D_y^p u\rangle + \|D_y^p u\|_0^2 \right). \tag{24}$$

Consider the error term first. Since

$$\begin{aligned}\|D_y^p u\|_0 \le \|D_y^{p-\frac{1}{\ell+1}} u\|_{\frac{1}{\ell+1}} &\le \delta_1^{\frac{p(\ell+1)}{p(\ell+1)-1}} \|D_y^p u\|_{\frac{1}{\ell+1}} + \delta_1^{-p(\ell+1)} \|u\|_{\frac{1}{\ell+1}} \\ &= \delta \|D_y^p u\|_{\frac{1}{\ell+1}} + C_\delta^p \|u\|_{\frac{1}{\ell+1}},\end{aligned} \tag{25}$$

the first term can be absorbed on the left of (24) if δ is chosen small. The other term is harmless in any Gevrey type estimate.

Consider now the term involving the scalar product in (24). We have

$$\begin{aligned}\langle P D_y^p u, D_y^p u\rangle &= \langle D_y^p P u, D_y^p u\rangle + \langle [P, D_y^p] u, D_y^p u\rangle \\ &= \langle D_y^p P u, D_y^p u\rangle + 2\langle [X_2, D_y^p] u, X_2^* D_y^p u\rangle - \langle [X_2, [X_2, D_y^p]] u, D_y^p u\rangle.\end{aligned}$$

For the first term above we use the fact that $Pu = g \in G_0^{s_0}$:

$$\left| \langle D_y^p g, D_y^p u\rangle \right| \le \|D_y^p u\|_0^2 + \left(C_g^{1+p} p!^{s_0} \right)^2.$$

Since $X_2^* = X_2 + f'$, using the same notations as in Sect. 2, and

$$[X_2, D_y^p] = -\sum_{k=1}^{p} \binom{p}{k} f^{(k)} D_y^{p-k+1},$$

we have

$$\langle [X_2, D_y^p]u, X_2^* D_y^p u\rangle = -\sum_{k=1}^{p} \binom{p}{k} \Big(\langle f^{(k)} D_y^{p-k+1} u, X_2 D_y^p u\rangle + \langle f^{(k)} D_y^{p-k+1} u, f' D_y^p u\rangle \Big).$$

The norm of the second factor in $\langle f^{(k)} D_y^{p-k+1} u, X_2 D_y^p u\rangle$ can be absorbed on the l.h.s. of (24), while the second factor of $\langle f^{(k)} D_y^{p-k+1} u, f' D_y^p u\rangle$ can be treated as in (25), so that we only need to estimate the norm $\| f^{(k)} D_y^{p-k+1} u\|_0$.

Modulo an inessential constant, keeping the binomial into account, we want to bound the quantity

$$p^k \|x^m y^{2a-k} D_y^{p-k+1} u\|_0. \tag{26}$$

We follow the same method we used in [8], but in this case it is simpler due to the lack of a cutoff function.

First observe that

$$p^k |y|^{2a-k} = p^{\theta k} |y|^{2a-k} p^{(1-\theta)k} \le C_0 \left((p^{\theta k} |y|^{2a-k})^{\lambda} + p^{(1-\theta)k\mu} \right),$$

where

$$\frac{1}{\lambda} + \frac{1}{\mu} = 1.$$

We point out explicitly that we are arguing in the case $k < 2a$. If $k = 2a$, we simply skip this step and go to the next. We omit the details. Choosing

$$\lambda = \frac{2a}{2a-k},$$

we have $\mu = \frac{2a}{k}$ and

$$p^k |y|^{2a-k} \le C_0 \left(p^{\theta \frac{2ak}{2a-k}} y^{2a} + p^{(1-\theta)2a} \right).$$

As a consequence

$$p^k \|x^m y^{2a-k} D_y^{p-k+1} u\|_0 \\ \leq C_0\Big(p^{\theta \frac{2ak}{2a-k}} \|x^m y^{2a} D_y^{p-k+1} u\|_0 + p^{(1-\theta)2a} \|x^m D_y^{p-k+1} u\|_0\Big). \tag{27}$$

Observe now that $|x|^m y^{2a} \leq |x|^m (y^{2a} + x^{\ell-m})$ because of assumption (3). Hence we get

$$p^k \|x^m y^{2a-k} D_y^{p-k+1} u\|_0 \\ \leq C_0\Big(p^{\theta \frac{2ak}{2a-k}} \|X_2 D_y^{p-k} u\|_0 + p^{(1-\theta)2a} \|x^m D_y^{p-k+1} u\|_0\Big). \tag{28}$$

The first term can be resubjected to the procedure. It gives a Gevrey

$$G^{\theta \frac{2a}{2a-k}}.$$

The above result may be established using an induction argument on the order of y powers, y-derivatives and powers of p. This is a classical technical step that, for the sake of brevity, we skip here and in what follows (see [8, 10] for details). We also point out that in the present case the constants are easy to control since they depend on the problem data and not on p due to the absence of cutoff functions.

Consider the second term. We write

$$p^{(1-\theta)2a} |x|^m = p^\rho \cdot p^{(1-\theta)2a-\rho} |x|^m,$$

and we use the same inequality used above:

$$p^\rho \cdot p^{(1-\theta)2a-\rho} |x|^m \leq C_0\Big(p^{\rho\sigma_1} + p^{((1-\theta)2a-\rho)\sigma_2} |x|^{m\sigma_2}\Big),$$

with

$$\frac{1}{\sigma_1} + \frac{1}{\sigma_2} = 1.$$

Choosing $\sigma_2 = \frac{\ell}{m}$, and hence $\sigma_1 = \frac{\ell}{\ell-m}$, we obtain

$$p^\rho \cdot p^{(1-\theta)2a-\rho} |x|^m \leq C_0\Big(p^{\rho \frac{\ell}{\ell-m}} + p^{((1-\theta)2a-\rho)\frac{\ell}{m}} |x|^\ell\Big).$$

Plugging this into the second term on the r.h.s. of (28), we get

$$
\begin{aligned}
p^{(1-\theta)2a}\|x^m D_y^{p-k+1}u\|_0 &\le C_0\Big(p^{\rho\frac{\ell}{\ell-m}}\|D_y^{p-k+1}u\|_0 \\
&\quad + p^{((1-\theta)2a-\rho)\frac{\ell}{m}}\|x^\ell D_y^{p-k+1}u\|_0\Big) \\
&\le C_0\Big(p^{\rho\frac{\ell}{\ell-m}}\|D_y^{p-k+1-\frac{1}{\ell+1}}u\|_{\frac{1}{\ell+1}} + p^{((1-\theta)2a-\rho)\frac{\ell}{m}}\|X_2 D_y^{p-k}u\|_0\Big). \qquad (29)
\end{aligned}
$$

The last term yields

$$
G^{((1-\theta)2a-\rho)\frac{\ell}{m}\frac{1}{k}},
$$

as Gevrey regularity, while the first gives (see Sect. 3 [10])

$$
G^{\rho\frac{\ell}{\ell-m}\left(k-\frac{\ell}{\ell+1}\right)^{-1}}.
$$

Next we compute the parameters θ, ρ in such a way that the three Gevrey regularities coincide, i.e.,

$$
\theta\frac{2a}{2a-k} = \rho\frac{\ell}{\ell-m}\left(k-\frac{\ell}{\ell+1}\right)^{-1} = ((1-\theta)2a-\rho)\frac{\ell}{m}\frac{1}{k}. \qquad (30)
$$

This yields

$$
\theta = \frac{2a-k}{2a}\rho\frac{\ell}{\ell-m}\left(k-\frac{\ell}{\ell+1}\right)^{-1},
$$

from which we can compute ρ in such a way that

$$
\rho\frac{\ell}{\ell-m}\left(k-\frac{\ell}{\ell+1}\right)^{-1} = \left(1-\frac{1}{2a}\frac{\ell-m}{\ell+1}\right)^{-1},
$$

which is the desired regularity.

The next and last term to examine is

$$
\langle [X_2, [X_2, D_y^p]]u, D_y^p u\rangle.
$$

Since

$$
[X_2, D_y^p] = -\sum_{k=1}^{p}\binom{p}{k} f^{(k)} D_y^{p-k+1},
$$

and

$$[X_2, f^{(k)} D_y^{p-k+1}] = [f D_y, f^{(k)} D_y^{p-k+1}]$$
$$= f f^{(k+1)} D_y^{p-k+1} - f^{(k)} \sum_{k_1=1}^{p-k+1} \binom{p-k+1}{k_1} f^{(k_1)} D_y^{p-(k+k_1)+2},$$

we may write

$$[X_2, [X_2, D_y^p]] = -\sum_{k=1}^{p} \binom{p}{k} f f^{(k+1)} D_y^{p-k+1}$$
$$+ \sum_{k=1}^{p} \sum_{k_1=1}^{p-k+1} \binom{p}{k} \binom{p-k+1}{k_1} f^{(k)} f^{(k_1)} D_y^{p-(k+k_1)+2}$$
$$= S_1 + S_2. \tag{31}$$

We recall that $f^{(k)} \not\equiv 0$ only if $0 \le k \le 2a$, so that it is enough to estimate each summand.

Hence

$$\left|\langle S_1 u, D_y^p u\rangle\right| \le \sum_{k=1}^{p} p^k \| f D_y^{p-k+1} u\|_0 \| f^{(k+1)} D_y^p u\|_0$$
$$\le C_1 \sum_{k=1}^{p} p^k \| X_2 D_y^{p-k} u\|_0 \| D_y^p u\|_0.$$

The second factor is treated as in (25), and the first has an analytic growth rate.

We recall here that only a finite number of y-derivatives of f appear, generating constants depending on the problem's data, but independent of p. We shall often use this fact without an explicit mention.

This sets S_1. Consider S_2. Again

$$\left|\langle S_2 u, D_y^p u\rangle\right|$$
$$\le \sum_{k=1}^{p} \sum_{k_1=1}^{p-k+1} \binom{p}{k} \binom{p-k+1}{k_1} \left|\langle f^{(k)} f^{(k_1)} D_y^{p-(k+k_1)+2} u, D_y^p u\rangle\right|$$
$$\le C_a \sum_{k=1}^{p} \sum_{k_1=1}^{p-k+1} p^{k+k_1} \left|\langle x^m y^{2a-(k+k_1-1)} D_y^{p-(k+k_1-1)+1} u, x^m y^{2a-1} D_y^p u\rangle\right|$$

$$\leq C_a \sum_{k=1}^{p} \sum_{k_1=1}^{p-k+1} p^{k+k_1-1} \|x^m y^{2a-(k+k_1-1)} D_y^{p-(k+k_1-1)+1} u\|_0$$

$$\cdot\, p\|x^m y^{2a-1} D_y^p u\|_0. \qquad (32)$$

Here C_a denotes a positive constant independent of p.

Next we estimate both terms

$$p^{k+k_1-1} \|x^m y^{2a-(k+k_1-1)} D_y^{p-(k+k_1-1)+1} u\|_0$$

and

$$p\|x^m y^{2a-1} D_y^p u\|_0,$$

with the same argument as that for (26).

This completes the proof of the theorem.

Appendix

For the sake of completeness we recall here some well-known facts used throughout the chapter.

Definition A.1 For any $m \in \mathbb{R}$, $\rho, \delta \in \mathbb{R}$ with $0 \leq \delta \leq \rho \leq 1$, $\delta < 1$, we denote by $S^m_{\rho,\delta}$ the set of all the functions $p(x, \xi) \in C^\infty(\mathbb{R}^{2n})$ such that for every multi-index α, β there exists a positive constant $C_{\alpha,\beta}$ for which

$$|\partial_\xi^\alpha \partial_x^\beta p(x, \xi)| \leq C_{\alpha,\beta} \langle \xi \rangle^{m-\rho|\alpha|+\delta|\beta|},$$

where $\langle \xi \rangle = (1 + |\xi|^2)^{\frac{1}{2}}$.

We denote by $OPS^m_{\rho,\delta}$ the class of the corresponding pseudodifferential operators $P = p(x, D)$.

It is trivial to see that the symbol class $S^m_{\rho,\delta}$ equipped with the seminorms

$$|p|_\ell^{(m)} = \max_{|\alpha+\beta| \leq \ell} \sup_{(x,\xi)} \{|\partial_\xi^\alpha \partial_x^\beta p(x, \xi)| \langle \xi \rangle^{-(m-\rho|\alpha|+\delta|\beta|)}\}, \quad \ell \in \mathbb{N}$$

is a Fréchet space.

The Calderón–Vaillancourt theorem shows the L^2-continuity properties of the pseudodifferential operators in the above classes (see [13] or, for a more general setting, [23] Chap. 7, Th.1.6). We state below a formulation of such a theorem for pseudodifferential operators of order zero.

Theorem A.2 (Calderón–Vaillancourt) *Let* $P = p(x, D) \in OPS^0_{\rho,\delta}$ *with* $0 \leq \delta \leq \rho, \delta < 1$. *Then there exist a positive integer* ℓ *and a positive constant* M *(depending only on* n*) such that*

$$\|Pu\| \leq M|p|^{(0)}_{\ell}\|u\|, \quad \textit{for every } u \in L^2(\mathbb{R}^n).$$

References

1. P. Albano, A. Bove and M. Mughetti, *Analytic Hypoellipticity for Sums of Squares and the Treves Conjecture*, J. Funct. Anal. **274** (2018), no. 10, 2725–2753.
2. M. S. Baouendi and Ch. Goulaouic, *Nonanalytic-hypoellipticity for some degenerate operators*, Bull. A. M. S., **78** (1972), 483–486.
3. A. Bove, *Analytic and Gevrey Regularity for Certain Sums of Two Squares in Two Variables*, 2023, to appear in Tohoku Math. J. **76** (2) (2024), no. 1, 1–85.
4. A. Bove, M. Mughetti and D.S. Tartakoff, *Hypoellipticity and nonhypoellipticity for sums of squares of complex vector fields*, Analysis and PDE **6** (2) (2013), 371–446.
5. A. Bove, M. Mughetti, *On a new method of proving Gevrey hypoellipticity for certain sums of squares*, Advances in Mathematics **293** (2016), 146–220.
6. A. Bove and M. Mughetti, *Analytic hypoellipticity for sums of squares in the presence of symplectic non Treves strata*, J. Inst. Math. Jussieu **19**(6) (2020), 1877–1888.
7. A. Bove and M. Mughetti, *Analytic Hypoellipticity for Sums of Squares and the Treves Conjecture, II*, Analysis and PDE **10** (7) (2017), 1613–1635.
8. A. Bove and M. Mughetti, *Gevrey Regularity for a Class of Sums of Squares of Monomial Vector Fields*, Advances in Math. **373** (2020), 35 pages.
9. A. Bove and M. Mughetti, *Analytic regularity for solutions to sums of squares: an assessment*, Complex Analysis and its Synergies **6** (2020), article no. 18.
10. A. Bove and M. Mughetti, *Minimal Gevrey Regularity for Hörmander Operators*, Int. Math. Res. Not. IMRN (2024), no. 4, 2790–2832. https://doi.org/10.1093/imrn/rnad055.
11. A. Bove and M. Mughetti, *Optimal Gevrey Regularity for Certain Sums of Squares in Two Variables*, 2023, to appear in Ann. S.N.S. Pisa.
12. A. Bove and F. Treves, *On the Gevrey hypo-ellipticity of sums of squares of vector fields*, Ann. Inst. Fourier (Grenoble) **54**(2004), 1443-1475.
13. A. P. Calderón and R. Vaillancourt, *A Class of Bounded Pseudo-Differential Operators*, Proc. Nat. Acad. Sci. USA **69** (1972), no. 5, 1185–1187.
14. G. Chinni, *On the partial and microlocal regularity for generalized Métivier operators*, J. Pseudo-Differ. Oper. Appl. **14** (2023), no. 3, 26 pp.
15. G. Chinni, *On the sharp Gevrey regularity for a generalization of the Métivier operator*, Math. Ann. **388** (2024), no.2, 1629–1675. https://doi.org/10.1007/s00208-022-02558-7.
16. P. D. Cordaro and N. Hanges, *A New Proof of Okaji's Theorem for a Class of Sum of Squares Operators*, Ann. Inst. Fourier (Grenoble) **59** (2009), no. 2, 595–619.
17. L. Ehrenpreis, *Solutions of some problems of division IV*, Amer. J. Math **82** (1960), 522–588.
18. M. Derridj and C. Zuily, *Sur la régularité Gevrey des opérateurs de Hörmander*, J. Math. Pures Appl. **52** (1973), 309–336.
19. V. V. Grušin, *On a class of elliptic pseudodifferential operators degenerate on a submanifold*, Mat. Sbornik **84**(2) (1971), 155–185.
20. L. Hörmander, *Hypoelliptic second order differential equations*, Acta Math. **119** (1967), 147–171.
21. L. Hörmander, *Uniqueness Theorems and Wave Front Sets for Solutions of Linear Differential Equations with Analytic Coefficients*, Communications Pure Appl. Math. **24** (1971), 671–704.

22. L. Hörmander, *The Analysis of Partial Differential Operators, III*, Springer Verlag, 1985.
23. H. Kumano-Go, *Pseudodifferential Operators* , MIT Press, Cambridge, Mass.–London, 1981.
24. G. Métivier, *Non-hypoellipticité Analytique pour* $D_x^2 + (x^2 + y^2)D_y^2$, C. R. Acad. Sci. Paris Sér. I Math. **292** (1981), no. 7, 401–404.
25. M. Mughetti, *Regularity properties of a double characteristics differential operator with complex lower order terms*, J. Pseudo-Differ. Oper. Appl. **5** (2014), no. 3, 343–358.
26. T. Ōkaji, *Analytic hypoellipticity for operators with symplectic characteristics*, J. Math. Kyoto Univ. **25** (1985), no. 3, 489–514.
27. O. A. Oleĭnik and E. V. Radkevič, *The analyticity of the solutions of linear partial differential equations*, (Russian) Mat. Sb. (N.S.), **90(132)** (1973), 592–606.
28. L. Rothschild and E. M. Stein, *Hypoelliptic differential operators and nilpotent groups*, Acta Math. **137** (3–4) (1976), 247–320.
29. D.S. Tartakoff, *On the Local Real Analyticity of Solutions to* $\Box_b$ *and the* $\bar{\partial}$*-Neumann Problem*, Acta Math. **145** (1980), 117–204.
30. F. Trèves, *Analytic hypo-ellipticity of a class of pseudodifferential operators with double characteristics and applications to the* $\bar{\partial}$*-Neumann problem*, Comm. Partial Differential Equations **3** (1978), no. 6–7, 475–642.
31. F. Treves, *Symplectic geometry and analytic hypo-ellipticity, in Differential equations,* La Pietra 1996 (Florence), Proc. Sympos. Pure Math. **65**, Amer. Math. Soc., Providence, RI, 1999, 201–219.

On the Local Regularity of the Gevrey Vectors for Hörmander's Operators

Gregorio Chinni and Makhlouf Derridj

In honor of Jorge Hounie for his many mathematical achievements

1 Introduction

In [17] we dealt with Hörmander's operators of the first kind (or commonly known as "sums of squares of vector fields") and considered the case of analytic vectors of operators with analytic coefficients. For that we used the method by F.B.I. transform taking advantage from the sharp subelliptic estimate obtained in [1], see also [27]. Here we study the general class of Hörmander's operators, meaning those of degenerate elliptic type or of degenerate parabolic type, i.e., operators of the form

$$P(x, D) = -\sum_{\ell=1}^{m} (X_\ell(x, D))^2 + i X_0(x, D) + c(x), \tag{1}$$

in Ω, open neighborhood of the origin in $\mathbb{R}^n$, where $c(x)$ is a Gevrey function of order s ($s \geq 1$) in Ω and

$$X_\ell(x, D) = \sum_{j=1}^{n} a_{j,\ell}(x) D_j, \quad \ell = 0, 1, \ldots, m, \qquad \left(D_j = \frac{1}{\sqrt{-1}} \frac{\partial}{\partial x_j}\right),$$

where $a_{j,\ell}(x)$ are real-valued Gevrey functions of order s on Ω, satisfying the Hörmander condition, see Definition 2.2.

G. Chinni
Dipartimento di Matematica, Universitá di Bologna, Bologna, Italy

M. Derridj (✉)
Rue de la Juviniére, Les Loges en Josas, France
e-mail: maklouf.derridj@numericable.fr

S. Berhanu et al. (eds.), *Geometric Analysis of PDEs and Several Complex Variables*, Latin American Mathematics Series – UFSCar subseries,
https://doi.org/10.1007/978-3-031-69702-9_5

This question of investigation of the Gevrey regularity of analytic-Gevrey vectors of partial differential operators goes back to the work by T. Kotake and M. Narasimhan ([37], 1962) where they considered analytic vectors of elliptic partial differential operators with analytic coefficients on an open set in R^n and showed there analyticity. This property on the iterates of elliptic operators was also called, after their work, "Kotake-Narasimhan property." (E. Nelson studied before analytic vectors of operators in the context of Lie group theory ([39], 1959.) Then after more and more works followed, generalizing their result in various directions (systems, s-Gevrey regularity for s greater or equal to 1, $s = 1$ corresponding to analytic case), non-elliptic operators, global, local, or even microlocal situations. Let us mention a few of the numerous results in the field, as a guide leading at our result (we mention for that two surveys, one quite complete on results obtained before 1987 [7] and one very short, mainly on some results obtained in the eighties until 2016 [22]). A result of G. Metivier obtained in 1978, [38], lightened the way in the study of the case of non-elliptic operators and s-Gevrey situation, s greater or equal to 1, showing that the Kotake–Narasimhan property in s-Gevrey class, $s \geq 1$, of a partial differential operator P in an open set characterizes ellipticity of P in this open set, letting of course open the question on iterates of non-elliptic operators in case $s = 1$. A few years after, M.S. Baouendi and G.Metivier proved that the hypoelliptic operators of principal type with analytic coefficients satisfy the Kotake–Narasimhan property and gave condition on s' such that s-Gevrey vectors are in s'-Gevrey space for $s > 1$ [2]. In the case of systems, let us just mention the case of vector fields: For real vector fields $(X_1, \ldots, X_r)$ satisfying the Hörmander's condition introduced in his famous paper on hypoelliptic second-order partial differential equations [33], two papers appeared in 1980, one by M.Damlakhi–B.Helffer proving "Kotake-Narasimhan property" for such systems with analytic coefficients [19], the other by B.Helffer–C.Mattera [30], proving analyticity for what they called reducible analytic vectors of the system (which is a less restrictive condition on the system); for complex vector fields, we mention two papers: in [3], 2011, and [12], 2013, for systems defining locally integrable structures. Since the paper of L. Hörmander appeared, works on analytic or Gevrey regularity of associated analytic-Gevrey vectors began many years after: The first one was in the global context for a subclass of Hörmander's operators on a product of two tori [11] 2016 (in the global context the "Kotake-Narasimhan property" was also studied in [13]). The local version of such result for the general class of Hörmander's operators was established by the second author in two articles ([23], 2018, [24], 2019), with an optimal version in the case of the so-called Sum of squares [25] and more general operators of second order introduced by O.A. Oleĭnik and E.V. Radkevič in [40, 41, 26]. The microlocal version of such work was done by the two authors in the case of "Sum of squares with analytic coefficients," [17], 2022. We recall that in the case of second-order operators of type Hörmander, subclass of that studied by Oleĭnik and Radkevič, the so-called Oleĭnik and Radkevič-H örmander condition is weaker than that of Hörmander, in the sense that it does not allow obtaining the optimal subellipticity index, see the examples discussed in [26] and [17], related to the operators studied in [6, 9, 10, 14, 15, 16, 20, 21]. We would like to mention some results using intensively the FBI transform (and generalization of it as in [4, 5, 31, 32, 43]) and studying mainly operators (in particular with constant coefficients or of elliptic type) in more and more general classes of ultradifferentiable functions (see [31, 29], where there are many references).

2 Notations, Definitions, Preliminary Facts, and Main Result

Before stating the main result, we recall some basic definitions and facts. First the definition of Gevrey class:

Definition 2.1 If $\Omega \subset \mathbb{R}^n$ is an open set, we say that the function u belongs to the Gevrey class of order $s \geq 1$, $G^s(\Omega)$, if $u \in C^\infty(\Omega)$, and for every compact set $K \subset U$, there exists a positive constant C_K such that

$$\sup_K |\partial_x^\alpha u(x)| \leq C_K^{|\alpha|+1} \alpha!^s,$$

with $\alpha = (\alpha_1, \dots, \alpha_n) \in \mathbb{N}^n$.

When $s = 1$, we get the class of real analytic functions, also denoted by $C^\omega(U)$.

Remark 2.2 The sup-norm in the above definition can be replaced by $L^p(K)$ norms, $p \geq 1$. In particular we have that $u \in G^s(\Omega)$ if and only if $u \in C^\infty(\Omega)$, and for every compact set $K \subset U$, there exists a positive constant $\widetilde{C}_K$ such that

$$\|\partial_x^\alpha u\|_{L^2(K)} \leq \widetilde{C}_K^{|\alpha|+1} \alpha!^s.$$

Let $X(x, D)$ and $Y(x, D)$ be two vector fields with smooth coefficients. We write $\text{ad}_X(Y) = [X, Y]$, where $[X, Y]$ is the Lie bracket or the commutator between $X(x, D)$ and $Y(x, D)$. Let $I = (i_1, i_2, \dots, i_k)$, $i_j \in \{0, 1, 2, \dots, m\}$, and we denote by

$$X_I = \text{ad}_{X_{i_1}} \text{ad}_{X_{i_2}} \cdots \text{ad}_{X_{i_{k-1}}} (X_{i_k}),$$

the k-times iterated commutator.

To the multi-index I, we associate the weight length $|I|$ defined as

$$|I| = \#\left\{i_j \mid 1 \leq i_j \leq m,\ j = 1, \dots, k\right\} + 2\#\left\{i_j \mid i_j = 0,\ j = 1, \dots, k\right\}. \quad (2)$$

Definition 2.2 We say that the system of vector fields $\{X_0(x, D), \dots, X_m(x, D)\}$ satisfies the Hörmander's condition in Ω, if for every $x_0 \in \Omega$, there is a neighborhood Ω_0 of x_0, $\Omega_0 \Subset \Omega$, and a positive integer r, depending on Ω_0, such that the vector fields $X_0, \dots, X_m$ and X_I, with $|I| \leq r$, $|I|$ as in (2), generate the n-dimensional Lie algebra on Ω_0. We say that the system, $\{X_0(x, D), \dots, X_m(x, D)\}$, satisfies Hörmader's condition at step $\widetilde{r}$ in Ω, if $\widetilde{r} = \sup_{x_0 \in \Omega}\{r_{x_0}\}$, where r_{x_0} is the infimum of the integers such that the Hörmander's condition is satisfied at the point x_0.

In [33], L. Hörmander showed that if the system of vector fields $\{X_0 \dots, X_m\}$ satisfies the Hörmander's condition in Ω, then P is C^∞-hypoelliptic. Moreover in [42], Rothschild and Stein showed that the following a priori estimate holds

$$\|u\|_{\frac{2}{r}} + \sum_{\ell,j=1}^{m} \|X_\ell X_j u\|_0 + \|X_0 u\|_0 \leq C\left(\|Pu\|_0 + \|u\|_0\right), \tag{3}$$

where $u \in \mathscr{D}(\Omega_0)$, $\Omega_0 \Subset \Omega$, r is as in Definition 2.2, and C is a suitable positive constant.

We recall the local notion of Gevrey vectors.

Definition 2.3 Let $P(x, D)$ a differential operator of order N with Gevrey coefficients of order $s \geq 1$ in Ω open subset of $\mathbb{R}^n$. We denote by $G^s(\Omega; P)$ the space of the Gevrey vectors of order $s \geq 1$, in Ω, of P, i.e., the set of all distributions $u \in \mathscr{D}'(\Omega)$ such that for any compact subset K of Ω and every $k \in \mathbb{N}$, $P^k u$ is in $L^2(K)$ and there is a positive constant C_K such that

$$\|P^k u\|_{L^2(K)} \leq C_K^{k+1} (Nk!)^s. \tag{4}$$

When $s = 1$, we set $G^1(\Omega; P) = \mathscr{A}(\Omega; P)$ the set of the analytic vectors of P in Ω.

We state now the main result of this chapter.

Theorem 2.1 *Let $P(x, D)$ be as in* (1). *Let x_0 in Ω, Ω_0 a neighborhood of x_0, where the Hörmander's condition is satisfied at the step r, Definition 2.2, and $u \in G^s(\Omega; P)$. Then $u \in G^{rs}(\Omega_0)$.*

3 Estimates of a Gevrey Vector of P

Let $x_0 \in \Omega$, Ω_0 a neighborhood of x_0, $\Omega_0 \Subset \Omega$, where the Hörmander's condition is satisfied at the step r, and $u \in G^s(\Omega_0; P)$. Let M a given fixed integer which will be determined at the end of the section, having the form $M = pn + q$, p and q suitable integers. Let $\psi_N(x)$ be an Ehrenpreis–Hörmander sequence ([28], see also [35, 44]) associated to the couple (Ω_1, Ω_2), $x_0 \in \Omega_1$. More precisely, $\psi_N(x) \equiv 1$ on Ω_1 and supported in Ω_2, with $\overline{\Omega_1} \Subset \Omega_2 \Subset \Omega_0$, and there is a positive constant C_ψ such that

$$|D^\alpha \psi_N| \leq C_\psi^{|\alpha|+1} N^{(|\alpha|-M)^+}, \tag{5}$$

for all $\alpha \in \mathbb{Z}_+^n$ such that $|\alpha| \leq N$.

We will use the following convention: The Latin alphabet letters in the upper index will denote the derivatives with respect to the corresponding direction, i.e., $a^{(k)}(x) = D_k a(x)$, and the Greek alphabet letters in the upper index will denote the usual multi-index derivatives, i.e., $a^{(\alpha)}(x) = D^\alpha a(x) = D_1^{\alpha_1} \cdots D_n^{\alpha_n} a(x)$, $\alpha = (\alpha_1, \ldots, \alpha_n) \in \mathbb{Z}_+^n$.

The purpose of the present section is to obtain a suitable estimate of $\|D^\alpha \psi_N^{(\beta)} P^k u\|_{p/r}^2$, $p = 1, \ldots, r$.[1]

3.1 Estimate in $H^{1/r}$ and $H^{2/r}$

Our goal is to obtain a suitable estimate for $\|D^\alpha \psi_N^{(\beta)} P^k u\|_{1/r}$. In order to obtain it, we use a modification of the Rothschild–Stein estimate (3) proved in [8]:

Lemma 3.1 (Lemma 2.1 in [8]) *The following a priori estimate holds*

$$\|v\|_{\frac{2}{r}} + \sum_{j=1}^{m} \left(\|\Lambda_r X_j v\|_0 + \|X_j \Lambda_r v\|_0\right) + \sum_{\ell,j=1}^{m} \|X_\ell X_j v\|_0 + \|X_0 v\|_0$$
$$\leq C \left(\|Pv\|_0 + \|v\|_0\right), \tag{6}$$

where $\Lambda_r = \left(1 + |D|^2\right)^{\frac{1}{2r}}$, $v \in \mathscr{D}(\Omega_0)$, $\Omega_0 \Subset \Omega$, *r is as in Definition 2.2, and C is a suitable positive constant.*

By the above Lemma, replacing v by $D^\alpha \psi_N^{(\beta)} P^k u$ in (6), the following estimate holds

$$\|D^\alpha \psi_N^{(\beta)} P^k u\|_{\frac{2}{r}} + \sum_{j=1}^{m} \|X_j D^\alpha \psi_N^{(\beta)} P^k u\|_{\frac{1}{r}}$$
$$+ \sum_{\ell,j=1}^{m} \|X_\ell X_j D^\alpha \psi_N^{(\beta)} P^k u\|_0 + \|X_0 D^\alpha \psi_N^{(\beta)} P^k u\|_0$$
$$\leq C \left(\|P D^\alpha \psi_N^{(\beta)} P^k u\|_0 + \|D^\alpha \psi_N^{(\beta)} P^k u\|_0\right). \tag{7}$$

[1] The choice to use the set Ω_2 related to the Ehrenpreis–Hörmander sequences and not Ω_0, and it will be more clear in the following when we will handle the case $p > 2$.

We begin to handle the first term on the right-hand side. We have

$$\begin{aligned}\|PD^\alpha\psi_N^{(\beta)}P^k u\|_0 &\le \|D^\alpha\psi_N^{(\beta)}P^{k+1}u\|_0 + \|[P, D^\alpha\psi_N^{(\beta)}]P^k u\|_0\\ &\le \|D^\alpha\psi_N^{(\beta)}P^{k+1}u\|_0 + \sum_{\ell=1}^{m}\|[X_\ell^2, D^\alpha\psi_N^{(\beta)}]P^k u\|_0 + \|[X_0, D^\alpha\psi_N^{(\beta)}]P^k u\|_0\\ &\quad + \|[c, D^\alpha\psi_N^{(\beta)}]P^k u\|_0. \end{aligned} \tag{8}$$

In order to estimate the above terms, we will take advantage from the following relation: Let $f \in G^s(\Omega)$, and then for every $\alpha \in \mathbb{Z}_+^n$, we have

$$[f(x), D^\gamma] = -\sum_{\substack{\nu\le\gamma,\\ \nu\neq 0}}\binom{\gamma}{\nu} f^{(\nu)}(x)D^{\gamma-\nu}. \tag{9}$$

Concerning the last term on the right-hand side (8), using the above formula with $f = c$ and $\gamma = \alpha$, we get

$$\begin{aligned}\|[c, D^\alpha\psi_N^{(\beta)}]P^k u\|_0 &\le \sum_{\substack{\nu\le\alpha,\\ \nu\neq 0}}\binom{\alpha}{\nu}\|c^{(\nu)}D^{\alpha-\nu}\psi_N^{(\beta)}P^k u\|_0\\ &\le \sum_{\substack{\nu\le\alpha,\\ \nu\neq 0}}\frac{\alpha!}{(\alpha-\nu)!}C_1^{|\nu|+1}(\nu!)^{s-1}\|D^{\alpha-\nu}\psi_N^{(\beta)}P^k u\|_0. \end{aligned} \tag{10}$$

We remark that when $\nu = \alpha$, we have

$$C_1^{|\alpha|+1}(\alpha!)^s\|\psi_N^{(\beta)}P^k u\|_0.$$

We recall the following result:

Lemma 3.2 ([18]) *For every $N \in \mathbb{N}^*$, one has*

$$k^j \le B^j N^{(k-M)^+} \text{ for } B = \sup(M, 3),\ j \le k \le N. \tag{11}$$

Moreover, the following elementary fact holds: If $p_1, \ldots, p_\ell$ and M are integers, then

$$N^{(p_1-M)^+}\ldots N^{(p_\ell-M)^+} \le N^{(p_1+\cdots+p_\ell-M)^+}.$$

Let K_0 be a compact set contained in Ω and containing all the supports of ψ_N, $\Omega_0 \subset K_0$. Since u is a G^s vector for P in Ω, we have $\|P^k u\|_{L^2(K_0)} \le C_{K_0}^{2k+1}(2k)^{2sk}$. Taking advantage from Lemma 3.2, we get

$$C_1^{|\alpha|+1}(\alpha!)^s \|\psi_N^{(\beta)} P^k u\|_0 \le C_1^{|\alpha|+1}(\alpha!)^s C_\psi^{|\beta|+1} N^{(|\beta|-M)^+} \|P^k u\|_{L^2(K_0)}$$
$$\le C_1^{|\alpha|+1} C_2^{|\beta|+2k+1} N^{s(|\alpha|+|\beta|+2k-M)^+}. \qquad (12)$$

Now, we consider the second to the last term on the right-hand side of (8):

$$\|[X_0, D^\alpha \psi_N^{(\beta)}] P^k u\|_0 \le \sum_{j=1}^n \|[a_{j,0}(x) D_j, D^\alpha \psi_N^{(\beta)}] P^k u\|_0$$
$$\le \sum_{j=1}^n \Big(\|a_{j,0}(x) D^\alpha \psi_N^{(\beta+j)} P^k u\|_0 + \|[a_{j,0}(x), D^\alpha] \psi_N^{(\beta+j)} P^k u\|_0$$
$$+ \|[a_{j,0}(x), D^\alpha] D_j \psi_N^{(\beta)} P^k u\|_0 \Big)$$
$$\le \sum_{j=1}^n \sum_{\substack{\nu\le\alpha,\\ \nu\ne 0}} C_1^{|\nu|+1} \frac{\alpha!\,(\nu!)^{s-1}}{(\alpha-\nu)!} \Big(\|D^{\alpha-\nu} \psi_N^{(\beta+j)} P^k u\|_0 + \|D^{\alpha-\nu+j} \psi_N^{(\beta)} P^k u\|_0 \Big)$$
$$+ C_1 \sum_{j=1}^n \|D^\alpha \psi_N^{(\beta+j)} P^k u\|_0, \qquad (13)$$

where we use (9) with $f(x) = a_{j,0}(x)$ and $\gamma = \alpha$ and the fact that

$$\sup_K |a_{\cdot,0}^{(\gamma)}(x)| \le C_1^{|\gamma|+1}(\gamma!)^s, \quad \gamma \in \mathbb{Z}_+^n,$$

where K is a compact set contained in Ω and containing all the supports of ψ_N.

We focus on the terms in the sum on the right-hand side of (8). We remark that

$$[X_\ell^2, D^\alpha \psi_N^{(\beta)}] = 2X_\ell [X_\ell, D^\alpha \psi_N^{(\beta)}] - [X_\ell, [X_\ell, D^\alpha \psi_N^{(\beta)}]], \qquad \ell = 1, 2, \ldots, m.$$

Then we have

$$\|[X_\ell^2, D^\alpha \psi_N^{(\beta)}] P^k u\|_0$$
$$\le 2\|X_\ell [X_\ell, D^\alpha \psi_N^{(\beta)}] P^k u\|_0 + \|[X_\ell, [X_\ell, D^\alpha \psi_N^{(\beta)}]] P^k u\|_0$$
$$= 2I_1 + I_2. \qquad (14)$$

We analyze separately these two terms.

Term I_1: We have

$$I_1 \le \|X_\ell[X_\ell, D^\alpha]\psi_N^{(\beta)} P^k u\|_0 + \|X_\ell D^\alpha [X_\ell, \psi_N^{(\beta)}] P^k u\|_0.$$

Now, by (9), where $f = a_{j,\ell}$ and $\gamma = \alpha$, $j = 1, \dots, n$, we have

$$[X_\ell, D^\alpha] = \sum_{j=1}^{n} [a_{j,\ell} D_j, D^\alpha] = - \sum_{j=1}^{n} \sum_{\substack{\nu \le \alpha, \\ \nu \ne 0}} \binom{\alpha}{\nu} a_{j,\ell}^{(\nu)}(x) D^{\alpha-\nu+j}, \tag{15}$$

and

$$D^\alpha [X_\ell, \psi_N^{(\beta)}] = \sum_{j=1}^{n} \left(a_{j,\ell}(x) D^\alpha \psi_N^{(\beta+j)} + \sum_{\substack{\nu \le \alpha, \\ \nu \ne 0}} \binom{\alpha}{\nu} a_{j,\ell}^{(\nu)}(x) D^{\alpha-\nu} \psi_N^{(\beta+j)} \right). \tag{16}$$

So, we obtain

$$\begin{aligned} I_1 \le \sum_{j=1}^{n} \sum_{\substack{\nu \le \alpha, \\ \nu \ne 0}} \binom{\alpha}{\nu} \Bigg(& \|a_{j,\ell}^{(\nu)} X_\ell D^{\alpha-\nu+j} \psi_N^{(\beta)} P^k u\|_0 \\ & + \sum_{i=1}^{n} \|a_{p,\ell} a_{j,\ell}^{(\nu+i)} D^{\alpha-\nu+j} \psi_N^{(\beta)} P^k u\|_0 \Bigg) \\ + \sum_{j=1}^{n} \Bigg(\|a_{j,\ell} X_\ell D^\alpha \psi_N^{(\beta+j)} P^k u\|_0 & + \sum_{i=1}^{n} \|a_{i,\ell} a_{j,\ell}^{(i)} D^\alpha \psi_N^{(\beta+j)} P^k u\|_0 \Bigg) \\ + \sum_{j=1}^{n} \sum_{\substack{\nu \le \alpha, \\ \nu \ne 0}} \binom{\alpha}{\nu} \Bigg(\|a_{j,\ell}^{(\nu)} X_\ell D^{\alpha-\nu} \psi_N^{(\beta+j)} P^k u\|_0 & + \sum_{i=1}^{n} \|a_{i,\ell} a_{j,\ell}^{(\nu+i)} D^{\alpha-\nu} \psi_N^{(\beta+j)} P^k u\|_0 \Bigg). \end{aligned}$$

Let K be a compact set contained in Ω and containing all the supports of ψ_N, and then the following estimate holds:

$$\begin{aligned} &\sup_K |a_{\cdot,\ell}^{(\gamma)}(x)| \le C_1^{|\gamma|+1} [(\gamma)!]^s, \\ &\sup_K |a_{\cdot,\ell}^{(\gamma+p)}(x)| \le C_1^{|\gamma|+2} [(\gamma+e_p)!]^s \le C_1^{|\gamma|+2} 2^{s[|\gamma|+1]} (\gamma!)^s \le C_2^{|\gamma|+2} (\gamma!)^s, \end{aligned} \tag{17}$$

for every $\gamma \in \mathbb{Z}_+^n$ and $1 \le p \le n$.

We conclude that

$$
\begin{aligned}
& I_1 \\
& \leq \sum_{j=1}^{n} \sum_{\substack{\nu\leq\alpha \\ \nu\neq 0}} C_1^{|\nu|+1} \frac{\alpha!\,(\nu!)^{s-1}}{(\alpha-\nu)!} \Big[\|X_\ell D^{\alpha-\nu} \psi_N^{(\beta+j)} P^k u\|_0 + \|X_\ell D^{\alpha-\nu+j} \psi_N^{(\beta)} P^k u\|_0 \Big] \\
& + n \sum_{j=1}^{n} \sum_{\substack{\nu\leq\alpha \\ \nu\neq 0}} C_1 C_2^{|\nu|+2} \frac{\alpha!\,(\nu!)^{s-1}}{(\alpha-\nu)!} \Big[\|D^{\alpha-\nu+j} \psi_N^{(\beta)} P^k u\|_0 + \|D^{\alpha-\nu} \psi_N^{(\beta+j)} P^k u\|_0 \Big] \\
& \qquad + \sum_{j=1}^{n} C_1 \Big[\|X_\ell D^{\alpha} \psi_N^{(\beta+j)} P^k u\|_0 + n C_1^2 \|D^{\alpha} \psi_N^{(\beta+j)} P^k u\|_0 \Big]. \qquad (18)
\end{aligned}
$$

Term I_2. We begin to observe that

$$
\begin{aligned}
& [X_\ell, [X_\ell, D^\alpha \psi_N^{(\beta)}]] \\
& = [X_\ell, [X_\ell, D^\alpha]] \psi_N^{(\beta)} + 2[X_\ell, D^\alpha][X_\ell, \psi_N^{(\beta)}] + D^\alpha [X_\ell, [X_\ell, \psi_N^{(\beta)}]]. \qquad (19)
\end{aligned}
$$

Using (9) with $f = a_{\cdot,\ell}$ or $f = a_{\cdot,\ell}^{(\cdot)}$ and $\gamma = \alpha$ or $\gamma = \alpha - \nu$, we get

$$
\begin{aligned}
[X_\ell, [X_\ell, D^\alpha]] \psi_N^{(\beta)} &= - \sum_{j=1}^{n} \sum_{i=1}^{n} \sum_{\substack{\nu\leq\alpha, \\ \nu\neq 0}} \binom{\alpha}{\nu} [a_{i,\ell} D_i, a_{j,\ell}^{(\nu+i)} D^{\alpha-\nu+j}] \psi_N^{(\beta)} \\
= - \sum_{j=1}^{n} \sum_{i=1}^{n} \sum_{\substack{\nu\leq\alpha, \\ \nu\neq 0}} \binom{\alpha}{\nu} a_{i,\ell}\, a_{j,\ell}^{(\nu+i)} D^{\alpha-\nu+j} \psi_N^{(\beta)} &+ \sum_{j=1}^{n} \sum_{i=1}^{n} a_{j,\ell}^{(\alpha)} a_{i,\ell}^{(j)} D_i \psi_N^{(\beta)} \\
& - \sum_{j=1}^{n} \sum_{i=1}^{n} \sum_{\substack{\nu<\alpha, \\ \nu\neq 0}} \binom{\alpha}{\nu} a_{j,\ell}^{(\nu)} [a_{i,\ell}, D^{\alpha-\nu+j}] D_i \psi_N^{(\beta)};
\end{aligned}
$$

since $[a_{i,\ell}, D^{\alpha-\nu+j}] = [a_{i,\ell}, D^{\alpha-\nu}] D_j - a_{i,\ell}^{(j)}, D^{\alpha-\nu} - [D^{\alpha-\nu}, a_{i,\ell}^{(j)}]$, we obtain

$$
\begin{aligned}
[X_\ell, [X_\ell, D^\alpha]] \psi_N^{(\beta)} &= - \sum_{j=1}^{n} \sum_{i=1}^{n} \sum_{\substack{\nu\leq\alpha, \\ \nu\neq 0}} \binom{\alpha}{\nu} a_{i,\ell}\, a_{j,\ell}^{(\nu+i)} D^{\alpha-\nu+j} \psi_N^{(\beta)} \\
+ \sum_{j,i=1}^{n} \sum_{\substack{\nu<\alpha, \\ \nu\neq 0}} \sum_{\substack{\mu\leq\alpha-\nu, \\ \mu\neq 0}} \binom{\alpha}{\nu} \binom{\alpha-\nu}{\mu} & a_{j,\ell}^{(\nu)} \left(a_{i,\ell}^{(\mu)} D^{\alpha-\nu-\mu+j+i} + a_{i,\ell}^{(\mu+j)} D^{\alpha-\nu-\mu+i} \right) \psi_N^{(\beta)}
\end{aligned}
$$

$$+\sum_{j,i=1}^{n}\sum_{\substack{\nu\leq\alpha,\\ \nu\neq 0}}\binom{\alpha}{\nu}a_{j,\ell}^{(\nu)}a_{i,\ell}^{(j)}D^{\alpha-\nu+i}\psi_N^{(\beta)}. \tag{20}$$

Acting in the same way, we have

$$\begin{aligned}
&2[X_\ell, D^\alpha][X_\ell, \psi_N^{(\beta)}]\\
&\quad = -2\sum_{j,i=1}^{n}\sum_{\substack{\nu\leq\alpha,\\ \nu\neq 0}}\binom{\alpha}{\nu}a_{j,\ell}^{(\nu)}\left(a_{i,\ell}D^{\alpha-\nu+j}+a_{i,\ell}^{(j)}D^{\alpha-\nu}\right)\psi_N^{(\beta+i)}\\
&\qquad -2\sum_{j,i=1}^{n}\sum_{\substack{\nu<\alpha,\\ \nu\neq 0}}\sum_{\substack{\mu\leq\alpha-\nu,\\ \mu\neq 0}}\binom{\alpha}{\nu}\binom{\alpha-\nu}{\mu}a_{j,\ell}^{(\nu)}\\
&\qquad \times\left(a_{i,\ell}^{(\mu+j)}D^{\alpha-\nu-\mu}+a_{i,\ell}^{(\mu)}D^{\alpha-\nu-\mu+j}\right)\psi_N^{(\beta+i)};
\end{aligned} \tag{21}$$

and

$$\begin{aligned}
D^\alpha[X_\ell,[X_\ell,\psi_N^{(\beta)}]] &= \sum_{j,i=1}^{n}a_{i,\ell}\left(a_{j,\ell}^{(i)}D^\alpha\psi_N^{(\beta+j)}+a_{j,\ell}D^\alpha\psi_N^{(\beta+j+i)}\right)\\
&+\sum_{j,i=1}^{n}\sum_{\substack{\nu\leq\alpha,\\ \nu\neq 0}}\binom{\alpha}{\nu}\Big[\left(a_{i,\ell}a_{j,\ell}^{(\nu+i)}+a_{i,\ell}^{(\nu)}a_{j,\ell}^{(i)}\right)D^{\alpha-\nu}\psi_N^{(\beta+j)}\\
&\qquad +\left(a_{i,\ell}a_{j,\ell}^{(\nu)}+a_{i,\ell}^{(\nu)}a_{j,\ell}\right)D^{\alpha-\nu}\psi_N^{(\beta+j+i)}\Big]\\
&+\sum_{j,i=1}^{n}\sum_{\substack{\nu<\alpha,\\ \nu\neq 0}}\sum_{\substack{\mu\leq\alpha-\nu,\\ \mu\neq 0}}\binom{\alpha}{\nu}\binom{\alpha-\nu}{\mu}a_{i,\ell}^{(\nu)}\Big[a_{j,\ell}^{(\mu+i)}D^{\alpha-\nu-\mu}\psi_N^{(\beta+j)}\\
&\qquad +a_{j,\ell}^{(\mu)}D^{\alpha-\nu-\mu}\psi_N^{(\beta+j+i)}\Big].
\end{aligned} \tag{22}$$

By (17) and the identities, (19), (20), (21), and (22), we obtain

$$\begin{aligned}
&I_2\\
&\leq 4n\sum_{j=1}^{n}\sum_{\substack{\nu\leq\alpha,\\ \nu\neq 0}}C_2^{|\nu|+3}\frac{\alpha!\,(\nu!)^{s-1}}{(\alpha-\nu)!}\left(\|D^{\alpha-\nu}\psi_N^{(\beta+j)}P^k u\|_0+\|D^{\alpha-\nu+j}\psi_N^{(\beta)}P^k u\|_0\right)
\end{aligned}$$

$$+2\sum_{i,j=1}^{n}\sum_{\substack{\nu\le\alpha,\\ \nu\ne0}} C_1^{|\nu|+2}\frac{\alpha!\,(\nu!)^{s-1}}{(\alpha-\nu)!}\Big(\|D^{\alpha-\nu}\psi_N^{(\beta+j+i)}P^k u\|_0+\|D^{\alpha-\nu+j}\psi_N^{(\beta+i)}P^k u\|_0\Big)$$

$$+3n\sum_{j=1}^{n}\sum_{\substack{\nu<\alpha,\\ \nu\ne0}}\sum_{\substack{\mu\le\alpha-\nu,\\ \mu\ne0}} C_1^{|\nu|+1}C_2^{|\mu|+2}\frac{\alpha!\,(\nu!)^{s-1}\,(\mu!)^{s-1}}{(\alpha-\nu-\mu)!}\Big[\|D^{\alpha-\nu-\mu}\psi_N^{(\beta+j)}P^k u\|_0 + \|D^{\alpha-\nu-\mu+j}\psi_N^{(\beta)}P^k u\|_0\Big]$$

$$+2\sum_{i,j=1}^{n}\sum_{\substack{\nu<\alpha,\\ \nu\ne0}}\sum_{\substack{\mu\le\alpha-\nu,\\ \mu\ne0}} C_1^{|\nu|+|\mu|+2}\frac{\alpha!\,(\nu!)^{s-1}\,(\mu!)^{s-1}}{(\alpha-\nu-\mu)!}\Big[\|D^{\alpha-\nu-\mu}\psi_N^{(\beta+j+i)}P^k u\|_0 + \|D^{\alpha-\nu-\mu+j}\psi_N^{(\beta+i)}P^k u\|_0+\|D^{\alpha-\nu-\mu+j+i}\psi_N^{(\beta)}P^k u\|_0\Big]$$

$$+nC_1^3\sum_{j=1}^{n}\|D^{\alpha}\psi_N^{(\beta+j)}P^k u\|_0+C_1^2\sum_{j=1}^{n}\sum_{i=1}^{n}\|D^{\alpha}\psi_N^{(\beta+j+i)}P^k u\|_0. \tag{23}$$

Summing up, by (7), (8), (10), (12), (13), (14), (18), and (23), we conclude that there are two new positive constants C_1 and C_2, independent of α, β, k and N, such that

$$\|D^{\alpha}\psi_N^{(\beta)}P^k u\|_{\frac{2}{r}}+\sum_{j=1}^{m}\|X_jD^{\alpha}\psi_N^{(\beta)}P^k u\|_{\frac{1}{r}} + \sum_{\ell,j=1}^{m}\|X_\ell X_j D^{\alpha}\psi_N^{(\beta)}P^k u\|_0+\|X_0D^{\alpha}\psi_N^{(\beta)}P^k u\|_0$$

$$\le C_2\Big[\|D^{\alpha}\psi_N^{(\beta)}P^{k+1}u\|_0+\|D^{\alpha}\psi_N^{(\beta)}P^k u\|_0+\sum_{\ell=1}^{m}\sum_{j=1}^{n}\|X_\ell D^{\alpha}\psi_N^{(\beta+j)}P^k u\|_0 + \sum_{j=1}^{n}\|D^{\alpha}\psi_N^{(\beta+j)}P^k u\|_0+\sum_{j=1}^{n}\sum_{i=1}^{n}\|D^{\alpha}\psi_N^{(\beta+j+i)}P^k u\|_0$$

$$+\sum_{\ell=1}^{m}\sum_{j=1}^{n}\sum_{\substack{\nu\le\alpha,\\ \nu\ne0}} C_1^{|\nu|+1}\frac{\alpha!\,(\nu!)^{s-1}}{(\alpha-\nu)!}\Big(\|X_\ell D^{\alpha-\nu}\psi_N^{(\beta+j)}P^k u\|_0 + \|X_\ell D^{\alpha-\nu+j}\psi_N^{(\beta)}P^k u\|_0\Big)$$

$$+\sum_{j=1}^{n}\sum_{\substack{\nu\le\alpha,\\ \nu\ne0}} C_1^{|\nu|+1}\frac{\alpha!\,(\nu!)^{s-1}}{(\alpha-\nu)!}\Big(\|D^{\alpha-\nu}\psi_N^{(\beta+j)}P^k u\|_0+\|D^{\alpha-\nu+j}\psi_N^{(\beta)}P^k u\|_0\Big)$$

$$+\sum_{j,i=1}^{n}\sum_{\substack{\nu\leq\alpha,\\ \nu\neq 0}} C_1^{|\nu|+1}\frac{\alpha!\,(\nu!)^{s-1}}{(\alpha-\nu)!}\Big(\|D^{\alpha-\nu}\psi_N^{(\beta+j+i)}P^k u\|_0+\|D^{\alpha-\nu+j}\psi_N^{(\beta+i)}P^k u\|_0\Big)$$

$$+\sum_{j=1}^{n}\sum_{\substack{\nu<\alpha,\\ \nu\neq 0}}\sum_{\substack{\mu\leq\alpha-\nu,\\ \mu\neq 0}} C_1^{|\nu|+|\mu|+1}\frac{\alpha!\,(\nu!)^{s-1}(\mu!)^{s-1}}{(\alpha-\nu-\mu)!}\Big(\|D^{\alpha-\nu-\mu}\psi_N^{(\beta+j)}P^k u\|_0 + \|D^{\alpha-\nu-\mu+j}\psi_N^{(\beta)}P^k u\|_0\Big)$$

$$+\sum_{i,j=1}^{n}\sum_{\substack{\nu<\alpha,\\ \nu\neq 0}}\sum_{\substack{\mu\leq\alpha-\nu,\\ \mu\neq 0}} C_1^{|\nu|+|\mu|+2}\frac{\alpha!\,(\nu!)^{s-1}(\mu!)^{s-1}}{(\alpha-\nu-\mu)!}\Big(\|D^{\alpha-\nu-\mu}\psi_N^{(\beta+j+i)}P^k u\|_0 + \|D^{\alpha-\nu-\mu+j}\psi_N^{(\beta+i)}P^k u\|_0 + \|D^{\alpha-\nu-\mu+j+i}\psi_N^{(\beta)}P^k u\|_0\Big)$$

$$+\sum_{\substack{\nu<\alpha,\\ \nu\neq 0}} C_1^{|\nu|+1}\frac{\alpha!\,(\nu!)^{s-1}}{(\alpha-\nu)!}\|D^{\alpha-\nu}\psi_N^{(\beta)}P^k u\|_0 + C_1^{|\alpha|+1}C_2^{|\beta|+2k}N^{s(|\alpha|+|\beta|+2k-M)^+}\Bigg]. \tag{24}$$

We remark that when $\nu=\alpha$ or $\mu=\alpha-\nu$, the first term in the multiple sums can be directly estimated. More precisely, setting $\nu=\alpha$ and using the same strategy adopted to estimate the last term in (10), see (12), the first term of second and third multiple sums can be bounded as follows:

$$\sum_{j=1}^{n} C_1^{|\alpha|+1}(\alpha!)^s\,\|\psi_N^{(\beta+j)}P^k u\|_0 \leq nC_1^{|\alpha|+1}\widetilde{C}_2^{|\beta|+2k+2}N^{s(|\alpha|+|\beta|+2k+1-M)^+};$$

$$\sum_{j=1}^{n}\sum_{i=1}^{n} C_1^{|\alpha|+1}(\alpha!)^s\,\|\psi_N^{(\beta+j+i)}P^k u\|_0 \leq n^2C_1^{|\alpha|+1}\widetilde{C}_2^{|\beta|+2k+3}N^{s(|\alpha|+|\beta|+2k+2-M)^+}. \tag{25}$$

In the same way, setting $\mu=\alpha-\nu$ in the first term of the fourth and fifth multiple sums involving μ, and recalling that the number of the multi-index ν with $|\nu|\leq|\alpha|$ is equal to $\binom{|\alpha|+n}{|\alpha|}$, smaller than $2^{|\alpha|+n}$, we have

$$\sum_{j=1}^{n}\sum_{\substack{\nu<\alpha,\\ \nu\neq 0}} C_1^{|\alpha|+1}\alpha!\,(\nu!)^{s-1}\,[(\alpha-\nu)!]^{s-1}\,\|\psi_N^{(\beta+j)}P^k u\|_0 \leq n2^{|\alpha|+n}C_1^{|\alpha|+1}\widetilde{C}_2^{|\beta|+2k+2}N^{s(|\alpha|+|\beta|+2k+1-M)^+};$$

$$\sum_{j=1}^{n}\sum_{i=1}^{n}\sum_{\substack{\nu<\alpha,\\ \nu\neq 0}} C_1^{|\alpha|+1}\alpha!\,(\nu!)^{s-1}\,[(\alpha-\nu)!]^{s-1}\,\|\psi_N^{(\beta+i+j)}P^k u\|_0 \\ \leq n^2 2^{|\alpha|+n} C_1^{|\alpha|+1}\widetilde{C}_2^{|\beta|+2k+3} N^{s(|\alpha|+|\beta|+2k+2-M)^+}. \qquad (26)$$

Concerning the first term in the first multiple sum on the right-hand side of (24), setting $\nu=\alpha$, it becomes

$$\sum_{\ell=1}^{m}\sum_{j=1}^{n} C_1^{|\alpha|+1}\,(\alpha!)^s\,\|X_\ell\psi_N^{(\beta+j)}P^k u\|_0. \qquad (27)$$

In order to handle it, we recall that the following a priori estimate holds

$$\sum_{\ell=1}^{m}\|X_\ell v\|_0^2 \leq \widetilde{C}\left(|\langle Pv, v\rangle| + \|v\|^2\right), \qquad \forall\, v\in\mathscr{D}(\Omega_0).$$

Replacing v by $\psi_N^{(\beta+j)}P^k u$, we obtain

$$\sum_{\ell=1}^{m}\|X_\ell\psi_N^{(\beta+j)}P^k u\|_0^2 \leq \widetilde{C}\left(|\langle P\psi_N^{(\beta+j)}P^k u, \psi_N^{(\beta+j)}P^k u\rangle| + \|\psi_N^{(\beta+j)}P^k u\|^2\right).$$

Since

$$P\psi_N^{(\beta+j)}P^k u = \psi_N^{(\beta+j)}P^{k+1}u + \sum_{\ell=1}^{m}\left(2X_\ell[X_\ell,\psi_N^{(\beta+j)}] - [X_\ell,[X_\ell,\psi_N^{(\beta+j)}]]\right) \\ + [X_0,\psi_N^{(\beta+j)}],$$

where

$$[X_\ell,\psi_N^{(\beta+j)}] = \sum_{i=1}^{n} a_{i,\ell}\psi_N^{(\beta+j+i)}, \qquad \ell = 0, 1, \ldots, m;$$

$$[X_\ell,[X_\ell,\psi_N^{(\beta+j)}]] = \sum_{i=1}^{n}\sum_{p=1}^{n} a_{p,\ell}\left(a_{i,\ell}^{(p)}\psi_N^{(\beta+j+i)} + a_{i,\ell}\psi_N^{(\beta+j+i+p)}\right),$$

$\ell = 1,\ldots,m,$

and keeping in mind that $X_\ell^* = X_\ell + \sum_{i=1}^{n} a_{i,\ell}^{(i)}$, we obtain

$$\sum_{\ell=1}^{m}\|X_\ell\psi_N^{(\beta+j)}P^k u\|_0^2 \leq \widetilde{C}_1\Big[\|\psi_N^{(\beta+j)}P^{k+1}u\|_0\|\psi_N^{(\beta+j)}P^k u\|_0 + \|\psi_N^{(\beta+j)}P^k u\|_0^2$$

$$+\sum_{\ell=0}^{m}\sum_{i=1}^{n}\|a_{i,\ell}\psi_N^{(\beta+j+i)}P^k u\|_0^2+\sum_{\ell=1}^{m}\sum_{i=1}^{n}\|a_{i,\ell}^{(i)}\psi_N^{(\beta+j)}P^k u\|_0^2$$
$$+\sum_{\ell=1}^{m}\sum_{i=1}^{n}\sum_{p=1}^{n}\Big(\|a_{p,\ell}a_{i,\ell}^{(p)}\psi_N^{(\beta+j+i)}P^k u\|_0$$
$$+\|a_{p,\ell}a_{i,\ell}\psi_N^{(\beta+j+i+p)}P^k u\|_0\Big)\,\|\psi_N^{(\beta+j)}P^k u\|_0\Big].$$

So, using the same strategy adopted to estimate the last term in (10) (see (12)), we conclude that

$$\sum_{\ell=1}^{m}\|X_\ell\psi_N^{(\beta+j)}P^k u\|_0^2\le \widetilde{C}_2^{2(|\beta|+2k+4)}N^{2s(|\beta|+2k+3-M)^+}.$$

The (27) can be bounded by

$$C_1^{|\alpha|+1}\widetilde{C}_3^{|\beta|+2k+4}N^{s(|\alpha|+|\beta|+2k+3-M)^+},\tag{28}$$

where $\widetilde{C}_3$ is a suitable positive constant independent of α, β, k, and N.

We end this subsection by recalling that by interpolation the following inequality holds:

$$\|D^\alpha\psi_N^{(\beta)}P^k u\|_{\frac{1}{r}}\le\left(\|D^\alpha\psi_N^{(\beta)}P^k u\|_{\frac{2}{r}}\right)^{1/2}\left(\|D^\alpha\psi_N^{(\beta)}P^k u\|_0\right)^{1/2}.\tag{29}$$

3.2 Estimate in $H^{p/r}$, $p = 3, \ldots, r$

The purpose of the present subsection is to obtain a suitable estimate of $\|v\|_{p/r}$, $\|X_j v\|_{(p-1)/r}$ and $\|X_0 v\|_{(p-2)/r}$, $j = 1, \ldots, m$, $p = 3, \ldots, r$, where $v = D^\alpha\psi_N^{(\beta)}P^k u$.

We denote by Λ_r^ℓ the pseudodifferential operator with symbol $(1+|\xi|^2)^{\ell/2r}$. Let $\widetilde{\psi}$ be in $\mathscr{D}(\Omega_3)$, $\overline{\Omega_3}\Subset\Omega_0$, such that $\widetilde{\psi}\equiv 1$ on Ω_2, $\overline{\Omega_2}\Subset\Omega_3$. We recall that $u\in G^s(\Omega_0; P)$, where Ω_0 is a neighborhood of x_0, $\Omega_0\Subset\Omega$, where the Hörmander condition is satisfied at the step r, and that ψ_N is the Ehrenpreis–Hörmander sequence associated to the couple (Ω_1, Ω_2), $x_0\in\Omega_1$, so $\widetilde{\psi}\psi_N=\psi_N$.

We have

$$\|D^\alpha\psi_N^{(\beta)}P^k u\|_{\frac{p}{r}}+\sum_{j=1}^{m}\|X_j D^\alpha\psi_N^{(\beta)}P^k u\|_{\frac{p-1}{r}}+\|X_0 D^\alpha\psi_N^{(\beta)}P^k u\|_{\frac{p-2}{r}}$$

$$+\sum_{\ell,j=1}^{m}\|X_\ell X_j D^\alpha\psi_N^{(\beta)}P^k u\|_{\frac{p-2}{r}}$$

$$= \|\widetilde{\psi}D^\alpha\psi_N^{(\beta)}P^k u\|_{\frac{p}{r}} + \sum_{j=1}^{m}\|\widetilde{\psi}X_j D^\alpha\psi_N^{(\beta)}P^k u\|_{\frac{p-1}{r}} + \|\widetilde{\psi}X_0 D^\alpha\psi_N^{(\beta)}P^k u\|_{\frac{p-2}{r}}$$

$$+\sum_{\ell,j=1}^{m}\|\widetilde{\psi}X_\ell X_j D^\alpha\psi_N^{(\beta)}P^k u\|_{\frac{p-2}{r}}. \tag{30}$$

We analyze separately the terms on the right-hand side. About the first term:

$$\|\widetilde{\psi}D^\alpha\psi_N^{(\beta)}P^k u\|_{\frac{p}{r}} = \|\Lambda_r^{p-2}\widetilde{\psi}D^\alpha\psi_N^{(\beta)}P^k u\|_{\frac{2}{r}}$$
$$\leq \|\widetilde{\psi}\Lambda_r^{p-2}D^\alpha\psi_N^{(\beta)}P^k u\|_{\frac{2}{r}} + \|[\widetilde{\psi},\Lambda_r^{p-2}]D^\alpha\psi_N^{(\beta)}P^k u\|_{\frac{2}{r}}.$$

Since $[\Lambda_r^{p-2},\widetilde{\psi}]$ is a pseudodifferential operator of order $\frac{p-2}{r}-1<0$, then the symbol associated to the operator $\Lambda_r^2[\Lambda_r^{p-2},\widetilde{\psi}]$ satisfies the assumptions of Theorem 18.1.1′ in [36], and we conclude that

$$\|\widetilde{\psi}D^\alpha\psi_N^{(\beta)}P^k u\|_{\frac{p}{r}} \leq \|\widetilde{\psi}\Lambda_r^{p-2}D^\alpha\psi_N^{(\beta)}P^k u\|_{\frac{2}{r}} + C_0\|D^\alpha\psi_N^{(\beta)}P^k u\|. \tag{31}$$

Before to trite the other terms on the right-hand side of (31), for technical reasons, i.e., in order to handle the commutators of X_j, $j=0,1,\dots,m$, with the pseudodifferential operator Λ_r^ℓ, we introduce a new Ehrenpreis–Hörmander sequence φ_N associated to the couple (Ω_3,Ω_4), $\overline{\Omega_3}\Subset\Omega_4\Subset\Omega_0$. So, φ_N is identically one on the supports of ψ_N and $\widetilde{\psi}$, $\varphi_N\psi_N=\psi_N$ and $\varphi_N\widetilde{\psi}=\widetilde{\psi}$. We set

$$\widetilde{X}_j(x,D)=\sum_{\ell=1}^{n}\widetilde{a}_{N,\ell,j}(x)D_\ell, \qquad j=0,1,\dots,m, \tag{32}$$

where $\widetilde{a}_{N,\ell,j}(x)=\varphi_N a_{\ell,j}(x)$; we remark: $\widetilde{\psi}\widetilde{X}_j=\widetilde{\psi}X_j$ and $[\widetilde{\psi},\widetilde{X}_j]=[\widetilde{\psi},X_j]$.

For every j, $j=1,\dots,m$, we have

$$\|\widetilde{\psi}X_j D^\alpha\psi_N^{(\beta)}P^k u\|_{\frac{p-1}{r}} = \|\widetilde{\psi}\widetilde{X}_j D^\alpha\psi_N^{(\beta)}P^k u\|_{\frac{p-1}{r}} = \|\Lambda_r^{p-2}\widetilde{\psi}\widetilde{X}_j D^\alpha\psi_N^{(\beta)}P^k u\|_{\frac{1}{r}}$$
$$\leq \|\widetilde{\psi}\widetilde{X}_j\Lambda_r^{p-2}D^\alpha\psi_N^{(\beta)}P^k u\|_{\frac{1}{r}} + \|[\Lambda_r^{p-2},\widetilde{\psi}\widetilde{X}_j]D^\alpha\psi_N^{(\beta)}P^k u\|_{\frac{1}{r}}$$
$$\leq \|\widetilde{X}_j\widetilde{\psi}\Lambda_r^{p-2}D^\alpha\psi_N^{(\beta)}P^k u\|_{\frac{1}{r}} + \|[\widetilde{\psi},\widetilde{X}_j]\Lambda_r^{p-2}D^\alpha\psi_N^{(\beta)}P^k u\|_{\frac{1}{r}}$$
$$+\|[\Lambda_r^{p-2},\widetilde{\psi}\widetilde{X}_j]D^\alpha\psi_N^{(\beta)}P^k u\|_{\frac{1}{r}}. \tag{33}$$

We set $[\widetilde{\psi}, \widetilde{X}_j] = \sum_{\ell=1}^n \widetilde{a}_{N,\ell,j} \widetilde{\psi}^{(\ell)} = f$; since

$$\left(1 + |\xi|^2\right)^{t/2} \leq (1 + |\eta|)^t \left(1 + |\xi - \eta|^2\right)^{t/2}, \qquad t \geq 0,$$

we have

$$\begin{aligned} \|[\widetilde{\psi}, \widetilde{X}_j]\Lambda_r^{p-2} D^\alpha \psi_N^{(\beta)} P^k u\|_{\frac{1}{r}} &= \|f \Lambda_r^{p-2} D^\alpha \psi_N^{(\beta)} P^k u\|_{\frac{1}{r}} \\ &\leq \left(\int (1 + |\xi|)^{\frac{1}{r}} \, |\widehat{f}(\xi)| \, d\xi\right) \|D^\alpha \psi_N^{(\beta)} P^k u\|_{\frac{p-1}{r}} \\ &\leq C^{n+3} N^{(n+2-M)^+} \|D^\alpha \psi_N^{(\beta)} P^k u\|_{\frac{p-1}{r}}. \end{aligned}$$

In the same way we obtain

$$\begin{aligned} \|\widetilde{X}_j \widetilde{\psi} \Lambda_r^{p-2} D^\alpha \psi_N^{(\beta)} P^k u\|_{\frac{1}{r}} &= \|\varphi_N X_j \widetilde{\psi} \Lambda_r^{p-2} D^\alpha \psi_N^{(\beta)} P^k u\|_{\frac{1}{r}} \\ &\leq \left(\int (1 + |\xi|)^{\frac{1}{r}} \, |\widehat{\varphi}_N(\xi)| \, d\xi\right) \|X_j \widetilde{\psi} \Lambda_r^{p-2} D^\alpha \psi_N^{(\beta)} P^k u\|_{\frac{1}{r}} \\ &\leq C^{n+3} N^{(n+2-M)^+} \|X_j \widetilde{\psi} \Lambda_r^{p-2} D^\alpha \psi_N^{(\beta)} P^k u\|_{\frac{1}{r}}. \end{aligned}$$

Now, we handle the last term on the right-hand side of (33):

$$[\Lambda_r^{p-2}, \widetilde{\psi} \widetilde{X}_j] = [\Lambda_r^{p-2}, \widetilde{\psi}]\widetilde{X}_j + \widetilde{\psi}[\Lambda_r^{p-2}, \widetilde{X}_j];$$

moreover

$$[\Lambda_r^{p-2}, \widetilde{X}_j] = \sum_{\ell_1=1}^n \sum_{\ell_2=1}^n \widetilde{a}_{N,\ell_1,j}^{(\ell_2)} (\Lambda_r^{p-2})^{(\ell_2)} D_{\ell_1} + \sum_{\ell_1=1}^n \mathscr{R}_1\left([\Lambda_r^{p-2}, \widetilde{a}_{N,\ell_1,j}]D_{\ell_1}\right), \tag{34}$$

where the symbol associated to $\mathscr{R}_1$ is

$$\begin{aligned} &\mathscr{R}_1\left([\Lambda_r^{p-2}, \widetilde{a}_{N,\ell_1,j}]D_{\ell_1}\right)(x, \xi) \\ &\quad = \frac{2}{(2\pi)^n} \sum_{|\mu|=2} \frac{1}{\mu!} \int e^{i(\eta-\xi)(x-y)} \widetilde{a}_{N,\ell_1,j}^{(\mu)}(y)\, \xi_{\ell_1} \\ &\qquad \int_0^1 (1-t)\left(\Lambda_r^{p-2}\right)^{(\mu)} (\xi + t(\eta - \xi))\, dt\, d\zeta\, dy. \end{aligned}$$

We remark that $\mathscr{R}_1$ is a pseudodifferential operator of order $\frac{p-2}{r} - 1$.

We conclude that

$$\|[\Lambda_r^{p-2}, \widetilde{\psi}\widetilde{X}_j]D^\alpha \psi_N^{(\beta)} P^k u\|_{\frac{1}{r}} \leq \|[\Lambda_r^{p-2}, \widetilde{\psi}]\widetilde{X}_j D^\alpha \psi_N^{(\beta)} P^k u\|_{\frac{1}{r}} + \|\widetilde{\psi}[\Lambda_r^{p-2}, \widetilde{X}_j]D^\alpha \psi_N^{(\beta)} P^k u\|_{\frac{1}{r}}$$

$$\leq C_1\|X_j D^\alpha \psi_N^{(\beta)} P^k u\|_{\frac{p-1}{r}-1} + C_2\|D^\alpha \psi_N^{(\beta)} P^k u\|_{\frac{p-1}{r}} + C_3\|\widetilde{\psi} D^\alpha \psi_N^{(\beta)} P^k u\|_{\frac{p-1}{r}-1}.$$

Without loss of generality we assumed that $M > n + 3$. By (33) and the above consideration, we obtain

$$\|\widetilde{\psi} X_j D^\alpha \psi_N^{(\beta)} P^k u\|_{\frac{p-1}{r}} \leq C\Big(\|X_j \widetilde{\psi} \Lambda_r^{p-2} D^\alpha \psi_N^{(\beta)} P^k u\|_{\frac{1}{r}} + \|D^\alpha \psi_N^{(\beta)} P^k u\|_{\frac{p-1}{r}} + \|X_j D^\alpha \psi_N^{(\beta)} P^k u\| + \|\widetilde{\psi} D^\alpha \psi_N^{(\beta)} P^k u\|\Big), \tag{35}$$

where C is a suitable positive constant.

Using the same strategy adopted above, we conclude that the third term on the right-hand side of (30) can be estimated as follows:

$$\|\widetilde{\psi} X_0 D^\alpha \psi_N^{(\beta)} P^k u\|_{\frac{p-2}{r}} \leq C\Big(\|X_0 \widetilde{\psi} \Lambda_r^{p-2} D^\alpha \psi_N^{(\beta)} P^k u\| + \|D^\alpha \psi_N^{(\beta)} P^k u\|_{\frac{p-2}{r}} + \|X_0 D^\alpha \psi_N^{(\beta)} P^k u\| + \|D^\alpha \psi_N^{(\beta)} P^k u\|\Big). \tag{36}$$

Lastly, we have to estimate the last term on the right-hand side of (30). For every j and ℓ, $1 \leq j, \ell \leq m$, we have

$$\begin{aligned}
\|\widetilde{\psi} X_\ell X_j D^\alpha \psi_N^{(\beta)} P^k u\|_{\frac{p-2}{r}} &= \|\Lambda_r^{p-2} \widetilde{\psi} \widetilde{X}_\ell \widetilde{X}_j D^\alpha \psi_N^{(\beta)} P^k u\| \\
&\leq \|[\Lambda_r^{p-2}, \widetilde{\psi}]\widetilde{X}_\ell \widetilde{X}_j D^\alpha \psi_N^{(\beta)} P^k u\| + \|\widetilde{\psi} \Lambda_r^{p-2} \widetilde{X}_\ell \widetilde{X}_j D^\alpha \psi_N^{(\beta)} P^k u\| \\
&\leq \|[\Lambda_r^{p-2}, \widetilde{\psi}]\widetilde{X}_\ell \widetilde{X}_j D^\alpha \psi_N^{(\beta)} P^k u\| + \|\widetilde{\psi}[\Lambda_r^{p-2}, \widetilde{X}_\ell]\widetilde{X}_j D^\alpha \psi_N^{(\beta)} P^k u\| \\
&\quad + \|\widetilde{\psi} \widetilde{X}_\ell \Lambda_r^{p-2} \widetilde{X}_j D^\alpha \psi_N^{(\beta)} P^k u\| \\
&\leq \|[\Lambda_r^{p-2}, \widetilde{\psi}]\widetilde{X}_\ell \widetilde{X}_j D^\alpha \psi_N^{(\beta)} P^k u\| + \|\widetilde{\psi}[\Lambda_r^{p-2}, \widetilde{X}_\ell]\widetilde{X}_j D^\alpha \psi_N^{(\beta)} P^k u\| \\
&\quad + \|\widetilde{X}_\ell \widetilde{X}_j \widetilde{\psi} \Lambda_r^{p-2} D^\alpha \psi_N^{(\beta)} P^k u\| + \|\widetilde{\psi}[\Lambda_r^{p-2}, \widetilde{X}_j]\widetilde{X}_\ell D^\alpha \psi_N^{(\beta)} P^k u\| \\
&\quad + \|\widetilde{\psi}[\widetilde{X}_\ell, [\Lambda_r^{p-2}, \widetilde{X}_j]]D^\alpha \psi_N^{(\beta)} P^k u\|.
\end{aligned}$$

Since $[\Lambda_r^{p-2}, \widetilde{\psi}]$ is a pseudodifferential operator of order $\frac{p-2}{r} - 1$, we have

$$\|[\Lambda_r^{p-2}, \widetilde{\psi}]\widetilde{X}_\ell \widetilde{X}_j D^\alpha \psi_N^{(\beta)} P^k u\| \leq C_0\|X_\ell X_j D^\alpha \psi_N^{(\beta)} P^k u\|.$$

Without fall in details, using the same strategy adopted previously to obtain (35), we conclude that

$$\|\widetilde{\psi}[\Lambda_r^{p-2}, \widetilde{X}_\ell]\widetilde{X}_j D^\alpha \psi_N^{(\beta)} P^k u\| \le C_1 \|\widetilde{X}_j D^\alpha \psi_N^{(\beta)} P^k u\|_{\frac{p-2}{r}} + C_2 \|\widetilde{X}_j D^\alpha \psi_N^{(\beta)} P^k u\|;$$

$$\|\widetilde{\psi}[\Lambda_r^{p-2}, \widetilde{X}_j]\widetilde{X}_\ell D^\alpha \psi_N^{(\beta)} P^k u\| \le C_1 \|\widetilde{X}_\ell D^\alpha \psi_N^{(\beta)} P^k u\|_{\frac{p-2}{r}} + C_2 \|\widetilde{X}_\ell D^\alpha \psi_N^{(\beta)} P^k u\|;$$

and

$$\|\widetilde{\psi}[\widetilde{X}_\ell, [\Lambda_r^{p-2}, \widetilde{X}_j]] D^\alpha \psi_N^{(\beta)} P^k u\| \le C_3 \|D^\alpha \psi_N^{(\beta)} P^k u\|.$$

So

$$\begin{aligned}
\|\widetilde{\psi} X_\ell X_j D^\alpha \psi_N^{(\beta)} P^k u\|_{\frac{p-2}{r}} \le C\Big(& \|X_\ell X_j \widetilde{\psi} \Lambda_r^{p-2} D^\alpha \psi_N^{(\beta)} P^k u\| \\
& + \|X_j D^\alpha \psi_N^{(\beta)} P^k u\|_{\frac{p-2}{r}} + \|X_\ell D^\alpha \psi_N^{(\beta)} P^k u\|_{\frac{p-2}{r}} + \|D^\alpha \psi_N^{(\beta)} P^k u\|_{\frac{p-2}{r}} \\
& + \|X_\ell X_j D^\alpha \psi_N^{(\beta)} P^k u\| + \|X_j D^\alpha \psi_N^{(\beta)} P^k u\| + \|X_\ell D^\alpha \psi_N^{(\beta)} P^k u\| \\
& + \|D^\alpha \psi_N^{(\beta)} P^k u\|\Big), \qquad (37)
\end{aligned}$$

where C is a suitable positive constant independent of α, β, k, and N.

Summing up by (30), (31), (35), (36), and (37), there is a suitable positive constant, C_1, independent of α, β, k, and N, such that

$$\begin{aligned}
& \|D^\alpha \psi_N^{(\beta)} P^k u\|_{\frac{p}{r}} + \sum_{j=1}^m \|X_j D^\alpha \psi_N^{(\beta)} P^k u\|_{\frac{p-1}{r}} + \|X_0 D^\alpha \psi_N^{(\beta)} P^k u\|_{\frac{p-2}{r}} \\
& \qquad + \sum_{\ell, j=1}^m \|X_\ell X_j D^\alpha \psi_N^{(\beta)} P^k u\|_{\frac{p-2}{r}} \\
& \le C_1\Big(\|\widetilde{\psi} \Lambda_r^{p-2} D^\alpha \psi_N^{(\beta)} P^k u\|_{\frac{2}{r}} + \sum_{j=1}^m \|X_j \widetilde{\psi} \Lambda_r^{p-2} D^\alpha \psi_N^{(\beta)} P^k u\|_{\frac{1}{r}} \\
& \quad + \|X_0 \widetilde{\psi} \Lambda_r^{p-2} D^\alpha \psi_N^{(\beta)} P^k u\| + \sum_{\ell, j=1}^m \|X_\ell X_j \widetilde{\psi} \Lambda_r^{p-2} D^\alpha \psi_N^{(\beta)} P^k u\| \\
& \quad + \|D^\alpha \psi_N^{(\beta)} P^k u\|_{\frac{p-1}{r}} + \|D^\alpha \psi_N^{(\beta)} P^k u\|_{\frac{p-2}{r}} + \sum_{j=1}^m \|X_j D^\alpha \psi_N^{(\beta)} P^k u\|_{\frac{p-2}{r}}
\end{aligned}$$

$$+\sum_{j=1}^{m}\|X_jD^\alpha\psi_N^{(\beta)}P^ku\|+\|X_0D^\alpha\psi_N^{(\beta)}P^ku\|+\sum_{\ell,j=1}^{m}\|X_\ell X_jD^\alpha\psi_N^{(\beta)}P^ku\|$$
$$+\|D^\alpha\psi_N^{(\beta)}P^ku\|\Big). \tag{38}$$

We have to estimate the first four terms on the right-hand side; the other terms are "good" terms for the induction argument used in the last section to gain the main result. Since $\widetilde{\psi}\Lambda_r^{p-2}D^\alpha\psi_N^{(\beta)}P^ku\in\mathscr{D}(\Omega_0)$, we take advantage from the estimate (6), Lemma 3.1, replacing v by $\widetilde{\psi}\Lambda_r^{p-2}D^\alpha\psi_N^{(\beta)}P^ku\in\mathscr{D}(\Omega_0)$. We have

$$\|\widetilde{\psi}\Lambda_r^{p-2}D^\alpha\psi_N^{(\beta)}P^ku\|_{\frac{2}{r}}+\sum_{j=1}^{m}\|X_j\widetilde{\psi}\Lambda_r^{p-2}D^\alpha\psi_N^{(\beta)}P^ku\|_{\frac{1}{r}}$$
$$+\|X_0\widetilde{\psi}\Lambda_r^{p-2}D^\alpha\psi_N^{(\beta)}P^ku\|+\sum_{\ell,j=1}^{m}\|X_\ell X_j\widetilde{\psi}\Lambda_r^{p-2}D^\alpha\psi_N^{(\beta)}P^ku\|$$
$$\leq C\left(\|P\widetilde{\psi}\Lambda_r^{p-2}D^\alpha\psi_N^{(\beta)}P^ku\|+\|\widetilde{\psi}\Lambda_r^{p-2}D^\alpha\psi_N^{(\beta)}P^ku\|\right). \tag{39}$$

We remark that

$$\|\widetilde{\psi}\Lambda_r^{p-2}D^\alpha\psi_N^{(\beta)}P^ku\|\leq\|D^\alpha\psi_N^{(\beta)}P^ku\|_{\frac{p-2}{r}}. \tag{40}$$

We focus on the first term on the right-hand side of (39). We proceed as in the case $H^{2/r}$ with the difference that in this case we have to handle new ingredients, in particular, the presence of the operator Λ_r^{p-2}. For technical reasons, we introduce

$$P_N=\sum_{\ell=1}^{m}\widetilde{X}_\ell^2+\widetilde{X}_0+\widetilde{c}_N, \tag{41}$$

where $\widetilde{X}_j$, $j=0,1,\ldots,m$, have the form described in (32). We point out that since φ_N is identically one on the support of $\widetilde{\psi}$ and φ_N, the identities $P_N(\widetilde{\psi}v)=P(\widetilde{\psi}v)$ and $\psi_N^{(\beta)}P_Nv=\psi_N^{(\beta)}Pv$ hold.

We have

$$\|P\widetilde{\psi}\Lambda_r^{p-2}D^\alpha\psi_N^{(\beta)}P^ku\|=\|P_N\widetilde{\psi}\Lambda_r^{p-2}D^\alpha\psi_N^{(\beta)}P^ku\|$$
$$\leq\|\widetilde{\psi}\Lambda_r^{p-2}D^\alpha\psi_N^{(\beta)}P^{k+1}u\|+\|[P_N,\widetilde{\psi}\Lambda_r^{p-2}D^\alpha\psi_N^{(\beta)}]P^ku\|$$
$$\leq\|D^\alpha\psi_N^{(\beta)}P^{k+1}u\|_{\frac{p-2}{r}}+\|[P_N,\widetilde{\psi}\Lambda_r^{p-2}]D^\alpha\psi_N^{(\beta)}P^ku\|$$
$$+\|\widetilde{\psi}\Lambda_r^{p-2}[P_N,D^\alpha\psi_N^{(\beta)}]P^ku\|. \tag{42}$$

We focus on the second term on the right-hand side:

$$[P_N, \widetilde{\psi}\Lambda_r^{p-2}] = \sum_{j=1}^{m}[\widetilde{X}_j^2, \widetilde{\psi}\Lambda_r^{p-2}] + [\widetilde{X}_0, \widetilde{\psi}\Lambda_r^{p-2}] + [\widetilde{c}_N, \widetilde{\psi}\Lambda_r^{p-2}]$$

$$= \sum_{j=1}^{m}\left(2[\widetilde{X}_j, \widetilde{\psi}\Lambda_r^{p-2}]\widetilde{X}_j + [\widetilde{X}_j, [\widetilde{X}_j, \widetilde{\psi}\Lambda_r^{p-2}]]\right) + [\widetilde{X}_0, \widetilde{\psi}\Lambda_r^{p-2}] + [\widetilde{c}_N, \widetilde{\psi}\Lambda_r^{p-2}].$$

So, we have

$$\begin{aligned}
&\|[P_N, \widetilde{\psi}\Lambda_r^{p-2}]D^\alpha\psi_N^{(\beta)}P^k u\| \\
&\leq \sum_{j=1}^{m}\left(2\|[\widetilde{X}_j, \widetilde{\psi}\Lambda_r^{p-2}]\widetilde{X}_j D^\alpha\psi_N^{(\beta)}P^k u\| + \|[\widetilde{X}_j, [\widetilde{X}_j, \widetilde{\psi}\Lambda_r^{p-2}]]D^\alpha\psi_N^{(\beta)}P^k u\|\right) \\
&\qquad + \|[\widetilde{X}_0, \widetilde{\psi}\Lambda_r^{p-2}]D^\alpha\psi_N^{(\beta)}P^k u\| + \|[\widetilde{c}_N, \widetilde{\psi}\Lambda_r^{p-2}]D^\alpha\psi_N^{(\beta)}P^k u\| \\
&= \sum_{k=1}^{4} I_k. \qquad (43)
\end{aligned}$$

We analyze separately the term on the right-hand side.

Term I_4: Since $[\widetilde{c}_N, \widetilde{\psi}\Lambda_r^{p-2}] = \widetilde{\psi}[\widetilde{c}_N, \Lambda_r^{p-2}]$ and $[\widetilde{c}_N, \Lambda_r^{p-2}]$ is a pseudodifferential operator of order $\frac{p-1}{r} - 1$, we have

$$I_4 \leq C_1\|D^\alpha\psi_N^{(\beta)}P^k u\|. \qquad (44)$$

The above estimate is a consequence of Theorem 18.1.1′ in [36], and we remark that since we are assuming that $M > n+2$, then C_1 is independent of N.

Term I_1: Since

$$[\widetilde{X}_j, \widetilde{\psi}\Lambda_r^{p-2}] = [\widetilde{X}_j, \widetilde{\psi}]\Lambda_r^{p-2} + \widetilde{\psi}[\widetilde{X}_j, \Lambda_r^{p-2}],$$

where $[\widetilde{X}_j, \Lambda_r^{p-2}] = -[\Lambda_r^{p-2}, \widetilde{X}_j]$ (explicitly given in (34)), we conclude that

$$I_1 \leq C_2\sum_{\ell=1}^{m}\left(\|X_j D^\alpha\psi_N^{(\beta)}P^k u\|_{\frac{p-2}{r}} + \|X_j D^\alpha\psi_N^{(\beta)}P^k u\|\right). \qquad (45)$$

Term I_3: It can be handled as the terms in I_1, and we have

$$I_3 \leq C_3\left(\|D^\alpha\psi_N^{(\beta)}P^k u\|_{\frac{p-2}{r}} + \|D^\alpha\psi_N^{(\beta)}P^k u\|\right). \qquad (46)$$

Term I_2: Observing that

$$[X_j, [X_j, \widetilde{\psi}\Lambda_r^{p-2}]] = [X_j, [X_j, \widetilde{\psi}]]\Lambda_r^{p-2} + 2[X_j, \widetilde{\psi}][\widetilde{X}_j, \Lambda_r^{p-2}] + \widetilde{\psi}[\widetilde{X}_j, [\widetilde{X}_j, \Lambda_r^{p-2}]],$$

where in particular

$$[\widetilde{X}_j, [\widetilde{X}_j, \Lambda_r^{p-2}]] = \sum_{\ell=1}^{n}\sum_{k=1}^{n} \Big(\widetilde{a}_{k,j}[\widetilde{a}_{\ell,j}^{(k)}, \Lambda_r^{p-2}]D_\ell + [\widetilde{a}_{k,j}, [\widetilde{a}_{\ell,j}, \Lambda_r^{p-2}]]D_k D_\ell$$
$$+[\widetilde{a}_{k,j}, [\widetilde{a}_{\ell,j}^{(k)}, \Lambda_r^{p-2}]]D_\ell + \widetilde{a}_{k,j}^{(\ell)}[\widetilde{a}_{\ell,j}, \Lambda_r^{p-2}]D_k - [\widetilde{a}_{k,j}^{(\ell)}, [\widetilde{a}_{\ell,j}, \Lambda_r^{p-2}]]D_k\Big),$$

with $[\widetilde{a}_{\ell,j}^{(k)}, \Lambda_r^{p-2}]$ and $[\widetilde{a}_{k,j}, [\widetilde{a}_{\ell,j}, \Lambda_r^{p-2}]]$, $[\widetilde{a}_{k,j}, [\widetilde{a}_{\ell,j}^{(k)}, \Lambda_r^{p-2}]]$ and $[\widetilde{a}_{k,j}^{(\ell)}, [\widetilde{a}_{\ell,j}, \Lambda_r^{p-2}]]$ pseudodifferential operators of order $\frac{p-2}{r} - 1$ and $\frac{p-2}{r} - 2$, respectively.

We have

$$\|[X_j, [X_j, \widetilde{\psi}]]\Lambda_r^{p-2} D^\alpha \psi_N^{(\beta)} P^k u\| \le C\|D^\alpha \psi_N^{(\beta)} P^k u\|_{\frac{p-2}{r}},$$

and, by (34) and the above considerations, that

$$\|[X_j, \widetilde{\psi}][\widetilde{X}_j, \Lambda_r^{p-2}]D^\alpha \psi_N^{(\beta)} P^k u\| + \|\widetilde{\psi}[\widetilde{X}_j, [\widetilde{X}_j, \Lambda_r^{p-2}]]D^\alpha \psi_N^{(\beta)} P^k u\|$$
$$\le C\left(\|D^\alpha \psi_N^{(\beta)} P^k u\|_{\frac{p-2}{r}} + \|D^\alpha \psi_N^{(\beta)} P^k u\|\right).$$

We conclude that

$$I_2 \le C_4\left(\|D^\alpha \psi_N^{(\beta)} P^k u\|_{\frac{p-2}{r}} + \|D^\alpha \psi_N^{(\beta)} P^k u\|\right). \tag{47}$$

Summing up, by (43), (44), (45), (46), and (47), we obtain

$$\|[P_N, \widetilde{\psi}\Lambda_r^{p-2}]D^\alpha \psi_N^{(\beta)} P^k u\|$$
$$\le C\Bigg(\sum_{\ell=1}^{m} \|X_j D^\alpha \psi_N^{(\beta)} P^k u\|_{\frac{p-2}{r}} + \|X_0 D^\alpha \psi_N^{(\beta)} P^k u\|_{\frac{p-2}{r}} + \|D^\alpha \psi_N^{(\beta)} P^k u\|_{\frac{p-2}{r}}$$
$$+\sum_{\ell=1}^{m} \|X_j D^\alpha \psi_N^{(\beta)} P^k u\| + \|X_0 D^\alpha \psi_N^{(\beta)} P^k u\| + \|D^\alpha \psi_N^{(\beta)} P^k u\|\Bigg), \tag{48}$$

where C is a new suitable constant independent of N.

Now, we have to estimate the last term on the right-hand side of (42):

$$\|\widetilde{\psi}\Lambda_r^{p-2}[P_N, D^\alpha \psi_N^{(\beta)}]P^k u\|.$$

This term is similar to that one treated in the previous subsection, and in order to make the paper more readable, we will cover it in quite detail anyway. Since

$$[P_N, D^\alpha \psi_N^{(\beta)}] = \sum_{j=1}^{m} \left(2\widetilde{X}_j[\widetilde{X}_j, D^\alpha \psi_N^{(\beta)}] - [\widetilde{X}_j, [\widetilde{X}_j, D^\alpha \psi_N^{(\beta)}]]\right)$$
$$+ [\widetilde{X}_0, D^\alpha \psi_N^{(\beta)}] + [\widetilde{c}_N, D^\alpha \psi_N^{(\beta)}],$$

we have

$$\|\widetilde{\psi}\Lambda_r^{p-2}[P_N, D^\alpha \psi_N^{(\beta)}]P^k u\| \le 2\sum_{j=1}^{m} \|\widetilde{\psi}\Lambda_r^{p-2}\widetilde{X}_j[\widetilde{X}_j, D^\alpha \psi_N^{(\beta)}]P^k u\|$$
$$+ \sum_{j=1}^{m} \|\widetilde{\psi}\Lambda_r^{p-2}[\widetilde{X}_j, [\widetilde{X}_j, D^\alpha \psi_N^{(\beta)}]]P^k u\| + \|\widetilde{\psi}\Lambda_r^{p-2}[\widetilde{X}_0, D^\alpha \psi_N^{(\beta)}]P^k u\|$$
$$+ \|\widetilde{\psi}\Lambda_r^{p-2}[\widetilde{c}_N, D^\alpha \psi_N^{(\beta)}]P^k u\| = H_{2,1} + H_{2,2} + H_{2,3} + H_{2,4}. \tag{49}$$

We analyze separately the terms on the right-hand side.

Term $H_{2,4}$. By (9), we have

$$[\widetilde{c}_N, D^\alpha \psi_N^{(\beta)}] = [\widetilde{c}_N, D^\alpha]\psi_N^{(\beta)} = -\sum_{\substack{\nu\le\alpha,\\ \nu\neq 0}} \binom{\alpha}{\nu} \widetilde{c}_N^{(\nu)} D^{\alpha-\nu}\psi_N^{(\beta)},$$

and then

$$\|\widetilde{\psi}\Lambda_r^{p-2}[\widetilde{c}_N, D^\alpha \psi_N^{(\beta)}]P^k u\|$$
$$\le 2^{\frac{p-2}{4r}} \sum_{\substack{\nu\le\alpha,\\ \nu\neq 0}} \binom{\alpha}{\nu} \left(\int (1+|\zeta|)^{\frac{p-2}{r}} |\widehat{\widetilde{c}_N^{(\nu)}}(\zeta)|\, d\zeta\right) \|D^{\alpha-\nu}\psi_N^{(\beta)} P^k u\|_{\frac{p-2}{r}}.$$

Let us proceed to estimate in detail the product of the first two factors in the sum. Later on we will use the same strategy several times, but we will omit the details in order to do not weigh down the discussion. Let q equal to 1 or 0 if n is odd or even. We have

$$\int (1+|\zeta|)^{\frac{p-2}{r}} |\widehat{\widetilde{c}_N^{(\nu)}}(\zeta)|\, d\zeta \le \int \frac{1}{(1+|\zeta|)^{n+1}} (1+|\zeta|)^{n+2+q} |\widehat{\widetilde{c}_N^{(\nu)}}(\zeta)|\, d\zeta$$
$$\le \int \frac{1}{(1+|\zeta|)^{n+1}} 2^{n+2+q}\left(1+|\zeta|^{n+2+q}\right) |\widehat{\widetilde{c}_N^{(\nu)}}(\zeta)|\, d\zeta$$

$$= \int \frac{1}{(1+|\zeta|)^{n+1}} 2^{n+2+q} \left(|\widehat{\widetilde{c}}_N^{(\nu)}(\zeta)| + \sum_{\mu=q'} \frac{q'!}{\mu_1! \cdots \mu_n!} |\widehat{\widetilde{c}}_N^{(\nu+2\mu)}(\zeta)| \right) d\zeta,$$

where $q' = n + 2 + q$. So

$$\binom{\alpha}{\nu} |\widetilde{c}_N^{(\nu+2\mu)}| \le \binom{\alpha}{\nu} \sum_{\gamma \le \nu+2\mu} \binom{\nu+2\mu}{\gamma} |\varphi_N^{(\nu+2\mu-\gamma)}| |c^{(\gamma)}|$$

$$\le 2^{|\nu|+n+2+q} (2\mu!) \frac{\alpha!}{(\alpha-\nu)!}$$

$$\times \sum_{\gamma \le \nu+2\mu} \frac{1}{(\nu+2\mu-\gamma)!} C_1^{|\nu|-|\gamma|+n+q+3} N^{(|\nu|+2|\mu|-|\gamma|-M)^+} C_2^{|\gamma|+1} (\gamma!)^{s-1}.$$

By the Young inequality for products, assuming that $M > 2|\mu|+1 = n+3+q$ and since the number of the multi-indexes γ with $|\gamma| \le |\nu| + 2|\mu|$ is smaller or equal to $2^{|\nu|+n+2+q}$, we obtain

$$\binom{\alpha}{\nu} |\widetilde{c}_N^{(\nu+2\mu)}| \le 2^{2(|\nu|+n+2+q)} (2\mu!) C_3^{|\nu|+n+q+3} \frac{\alpha!}{(\alpha-\nu)!} \left(N^{|\nu|-1} + |\nu|^{(s-1)|\nu|} \right)$$

and

$$\binom{\alpha}{\nu} |\widetilde{c}_N^{(\nu)}| \le 2^{2(|\nu|)} C_3^{|\nu|+1} \frac{\alpha!}{(\alpha-\nu)!} \left(N^{|\nu|-1} + |\nu|^{(s-1)|\nu|} \right).$$

Since $\sum_{\mu=q'} \frac{q'!}{\mu_1! \cdots \mu_n!} = n^{n+2+q}$, we have

$$\binom{\alpha}{\nu} |\int (1+|\zeta|)^{\frac{p-2}{r}} |\widehat{\widetilde{c}_N^{(\nu)}}(\zeta)| \, d\zeta \le C_4^{|\nu|+1} \frac{\alpha!}{(\alpha-\nu)!} \left(N^{|\nu|-1} + |\nu|^{(s-1)|\nu|} \right), \tag{50}$$

where C_4 is a suitable positive constant independent of ν and N.

We conclude that

$$H_{2,4} = \|\widetilde{\psi} \Lambda_r^{p-2} [\widetilde{c}_N, D^\alpha \psi_N^{(\beta)}] P^k u\|$$

$$\le \sum_{\substack{\nu \le \alpha, \\ \nu \ne 0}} C_5^{|\nu|+1} \frac{\alpha!}{(\alpha-\nu)!} \left(N^{|\nu|-1} + |\nu|^{(s-1)|\nu|} \right) \|D^{\alpha-\nu} \psi_N^{(\beta)} P^k u\|_{\frac{p-2}{r}}. \tag{51}$$

Term $H_{2,3}$. By (9), we have

$$[\widetilde{X}_0, D^\alpha \psi_N^{(\beta)}] = [\widetilde{X}_0, D^\alpha]\psi_N^{(\beta)} + D^\alpha[\widetilde{X}_0, \psi_N^{(\beta)}]$$
$$= \sum_{\ell=1}^{n} \sum_{\substack{\nu\le\alpha,\\ \nu\neq 0}} \binom{\alpha}{\nu} \left(\widetilde{a}_{N,\ell,0}^{(\nu)} D^{\alpha-\nu}\psi_N^{(\beta+\ell)} - \widetilde{a}_{N,\ell,0}^{(\nu)} D^{\alpha-\nu+\ell}\psi_N^{(\beta)}\right) + \sum_{\ell=1}^{n} \widetilde{a}_{N,\ell,0} D^\alpha \psi_N^{(\beta+\ell)},$$

so

$$H_{2,3} \le \sum_{\ell=1}^{n} \|\widetilde{\psi}\Lambda_r^{p-2}\widetilde{a}_{N,\ell,0} D^\alpha \psi_N^{(\beta+\ell)} P^k u\|$$
$$+ \sum_{\ell=1}^{n} \sum_{\substack{\nu\le\alpha,\\ \nu\neq 0}} \binom{\alpha}{\nu} \Big(\|\widetilde{\psi}\Lambda_r^{p-2}\widetilde{a}_{N,\ell,0}^{(\nu)} D^{\alpha-\nu+\ell}\psi_N^{(\beta)} P^k u\|$$
$$+ \|\widetilde{\psi}\Lambda_r^{p-2}\widetilde{a}_{N,\ell,0}^{(\nu)} D^{\alpha-\nu}\psi_N^{(\beta+\ell)} P^k u\|\Big)$$
$$\le 2^{\frac{p-2}{4r}} \sum_{\ell=1}^{n} \left(\int (1+|\zeta|)^{\frac{p-2}{r}} |\widehat{\widetilde{a}_{N,\ell,0}}(\zeta)|\, d\zeta \right) \|D^\alpha \psi_N^{(\beta+\ell)} P^k u\|_{\frac{p-2}{r}}$$
$$+ 2^{\frac{p-2}{4r}} \sum_{\ell=1}^{n} \sum_{\substack{\nu\le\alpha,\\ \nu\neq 0}} \binom{\alpha}{\nu} \left(\int (1+|\zeta|)^{\frac{p-2}{r}} |\widehat{\widetilde{a}_{N,\ell,0}^{(\nu)}}(\zeta)|\, d\zeta \right)$$
$$\times \left(\|D^{\alpha-\nu+\ell}\psi_N^{(\beta)} P^k u\|_{\frac{p-2}{r}} + \|D^{\alpha-\nu}\psi_N^{(\beta+\ell)} P^k u\|_{\frac{p-2}{r}} \right).$$

Using the same strategy adopted to obtain (50), we conclude that

$$H_{2,3} = \|\widetilde{\psi}\Lambda_r^{p-2}[\widetilde{X}_0, D^\alpha \psi_N^{(\beta)}] P^k u\|$$
$$\le \sum_{\ell=1}^{n} C_1^{n+3} N^{s(n+2-M)^+} \|D^\alpha \psi_N^{(\beta+\ell)} P^k u\|_{\frac{p-2}{r}}$$
$$+ \sum_{\ell=1}^{n} \sum_{\substack{\nu\le\alpha,\\ \nu\neq 0}} C_1^{|\nu|+1} \frac{\alpha!}{(\alpha-\nu)!} \left(N^{|\nu|-1} + |\nu|^{(s-1)|\nu|} \right)$$
$$\times \left(\|D^{\alpha-\nu+\ell}\psi_N^{(\beta)} P^k u\|_{\frac{p-2}{r}} + \|D^{\alpha-\nu}\psi_N^{(\beta+\ell)} P^k u\|_{\frac{p-2}{r}} \right). \tag{52}$$

Term $H_{2,1}$. For every $j = 1, \dots, m$, we observe that

$$\widetilde{X}_j[\widetilde{X}_j, D^\alpha \psi_N^{(\beta)}]$$

$$= \sum_{\ell=1}^{n} \sum_{\substack{\nu \le \alpha, \\ \nu \ne 0}} \binom{\alpha}{\nu} \left(\widetilde{a}^{(\nu)}_{N,\ell,j} \widetilde{X}_j D^{\alpha-\nu} \psi_N^{(\beta+\ell)} - \widetilde{a}^{(\nu)}_{N,\ell,j} \widetilde{X}_j D^{\alpha-\nu+\ell} \psi_N^{(\beta)} \right)$$

$$+ \sum_{\ell=1}^{n} \sum_{i=1}^{n} \sum_{\substack{\nu \le \alpha, \\ \nu \ne 0}} \binom{\alpha}{\nu} \left(\widetilde{a}_{N,i,j} \widetilde{a}^{(\nu+i)}_{N,\ell,j} D^{\alpha-\nu} \psi_N^{(\beta+\ell)} - \widetilde{a}_{N,i,j} \widetilde{a}^{(\nu+i)}_{N,\ell,j} D^{\alpha-\nu+\ell} \psi_N^{(\beta)} \right)$$

$$+ \sum_{\ell=1}^{n} \widetilde{a}_{N,\ell,j} \widetilde{X}_j D^{\alpha} \psi_N^{(\beta+\ell)} + \sum_{\ell=1}^{n} \sum_{i=1}^{n} \widetilde{a}_{N,i,j} \widetilde{a}^{(i)}_{N,\ell,j} D^{\alpha} \psi_N^{(\beta+\ell)}.$$

So, we have

$$\frac{1}{2} H_{2,1} \le \sum_{j=1}^{m} \sum_{\ell=1}^{n} \sum_{\substack{\nu \le \alpha, \\ \nu \ne 0}} \binom{\alpha}{\nu} \Big(\| \widetilde{\psi} \Lambda_r^{p-2} \widetilde{a}^{(\nu)}_{N,\ell,j} \widetilde{X}_j D^{\alpha-\nu+\ell} \psi_N^{(\beta)} P^k u \|$$

$$+ \| \widetilde{\psi} \Lambda_r^{p-2} \widetilde{a}^{(\nu)}_{N,\ell,j} \widetilde{X}_j D^{\alpha-\nu} \psi_N^{(\beta+\ell)} P^k u \| \Big)$$

$$+ \sum_{j=1}^{m} \sum_{i,\ell=1}^{n} \sum_{\substack{\nu \le \alpha, \\ \nu \ne 0}} \binom{\alpha}{\nu} \Big(\| \widetilde{\psi} \Lambda_r^{p-2} \widetilde{a}_{N,i,j} \widetilde{a}^{(\nu+i)}_{N,\ell,j} D^{\alpha-\nu+\ell} \psi_N^{(\beta)} P^k u \|$$

$$+ \| \widetilde{\psi} \Lambda_r^{p-2} \widetilde{a}_{N,i,j} \widetilde{a}^{(\nu+i)}_{N,\ell,j} D^{\alpha-\nu} \psi_N^{(\beta+\ell)} P^k u \| \Big)$$

$$+ \sum_{j=1}^{m} \sum_{\ell=1}^{n} \Big(\| \widetilde{\psi} \Lambda_r^{p-2} \widetilde{a}_{N,\ell,j} \widetilde{X}_j D^{\alpha} \psi_N^{(\beta+\ell)} P^k u \|$$

$$+ \sum_{i=1}^{n} \| \widetilde{\psi} \Lambda_r^{p-2} \widetilde{a}_{N,i,j} \widetilde{a}^{(i)}_{N,\ell,j} D^{\alpha} \psi_N^{(\beta+\ell)} P^k u \| \Big)$$

$$\le \sum_{j=1}^{m} \sum_{\ell=1}^{n} \sum_{\substack{\nu \le \alpha, \\ \nu \ne 0}} \binom{\alpha}{\nu} \left(\int (1+|\zeta|)^{\frac{p-2}{r}} |\widehat{\widetilde{a}^{(\nu)}_{N,\ell,j}}(\zeta)| \, d\zeta \right) \left(\int (1+|\zeta|)^{\frac{p-2}{r}} |\widehat{\varphi}_N(\zeta)| \, d\zeta \right)$$

$$\times \left[\| X_j D^{\alpha-\nu+\ell} \psi_N^{(\beta)} P^k u \|_{\frac{p-2}{r}} + \| X_j D^{\alpha-\nu} \psi_N^{(\beta+\ell)} P^k u \|_{\frac{p-2}{r}} \right]$$

$$+ \sum_{j=1}^{m} \sum_{i,\ell=1}^{n} \sum_{\substack{\nu \le \alpha, \\ \nu \ne 0}} \binom{\alpha}{\nu} \left(\int (1+|\zeta|)^{\frac{p-2}{r}} |\widehat{\widetilde{a}^{(\nu+i)}_{N,\ell,j}}(\zeta)| \, d\zeta \right) \left(\int (1+|\zeta|)^{\frac{p-2}{r}} |\widehat{\widetilde{a}_{N,i,j}}(\zeta)| \, d\zeta \right)$$

$$\times \left[\| D^{\alpha-\nu+\ell} \psi_N^{(\beta)} P^k u \|_{\frac{p-2}{r}} + \| D^{\alpha-\nu} \psi_N^{(\beta+\ell)} P^k u \|_{\frac{p-2}{r}} \right]$$

$$+\sum_{j=1}^{m}\sum_{\ell=1}^{n}\left(\int (1+|\zeta|)^{\frac{p-2}{r}}\,|\widehat{\widetilde{a}_{N,\ell,j}}(\zeta)|\,d\zeta\right)\left(\int (1+|\zeta|)^{\frac{p-2}{r}}\,|\widehat{\varphi}_N(\zeta)|\,d\zeta\right)$$
$$\times \|X_j D^\alpha \psi_N^{(\beta+\ell)} P^k u\|_{\frac{p-2}{r}}$$

$$+\sum_{j=1}^{m}\sum_{i,\ell=1}^{n}\left(\int (1+|\zeta|)^{\frac{p-2}{r}}\,|\widehat{\widetilde{a}^{(i)}_{N,\ell,j}}(\zeta)|\,d\zeta\right)\left(\int (1+|\zeta|)^{\frac{p-2}{r}}\,|\widehat{\widetilde{a}_{N,i,j}}(\zeta)|\,d\zeta\right)$$
$$\times \|D^\alpha \psi_N^{(\beta+\ell)} P^k u\|_{\frac{p-2}{r}}.$$

Now, using suitably the same strategy adopted to obtain (50), we conclude that

$$H_{2,1} = 2\sum_{j=1}^{m}\|\widetilde{\psi}\Lambda_r^{p-2}\widetilde{X}_j[\widetilde{X}_j, D^\alpha\psi_N^{(\beta)}]P^k u\|$$
$$\leq 2\sum_{j=1}^{m}\sum_{\ell=1}^{n}\sum_{\substack{\nu\leq\alpha,\\ \nu\neq 0}} C_1^{|\nu|+1}\frac{\alpha!}{(\alpha-\nu)!}\left(N^{|\nu|-1}+|\nu|^{(s-1)|\nu|}\right)N^{s(n+3-M)^+}$$
$$\times\left[\|X_j D^{\alpha-\nu+\ell}\psi_N^{(\beta)}P^k u\|_{\frac{p-2}{r}}+\|X_j D^{\alpha-\nu}\psi_N^{(\beta+\ell)}P^k u\|_{\frac{p-2}{r}}\right]$$
$$+2\sum_{\ell=1}^{n}\sum_{\substack{\nu\leq\alpha,\\ \nu\neq 0}} C_1^{|\nu|+1}\frac{\alpha!}{(\alpha-\nu)!}\left(N^{|\nu|-1}+|\nu|^{(s-1)|\nu|}\right)N^{s(n+3-M)^+}$$
$$\times\left[\|D^{\alpha-\nu+\ell}\psi_N^{(\beta)}P^k u\|_{\frac{p-2}{r}}+\|D^{\alpha-\nu}\psi_N^{(\beta+\ell)}P^k u\|_{\frac{p-2}{r}}\right]$$
$$+2C_1^{2n+6}N^{s(2n+4-M)^+}\sum_{j=1}^{m}\sum_{\ell=1}^{n}\|X_j D^\alpha\psi_N^{(\beta+\ell)}P^k u\|_{\frac{p-2}{r}}$$
$$+2C_1^{2n+7}N^{s(2n+5-M)^+}\sum_{\ell=1}^{n}\|D^\alpha\psi_N^{(\beta+\ell)}P^k u\|_{\frac{p-2}{r}}. \tag{53}$$

Term $H_{2,2}$. For every $j=1,\ldots,m$, we have

$$[\widetilde{X}_j,[\widetilde{X}_j, D^\alpha\psi_N^{(\beta)}]] = [\widetilde{X}_j,[\widetilde{X}_j, D^\alpha]\psi_N^{(\beta)}]+[\widetilde{X}_j, D^\alpha[\widetilde{X}_j,\psi_N^{(\beta)}]]$$
$$=[\widetilde{X}_j,[\widetilde{X}_j,D^\alpha]]\psi_N^{(\beta)}+2[\widetilde{X}_j,D^\alpha][\widetilde{X}_j,\psi_N^{(\beta)}]+D^\alpha[\widetilde{X}_j,[\widetilde{X}_j,\psi_N^{(\beta)}]].$$

We obtain

$$H_{2,2}=\sum_{j=1}^{m}\|\widetilde{\psi}\Lambda_r^{p-2}[\widetilde{X}_j,[\widetilde{X}_j,D^{\alpha}\psi_N^{(\beta)}]]P^k u\|$$

$$\leq\sum_{j=1}^{m}\Big(\|\widetilde{\psi}\Lambda_r^{p-2}[\widetilde{X}_j,[\widetilde{X}_j,D^{\alpha}]]\psi_N^{(\beta)}P^k u\|+2\|\widetilde{\psi}\Lambda_r^{p-2}[\widetilde{X}_j,D^{\alpha}][\widetilde{X}_j,\psi_N^{(\beta)}]P^k u\|$$

$$+\|\widetilde{\psi}D^{\alpha}[\widetilde{X}_j,[\widetilde{X}_j,\psi_N^{(\beta)}]]\Lambda_r^{p-2}P^k u\|\Big)$$

$$=H_{2,2,1}+H_{2,2,2}+H_{2,2,3}. \tag{54}$$

We proceed to estimate the three terms separately.

Term $H_{2,2,1}$. Using the same steps to obtain (20), we have

$$[\widetilde{X}_j,[\widetilde{X}_j,D^{\alpha}]]=-\sum_{\ell,i=1}^{n}\sum_{\substack{\nu\leq\alpha,\\ \nu\neq 0}}\binom{\alpha}{\nu}\widetilde{a}_{N,i,j}\,\widetilde{a}_{N,\ell,j}^{(\nu+i)}D^{\alpha-\nu+\ell}$$

$$+\sum_{\ell,i=1}^{n}\sum_{\substack{\nu\leq\alpha,\\ \nu\neq 0}}\sum_{\substack{\mu\leq\alpha-\nu\\ \mu\neq 0}}\binom{\alpha}{\nu}\binom{\alpha-\nu}{\mu}\widetilde{a}_{N,\ell,j}^{(\nu)}\left(\widetilde{a}_{N,i,j}^{(\mu)}D^{\alpha-\nu-\mu+\ell+i}+\widetilde{a}_{N,i,j}^{(\mu+\ell)}D^{\alpha-\nu-\mu+i}\right)$$

$$+\sum_{\ell,i=1}^{n}\sum_{\substack{\nu\leq\alpha,\\ \nu\neq 0}}\binom{\alpha}{\nu}\widetilde{a}_{N,\ell,j}^{(\nu)}\widetilde{a}_{N,i,j}^{(\ell)}D^{\alpha-\nu+i}.$$

So, using the strategy adopted to handle the previous terms, in particular the strategy to obtain (50), and assuming that M is greater than $n+6$, we have

$$H_{2,2,1}=\sum_{j=1}^{m}\|\widetilde{\psi}\Lambda_r^{p-2}[\widetilde{X}_j,[\widetilde{X}_j,D^{\alpha}]]\psi_N^{(\beta)}P^k u\|$$

$$\leq\sum_{j=1}^{m}\sum_{\ell,i=1}^{n}\sum_{\substack{\nu\leq\alpha,\\ \nu\neq 0}}\binom{\alpha}{\nu}\left(\int(1+|\zeta|)^{\frac{p-2}{r}}|\widehat{\widetilde{a}}_{N,i,j}(\zeta)|\,d\zeta\right)$$

$$\times\left(\int(1+|\zeta|)^{\frac{p-2}{r}}|\widehat{\widetilde{a}}_{N,\ell,j}^{(\nu+i)}(\zeta)|\,d\zeta\right)\|D^{\alpha-\nu+\ell}\psi_N^{(\beta)}P^k u\|_{\frac{p-2}{r}}$$

$$+\sum_{j=1}^{m}\sum_{\ell,i=1}^{n}\sum_{\substack{\nu\le\alpha,\\ \nu\neq 0}}\binom{\alpha}{\nu}\left(\int(1+|\zeta|)^{\frac{p-2}{r}}\,|\widehat{\widetilde{a}}^{(\nu)}_{N,\ell,j}(\zeta)|\,d\zeta\right)$$

$$\times\left(\int(1+|\zeta|)^{\frac{p-2}{r}}\,|\widehat{\widetilde{a}}^{(\ell)}_{N,i,j}(\zeta)|\,d\zeta\right)\|D^{\alpha-\nu+i}\psi_N^{(\beta)}P^k u\|_{\frac{p-2}{r}}$$

$$+\sum_{j=1}^{m}\sum_{\ell,i=1}^{n}\sum_{\substack{\nu\le\alpha,\\ \nu\neq 0}}\sum_{\substack{\mu\le\alpha-\nu\\ \mu\neq 0}}\binom{\alpha}{\nu}\binom{\alpha-\nu}{\mu}\Bigg[\left(\int(1+|\zeta|)^{\frac{p-2}{r}}\,|\widehat{\widetilde{a}}^{(\nu)}_{N,\ell,j}(\zeta)|\,d\zeta\right)$$

$$\times\left(\int(1+|\zeta|)^{\frac{p-2}{r}}\,|\widehat{\widetilde{a}}^{(\mu+\ell)}_{N,i,j}(\zeta)|\,d\zeta\right)\|D^{\alpha-\nu-\mu+i}\psi_N^{(\beta)}P^k u\|_{\frac{p-2}{r}}$$

$$+\left(\int(1+|\zeta|)^{\frac{p-2}{r}}\,|\widehat{\widetilde{a}}^{(\nu)}_{N,\ell,j}(\zeta)|\,d\zeta\right)\left(\int(1+|\zeta|)^{\frac{p-2}{r}}\,|\widehat{\widetilde{a}}^{(\mu)}_{N,i,j}(\zeta)|\,d\zeta\right)$$

$$\times\|D^{\alpha-\nu-\mu+\ell+i}\psi_N^{(\beta)}P^k u\|_{\frac{p-2}{r}}\Bigg]$$

$$\le\sum_{\ell=1}^{n}\sum_{\substack{\nu\le\alpha,\\ \nu\neq 0}}C_1^{|\nu|+2}\frac{\alpha!}{(\alpha-\nu)!}\left(N^{|\nu|-1}+|\nu|^{(s-1)|\nu|}\right)\|D^{\alpha-\nu+\ell}\psi_N^{(\beta)}P^k u\|_{\frac{p-2}{r}}$$

$$+\sum_{i=1}^{n}\sum_{\substack{\nu\le\alpha,\\ \nu\neq 0}}\sum_{\substack{\mu\le\alpha-\nu\\ \mu\neq 0}}C_1^{|\nu|+|\mu|+2}\frac{\alpha!}{(\alpha-\nu-\mu)!}\left(N^{|\nu|-1}+|\nu|^{(s-1)|\nu|}\right)$$

$$\times\left(N^{|\mu|-1}+|\mu|^{(s-1)|\mu|}\right)\|D^{\alpha-\nu-\mu+i}\psi_N^{(\beta)}P^k u\|_{\frac{p-2}{r}}$$

$$+\sum_{\ell,i=1}^{n}\sum_{\substack{\nu\le\alpha,\\ \nu\neq 0}}\sum_{\substack{\mu\le\alpha-\nu\\ \mu\neq 0}}C_1^{|\nu|+|\mu|+1}\frac{\alpha!}{(\alpha-\nu-\mu)!}\left(N^{|\nu|-1}+|\nu|^{(s-1)|\nu|}\right)$$

$$\times\left(N^{|\mu|-1}+|\mu|^{(s-1)|\mu|}\right)\|D^{\alpha-\nu-\mu+i+\ell}\psi_N^{(\beta)}P^k u\|_{\frac{p-2}{r}}. \tag{55}$$

Term $H_{2,2,2}$. By (21), for every j, $j=1,\dots,m$, we have

$$2[\widetilde{X}_j,D^\alpha][\widetilde{X}_j,\psi_N^{(\beta)}]$$

$$=-2\sum_{\ell,i=1}^{n}\sum_{\substack{\nu\le\alpha,\\ \nu\neq 0}}\binom{\alpha}{\nu}\widetilde{a}^{(\nu)}_{N,\ell,j}\left(\widetilde{a}_{N,i,j}D^{\alpha-\nu+\ell}+\widetilde{a}^{(\ell)}_{N,i,j}D^{\alpha-\nu}\right)\psi_N^{(\beta+i)}$$

$$-2\sum_{\ell,i=1}^{n}\sum_{\substack{\nu\le\alpha,\\ \nu\neq 0}}\sum_{\substack{\mu\le\alpha-\nu+\ell\\ \mu\neq 0}}\binom{\alpha}{\nu}\binom{\alpha-\nu}{\mu}\widetilde{a}^{(\nu)}_{N,\ell,j}\left(\widetilde{a}^{(\mu+\ell)}_{N,i,j}D^{\alpha-\nu-\mu}\right.$$
$$\left.+\widetilde{a}^{(\mu)}_{N,i,j}D^{\alpha-\nu-\mu+\ell}\right)\psi_N^{(\beta+i)}.$$

We conclude that

$$H_{2,2,2}=2\sum_{j=1}^{m}\|\widetilde{\psi}\Lambda_r^{p-2}[\widetilde{X}_j,D^\alpha][\widetilde{X}_j,\psi_N^{(\beta)}]P^ku\|$$
$$\le\sum_{i=1}^{n}\sum_{\substack{\nu\le\alpha,\\ \nu\neq 0}}C_1^{|\nu|+2}\frac{\alpha!}{(\alpha-\nu)!}\left(N^{|\nu|-1}+|\nu|^{(s-1)|\nu|}\right)\|D^{\alpha-\nu}\psi_N^{(\beta+i)}P^ku\|_{\frac{p-2}{r}}$$
$$+\sum_{\ell,i=1}^{n}\sum_{\substack{\nu\le\alpha,\\ \nu\neq 0}}C_1^{|\nu|+1}\frac{\alpha!}{(\alpha-\nu)!}\left(N^{|\nu|-1}+|\nu|^{(s-1)|\nu|}\right)\|D^{\alpha-\nu+\ell}\psi_N^{(\beta+i)}P^ku\|_{\frac{p-2}{r}}$$
$$+\sum_{i=1}^{n}\sum_{\substack{\nu\le\alpha,\\ \nu\neq 0}}\sum_{\substack{\mu\le\alpha-\nu\\ \mu\neq 0}}C_1^{|\nu|+|\mu|+3}\frac{\alpha!}{(\alpha-\nu-\mu)!}\left(N^{|\nu|-1}+|\nu|^{(s-1)|\nu|}\right)$$
$$\times\left(N^{|\mu|-1}+|\mu|^{(s-1)|\mu|}\right)\|D^{\alpha-\nu-\mu}\psi_N^{(\beta+i)}P^ku\|_{\frac{p-2}{r}}$$
$$+\sum_{\ell,i=1}^{n}\sum_{\substack{\nu\le\alpha,\\ \nu\neq 0}}\sum_{\substack{\mu\le\alpha-\nu\\ \mu\neq 0}}C_1^{|\nu|+|\mu|+2}\frac{\alpha!}{(\alpha-\nu-\mu)!}\left(N^{|\nu|-1}+|\nu|^{(s-1)|\nu|}\right)$$
$$\times\left(N^{|\mu|-1}+|\mu|^{(s-1)|\mu|}\right)\|D^{\alpha-\nu-\mu+\ell}\psi_N^{(\beta+i)}P^ku\|_{\frac{p-2}{r}}.\tag{56}$$

Term $H_{2,2,3}$. Taking advantage from (22), for every j, $j=1,\dots,m$, we have

$$D^\alpha[\widetilde{X}_j,[\widetilde{X}_j,\psi_N^{(\beta)}]]=\sum_{j,i=1}^{n}\widetilde{a}_{N,i,j}\left(\widetilde{a}^{(i)}_{N,\ell,j}D^\alpha\psi_N^{(\beta+\ell)}+\widetilde{a}_{N,\ell,j}D^\alpha\psi_N^{(\beta+\ell+i)}\right)$$
$$+\sum_{\ell,i=1}^{n}\sum_{\substack{\nu\le\alpha\\ \nu\neq 0}}\binom{\alpha}{\nu}\left[\left(\widetilde{a}_{N,i,j}\widetilde{a}^{(\nu+i)}_{N,\ell,j}+\widetilde{a}^{(\nu)}_{N,i,j}\widetilde{a}^{(i)}_{N,\ell,j}\right)D^{\alpha-\nu}\psi_N^{(\beta+\ell)}\right.$$
$$\left.+\left(\widetilde{a}_{N,i,j}\widetilde{a}^{(\nu)}_{N,\ell,j}+\widetilde{a}^{(\nu)}_{N,i,j}\widetilde{a}_{N,\ell,j}\right)D^{\alpha-\nu}\psi_N^{(\beta+\ell+i)}\right]$$

$$+\sum_{\ell,i=1}^{n}\sum_{\substack{\nu<\alpha\\ \nu\neq 0}}\sum_{\substack{\mu\leq\alpha-\nu\\ \mu\neq 0}}\binom{\alpha}{\nu}\binom{\alpha-\nu}{\mu}\widetilde{a}_{N,i,j}^{(\nu)}\left[\widetilde{a}_{N,\ell,j}^{(\mu+i)}D^{\alpha-\nu-\mu}\psi_N^{(\beta+\ell)}\right.$$
$$\left.+\widetilde{a}_{N,\ell,j}^{(\mu)}D^{\alpha-\nu-\mu}\psi_N^{(\beta+\ell+i)}\right].$$

From this identity, we obtain

$$\begin{aligned}H_{2,2,3}&=\sum_{j=1}^{m}\|\widetilde{\psi}\Lambda_r^{p-2}D^{\alpha}[\widetilde{X}_j,[\widetilde{X}_j,\psi_N^{(\beta)}]]P^k u\|\\&\leq\sum_{\ell=1}^{n}C_1^3\|D^{\alpha}\psi_N^{(\beta+\ell)}P^k u\|_{\frac{p-2}{r}}+\sum_{j,i=1}^{n}C_1^2\|D^{\alpha}\psi_N^{(\beta+\ell+i)}P^k u\|_{\frac{p-2}{r}}\\&+\sum_{\ell=1}^{n}\sum_{\substack{\nu\leq\alpha\\ \nu\neq 0}}C_1^{|\nu|+3}\frac{\alpha!}{(\alpha-\nu)!}\left(N^{|\nu|-1}+|\nu|^{(s-1)|\nu|}\right)\|D^{\alpha-\nu}\psi_N^{(\beta+\ell)}P^k u\|_{\frac{p-2}{r}}\\&+\sum_{\ell,i=1}^{n}\sum_{\substack{\nu\leq\alpha\\ \nu\neq 0}}C_1^{|\nu|+2}\frac{\alpha!}{(\alpha-\nu)!}\left(N^{|\nu|-1}+|\nu|^{(s-1)|\nu|}\right)\|D^{\alpha-\nu}\psi_N^{(\beta+\ell+i)}P^k u\|_{\frac{p-2}{r}}\\&+\sum_{\ell=1}^{n}\sum_{\substack{\nu<\alpha\\ \nu\neq 0}}\sum_{\substack{\mu\leq\alpha-\nu\\ \mu\neq 0}}C_1^{|\nu|+|\mu|+3}\frac{\alpha!}{(\alpha-\nu-\mu)!}\left(N^{|\nu|-1}+|\nu|^{(s-1)|\nu|}\right)\\&\times\left(N^{|\mu|-1}+|\mu|^{(s-1)|\mu|}\right)\|D^{\alpha-\nu-\mu}\psi_N^{(\beta+\ell)}P^k u\|_{\frac{p-2}{r}}\\&+\sum_{\ell,i=1}^{n}\sum_{\substack{\nu<\alpha\\ \nu\neq 0}}\sum_{\substack{\mu\leq\alpha-\nu\\ \mu\neq 0}}C_1^{|\nu|+|\mu|+2}\frac{\alpha!}{(\alpha-\nu-\mu)!}\left(N^{|\nu|-1}+|\nu|^{(s-1)|\nu|}\right)\\&\times\left(N^{|\mu|-1}+|\mu|^{(s-1)|\mu|}\right)\|D^{\alpha-\nu-\mu}\psi_N^{(\beta+\ell+i)}P^k u\|_{\frac{p-2}{r}}.\end{aligned}\tag{57}$$

By (54), (55),(56), and(57), we have

$$\begin{aligned}H_{2,2}&=\sum_{j=1}^{m}\|\widetilde{\psi}\Lambda_r^{p-2}[\widetilde{X}_j,[\widetilde{X}_j,D^{\alpha}\psi_N^{(\beta)}]]P^k u\|\\&\leq\sum_{\ell=1}^{n}C_1^3\|D^{\alpha}\psi_N^{(\beta+\ell)}P^k u\|_{\frac{p-2}{r}}+\sum_{j,i=1}^{n}C_1^2\|D^{\alpha}\psi_N^{(\beta+\ell+i)}P^k u\|_{\frac{p-2}{r}}\end{aligned}$$

$$+\sum_{\ell=1}^{n}\sum_{\substack{\nu\le\alpha\\ \nu\neq 0}} C_1^{|\nu|+3}\frac{\alpha!}{(\alpha-\nu)!}\left(N^{|\nu|-1}+|\nu|^{(s-1)|\nu|}\right)\Bigg[\|D^{\alpha-\nu+\ell}\psi_N^{(\beta)}P^k u\|_{\frac{p-2}{r}}$$

$$+\|D^{\alpha-\nu}\psi_N^{(\beta+\ell)}P^k u\|_{\frac{p-2}{r}}+\sum_{i=1}^{n}\Bigg(\|D^{\alpha-\nu+\ell}\psi_N^{(\beta+i)}P^k u\|_{\frac{p-2}{r}}$$

$$+\|D^{\alpha-\nu}\psi_N^{(\beta+\ell+i)}P^k u\|_{\frac{p-2}{r}}\Bigg)\Bigg]$$

$$+\sum_{\ell=1}^{n}\sum_{\substack{\nu<\alpha\\ \nu\neq 0}}\sum_{\substack{\mu\le\alpha-\nu\\ \mu\neq 0}} C_1^{|\nu|+|\mu|+3}\frac{\alpha!}{(\alpha-\nu-\mu)!}\left(N^{|\nu|-1}+|\nu|^{(s-1)|\nu|}\right)$$

$$\times\left(N^{|\mu|-1}+|\mu|^{(s-1)|\mu|}\right)\Bigg[\|D^{\alpha-\nu-\mu+\ell}\psi_N^{(\beta)}P^k u\|_{\frac{p-2}{r}}+\|D^{\alpha-\nu-\mu}\psi_N^{(\beta+\ell)}P^k u\|_{\frac{p-2}{r}}$$

$$+\sum_{i=1}^{n}\Bigg(\|D^{\alpha-\nu-\mu+\ell+i}\psi_N^{(\beta)}P^k u\|_{\frac{p-2}{r}}+\|D^{\alpha-\nu-\mu+\ell}\psi_N^{(\beta+i)}P^k u\|_{\frac{p-2}{r}}$$

$$+\|D^{\alpha-\nu-\mu}\psi_N^{(\beta+\ell+i)}P^k u\|_{\frac{p-2}{r}}\Bigg)\Bigg]. \tag{58}$$

Summing up, by (38), (39), (40), (42), (48), (49), (51), (52), (53), and (58), we conclude that there are suitable constants, C_1 and C_2, independent of α, β, and N, such that

$$\|D^\alpha\psi_N^{(\beta)}P^k u\|_{\frac{p}{r}}+\sum_{j=1}^{m}\|X_jD^\alpha\psi_N^{(\beta)}P^k u\|_{\frac{p-1}{r}}+\|X_0D^\alpha\psi_N^{(\beta)}P^k u\|_{\frac{p-2}{r}}$$

$$+\sum_{\ell,j=1}^{m}\|X_\ell X_jD^\alpha\psi_N^{(\beta)}P^k u\|_{\frac{p-2}{r}}$$

$$\le C_2\Bigg\{\|D^\alpha\psi_N^{(\beta)}P^{k+1}u\|_{\frac{p-2}{r}}+\sum_{j=1}^{m}\|X_jD^\alpha\psi_N^{(\beta)}P^k u\|_{\frac{p-2}{r}}$$

$$+\|D^\alpha\psi_N^{(\beta)}P^k u\|_{\frac{p-1}{r}}+\|D^\alpha\psi_N^{(\beta)}P^k u\|_{\frac{p-2}{r}}+\|D^\alpha\psi_N^{(\beta)}P^k u\|$$

$$+\sum_{j=1}^{m}\|X_jD^\alpha\psi_N^{(\beta)}P^k u\|+\|X_0D^\alpha\psi_N^{(\beta)}P^k u\|+\sum_{\ell,j=1}^{m}\|X_\ell X_jD^\alpha\psi_N^{(\beta)}P^k u\|$$

$$+C_1N^{s(2n+5-M)^+}\Bigg(\sum_{\ell=1}^{n}\|D^\alpha\psi_N^{(\beta+\ell)}P^k u\|_{\frac{p-2}{r}}+\sum_{\ell,i=1}^{n}\|D^\alpha\psi_N^{(\beta+\ell+i)}P^k u\|_{\frac{p-2}{r}}$$

$$+\sum_{j=1}^{m}\sum_{\ell=1}^{n}\|X_j D^{\alpha}\psi_N^{(\beta+\ell)}P^k u\|_{\frac{p-2}{r}}\Bigg)$$

$$+\sum_{j=1}^{m}\sum_{\ell=1}^{n}\sum_{\substack{\nu\leq\alpha\\ \nu\neq 0}} C_1^{|\nu|+1}\frac{\alpha!}{(\alpha-\nu)!}\left(N^{|\nu|-1}+|\nu|^{(s-1)|\nu|}\right)N^{s(n+3-M)^+}$$
$$\times\left(\|X_j D^{\alpha-\nu+\ell}\psi_N^{(\beta)}P^k u\|_{\frac{p-2}{r}}+\|X_j D^{\alpha-\nu}\psi_N^{(\beta+\ell)}P^k u\|_{\frac{p-2}{r}}\right)$$

$$+\sum_{\substack{\nu\leq\alpha\\ \nu\neq 0}} C_1^{|\nu|+1}\frac{\alpha!}{(\alpha-\nu)!}\left(N^{|\nu|-1}+|\nu|^{(s-1)|\nu|}\right)\|D^{\alpha-\nu}\psi_N^{(\beta)}P^k u\|_{\frac{p-2}{r}}$$

$$+\sum_{\ell=1}^{n}\sum_{\substack{\nu\leq\alpha\\ \nu\neq 0}} C_1^{|\nu|+1}\frac{\alpha!}{(\alpha-\nu)!}\left(N^{|\nu|-1}+|\nu|^{(s-1)|\nu|}\right)N^{s(n+3-M)^+}$$
$$\times\left(\|D^{\alpha-\nu+\ell}\psi_N^{(\beta)}P^k u\|_{\frac{p-2}{r}}+\|D^{\alpha-\nu}\psi_N^{(\beta+\ell)}P^k u\|_{\frac{p-2}{r}}\right)$$

$$+\sum_{\ell,i=1}^{n}\sum_{\substack{\nu\leq\alpha\\ \nu\neq 0}} C_1^{|\nu|+1}\frac{\alpha!}{(\alpha-\nu)!}\left(N^{|\nu|-1}+|\nu|^{(s-1)|\nu|}\right)$$
$$\times\left(\|D^{\alpha-\nu+\ell}\psi_N^{(\beta+i)}P^k u\|_{\frac{p-2}{r}}+\|D^{\alpha-\nu}\psi_N^{(\beta+\ell+i)}P^k u\|_{\frac{p-2}{r}}\right)$$

$$+\sum_{\ell=1}^{n}\sum_{\substack{\nu<\alpha\\ \nu\neq 0}}\sum_{\substack{\mu\leq\alpha-\nu\\ \mu\neq 0}} C_1^{|\nu|+|\mu|+3}\frac{\alpha!}{(\alpha-\nu-\mu)!}\left(N^{|\nu|-1}+|\nu|^{(s-1)|\nu|}\right)$$
$$\times\left(N^{|\mu|-1}+|\mu|^{(s-1)|\mu|}\right)\Bigg[\|D^{\alpha-\nu-\mu+\ell}\psi_N^{(\beta)}P^k u\|_{\frac{p-2}{r}}+\|D^{\alpha-\nu-\mu}\psi_N^{(\beta+\ell)}P^k u\|_{\frac{p-2}{r}}$$
$$+\sum_{i=1}^{n}\Bigg(\|D^{\alpha-\nu-\mu+\ell+i}\psi_N^{(\beta)}P^k u\|_{\frac{p-2}{r}}+\|D^{\alpha-\nu-\mu+\ell}\psi_N^{(\beta+i)}P^k u\|_{\frac{p-2}{r}}$$
$$+\|D^{\alpha-\nu-\mu}\psi_N^{(\beta+\ell+i)}P^k u\|_{\frac{p-2}{r}}\Bigg)\Bigg]\Bigg\}. \tag{59}$$

We point out that now we can choose M equal to $2n+7$.

4 Local Gevrey Regularity of a Gevrey Vector of P

In this section we prove our main theorem concerning the regularity of the Gevrey vectors of $P(x, D)$, (1). About that, we begin to prove a couple of results consequence of the estimates established in the previous section.

Proposition 4.1 *Let P be as in* (1), *x_0 in Ω and Ω_0 a neighborhood of x_0 where the Hörmander's condition is satisfied at the step r, Definition 2.2. Let ψ_N be as in the previous section,* (5), *and u a Gevrey vector of order s for P. There exist constants A_1 and B_1 such that, if:*

$$(1)_0 \begin{cases} \|D^\alpha \psi_N^{(\beta)} P^k u\| \leq A_1^{|\beta|+2k+1} B_1^{2r|\alpha|+1} N^{s[r|\alpha|+|\beta|+2k]}, \\ \|X_\ell D^\alpha \psi_N^{(\beta)} P^k u\| \leq A_1^{|\beta|+2k+1} B_1^{2r(|\alpha|+(2r)^{-1})+1} N^{s[r(|\alpha|+r^{-1})+|\beta|+2k]}, \\ \ell = 1, \ldots, m, \text{ for } 2r|\alpha| + (|\beta| - M) + 2k \leq N. \end{cases} \tag{60}$$

Then, one has for $p = 1$

$$(1)_1 \begin{cases} \|D^\alpha \psi_N^{(\beta)} P^k u\|_{\frac{1}{r}} \leq A_1^{|\beta|+2k+1+1} B_1^{2r|\alpha|+1+1} N^{s[r|\alpha|+|\beta|+2k+1]}, \\ \text{for } 2r|\alpha| + (|\beta| - M) + 2k \leq N - 4, \end{cases} \tag{61}$$

and for $2 \leq p \leq r$

$$(1)_p \begin{cases} \|D^\alpha \psi_N^{(\beta)} P^k u\|_{\frac{p}{r}} \leq A_1^{|\beta|+2k+p+1} B_1^{2r|\alpha|+p+1} N^{s[r|\alpha|+|\beta|+2k+p]}, \\ \|X_\ell D^\alpha \psi_N^{(\beta)} P^k u\|_{\frac{p-1}{r}} \leq A_1^{|\beta|+2k+p+1} B_1^{2r|\alpha|+p+1} N^{s[r|\alpha|+|\beta|+2k+p]}, \\ \|X_0 D^\alpha \psi_N^{(\beta)} P^k u\|_{\frac{p-2}{r}} \leq A_1^{|\beta|+2k+p+1} B_1^{2r|\alpha|+p+1} N^{s[r|\alpha|+|\beta|+2k+p]}, \\ \|X_j X_\ell D^\alpha \psi_N^{(\beta)} P^k u\|_{\frac{p-2}{r}} \leq A_1^{|\beta|+2k+p+1} B_1^{2r|\alpha|+p+1} N^{s[r|\alpha|+|\beta|+2k+p]}, \\ \ell, j \in \{1, \ldots, m\}, \text{ for } 2r|\alpha| + (|\beta| - M) + 2k \leq N - 2p. \end{cases} \tag{62}$$

We recall that the constant M is fixed, more precisely $M = 2n + 7$.

Proof The result is obtained by induction on p. The main tools are the estimates (24) and (59) obtained in the previous section.

Step $p = 1, 2$. We want to show (61) and (62) and cases $p = 1$ and $p = 2$, respectively. We begin from the case $p = 2$; the case $p = 1$ will be obtained using (29).

We recall the estimate (24) slightly modified in view of the considerations made at the end of the subsection "Estimate in $H^{1/r}$ and $H^{2/r}$":

$$
\begin{aligned}
&\|D^\alpha\psi_N^{(\beta)}P^k u\|_{\frac{2}{r}}+\sum_{j=1}^{m}\|X_jD^\alpha\psi_N^{(\beta)}P^k u\|_{\frac{1}{r}}\\
&\qquad+\sum_{\ell,j=1}^{m}\|X_\ell X_jD^\alpha\psi_N^{(\beta)}P^k u\|_0+\|X_0D^\alpha\psi_N^{(\beta)}P^k u\|_0\\
&\leq C_2\Big\{\|D^\alpha\psi_N^{(\beta)}P^{k+1}u\|_0+\|D^\alpha\psi_N^{(\beta)}P^k u\|_0+\sum_{\ell=1}^{m}\sum_{j=1}^{n}\|X_\ell D^\alpha\psi_N^{(\beta+j)}P^k u\|_0\\
&\qquad+\sum_{j=1}^{n}\|D^\alpha\psi_N^{(\beta+j)}P^k u\|_0+\sum_{j=1}^{n}\sum_{i=1}^{n}\|D^\alpha\psi_N^{(\beta+j+i)}P^k u\|_0\\
&\quad+\sum_{\ell=1}^{m}\sum_{j=1}^{n}\sum_{\substack{\nu<\alpha,\\ \nu\neq 0}}C_1^{|\nu|+1}\frac{\alpha!\,(\nu!)^{s-1}}{(\alpha-\nu)!}\Big(\|X_\ell D^{\alpha-\nu}\psi_N^{(\beta+j)}P^k u\|_0\\
&\qquad+\|X_\ell D^{\alpha-\nu+j}\psi_N^{(\beta)}P^k u\|_0\Big)\\
&+\sum_{j=1}^{n}\sum_{\substack{\nu<\alpha,\\ \nu\neq 0}}C_1^{|\nu|+1}\frac{\alpha!\,(\nu!)^{s-1}}{(\alpha-\nu)!}\Big(\|D^{\alpha-\nu}\psi_N^{(\beta+j)}P^k u\|_0+\|D^{\alpha-\nu+j}\psi_N^{(\beta)}P^k u\|_0\Big)\\
&\quad+\sum_{j=1}^{n}\sum_{i=1}^{n}\sum_{\substack{\nu<\alpha,\\ \nu\neq 0}}C_1^{|\nu|+1}\frac{\alpha!\,(\nu!)^{s-1}}{(\alpha-\nu)!}\Big(\|D^{\alpha-\nu}\psi_N^{(\beta+j+i)}P^k u\|_0\\
&\qquad+\|D^{\alpha-\nu+j}\psi_N^{(\beta+i)}P^k u\|_0\Big)\\
&+\sum_{j=1}^{n}\sum_{\substack{\nu<\alpha,\\ \nu\neq 0}}\sum_{\substack{\mu<\alpha-\nu,\\ \mu\neq 0}}C_1^{|\nu|+|\mu|+2}\frac{\alpha!\,(\nu!)^{s-1}\,(\mu!)^{s-1}}{(\alpha-\nu-\mu)!}\Big[\|D^{\alpha-\nu-\mu}\psi_N^{(\beta+j)}P^k u\|_0\\
&\quad+\|D^{\alpha-\nu-\mu+j}\psi_N^{(\beta)}P^k u\|_0+\sum_{i=1}^{n}\Big(\|D^{\alpha-\nu-\mu}\psi_N^{(\beta+j+i)}P^k u\|_0\\
&\qquad+\|D^{\alpha-\nu-\mu+j}\psi_N^{(\beta+i)}P^k u\|_0+\|D^{\alpha-\nu-\mu+j+i}\psi_N^{(\beta)}P^k u\|_0\Big)\Big]\\
&\quad+C_1^{|\alpha|+1}(\alpha!)^s\sum_{j=1}^{n}\Big[\sum_{\ell=1}^{m}\|X_\ell D_j\psi_N^{(\beta)}P^k u\|_0+\Big(2^{|\alpha|+n}+1\Big)\|D_j\psi_N^{(\beta)}P^k u\|_0
\end{aligned}
$$

$$+\left(2^{|\alpha|+n}+1\right)\sum_{i=1}^{n}\left(\|D_j\psi_N^{(\beta+i)}P^k u\|_0+\|D_jD_i\psi_N^{(\beta)}P^k u\|_0\right)\Bigg]$$
$$+\sum_{\substack{\nu<\alpha,\\ \nu\neq 0}} C_1^{|\nu|+1}\frac{\alpha!\,(\nu!)^{s-1}}{(\alpha-\nu)!}\|D^{\alpha-\nu}\psi_N^{(\beta)}P^k u\|_0+C_1^{|\alpha|+1}C_2^{|\beta|+2k}N^{s(|\alpha|+|\beta|+2k)}\Bigg\}. \tag{63}$$

We remark that all the $2n+1$-tuples of the form $(\alpha', \beta', k') \in \mathbb{N}^{2n+1}$ associated to each term on the right-hand side of (63) satisfy the condition (60).

In order to make more readable and flowing the writing of the proof, we introduce the following notation: Let us denote the powers of A_1, B_1 and N appearing in (60), (61), and (62), respectively by:

$$\begin{aligned} S_p &\doteq S_p(|\beta|, k) = |\beta| + 2k + p + 1;\\ T_p &\doteq T_p(|\alpha|) = 2r|\alpha| + p + 1;\\ U_p &\doteq U_p(|\alpha|, |\beta|, k) = s\,[r|\alpha| + |\beta| + 2k + p]\,. \end{aligned} \tag{64}$$

We point out that $S_{p+1} = S_p + 1$, $T_{p+1} = T_p + 1$, and $U_{p+1} = U_p + s$. With this notation the assumption (60) can be rewritten as

$$(1)_0 \begin{cases} \|D^\alpha\psi_N^{(\beta)}P^k u\| \le A_1^{S_0}B_1^{T_0}N^{U_0},\\ \|X_\ell D^\alpha\psi_N^{(\beta)}P^k u\| \le A_1^{S_0}B_1^{T_1}N^{U_1},\\ \ell = 1, \dots, m, \text{ for } T_0 + S_0 - M \le N + 2. \end{cases} \tag{65}$$

So, at this step, $p = 2$, our purpose is to show that for $T_2 + S_2 - M \le N + 2$ the left-hand side of (63) is bounded by $A_1^{S_2}B_1^{T_2}N^{U_2}$. This will be gain showing that all the terms on the right-hand side of (63) are smaller than $A_1^{S_2}B_1^{T_2}N^{U_2}$ times a factor depending on negative power of A_1 or B_1 or N. A suitable choice of A_1 and B_1 will yield the summand of all these factors less than one.

We begin to estimate the first five terms on the right-hand side of (63):

$$C_2\left(\|D^\alpha\psi_N^{(\beta)}P^{k+1}u\|_0+\|D^\alpha\psi_N^{(\beta)}P^k u\|_0+\sum_{\ell=1}^{m}\sum_{j=1}^{n}\|X_\ell D^\alpha\psi_N^{(\beta+j)}P^k u\|_0\right.$$
$$\left.+\sum_{j=1}^{n}\|D^\alpha\psi_N^{(\beta+j)}P^k u\|_0+\sum_{j=1}^{n}\sum_{i=1}^{n}\|D^\alpha\psi_N^{(\beta+j+i)}P^k u\|_0\right)$$
$$\le C_2\left(A_1^{S_2}B_1^{T_0}N^{U_2}+A_1^{S_0}B_1^{T_0}N^{U_0}+nmA_1^{S_1}B_1^{T_1}N^{U_2}+nA_1^{S_1}B_1^{T_0}N^{U_1}\right.$$

$$
\begin{aligned}
&\qquad\qquad +n^2 A_1^{S_2} B_1^{T_0} N^{U_2}\Big) \\
&\le A_1^{S_2} B_1^{T_2} N^{U_2} \times C_2 \left(B_1^{-2} + A_1^{-2} B_1^{-2} N^{-2s} \right.\\
&\qquad \left. + nm A_1^{-1} B_1^{-1} + n A_1^{-1} B_1^{-2} N^{-s} + n^2 B_1^{-2} \right). \qquad (66)
\end{aligned}
$$

Now, we handle the multiple sums in ν and μ. We stress that $|\nu|$ and $|\mu|$ are strictly greater than 0. We have

$$
\begin{aligned}
&C_2 \sum_{\ell=1}^{m} \sum_{j=1}^{n} \sum_{\substack{\nu<\alpha,\\ \nu\neq 0}} C_1^{|\nu|+1} \frac{\alpha!\,(\nu!)^{s-1}}{(\alpha-\nu)!} \Big(\| X_\ell D^{\alpha-\nu} \psi_N^{(\beta+j)} P^k u \|_0 \\
&\qquad\qquad + \| X_\ell D^{\alpha-\nu+j} \psi_N^{(\beta)} P^k u \|_0 \Big) \\
&\le nm C_2 C_1 \sum_{\substack{\nu<\alpha,\\ \nu\neq 0}} C_1^{|\nu|} \frac{\alpha!\,(\nu!)^{s-1}}{(\alpha-\nu)!} \Big(A_1^{S_2-1} B_1^{T_2-1-2r|\nu|} N^{U_2-sr|\nu|} \\
&\qquad\qquad + A_1^{S_2-2} B_1^{T_2-2r(|\nu|-1} N^{U_2-s[r(|\nu|-1)+2]} \Big).
\end{aligned}
$$

Since $\frac{\alpha!}{(\alpha-\nu)!} \le |\alpha|^{|\nu|} \le N^{|\nu|}$ and $(\nu!)^{s-1} \le N^{(s-1)|\nu|}$, the right-hand side of the above estimate can be bounded as follows:

$$
\begin{aligned}
&A_1^{S_2} B_1^{T_2} N^{U_2} \Bigg(nm C_2 C_1 A_1^{-1} B^{-1} \sum_{p=1}^{|\alpha|-1} \sum_{|\nu|=p} C_1^{|\nu|} B_1^{-2r|\nu|} N^{-s(r-1)|\nu|} \\
&\qquad + nm C_2 C_1^2 A_1^{-2} \sum_{p=1}^{|\alpha|-1} \sum_{|\nu|=p} C_1^{|\nu|-1} B_1^{-2r(|\nu|-1)} N^{-s(|\nu|-1)(r-1)-1} \Bigg).
\end{aligned}
$$

Now, $|\nu| \ge 1$ and $r > 1$, so $-s(r-1)|\nu|$ and $-s(|\nu|-1)(r-1)-1$ are less than zero; moreover, the number of multi-indexes ν with $|\nu| = p$ is given by $\binom{p+n-1}{n-1}$, bounded by $2^{|p|+n-1}$. The above quantity can be estimated by

$$
\begin{aligned}
&A_1^{S_2} B_1^{T_2} N^{U_2} \Bigg(nm 2^{n-1} C_2 C_1 A_1^{-1} B^{-1} \sum_{p=1}^{|\alpha|-1} \left(2 C_1 B_1^{-2r} \right)^p \\
&\qquad + nm 2^n C_2 C_1^2 A_1^{-2} \sum_{p=1}^{|\alpha|-1} \left(2 C_1 B_1^{-2r} \right)^{p-1} \Bigg).
\end{aligned}
$$

Choosing B_1 such that $2C_1 B_1^{-2r} < 2^{-1}$, we have

$$\sum_{p=1}^{|\alpha|-1} \left(2C_1 B_1^{-2r}\right)^p \leq 1 \quad \text{and} \quad \sum_{p=1}^{|\alpha|-1} \left(2C_1 B_1^{-2r}\right)^{p-1} \leq 2.$$

We obtain

$$C_2 \sum_{\ell=1}^{m} \sum_{j=1}^{n} \sum_{\substack{\nu<\alpha, \\ \nu\neq 0}} C_1^{|\nu|+1} \frac{\alpha!\,(\nu!)^{s-1}}{(\alpha-\nu)!} \Big(\|X_\ell D^{\alpha-\nu} \psi_N^{(\beta+j)} P^k u\|_0 + \|X_\ell D^{\alpha-\nu+j} \psi_N^{(\beta)} P^k u\|_0 \Big)$$

$$\leq A_1^{S_2} B_1^{T_2} N^{U_2} \times \left(2^{n-1} nm C_2 C_1\right) A_1^{-1} \left(B^{-1} + 4C_1 A_1^{-1}\right). \tag{67}$$

In the same way, we conclude that

$$C_2 \sum_{j=1}^{n} \sum_{\substack{\nu<\alpha, \\ \nu\neq 0}} C_1^{|\nu|+1} \frac{\alpha!\,(\nu!)^{s-1}}{(\alpha-\nu)!} \Big(\|D^{\alpha-\nu} \psi_N^{(\beta+j)} P^k u\|_0 + \|D^{\alpha-\nu+j} \psi_N^{(\beta)} P^k u\|_0 \Big)$$

$$\leq A_1^{S_2} B_1^{T_2} N^{U_2} \times \left(2^{n-1} n C_2 C_1\right) \left(N^{-s} + 4C_1\right) A_1^{-2} B^{-2} \tag{68}$$

and

$$C_2 \sum_{j=1}^{n} \sum_{i=1}^{n} \sum_{\substack{\nu<\alpha, \\ \nu\neq 0}} C_1^{|\nu|+1} \frac{\alpha!\,(\nu!)^{s-1}}{(\alpha-\nu)!} \Big(\|D^{\alpha-\nu} \psi_N^{(\beta+j+i)} P^k u\|_0 + \|D^{\alpha-\nu+j} \psi_N^{(\beta+i)} P^k u\|_0 \Big)$$

$$\leq A_1^{S_2} B_1^{T_2} N^{U_2} \times \left(2^{n-1} n^2 C_2 C_1\right) \left(B_1^{-2} + +4C_1 A_1^{-1}\right). \tag{69}$$

Concerning the other multiple sums on the right-hand side of (63) in ν and μ, we proceed analogously.

More precisely: since

$$\frac{\alpha!}{(\alpha-\nu-\mu)!} \leq N^{|\nu|+|\mu|}$$

and $(r-1)(|\nu|+|\mu|) > 0$, $(r-1)(|\nu|-1)+(r-1)|\mu| > 0$ and $(r-1)(|\nu|+|\mu|-2) \geq 0$, $r \geq 2$ and $|\nu|$, $|\mu| \geq 1$, we obtain

$$
\begin{aligned}
&C_2 \sum_{j=1}^{n} \sum_{\substack{\nu<\alpha,\\ \nu\neq 0}} \sum_{\substack{\mu<\alpha-\nu,\\ \mu\neq 0}} C_1^{|\nu|+|\mu|+2} \frac{\alpha!\,(\nu!)^{s-1}\,(\mu!)^{s-1}}{(\alpha-\nu-\mu)!} \Bigg[\| D^{\alpha-\nu-\mu} \psi_N^{(\beta+j)} P^k u \|_0 \\
&\qquad + \| D^{\alpha-\nu-\mu+j} \psi_N^{(\beta)} P^k u \|_0 + \sum_{i=1}^{n} \Big(\| D^{\alpha-\nu-\mu} \psi_N^{(\beta+j+i)} P^k u \|_0 \\
&\qquad\qquad + \| D^{\alpha-\nu-\mu+j} \psi_N^{(\beta+i)} P^k u \|_0 + \| D^{\alpha-\nu-\mu+j+i} \psi_N^{(\beta)} P^k u \|_0 \Big) \Bigg] \\
&\leq A_1^{S_2} B_1^{T_2} N^{U_2} \times \left(2^{2(n-1)} n C_2 C_1^2 B_1^{-2} \right) \left(n + (2n+1) A_1^{-1} + 2(2n+1) A_1^{-2} \right).
\end{aligned} \tag{70}
$$

We point out that, as before, we have chosen B_1 large enough in order that $2C_1 B_1^{-2r} < 2$.

Concerning the last terms on the right-hand side of (63), we have

$$
\begin{aligned}
&C_2 \Bigg\{ C_1^{|\alpha|+1} (\alpha!)^s \sum_{j=1}^{n} \Bigg[\sum_{\ell=1}^{m} \| X_\ell D_j \psi_N^{(\beta)} P^k u \|_0 + \left(2^{|\alpha|+n} + 1 \right) \| D_j \psi_N^{(\beta)} P^k u \|_0 \\
&\qquad + \left(2^{|\alpha|+n} + 1 \right) \sum_{i=1}^{n} \left(\| D_j \psi_N^{(\beta+i)} P^k u \|_0 + \| D_j D_i \psi_N^{(\beta)} P^k u \|_0 \right) \Bigg] \\
&\quad + \sum_{\substack{\nu<\alpha,\\ \nu\neq 0}} C_1^{|\nu|+1} \frac{\alpha!\,(\nu!)^{s-1}}{(\alpha-\nu)!} \| D^{\alpha-\nu} \psi_N^{(\beta)} P^k u \|_0 + C_1^{|\alpha|+1} C_2^{|\beta|+2k} N^{s(|\alpha|+|\beta|+2k)} \Bigg\}. \\
&\quad \leq A_1^{S_2} B_1^{T_2} N^{U_2} \times \Bigg\{ n C_2 C_1^2 A_1^{-1} B_1^{-1} \Bigg[m A_1^{-1} \left(C_1 B^{-2r} \right)^{|\alpha|-1} N^{-s(|\alpha|-1)(r-1)} \\
&\qquad + 2^{n+1} (n + N^{-s}) A_1^{-1} B_1^{-1} \left(2 C_1 B^{-2r} \right)^{|\alpha|-1} N^{-s(|\alpha|-1)(r-1)} \\
&\quad + 2^{n+1} n C_1 (A_1 B_1)^{-1} \left(2 C_1 B^{-2r} \right)^{|\alpha|-2} N^{-s(|\alpha|-2)(r-1)} \Bigg] + C_2 C_1 2^{n-1} (A_1 B_1)^{-2} \\
&\qquad + C_1 B_1^{-3} A_1^{-2} \left(C_1 B_1^{-2r} \right)^{|\alpha|} \left(C_2 A_1^{-1} \right)^{|\beta|+2k+1} N^{-s[|\alpha|(r-1)+2]} \Bigg\}.
\end{aligned} \tag{71}
$$

We point out that since $r \geq 2$ and we are assuming $|\alpha| \geq 2$ (as ν is nonzero and ν is strictly less than α, we get $|\alpha|$ greater or equal to 2), then $(|\alpha| - 1)(r - 1)$ and $(|\alpha| - 2)(r - 1)$ are greater or equal to zero.

Summing up, if A_1 and B_1 are chosen large enough, the sum of the second factor on the right-hand side of (66), (67), (68), (69), (70), and (71) can be made smaller than one. By (63) and above considerations, we obtain the desired estimate:

$$\|D^\alpha \psi_N^{(\beta)} P^k u\|_{\frac{2}{r}} + \sum_{j=1}^{m} \|X_j D^\alpha \psi_N^{(\beta)} P^k u\|_{\frac{1}{r}} + \sum_{\ell,j=1}^{m} \|X_\ell X_j D^\alpha \psi_N^{(\beta)} P^k u\|_0 + \|X_0 D^\alpha \psi_N^{(\beta)} P^k u\|_0 \le A_1^{S_2} B_1^{T_2} N^{U_2}. \tag{72}$$

We then obtained (62) in the case $p = 2$.

Step $p = 1$. We want to show that $\|v\|_{1/r} \le A_1^{S_1} B_1^{T_1} N^{U_1}$ for $T_1 + S_1 \le N$, $v = D^\alpha \psi_N^{(\beta)} P^k u$. We recall the estimate (29):

$$\|D^\alpha \psi_N^{(\beta)} P^k u\|_{\frac{1}{r}} \le \left(\|D^\alpha \psi_N^{(\beta)} P^k u\|_{\frac{2}{r}}\right)^{1/2} \left(\|D^\alpha \psi_N^{(\beta)} P^k u\|_0\right)^{1/2}.$$

We point out that in this case, since $T_1+S_1 \le N$, i.e., $2r|\alpha|+(|\beta|-M)+2k \le N-4$, the $(2n+1)$-tuple (α, β, k) satisfies the assumptions in (62), case $p = 2$, and those in (60).

By (72), we obtain

$$\begin{aligned}
\|D^\alpha \psi_N^{(\beta)} P^k u\|_{\frac{1}{r}} &\le \left(\|D^\alpha \psi_N^{(\beta)} P^k u\|_{\frac{2}{r}}\right)^{1/2} \left(\|D^\alpha \psi_N^{(\beta)} P^k u\|_0\right)^{1/2} \\
&\le \left(A_1^{S_2} B_1^{T_2} N^{U_2}\right)^{1/2} \left(A_1^{S_0} B_1^{T_0} N^{U_0}\right)^{1/2} \\
&= \left(A_1^{|\beta|+2k+2+1} B_1^{2r|\alpha|+2+1} N^{s[r|\alpha|+|\beta|+2k+2]}\right)^{\frac{1}{2}} \\
&\qquad \times \left(A_1^{|\beta|+2k+1} B_1^{2r|\alpha|+1} N^{s[r|\alpha|+|\beta|+2k]}\right)^{\frac{1}{2}} \\
&= A_1^{|\beta|+2k+1+1} B_1^{2r|\alpha|+1+1} N^{s[r|\alpha|+|\beta|+2k+1]} = A_1^{S_1} B_1^{T_1} N^{U_1}.
\end{aligned} \tag{73}$$

So, we showed (61).

Step $p \geq 3$. Using the notation introduced in (64), we assume

$$(1)_{p-1}\begin{cases}\|D^\alpha \psi_N^{(\beta)} P^k u\|_{\frac{p-1}{r}} \leq A_1^{S_{p-1}} B_1^{T_{p-1}} N^{U_{p-1}},\\ \|X_\ell D^\alpha \psi_N^{(\beta)} P^k u\|_{\frac{p-2}{r}} \leq A_1^{S_{p-1}} B_1^{T_{p-1}} N^{U_{p-1}},\\ \|X_0 D^\alpha \psi_N^{(\beta)} P^k u\|_{\frac{p-3}{r}} \leq A_1^{S_{p-1}} B_1^{T_{p-1}} N^{U_{p-1}},\\ \|X_j X_\ell D^\alpha \psi_N^{(\beta)} P^k u\|_{\frac{p-3}{r}} \leq A_1^{S_{p-1}} B_1^{T_{p-1}} N^{U_{p-1}},\\ \ell, j \in \{1, \dots, m\}, \text{ for } S_{p-1} + T_{p-1} - M \leq N + 2.\end{cases} \tag{74}$$

We want to show, via the estimate (59), that $(1)_p$ holds. We recall the estimate (59):

$$\begin{aligned}
&\|D^\alpha \psi_N^{(\beta)} P^k u\|_{\frac{p}{r}} + \sum_{j=1}^m \|X_j D^\alpha \psi_N^{(\beta)} P^k u\|_{\frac{p-1}{r}} + \|X_0 D^\alpha \psi_N^{(\beta)} P^k u\|_{\frac{p-2}{r}}\\
&\qquad + \sum_{\ell,j=1}^m \|X_\ell X_j D^\alpha \psi_N^{(\beta)} P^k u\|_{\frac{p-2}{r}}\\
&\leq C_2 \Bigg\{ \|D^\alpha \psi_N^{(\beta)} P^{k+1} u\|_{\frac{p-2}{r}} + \sum_{j=1}^m \|X_j D^\alpha \psi_N^{(\beta)} P^k u\|_{\frac{p-2}{r}} + \|D^\alpha \psi_N^{(\beta)} P^k u\|_{\frac{p-2}{r}}\\
&\qquad + \|D^\alpha \psi_N^{(\beta)} P^k u\| + \sum_{j=1}^m \|X_j D^\alpha \psi_N^{(\beta)} P^k u\| + \|X_0 D^\alpha \psi_N^{(\beta)} P^k u\|\\
&\qquad + \sum_{\ell,j=1}^m \|X_\ell X_j D^\alpha \psi_N^{(\beta)} P^k u\| + \|D^\alpha \psi_N^{(\beta)} P^k u\|_{\frac{p-1}{r}}\\
&\qquad + C_1 \sum_{\ell=1}^n \Bigg(\|D^\alpha \psi_N^{(\beta+\ell)} P^k u\|_{\frac{p-2}{r}} + \sum_{i=1}^n \|D^\alpha \psi_N^{(\beta+\ell+i)} P^k u\|_{\frac{p-2}{r}}\\
&\qquad\qquad + \sum_{j=1}^m \|X_j D^\alpha \psi_N^{(\beta+\ell)} P^k u\|_{\frac{p-2}{r}} \Bigg)\\
&\qquad + \sum_{j=1}^m \sum_{\ell=1}^n \sum_{\substack{\nu\leq\alpha\\ \nu\neq 0}} C_1^{|\nu|+1} \frac{\alpha!}{(\alpha-\nu)!} \left(N^{|\nu|-1} + |\nu|^{(s-1)|\nu|} \right)\\
&\qquad\qquad \times \left(\|X_j D^{\alpha-\nu+\ell} \psi_N^{(\beta)} P^k u\|_{\frac{p-2}{r}} + \|X_j D^{\alpha-\nu} \psi_N^{(\beta+\ell)} P^k u\|_{\frac{p-2}{r}} \right)
\end{aligned}$$

$$+\sum_{\substack{\nu\le\alpha\\ \nu\neq 0}} C_1^{|\nu|+1}\frac{\alpha!}{(\alpha-\nu)!}\left(N^{|\nu|-1}+|\nu|^{(s-1)|\nu|}\right)\Bigg[\|D^{\alpha-\nu}\psi_N^{(\beta)}P^k u\|_{\frac{p-2}{r}}$$

$$+\sum_{\ell=1}^{n}\left(\|D^{\alpha-\nu+\ell}\psi_N^{(\beta)}P^k u\|_{\frac{p-2}{r}}+\|D^{\alpha-\nu}\psi_N^{(\beta+\ell)}P^k u\|_{\frac{p-2}{r}}\right)$$

$$+\sum_{\ell,i=1}^{n}\left(\|D^{\alpha-\nu+\ell}\psi_N^{(\beta+i)}P^k u\|_{\frac{p-2}{r}}+\|D^{\alpha-\nu}\psi_N^{(\beta+\ell+i)}P^k u\|_{\frac{p-2}{r}}\right)\Bigg]$$

$$+\sum_{\ell=1}^{n}\sum_{\substack{\nu<\alpha\\ \nu\neq 0}}\sum_{\substack{\mu\le\alpha-\nu\\ \mu\neq 0}} C_1^{|\nu|+|\mu|+3}\frac{\alpha!}{(\alpha-\nu-\mu)!}\left(N^{|\nu|-1}+|\nu|^{(s-1)|\nu|}\right)$$

$$\times\left(N^{|\mu|-1}+|\mu|^{(s-1)|\mu|}\right)\Bigg[\|D^{\alpha-\nu-\mu+\ell}\psi_N^{(\beta)}P^k u\|_{\frac{p-2}{r}}+\|D^{\alpha-\nu-\mu}\psi_N^{(\beta+\ell)}P^k u\|_{\frac{p-2}{r}}$$

$$+\sum_{i=1}^{n}\Bigg(\|D^{\alpha-\nu-\mu+\ell+i}\psi_N^{(\beta)}P^k u\|_{\frac{p-2}{r}}+\|D^{\alpha-\nu-\mu+\ell}\psi_N^{(\beta+i)}P^k u\|_{\frac{p-2}{r}}$$

$$+\|D^{\alpha-\nu-\mu}\psi_N^{(\beta+\ell+i)}P^k u\|_{\frac{p-2}{r}}\Bigg)\Bigg]\Bigg\}. \tag{75}$$

The purpose is to show that, for $T_p+S_p\le N+2$, all the terms on the right-hand side of (75) are smaller than $A^{S_p}B_1^{T_p}N^{U_p}$ times a factor depending on negative powers of A_1 or B_1. As before, a suitable choice of the constants A_1 and B_1 will yield the sum of all these factors less than one. We point out that the $2n+1$-tuples of the form $(\alpha',\beta',k')\in\mathbb{N}^{2n+1}$ associated to each term on the right-hand side of (75) satisfy the condition in (74). To gain the result, we will essentially use the same strategies adopted in the case $p=2$. We begin to estimate the terms in first three lines on the right-hand side of (75):

$$\|D^\alpha\psi_N^{(\beta)}P^{k+1}u\|_{\frac{p-2}{r}}+\sum_{j=1}^{n}\|X_jD^\alpha\psi_N^{(\beta)}P^k u\|_{\frac{p-2}{r}}$$

$$+\|D^\alpha\psi_N^{(\beta)}P^k u\|+\sum_{j=1}^{n}\|X_jD^\alpha\psi_N^{(\beta)}P^k u\|+\|X_0D^\alpha\psi_N^{(\beta)}P^k u\|$$

$$+\sum_{\ell,j=1}^{m}\|X_\ell X_jD^\alpha\psi_N^{(\beta)}P^k u\|+\|D^\alpha\psi_N^{(\beta)}P^k u\|_{\frac{p-1}{r}}+\|D^\alpha\psi_N^{(\beta)}P^k u\|_{\frac{p-2}{r}}$$

$$\le A_1^{S_p}B_1^{T_p}N^{U_p}\times\Big(B_1^{-2}+A_1^{-1}B_1^{-1}N^{-s}\left(1+3A_1^{-1}B_1^{-1}N^{-s}\right)$$

$$+A_1^{-p+1}B_1^{-p+1}N^{-s(p-1)}\left(1+2B_1^{-1}N^{-s}\right)\Big). \tag{76}$$

Moreover, about the terms in the fourth line, we have

$$C_1 \sum_{\ell=1}^{n} \Bigg(\|D^\alpha \psi_N^{(\beta+\ell)} P^k u\|_{\frac{p-2}{r}} + \sum_{i=1}^{n} \|D^\alpha \psi_N^{(\beta+\ell+i)} P^k u\|_{\frac{p-2}{r}} + \sum_{j=1}^{m} \|X_j D^\alpha \psi_N^{(\beta+\ell)} P^k u\|_{\frac{p-2}{r}} \Bigg)$$
$$\leq A_1^{S_p} B_1^{T_p} N^{U_p} \times n C_1 B_1^{-2} \left(A_1^{-1} N^{-s} + n + m A_1^{-1} N^{-s} \right). \tag{77}$$

Now, we analyze the multi-index sums in ν and μ. We adopt the same strategy used to obtain (67), and we consider

$$\sum_{j=1}^{m} \sum_{\ell=1}^{n} \sum_{\substack{\nu \leq \alpha \\ \nu \neq 0}} C_1^{|\nu|+1} \frac{\alpha!}{(\alpha-\nu)!} \left(N^{|\nu|-1} + |\nu|^{(s-1)|\nu|} \right)$$
$$\times \left(\|X_j D^{\alpha-\nu+\ell} \psi_N^{(\beta)} P^k u\|_{\frac{p-2}{r}} + \|X_j D^{\alpha-\nu} \psi_N^{(\beta+\ell)} P^k u\|_{\frac{p-2}{r}} \right). \tag{78}$$

By inductive hypothesis, we have

$$\|X_j D^{\alpha-\nu+\ell} \psi_N^{(\beta)} P^k u\|_{\frac{p-2}{r}} \leq A_1^{S_p} B_1^{T_p} N^{U_p} \left(A_1^{-2} B_1^{-2-2r(|\nu|-1)} N^{-2s-sr(|\nu|-1)} \right);$$
$$\|X_j D^{\alpha-\nu} \psi_N^{(\beta+\ell)} P^k u\|_{\frac{p-2}{r}} \leq A_1^{S_p} B_1^{T_p} N^{U_p} \left(A_1^{-1} B_1^{-2-2r|\nu|} N^{-s-sr|\nu|} \right).$$

Since $\frac{\alpha}{(\alpha-\nu)!} \leq |\alpha|^{|\nu|} \leq N^{|\nu|}$, we have

$$\frac{\alpha!}{(\alpha-\nu)!} \left(N^{|\nu|-1} + |\nu|^{(s-1)|\nu|} \right) N^{-2s-sr(|\nu|-1)}$$
$$\leq N^{-(sr-2)(|\nu|-1)-2s+1} + N^{-s(r-1)(|\nu|-1)-s}.$$

Moreover, we observe that since $s \geq 1$ and $r \geq 2$, then $-(sr-2)(|\nu|-1)-2s+1 < 0$ and $-s(r-1)(|\nu|-1)-s < 0$. The right-hand side of the above inequality can be bounded by 2. Modulo $A_1^{S_p} B_1^{T_p} N^{U_p}$, (78) can be bounded by

$$2nmC_1 A_1^{-1} B_1^{-2} \left(A^{-1} \sum_{\substack{\nu \leq \alpha \\ \nu \neq 0}} (C_1 B_1^{-2r})^{|\nu|-1} + \sum_{\substack{\nu \leq \alpha \\ \nu \neq 0}} (C_1 B_1^{-2r})^{|\nu|} \right)$$

$$= 2nmC_1A^{-1}B_1^{-2}\left(A^{-1}\sum_{p>0}\sum_{|\nu|=p}(C_1B_1^{-2r})^{|p|-1}+\sum_{p>0}\sum_{|\nu|=p}(C_1B_1^{-2r})^{|p|}\right)$$

$$= 2nmC_1A^{-1}B_1^{-2}\left(A_1^{-1}\sum_{p>0}\binom{p+n-1}{n-1}(C_1B_1^{-2r})^{|p|-1}\right.$$

$$\left.+\sum_{p>0}\binom{p+n-1}{n-1}(C_1B_1^{-2r})^{|p|}\right)$$

$$\leq 2^{n-1}nmC_1A_1^{-1}B_1^{-2}\left(2A_1^{-1}\sum_{p\geq 0}(2C_1B_1^{-2r})^{|p|}+\sum_{p>0}(2C_1B_1^{-2r})^{|p|}\right).$$

Choosing B_1 such that $2C_1B_1^{-2r}\leq 2^{-1}$, we conclude that (78) is estimated by

$$A_1^{S_p}B_1^{T_p}N^{U_p}\times 2^{n-1}nmC_1A_1^{-1}B_1^{-2}\left(4A_1^{-1}+1\right). \tag{79}$$

In order to handle the other terms on the right-hand side of (75), we adopt essentially the same strategy, and in order to not bore the reader, we will omit the details.

We have

$$\sum_{\substack{\nu\leq\alpha\\ \nu\neq 0}} C_1^{|\nu|+1}\frac{\alpha!}{(\alpha-\nu)!}\left(N^{|\nu|-1}+|\nu|^{(s-1)|\nu|}\right)$$

$$\times\left[\|D^{\alpha-\nu}\psi_N^{(\beta)}P^ku\|_{\frac{p-2}{r}}+\sum_{\ell=1}^{n}\left(\|D^{\alpha-\nu+\ell}\psi_N^{(\beta)}P^ku\|_{\frac{p-2}{r}}+\|D^{\alpha-\nu}\psi_N^{(\beta+\ell)}P^ku\|_{\frac{p-2}{r}}\right)\right.$$

$$\left.\sum_{\ell=1}^{n}\sum_{i=1}^{n}\left(\|D^{\alpha-\nu+\ell}\psi_N^{(\beta+i)}P^ku\|_{\frac{p-2}{r}}+\|D^{\alpha-\nu}\psi_N^{(\beta+\ell+i)}P^ku\|_{\frac{p-2}{r}}\right)\right]$$

$$\leq A_1^{S_p}B_1^{T_p}N^{U_p}$$

$$\times\sum_{\substack{\nu\leq\alpha\\ \nu\neq 0}} C_1^{|\nu|+1}\left[A_1^{-2}B_1^{-2-2r|\nu|}\left(N^{-(sr-2)|\nu|-1-2s}+N^{-s(r-1)|\nu|-2s}\right)\right.$$

$$+nA_1^{-2}B_1^{-2-2r(|\nu|-1)}\left(N^{-(sr-2)(|\nu|-1)-2s+1}+N^{-s(r-1)(|\nu|-1)-s}\right)$$

$$+nA_1^{-1}B_1^{-2-2r|\nu|}\left(N^{-(sr-2)|\nu|-s-1}+N^{-s(r-1)|\nu|-s}\right)$$

$$+ n^2 A_1^{-1} B_1^{-2-2r(|\nu|-1)} \left(N^{-(sr-2)(|\nu|-1)-s+1} + N^{-s(r-1)(|\nu|-1)} \right)$$
$$+ n^2 B_1^{-2-2r|\nu|} \left(N^{-(sr-2)|\nu|-1} + N^{-s(r-1)|\nu|} \right) \Bigg]. \quad (80)$$

As $s \geq 1, r \geq 2$, and $|\nu| \geq 1$, all the exponents of N are less or equal to zero. Using the same strategy adopted to handle (78), we conclude that the right-hand side of (80) is bounded by

$$A_1^{S_p} B_1^{T_p} N^{U_p}$$
$$\times 2^n C_1 B_1^{-2} \left(A_1^{-2} + 2nC_1 A_1^{-2} + nA_1^{-1} + 2n^2 C_1 A_1^{-1} + 2n^2 A_1^{-1} + n \right). \quad (81)$$

Concerning the last multi-indexes sum on the right-hand side of (75), remarking that $\frac{\alpha!}{(\alpha-\nu-\mu)!} \leq N^{|\nu|+\mu}$, we get

$$\sum_{\ell=1}^{n} \sum_{\substack{\nu<\alpha \\ \nu\neq 0}} \sum_{\substack{\mu\leq\alpha-\nu \\ \mu\neq 0}} C_1^{|\nu|+|\mu|+3} \frac{\alpha!}{(\alpha-\nu-\mu)!} \left(N^{|\nu|-1} + |\nu|^{(s-1)|\nu|} \right)$$
$$\times \left(N^{|\mu|-1} + |\mu|^{(s-1)|\mu|} \right) \Bigg[\| D^{\alpha-\nu-\mu+\ell} \psi_N^{(\beta)} P^k u \|_{\frac{p-2}{r}} + \| D^{\alpha-\nu-\mu} \psi_N^{(\beta+\ell)} P^k u \|_{\frac{p-2}{r}}$$
$$+ \sum_{i=1}^{n} \Bigg(\| D^{\alpha-\nu-\mu+\ell+i} \psi_N^{(\beta)} P^k u \|_{\frac{p-2}{r}} + \| D^{\alpha-\nu-\mu+\ell} \psi_N^{(\beta+i)} P^k u \|_{\frac{p-2}{r}}$$
$$+ \| D^{\alpha-\nu-\mu} \psi_N^{(\beta+\ell+i)} P^k u \|_{\frac{p-2}{r}} \Bigg) \Bigg]$$
$$\leq A_1^{S_p} B_1^{T_p} N^{U_p} \times n \sum_{\substack{\nu<\alpha \\ \nu\neq 0}} \sum_{\substack{\mu\leq\alpha-\nu \\ \mu\neq 0}} C_1^{|\nu|+|\mu|+3}$$
$$\times \Bigg[A_1^{-2} B_1^{-2-2r(|\nu|-1)-2r|\mu|} \left(N^{-(sr-2)|\mu|-1} + N^{-s(r-1)|\mu|} \right)$$
$$\times \left(N^{-(sr-2)(|\nu|-1)-s+1} + N^{-s(r-1)(|\nu|-1)} \right) + A_1^{-1} B_1^{-2-2r|\nu|-2r|\mu|} N^{-s}$$
$$\times \left(N^{-(sr-2)|\mu|-1} + N^{-s(r-1)|\mu|} \right) \left(N^{-(sr-2)|\nu|-1} + N^{-s(r-1)|\nu|} \right)$$
$$+ nA_1^{-1} B_1^{-2-2r(|\nu|-1)-2r(|\mu|-1)} \left(N^{-(sr-2)(|\nu|-1)-s+1} + N^{-s(r-1)(|\nu|-1)} \right)$$
$$\times \left(N^{-(sr-2)(|\mu|-1)-s+1} + N^{-s(r-1)(|\mu|-1)} \right) + nA_1^{-1} B_1^{-2-2r(|\nu|-1)-2r|\mu|}$$

$$\times\left(N^{-(sr-2)(|\nu|-1)-s+1}+N^{-s(r-1)(|\nu|-1)}\right)\left(N^{-(sr-2)|\mu|-1}+N^{-s(r-1)|\mu|}\right)$$
$$+nA_1^{-1}B_1^{-2-2r|\nu|-2r|\mu|}N^{-s}\left(N^{-(sr-2)|\mu|-1}+N^{-s(r-1)|\mu|}\right)$$
$$\times\left(N^{-(sr-2)|\nu|-1}+N^{-s(r-1)|\nu|}\right)\Bigg]. \tag{82}$$

Since $|\nu|$, $|\mu|$ and s are greater than 1, and r is greater than 2, all the exponents of N are less than or equal to zero. Applying the same strategies to handle the previous multi-indexes sum, we obtain that the right-hand side of the above inequality is bounded by

$$A_1^{S_p}B_1^{T_p}N^{U_p}\le 2^{2n+2}n^2B_1^{-2}$$
$$\times\left[(2nC_1)^{-1}A_1^{-2}+\left(4nC_1^2\right)^{-1}A_1^{-1}+A_1^{-1}+(2C_1)^{-1}A_1^{-1}+(2C_1)^{-2}\right]. \tag{83}$$

Summing up, enlarging A_1 and B_1 if necessary, the summand of the second factor of (79), (81), and (83) and the second factor on the right-hand sides of (76) and (77) can be made smaller than C_2^{-1}, we conclude

$$\|D^\alpha\psi_N^{(\beta)}P^ku\|_{\frac{p}{r}}+\sum_{j=1}^m\|X_jD^\alpha\psi_N^{(\beta)}P^ku\|_{\frac{p-1}{r}}+\|X_0D^\alpha\psi_N^{(\beta)}P^ku\|_{\frac{p-2}{r}}$$
$$+\sum_{\ell,j=1}^m\|X_\ell X_jD^\alpha\psi_N^{(\beta)}P^ku\|_{\frac{p-2}{r}}\le A_1^{S_p}B_1^{T_p}N^{U_p}. \tag{84}$$

We obtained (62).

This concludes the proof of Proposition (4.1).

Proposition 4.2 *There exist positive constants* A_1 *and* B_1 *such that* (60) *in Proposition 4.1 is true.*

Proof We use the induction on $|\alpha|$. Case $|\alpha|=0$. Let K_0 be a compact set contained in Ω and containing all the supports of ψ_N, $\Omega_0\subset K_0$. Since u is a G^s vector for P in Ω, we have $\|P^ku\|_{L^2(K_0)}\le C_{K_0}^{2k+1}k^{2sk}$. Then, taking advantage from Lemma 3.2, we obtain

$$\|\psi_N^{(\beta)}P^ku\|_0\le C_\psi^{|\beta|+1}N^{(|\beta|-M)^+}\|P^ku\|_{L^2(K_0)}$$
$$\le C_1^{|\alpha|+1}C_2^{|\beta|+2k+1}N^{s(|\alpha|+|\beta|+2k-M)^+}.$$

Since M is a fixed constant depending only on n and greater than one, we conclude that there are two positive constants A_1 and B_1, independent of β and k, such that

$$\|\psi_N^{(\beta)} P^k u\| \leq A_1^{|\beta|+2k+1} B_1 N^{s||\beta|+2k|}. \tag{85}$$

In order to gain the second assertion in (60), we use the following estimate:

$$\sum_{\ell=1}^{m} \|X_\ell v\|_0^2 \leq \widetilde{C}\left(|\langle Pv, v\rangle| + \|v\|^2\right), \qquad \forall\, v \in \mathscr{D}(\Omega_0)\,. \tag{86}$$

Replacing v by $\psi_N^{(\beta)} P^k u$, we have

$$\sum_{\ell=1}^{m} \|X_\ell \psi_N^{(\beta)} P^k u\|_0^2 \leq \widetilde{C}\left(|\langle P\psi_N^{(\beta)} P^k u, \psi_N^{(\beta)} P^k u\rangle| + \|\psi_N^{(\beta)} P^k u\|^2\right).$$

Since

$$P\psi_N^{(\beta)} P^k u = \psi_N^{(\beta)} P^{k+1} u + \sum_{\ell=1}^{m}\left(2X_\ell[X_\ell, \psi_N^{(\beta)}] - [X_\ell, [X_\ell, \psi_N^{(\beta)}]]\right) + [X_0, \psi_N^{(\beta)}],$$

where

$$[X_\ell, \psi_N^{(\beta)}] = \sum_{i=1}^{n} a_{i,\ell}\psi_N^{(\beta+i)}, \qquad \ell = 0, 1, \dots, m;$$

$$[X_\ell, [X_\ell, \psi_N^{(\beta)}]] = \sum_{i=1}^{n}\sum_{j=1}^{n} a_{j,\ell}\left(a_{i,\ell}^{(j)}\psi_N^{(\beta+i)} + a_{i,\ell}\psi_N^{(\beta+i+j)}\right), \qquad \ell = 1, \dots, m,$$

and keeping in mind that $X_\ell^* = X_\ell - i\sum_{i=1}^{n} a_{i,\ell}^{(i)}$, we obtain

$$\begin{aligned}\sum_{\ell=1}^{m} \|X_\ell \psi_N^{(\beta)} P^k u\|_0^2 \leq \widetilde{C}_1\Bigg[& \|\psi_N^{(\beta)} P^{k+1} u\|_0 \|\psi_N^{(\beta)} P^k u\|_0 + \|\psi_N^{(\beta)} P^k u\|_0^2 \\ & + \sum_{\ell=0}^{m}\sum_{i=1}^{n} \|a_{i,\ell}\psi_N^{(\beta+i)} P^k u\|_0^2 + \sum_{\ell=1}^{m}\sum_{i=1}^{n} \|a_{i,\ell}^{(i)}\psi_N^{(\beta)} P^k u\|_0^2 \\ + \sum_{\ell=1}^{m}\sum_{i=1}^{n}\sum_{p=1}^{n} & \left(\|a_{p,\ell} a_{i,\ell}^{(p)}\psi_N^{(\beta+i)} P^k u\|_0 + \|a_{p,\ell} a_{i,\ell}\psi_N^{(\beta+i+p)} P^k u\|_0\right) \|\psi_N^{(\beta)} P^k u\|_0 \Bigg].\end{aligned}$$

So, there is a new positive constant $\widetilde{C}_2$ such that

$$\sum_{\ell=1}^{m} \|X_\ell \psi_N^{(\beta)} P^k u\|_0^2 \le \widetilde{C}_2 \Big[\|\psi_N^{(\beta)} P^{k+1} u\|_0 \|\psi_N^{(\beta)} P^k u\|_0 + \|\psi_N^{(\beta)} P^k u\|_0^2$$

$$+ \sum_{i=1}^{n} \|\psi_N^{(\beta+i)} P^k u\|_0^2 + \sum_{i=1}^{n} \|\psi_N^{(\beta+i)} P^k u\|_0 \|\psi_N^{(\beta)} P^k u\|_0$$

$$+ \sum_{i=1}^{n}\sum_{j=1}^{n} \|\psi_N^{(\beta+i+j)} P^k u\|_0 \|\psi_N^{(\beta)} P^k u\|_0 \Big].$$

From these estimates and the (85), for every $\ell \in \{1, \ldots, m\}$, we obtain

$$\|X_\ell \psi_N^{(\beta)} P^k u\|_0 \le A_1^{|\beta|+2k+1} B_1^2 N^{s(1+|\beta|+2k)}$$

$$\times \widetilde{C}_2^{\frac{1}{2}} B_1^{-1} \Big(A_1 + N^{-s} + \sqrt{n} A_1 + \sqrt{n} A_1^{\frac{1}{2}} N^{-\frac{s}{2}} + n A_1 \Big).$$

Taking B_1 large enough compared to A_1, we have

$$\widetilde{C}_2^{\frac{1}{2}} B_1^{-1} \Big(A_1 + N^{-s} + \sqrt{n} A_1 + \sqrt{n} A_1^{\frac{1}{2}} N^{-\frac{s}{2}} + n A_1 \Big) \le 1.$$

We conclude that

$$\|X_\ell \psi_N^{(\beta)} P^k u\|_0 \le A_1^{|\beta|+2k+1} B_1^2 N^{s(1+|\beta|+2k)}, \qquad \ell = 1, \ldots, m,$$

i.e., the second assertion in (60).

Let $m_0 \in \mathbb{N}$, and we assume that

$$\begin{cases} \|D^\alpha \psi_N^{(\beta)} P^k u\| \le A_1^{|\beta|+2k+1} B_1^{2r|\alpha|+1} N^{s[r|\alpha|+|\beta|+2k]}, \\ \|X_\ell D^\alpha \psi_N^{(\beta)} P^k u\| \le A_1^{|\beta|+2k+1} B_1^{2r(|\alpha|+(2r)^{-1})+1} N^{s[r(|\alpha|+r^{-1})+|\beta|+2k]}, \ \ell=1, \ldots, m, \end{cases}$$

for every α such that $|\alpha| \le m_0$ and $2r|\alpha| + (|\beta| - M) + 2k \le N$. Our goal is to show that the above inequalities are true for α such that $|\alpha| = m_0 + 1$.

By Proposition (4.1) we know that

$$\|D^\alpha \psi_N^{(\beta)} P^k u\|_{\frac{p}{r}} \le A_1^{|\beta|+2k+p+1} B_1^{2r|\alpha|+p+1} N^{s[r|\alpha|+|\beta|+2k+p]}$$

holds for $2 \le p \le r$, $|\alpha| \le m_0$ and $2r|\alpha| + (|\beta| - M) + 2k \le N - 2p$. In particular it holds when $p = r$:

$$\|D^\alpha \psi_N^{(\beta)} P^k u\|_1 \le A_1^{|\beta|+2k+r+1} B_1^{2rm_0+r+1} N^{s[rm_0+|\beta|+2k+r]}.$$

Let $(\alpha, \beta, k) \in \mathbb{N}^{2n+1}$, with $|\alpha| = m_0 + 1$, such that $2r|\alpha| + (|\beta| - M) + 2k \leq N$, and then $2r(m_0+1)+(|\beta|-M)+2k \leq N$, i.e., $2rm_0+(|\beta|-M)+2k \leq N-2r$. So $\alpha = \alpha_0 + e_j$, $e_j = (0, \ldots, 0, 1, 0, \ldots, 0)$, with $|\alpha_0| = m_0$. Since

$$2rm_0 + (|\beta| - M) + 2k = 2r(m_0 + 1) + (|\beta| - M) + 2k \leq N - 2r,$$

by inductive hypothesis, (60), in Proposition 4.1, is true for (α_0, β, k), so by Proposition 4.1, (62) holds for $p = r$ and (α_0, β, k), and we conclude that

$$\begin{aligned}\|D^\alpha \psi_N^{(\beta)} P^k u\|_0 &= \|D_j D^{\alpha_0} \psi_N^{(\beta)} P^k u\|_0 \\ &\leq \|D^{\alpha_0} \psi_N^{(\beta)} P^k u\|_1 \leq A_1^{|\beta|+2k+r+1} B_1^{2rm_0+r+1} N^{s[rm_0+|\beta|+2k+r]} \\ &= A_1^{|\beta|+2k+1} B_1^{2r|\alpha|+1} N^{s[r|\alpha|+|\beta|+2k]} \times (A_1 B_1^{-1})^r. \end{aligned} \tag{87}$$

As before, assuming B_1 large compared to A_1, the factor $A_1 B_1^{-1}$ is less than one. We have obtained what we want.

Now, we have to show the second assertion, i.e., that

$$\|X_\ell D^\alpha \psi_N^{(\beta)} P^k u\| \leq A_1^{|\beta|+2k+1} B_1^{2r(|\alpha|+(2r)^{-1})+1} N^{s[r(|\alpha|+r^{-1})+|\beta|+2k]}$$

is true for $|\alpha| = m_0 + 1$.

By (86), replacing v by $D^\alpha \psi_N^{(\beta)} P^k u$, we obtain the following estimate:

$$\begin{aligned}\sum_{\ell=1}^m \|X_\ell D^\alpha \psi_N^{(\beta)} P^k u\|_0^2 &\leq \widetilde{C}\Big[\|D^\alpha \psi_N^{(\beta)} P^{k+1} u\|_0 \|D^\alpha \psi_N^{(\beta)} P^k u\|_0 \\ &\quad + \|D^\alpha \psi_N^{(\beta)} P^k u\|_0^2 + |\langle [P, D^\alpha \psi_N^{(\beta)}] P^k u, D^\alpha \psi_N^{(\beta)} P^k u\rangle| \Big] \\ &\leq \widetilde{C}\Big[\|D^\alpha \psi_N^{(\beta)} P^{k+1} u\|_0 \|D^\alpha \psi_N^{(\beta)} P^k u\|_0 + \|D^\alpha \psi_N^{(\beta)} P^k u\|_0^2 \\ &\quad + \sum_{\ell=1}^m \Big(2|\langle X_\ell [X_\ell, D^\alpha \psi_N^{(\beta)}] P^k u, D^\alpha \psi_N^{(\beta)} P^k u\rangle| \\ &\qquad + |\langle [X_\ell, [X_\ell, D^\alpha \psi_N^{(\beta)}]] P^k u, D^\alpha \psi_N^{(\beta)} P^k u\rangle| \Big) \\ &\quad + |\langle [X_0, D^\alpha \psi_N^{(\beta)}] P^k u, D^\alpha \psi_N^{(\beta)} P^k u\rangle| + |\langle [c, D^\alpha] \psi_N^{(\beta)} P^k u, D^\alpha \psi_N^{(\beta)} P^k u\rangle| \Big]. \end{aligned} \tag{88}$$

Now, since $X_\ell^* = X_\ell + f_\ell$, $f_\ell = \sum_{j=1}^n a_{j,\ell}^{(j)}$, and

$$\sum_{\ell=1}^{m} 2|\langle X_\ell[X_\ell, D^\alpha \psi_N^{(\beta)}]P^k u, D^\alpha \psi_N^{(\beta)} P^k u\rangle| \le \frac{1}{2\widetilde{C}} \sum_{\ell=1}^{m} \|X_\ell D^\alpha \psi_N^{(\beta)} P^k u\|_0^2$$

$$+ 4\widetilde{C} \sum_{\ell=1}^{m} \|[X_\ell, D^\alpha \psi_N^{(\beta)}]P^k u\|_0^2 + 2\sum_{\ell=1}^{m} |\langle [X_\ell, D^\alpha \psi_N^{(\beta)}]P^k u, f_\ell D^\alpha \psi_N^{(\beta)} P^k u\rangle|,$$

then (88) can be rewritten as follows:

$$\sum_{\ell=1}^{m} \|X_\ell D^\alpha \psi_N^{(\beta)} P^k u\|_0^2 \le 2\widetilde{C}\Big[\|D^\alpha \psi_N^{(\beta)} P^{k+1} u\|_0 \|D^\alpha \psi_N^{(\beta)} P^k u\|_0$$

$$+ \|D^\alpha \psi_N^{(\beta)} P^k u\|_0^2 + 2\sum_{\ell=1}^{m} |\langle [X_\ell, D^\alpha \psi_N^{(\beta)}]P^k u, f_\ell D^\alpha \psi_N^{(\beta)} P^k u\rangle|$$

$$+ 4\widetilde{C} \sum_{\ell=1}^{m} \|[X_\ell, D^\alpha \psi_N^{(\beta)}]P^k u\|_0^2 + \sum_{\ell=1}^{m} |\langle [X_\ell, [X_\ell, D^\alpha \psi_N^{(\beta)}]]P^k u, D^\alpha \psi_N^{(\beta)} P^k u\rangle|$$

$$+ |\langle [X_0, D^\alpha \psi_N^{(\beta)}]P^k u, D^\alpha \psi_N^{(\beta)} P^k u\rangle| + |\langle [c, D^\alpha]\psi_N^{(\beta)} P^k u, D^\alpha \psi_N^{(\beta)} P^k u\rangle| \Big]. \tag{89}$$

We analyze separately the term on the right-hand side.

We begin to remark that by (87), we know that

$$\|D^\alpha \psi_N^{(\beta)} P^k u\|_0^2$$
$$\le \left(A_1^{|\beta|+2k+1} B_1^{2r(|\alpha|+(2r)^{-1})+1} N^{s[r(|\alpha|+r^{-1})+|\beta|+2k]} \right)^2 \times B_1^{-2} N^{-2s}. \tag{90}$$

Concerning the first term on the right-hand side, we have

$$\|D^\alpha \psi_N^{(\beta)} P^{k+1} u\|_0 \|D^\alpha \psi_N^{(\beta)} P^k u\|_0 \le \|D^{\alpha_0} \psi_N^{(\beta)} P^{k+1} u\|_1 \|D^\alpha \psi_N^{(\beta)} P^k u\|_0,$$

and now we remark that

$$2r|\alpha_0| + |\beta| + 2k + 2 = 2r(|\alpha| - 1) + |\beta| + 2k + 2 \le 2r|\alpha| + |\beta| + 2k;$$

$2(1-r)$ is negative as $r \ge 2$. Since $|\alpha_0| = m_0$, $(\alpha_0, \beta, k+1)$ satisfies the condition in (60), and then by Proposition 4.1, (62) is true for $p = r$ for the $2n+1$-tuples $(\alpha_0, \beta, k+1)$; we have

$$\|D^{\alpha_0} \psi_N^{(\beta)} P^{k+1} u\|_1 \le A_1^{|\beta|+2k+2+r+1} B_1^{2r|\alpha_0|+r+1} N^{s[r|\alpha_0|+|\beta|+2k+2+r]}.$$

By (90) and the above inequality, we conclude that

$$
\begin{aligned}
&\| D^{\alpha} \psi_N^{(\beta)} P^{k+1} u \|_0 \| D^{\alpha} \psi_N^{(\beta)} P^k u \|_0 \\
&\leq \left(A_1^{|\beta|+2k+1} B_1^{2r(|\alpha|+(2r)^{-1})+1} N^{s[r(|\alpha|+r^{-1})+|\beta|+2k]} \right)^2 \times A_1^{r+2} B_1^{-r-2}. \qquad (91)
\end{aligned}
$$

We focus on the third term on the right-hand side of (89). By (9), (15), (16), and (17), we have

$$
\begin{aligned}
&\sum_{\ell=1}^{m} |\langle [X_\ell, D^{\alpha} \psi_N^{(\beta)}] P^k u, f_\ell D^{\alpha} \psi_N^{(\beta)} P^k u \rangle| \\
&\qquad \leq \sum_{\ell=1}^{m} \sum_{j=1}^{n} C_1^3 \| D^{\alpha_0} \psi_N^{(\beta+j)} P^k u \|_1 \| D^{\alpha} \psi_N^{(\beta)} P^k u \|_0 \\
&+ \sum_{\ell=1}^{m} \sum_{j=1}^{n} \sum_{\substack{\nu \leq \alpha \\ \nu \neq 0}} C_1^{|\nu|+3} \frac{\alpha!}{(\alpha-\nu)!} (\nu!)^s \| D^{\alpha-\nu+j} \psi_N^{(\beta)} P^k u \|_0 \| D^{\alpha} \psi_N^{(\beta)} P^k u \|_0 \\
&+ \sum_{\ell=1}^{m} \sum_{j=1}^{n} \sum_{\substack{\nu \leq \alpha \\ \nu \neq 0}} C_1^{|\nu|+3} \frac{\alpha!}{(\alpha-\nu)!} (\nu!)^s \| D^{\alpha-\nu} \psi_N^{(\beta+j)} P^k u \|_0 \| D^{\alpha} \psi_N^{(\beta)} P^k u \|_0 .
\end{aligned}
\qquad (92)
$$

In order to handle the second factor in the first sum we remark that

$$
2r|\alpha_0| + |\beta| + 2k + 1 = 2r(|\alpha| - 1) + |\beta| + 2k + 1 \leq 2r|\alpha| + |\beta| + 2k;
$$

$1 - 2r$ is negative, $r \geq 2$. Since $|\alpha_0| = m_0$ and $(\alpha_0, \beta + j, k)$ satisfies the condition in (60), then by Proposition 4.1, (62) is true for $p = r$ for the $2n+1$-tuples $(\alpha_0, \beta + j, k)$; we have

$$
\| D^{\alpha_0} \psi_N^{(\beta+j)} P^k u \|_1 \leq A_1^{|\beta|+1+2k+r+1} B_1^{2r|\alpha_0|+r+1} N^{s[r|\alpha_0|+|\beta|+1+2k+r]}.
$$

By (90) and the above inequality, we obtain

$$
\begin{aligned}
&\| D^{\alpha_0} \psi_N^{(\beta+j)} P^k u \|_1 \| D^{\alpha} \psi_N^{(\beta)} P^k u \|_0 \\
&\leq \left(A_1^{|\beta|+2k+1} B_1^{2r(|\alpha|+(2r)^{-1})+1} N^{s[r(|\alpha|+r^{-1})+|\beta|+2k]} \right)^2 \times A_1^{r+1} B_1^{-r-2} N^{-s}.
\end{aligned}
$$

In order to handle the terms in the second multiple sum, we have to distinguish two cases, $|\nu| = 1$ and $|\nu| > 1$; we remark that in the second case, we have that

$|\alpha - \nu + j| \leq m_0$, so we can use directly the inductive hypothesis. Concerning the case $|\nu| = 1$, we have

$$\|D^{\alpha-\nu+j}\psi_N^{(\beta)}P^k u\|_0 \leq \|D^{\alpha_0-\nu+j}\psi_N^{(\beta)}P^k u\|_1.$$

Now

$$2r(|\alpha_0| - |\nu| + 1) + |\beta| + 2k = 2r(|\alpha| - |\nu|) + |\beta| + 2k < 2r|\alpha| + |\beta| + 2k;$$

$|\nu| = 1$. Since $|\alpha_0| = m_0$ and $(\alpha_0 - \nu + j, \beta, k)$ satisfies the condition in (60), then by Proposition 4.1, (62) is true for $p = r$ for the $2n + 1$-tuples $(\alpha_0, \beta + j, k)$. We have

$$\|D^{\alpha_0-\nu+j}\psi_N^{(\beta)}P^k u\|_1 \leq A_1^{|\beta|+2k+r+1}B_1^{2r|\alpha_0|-2r|\nu|+2r+r+1}N^{s[r|\alpha_0|-r|\nu|+r+|\beta|+2k+r]}.$$

Recalling that $|\nu| = 1$, by (90) and the above inequality, we obtain

$$\begin{aligned}&\|D^{\alpha_0-\nu+j}\psi_N^{(\beta)}P^k u\|_1\|D^{\alpha}\psi_N^{(\beta)}P^k u\|_0\\ &\quad\leq \left(A_1^{|\beta|+2k+1}B_1^{2r(|\alpha|+(2r)^{-1})+1}N^{s[r(|\alpha|+r^{-1})+|\beta|+2k]}\right)^2 \times A_1^r B_1^{-r-2}N^{-2s}.\end{aligned}$$

As $\frac{\alpha!}{(\alpha-\nu)!} \leq |\alpha|^{|\nu|} \leq N^{|\nu|}$, we have

$$\begin{aligned}&\sum_{\ell=1}^{m}\sum_{j=1}^{n}\sum_{\substack{\nu\leq\alpha\\ \nu\neq 0}} C_1^{|\nu|+3}\frac{\alpha!}{(\alpha-\nu)!}(\nu!)^s\|D^{\alpha-\nu+j}\psi_N^{(\beta)}P^k u\|_0\|D^{\alpha}\psi_N^{(\beta)}P^k u\|_0\\
&\qquad\leq \sum_{\ell=1}^{m}\sum_{j=1}^{n}\sum_{|\nu|=1} C_1^4 N\|D^{\alpha_0-\nu+j}\psi_N^{(\beta)}P^k u\|_1\|D^{\alpha}\psi_N^{(\beta)}P^k u\|_0\\
&\qquad+\sum_{\ell=1}^{m}\sum_{j=1}^{n}\sum_{\substack{\nu\leq\alpha\\ |\nu|\geq 2}} C_1^{|\nu|+3}N^{s|\nu|}\|D^{\alpha-\nu+j}\psi_N^{(\beta)}P^k u\|_0\|D^{\alpha}\psi_N^{(\beta)}P^k u\|_0\\
&\qquad\leq \left(A_1^{|\beta|+2k+1}B_1^{2r(|\alpha|+(2r)^{-1})+1}N^{s[r(|\alpha|+r^{-1})+|\beta|+2k]}\right)^2\\
&\times nmC_1^4\left[A_1^r B_1^{-r-2}N^{-2s} + B_1^{-1}N^{-s}\sum_{\substack{\nu\leq\alpha\\ |\nu|\geq 2}}\left(C_1 B_1^{-2r}\right)^{|\nu|-1}N^{-s(r-1)(|\nu|-1)}\right].\end{aligned}$$

Since we have chosen B_1 large enough in order that $2C_1 B_1^{-2r} < 2$ and $r \geq 2$, we conclude that

$$\sum_{\ell=1}^{m}\sum_{j=1}^{n}\sum_{\substack{\nu\le\alpha\\ \nu\neq 0}} C_1^{|\nu|+3}\frac{\alpha!}{(\alpha-\nu)!}(\nu!)^s\|D^{\alpha-\nu+j}\psi_N^{(\beta)}P^k u\|_0\|D^\alpha\psi_N^{(\beta)}P^k u\|_0$$
$$\le\left(A_1^{|\beta|+2k+1}B_1^{2r(|\alpha|+(2r)^{-1})+1}N^{s[r(|\alpha|+r^{-1})+|\beta|+2k]}\right)^2$$
$$\times\, nmC_1^4\left[A_1^rB_1^{-r-2}N^{-2s}+2^{n-1}B_1^{-1}N^{-s}\right].$$

Concerning the last multiple sum on the right-hand side of (92), we observe that since $|\nu|\ge 1$, we can use directly the inductive hypothesis. Using the same strategy adopted above, we obtain

$$\sum_{\ell=1}^{m}\sum_{j=1}^{n}\sum_{\substack{\nu\le\alpha\\ \nu\neq 0}} C_1^{|\nu|+3}\frac{\alpha!}{(\alpha-\nu)!}(\nu!)^s\|D^{\alpha-\nu}\psi_N^{(\beta+j)}P^k u\|_0\|D^\alpha\psi_N^{(\beta)}P^k u\|_0$$
$$\le\left(A_1^{|\beta|+2k+1}B_1^{2r(|\alpha|+(2r)^{-1})+1}N^{s[r(|\alpha|+r^{-1})+|\beta|+2k]}\right)^2$$
$$\times\, 2^{n-1}nmC_1^3A_1B_1^{-2}N^{-s}.$$

Summing up, by the above consideration, we conclude that

$$\sum_{\ell=1}^{m}|\langle[X_\ell, D^\alpha\psi_N^{(\beta)}]P^k u, f_\ell D^\alpha\psi_N^{(\beta)}P^k u\rangle|$$
$$\le\left(A_1^{|\beta|+2k+1}B_1^{2r(|\alpha|+(2r)^{-1})+1}N^{s[r(|\alpha|+r^{-1})+|\beta|+2k]}\right)^2$$
$$\times\left[A_1^rB_1^{-r-2}N^{-2s}+nm2^{n-1}C_1^4B_1^{-1}N^{-s}\left(A_1^rB_1^{-r-1}2^{-n+1}N^{-s}+1+A_1B_1^{-1}\right)\right]. \tag{93}$$

Concerning the second to last term on the right-hand side of (89): Using the same strategy adopted to handle the previous term, we have

$$|\langle[X_0, D^\alpha\psi_N^{(\beta)}]P^k u, D^\alpha\psi_N^{(\beta)}P^k u\rangle|$$
$$\le\left(A_1^{|\beta|+2k+1}B_1^{2r(|\alpha|+(2r)^{-1})+1}N^{s[r(|\alpha|+r^{-1})+|\beta|+2k]}\right)^2$$

$$\times\left[A_1^r B_1^{-r-2} N^{-2s} + nm2^{n-1}C_1^2 B_1^{-1} N^{-s}\left(A_1^r B_1^{-r-1} 2^{-n+1} N^{-s} + 1 + A_1 B_1^{-1}\right)\right]. \tag{94}$$

We handle now the last term on the right-hand side of (89). By (10) we have

$$|\langle [c, D^\alpha]\psi_N^{(\beta)} P^k u, D^\alpha \psi_N^{(\beta)} P^k u\rangle|$$
$$\leq \sum_{\substack{\nu\leq\alpha,\\ \nu\neq 0}} C_1^{|\nu|+1} \frac{\alpha!}{(\alpha-\nu)!}(\nu!)^{s-1} \| D^{\alpha-\nu}\psi_N^{(\beta)} P^k u\|_0 \| D^\alpha \psi_N^{(\beta)} P^k u\|_0.$$

We observe that since $|\nu| \geq 1$, $|\alpha - \nu| \leq m_0$. We can use directly the inductive hypothesis. Arguing as before to handle the other multi-index sums, we obtain

$$|\langle [c, D^\alpha]\psi_N^{(\beta)} P^k u, D^\alpha \psi_N^{(\beta)} P^k u\rangle|$$
$$\leq \left(A_1^{|\beta|+2k+1} B_1^{2r(|\alpha|+(2r)^{-1})+1} N^{s[r(|\alpha|+r^{-1})+|\beta|+2k]}\right)^2$$
$$\times \left(2^{n-1} C_1 B_1^{-2} N^{-2s}\right). \tag{95}$$

There remains one last term to handle, the third to last term on the right-hand side of (89). By (17) and the identities (19), (20), (21), and (22), we have

$$\sum_{\ell=1}^{m} |\langle [X_\ell, [X_\ell, D^\alpha \psi_N^{(\beta)}]] P^k u, D^\alpha \psi_N^{(\beta)} P^k u\rangle|$$
$$\leq \Bigg\{ 4n \sum_{j=1}^{n} \sum_{\substack{\nu\leq\alpha,\\ \nu\neq 0}} C_1^{|\nu|+3} \frac{\alpha!\,(\nu!)^{s-1}}{(\alpha-\nu)!} \left(\| D^{\alpha-\nu}\psi_N^{(\beta+j)} P^k u\|_0 + \| D^{\alpha-\nu+j}\psi_N^{(\beta)} P^k u\|_0\right)$$
$$+ 2\sum_{i,j=1}^{n} \sum_{\substack{\nu\leq\alpha,\\ \nu\neq 0}} C_1^{|\nu|+2} \frac{\alpha!\,(\nu!)^{s-1}}{(\alpha-\nu)!} \left(\| D^{\alpha-\nu}\psi_N^{(\beta+j+i)} P^k u\|_0 + \| D^{\alpha-\nu+j}\psi_N^{(\beta+i)} P^k u\|_0\right)$$
$$+ 3n \sum_{j=1}^{n} \sum_{\substack{\nu<\alpha,\\ \nu\neq 0}} \sum_{\substack{\mu\leq\alpha-\nu,\\ \mu\neq 0}} C_1^{|\nu|+|\mu|+3} \frac{\alpha!\,(\nu!)^{s-1}\,(\mu!)^{s-1}}{(\alpha-\nu-\mu)!} \Big[\| D^{\alpha-\nu-\mu}\psi_N^{(\beta+j)} P^k u\|_0$$
$$+ \| D^{\alpha-\nu-\mu+j}\psi_N^{(\beta)} P^k u\|_0\Big]$$

$$+2\sum_{i,j=1}^{n}\sum_{\substack{\nu<\alpha,\\ \nu\neq 0}}\sum_{\substack{\mu\leq\alpha-\nu,\\ \mu\neq 0}} C_1^{|\nu|+|\mu|+2}\frac{\alpha!\,(\nu!)^{s-1}\,(\mu!)^{s-1}}{(\alpha-\nu-\mu)!}\Big[\|D^{\alpha-\nu-\mu}\psi_N^{(\beta+j+i)}P^k u\|_0$$

$$+\|D^{\alpha-\nu-\mu+j}\psi_N^{(\beta+i)}P^k u\|_0+\|D^{\alpha-\nu-\mu+j+i}\psi_N^{(\beta)}P^k u\|_0\Big]$$

$$+\,nC_1^3\sum_{j=1}^{n}\|D^{\alpha}\psi_N^{(\beta+j)}P^k u\|_0+C_1^2\sum_{j,i=1}^{n}\|D^{\alpha}\psi_N^{(\beta+j+i)}P^k u\|_0\Big\}\|D^{\alpha}\psi_N^{(\beta)}P^k u\|_0. \tag{96}$$

We begin to analyze the terms in the last two sums. We have

$$C_1^2\Big(nC_1\sum_{j=1}^{n}\|D^{\alpha}\psi_N^{(\beta+j)}P^k u\|_0+\sum_{j,i=1}^{n}\|D^{\alpha}\psi_N^{(\beta+j+i)}P^k u\|_0\Big)\|D^{\alpha}\psi_N^{(\beta)}P^k u\|_0$$

$$\leq nC_1^3\Big(\sum_{j=1}^{n}\|D^{\alpha_0}\psi_N^{(\beta+j)}P^k u\|_1+\sum_{j,i=1}^{n}\|D^{\alpha_0}\psi_N^{(\beta+j+i)}P^k u\|_1\Big)\|D^{\alpha}\psi_N^{(\beta)}P^k u\|_0.$$

We remark that

$$2r|\alpha_0|+|\beta|+2k+1=2r(|\alpha|-1)+|\beta|+2k+1\leq 2r|\alpha|+|\beta|+2k;$$

$$2r|\alpha_0|+|\beta|+2k+2=2r(|\alpha|-1)+|\beta|+2k+2\leq 2r|\alpha|+|\beta|+2k;$$

$1-2r$ and $2(1-r)$ are negative, $r\geq 2$. Since $|\alpha_0|=m_0$ and $(\alpha_0,\beta+j,k)$, and $(\alpha_0,\beta+j+i,k)$ satisfy the condition in (60), then by Proposition 4.1, (62) is true for $p=r$ for these two $2n+1$-tuples. We have

$$\|D^{\alpha_0}\psi_N^{(\beta+i)}P^k u\|_1\leq A_1^{|\beta|+1+2k+r+1}B_1^{2r|\alpha_0|+r+1}N^{s[r|\alpha_0|+|\beta|+1+2k+r]};$$

$$\|D^{\alpha_0}\psi_N^{(\beta+i+j)}P^k u\|_1\leq A_1^{|\beta|+2+2k+r+1}B_1^{2r|\alpha_0|+r+1}N^{s[r|\alpha_0|+|\beta|+2+2k+r]}.$$

We obtain

$$C_1^2\Big(nC_1\sum_{j=1}^{n}\|D^{\alpha}\psi_N^{(\beta+j)}P^k u\|_0+\sum_{j,i=1}^{n}\|D^{\alpha}\psi_N^{(\beta+j+i)}P^k u\|_0\Big)\|D^{\alpha}\psi_N^{(\beta)}P^k u\|_0$$

$$\leq\Big(A_1^{|\beta|+2k+1}B_1^{2r(|\alpha|+(2r)^{-1})+1}N^{s[r(|\alpha|+r^{-1})+|\beta|+2k]}\Big)^2$$

$$\times\, n^3C_1^3A_1^{r+2}B_1^{-r-2}\Big(A_1^{-1}N^{-s}+1\Big).$$

About the other terms on the right-hand side of (95), we remark that, except the second term in the first two sums, when $|\nu| = 1$, and the third term in the last multi-indexes sum, when $|\nu| = |\mu| = 1$, since $|\nu|$, $|\mu|$ are greater than one, we can use the inductive hypothesis to handle all the terms in multi-indexes sums. We point out that in the cases $|\nu| = 1$ and $|\nu| = |\mu| = 1$, we have to cross the same problem run into to handle the first term in the first multi-index sum on the right-hand side of (92).

Case $|\nu| = 1$. By (87), we have

$$\|D^{\alpha-\nu+j}\psi_N^{(\beta)}P^k u\|_0 \le A_1^{|\beta|+2k+1}B_1^{2r|\alpha|+1}N^{s[r|\alpha|+|\beta|+2k]} \times (A_1B_1^{-1})^r;$$

then

$$\begin{aligned}&\|D^{\alpha-\nu+j}\psi_N^{(\beta)}P^k u\|_0\|D^{\alpha}\psi_N^{(\beta)}P^k u\|_0\\ &\quad\le \left(A_1^{|\beta|+2k+1}B_1^{2r(|\alpha|+(2r)^{-1})+1}N^{s[r(|\alpha|+r^{-1})+|\beta|+2k]}\right)^2 \times A_1^r B_1^{-r-2}N^{-2s}.\end{aligned}$$

Concerning the second term in the second multi-index sum, we have

$$\begin{aligned}&\sum_{j=1}^{n}\|D^{\alpha-\nu+j}\psi_N^{(\beta+i)}P^k u\|_0\|D^{\alpha}\psi_N^{(\beta)}P^k u\|_0\\ &\qquad= \|D^{\alpha-\nu}\psi_N^{(\beta+i)}P^k u\|_1\|D^{\alpha}\psi_N^{(\beta)}P^k u\|_0,\end{aligned}$$

where $|\alpha-\nu|$ is equal to m_0. Since

$$2rm_0 + (|\beta| - M) + 2k = 2r(m_0+1) + (|\beta| - M) + 2k \le N - 2r,$$

(60), in Proposition 4.1, is true for $(\alpha-\nu, \beta+j, k)$, so by Proposition 4.1, (62) holds for $p = r$ and $(\alpha-\nu, \beta+j, k)$. We have

$$\|D^{\alpha-\nu}\psi_N^{(\beta+i)}P^k u\|_1 \le A_1^{|\beta|+2k+1}B_1^{2r|\alpha|+1}N^{s[r|\alpha|+|\beta|+2k+1]} \times \left(A_1^{r+1}B_1^{-r-1}\right).$$

We conclude that

$$\begin{aligned}&\|D^{\alpha-\nu}\psi_N^{(\beta+i)}P^k u\|_1\|D^{\alpha}\psi_N^{(\beta)}P^k u\|_0\\ &\quad\le \left(A_1^{|\beta|+2k+1}B_1^{2r(|\alpha|+(2r)^{-1})+1}N^{s[r(|\alpha|+r^{-1})+|\beta|+2k]}\right)^2 \times A_1^{2r+1}B_1^{-2r-2}N^{-s}.\end{aligned}$$

Case $|\nu| = |\mu| = 1$:

$$\sum_{i=1}^{n} \| D^{\alpha-\nu-\mu+j+i} \psi_N^{(\beta)} P^k u\|_0 \| D^\alpha \psi_N^{(\beta)} P^k u\|_0$$
$$= \| D^{\alpha-\nu-\mu+j} \psi_N^{(\beta)} P^k u\|_1 \| D^\alpha \psi_N^{(\beta)} P^k u\|_0,$$

where $|\alpha - \nu - \mu + j|$ is equal to m_0. Using the same argument adopted above, we have

$$\| D^{\alpha-\nu-\mu+j} \psi_N^{(\beta)} P^k u\|_1 \| D^\alpha \psi_N^{(\beta)} P^k u\|_0$$
$$\leq \left(A_1^{|\beta|+2k+1} B_1^{2r(|\alpha|+(2r)^{-1})+1} N^{s[r(|\alpha|+r^{-1})+|\beta|+2k]} \right)^2 \times A_1^{2r} B_1^{-3r-2} N^{-s}.$$

So, by the above considerations and using the same strategies adopted, several times, previously to handle the multi-index sums, we conclude that

$$\sum_{\ell=1}^{m} |\langle [X_\ell, [X_\ell, D^\alpha \psi_N^{(\beta)}]] P^k u, D^\alpha \psi_N^{(\beta)} P^k u\rangle|$$
$$\leq \left(A_1^{|\beta|+2k+1} B_1^{2r(|\alpha|+(2r)^{-1})+1} N^{s[r(|\alpha|+r^{-1})+|\beta|+2k]} \right)^2$$
$$\times 2^{2n} n^2 C_1^4 A_1^{r+1} B_1^{-r-1} \left[2B_1^{-1} + 4A_1^{-1} B_1^{-1} + 2A_1^r B_1^{-r-1} + C_1^{-4} + C_1^{-1} A B^{-1} \right]. \tag{97}$$

Enlarging B_1 if necessary and taking it large enough compared to A_1, the second factor of (90), (91), (93), (94), (95), and (96) times $4\widetilde{C}^2$ can be made strictly less than 6^{-1}. We conclude that

$$\| X_\ell D^\alpha \psi_N^{(\beta)} P^k u\|_0 \leq A_1^{|\beta|+2k+1} B_1^{2r|\alpha|+1} N^{s[r|\alpha|+1+|\beta|+2k]},$$

where $|\alpha| = m_0 + 1$.

By induction we have obtained that (60) is true for $m \in \mathbb{N}$ and $(\alpha, \beta, k) \in \mathbb{N}^{2n+1}$ such that $2r|\alpha| + (|\beta| - M) + 2k \leq N$.

This concludes the proof of the proposition.

Corollary 4.1 *The following estimate holds*

$$\| D^\alpha \psi_N u\|_0 \leq B^{|\alpha|+1} N^{sr|\alpha|}, \tag{98}$$

where B is independent of α and N.

We recall the following result due to Hörmander [34].

Proposition 4.3 *Let* $x_0 \in \Omega_0 \subset \mathbb{R}^n$ *and* $u \in \mathscr{D}'(\Omega)$. *Then* $u \in G^s$, $s \geq 1$, *in a neighborhood of* x_0 *if and only if for some* Ω_1 *of* x_0 *there is a sequence* $u_N \in \mathscr{E}'(\Omega_0)$ *which is equal to* u *in* Ω_1 *and satisfies the estimates*

$$|\widehat{u}_N(\xi)| \leq C \left(CN^s/|\xi|\right)^N, \qquad |\xi| \geq 1, \qquad N = 1, 2, \ldots, \tag{99}$$

for some constant C.

Setting $u_N = \psi_N u$ and $N = 2r|\alpha|$ the estimate (98) is equivalent to (99) in the above Proposition, we conclude that $u \in G^{sr}(\Omega_0)$. We obtained Theorem 2.1.

References

1. P. Albano, A. Bove, G. Chinni, *Minimal Microlocal Gevrey Regularity for "Sums of Squares"* , Int. Math. Res. Notices, **12** (2009), 2275-2302.
2. M. S. Baouendi and G. Métivier, *Analytic vectors of hypoelliptic operators of principal type*, Amer. J. Math. **104**(2)(1982), 287–319.
3. R.F. Barostichi, P.D. Cordaro and G. Petronilho, *Analytic vectors in locally integrable structures*. Geometric analysis of several complex variables and related topics, pp. 1–14, Contemp. Math., **550**, Amer. Math. Soc., Providence, RI, 2011.
4. S. Berhanu and J. Hounie, *A Class of FBI Transforms* Commun. Partial Diff. Eq. **37**(1) (2012), 38–57.
5. S. Berhanu and A. Hailu, *Characterization of Gevrey regularity by a class of FBI transforms*, In: Pesenson I., Le Gia Q., Mayeli A., Mhaskar H., Zhou DX. (eds) Recent Applications of Harmonic Analysis to Function Spaces, Differential Equations, and Data Science; Applied and Numerical Harmonic Analysis; Birkhäuser, Cham (2017), 451–482.
6. P. Bolley and J. Camus, *Regularite Gevrey et iteres pour une classe d'operatours hypoelliptiques* Commun. Partial Diff. Eq. **6**(10) (1981), 1057–1110.
7. P. Bolley, J. Camus and L. Rodino, *Hypoellipticité Analytique-Gevrey et Iterés d'opérateurs*, Rend. Sem. Mat. Univ. Politecn. Torino, **45** (1987) 1–61.
8. A.Bove and M.Mughetti *Minimal Gevrey Regularity for Hörmander Operators*, International Mathematics Research Notices, 2023; https://doi.org/10.1093/imrn/rnad055.
9. A. Bove and G. Chinni *On a class of globally analytic hypoelliptic sums of squares*, J. Diff. Eq., **327**(2022), 109–126.
10. A. Bove, G. Chinni and M. Mughetti, *On a class of sums of squares related to Hamiltonians with a non periodic magnetic field*, Journal of Mathematical Analysis and Applications, **526**(2)(2023), doi: https://doi.org/10.1016/j.jmaa.2023.127303.
11. N. Braun Rodrigues, G. Chinni, P. D. Cordaro and M. R. Jahnke, *Lower order perturbation and global analytic vectors for a class of globally analytic hypoelliptic operators*, Proc. Amer. Math. Soc. **144** (2016), no. 12, 5159–5170.
12. J.E. Castellanos, P.D. Cordaro and G. Petronilho, *Gevrey vectors in involutive tube structures and Gevrey regularity for the solutions of certain classes of semilinear systems*, JAMA **119**(2013), 333–364.
13. G. Chinni and P. D. Cordaro, *On global analytic and Gevrey hypoellipticity on the torus and the Métivier inequality*, Communications in P.D.E. **42**(1) (2017), 121–141.
14. G. Chinni, *On the Gevrey regularity for sums of squares of vector fields, study of some models*, Journal of Differential Equations **265** (3) (2018), 906–920.

15. G. Chinni, *On the sharp Gevrey regularity for a generalization of the Métivier operator*, Math. Ann. (2023). https://doi.org/10.1007/s00208-022-02558-7.
16. G. Chinni, *On the partial and microlocal regularity for generalized Métivier operators*, J. Pseudo-Differ. Oper. Appl. 14, 38 (2023). https://doi.org/10.1007/s11868-023-00534-6.
17. G. Chinni and M. Derridj, *On the microlocal regularity of analytic vectors for "sums of squares" of vector fields*, Math. Z. **302**, 1983–2003 (2022). https://doi.org/10.1007/s00209-022-03129-x
18. G. Chinni and M. Derridj, *On the Microlocal Regularity of the Gevrey Vectors for second order partial differential operators with non negative characteristic form of first kind*, Preprint.
19. M. Damlakhi and B. Helffer, *Analyticité et itères d'un système de champs non elliptique*, Ann. scient. Éc. Norm. Sup., 4e série, 13 (1980), 397–403.
20. M. Derridj and C. Zuily, *Régularité analytique et Gevrey d'opérateurs elliptiques dégénérés*, J. Math. Pures Appl. **52** (1973), 65–80.
21. M. Derridj and C. Zuily, *Sur la régularité Gevrey des opérateurs de Hörmander*, J. Math. Pures Appl. **52** (1973), 309–336.
22. M. Derridj, *On the Gevrey vectors for some partial differential operators*, Complex Var. Ellip. Equ. **62**(10) (2017), 1474–1491.
23. M. Derridj, *On Gevrey vectors of L. Hörmander's operators* Trans. Amer. Math. Soc. 372 (2019), 3845–3865.
24. M. Derridj, *Local estimates for Hörmander's operators with Gevrey coefficients and application to the regularity of their Gevrey vectors* Tunisian Journal of Mathematics, Vol. **1**, No. 3 (2019), 321–345.
25. M. Derridj, *Local estimates for Hörmander's operators of first kind with analytic Gevrey coefficients and application to the regularity of their Gevrey vectors* Pacific J. Math., Vol. **302**, (2019), 511–543.
26. M.Derridj, *Gevrey regularity of Gevrey vectors of second order partial differential operators with non negative characteristic form*, Complex Anal Synerg **6**, 10 (2020), https://doi.org/10.1007/s40627-020-00047-8.
27. A. Grigis and J. Sjöstrand, *Front d'onde analytique et somme de carrès de champs de vecteurs*, Duke Math. J. **52** (1985), 35–51.
28. Ehrenpreis, L., *Solution of Some Problems of Division. Part IV. Invertible and Elliptic Operators*, American Journal of Mathematics, vol. **82**,no. 3, 1960, 522–588.
29. S. Fürdös *The Kotake-Narasimhan Theorem in general ultradifferentiable classes* (Preprint) https://arxiv.org/abs/2212.11905.
30. B. Helffer and Cl. Mattera, *Analyticite et iteres reduits d'un systeme de champs de vecteurs*, Communications in Partial Differential Equations, Vol. **5**, No. 10 (1980), 1065–1072.
31. G. Hoepfner and R. Medrado, *The FBI transforms and their use in microlocal analysis*, Journal of Functional Analysis, **275**(5), 1208–1258(2018).
32. G. Hoepfner and A. Raich, *Global L^q Gevrey Functions, Paley-Weiner Theorems, and the FBI Transform*, Indiana Univ. Math. J. **68**(3) (2019), 967–1002.
33. L. Hörmander, *Hypoelliptic second order differential equations*, Acta Math. **119** (1967), 147-171.
34. L. Hörmander, *Uniqueness Theorems and Wave Front Sets for Solutions of Linear Differential Equations with Analytic Coefficients*, Communications Pure Appl. Math. **24** (1971), 671–704.
35. L. Hörmander *The Analysis of Linear Partial Differential Operators I*, Springer-Verlag Berlin Heidelberg (1990).
36. L. Hörmander *The Analysis of Linear Partial Differential Operators III*, Springer-Verlag Berlin Heidelberg (1983).
37. T. Kotake and M. Narasimhan, *Regularity theorems for fractional powers of a linear elliptic operator*, Bull. Soc. Math. France **90** (1962), 449–471
38. G. Métivier, *Propriété des itérés et ellipticité*, Comm. Partial Differential Equations **3**(9) (1978), 827–876.
39. E. Nelson, *Analytic vectors*, Ann. Math. **70**(3), (1959), 572–615.

40. O.A. Oleĭnik and E.V. Radkevič, *Second order equations with nonnegative characteristic form* Itogi Nauki. Ser. Matematika. Mat. Anal. 1969 (Russian) 1971, 7–252.
41. E.V. Radkevič, *Equations with nonnegative characteristic form. I.* J. Math. Sci. **158**, 297–452 (2009).
42. L. Preiss Rothschild and E. M. Stein, *Hypoelliptic differential operators and nilpotent groups*, Acta Math. **137** (1976), 247-320.
43. J. Sjöstrand, *Singularités analytiques microlocales*, Astérisque **95** (1982).
44. F. Treves, *Analytic Partial Differential Equations*, Grundlehren der Mathematischen Wissenschaften, vol.**359**, Springer Verlag, 2022.

Bergman Logarithmically Flat and Obstruction Flat Hypersurfaces and Their CR Structures

Peter Ebenfelt and Ming Xiao

This paper is dedicated to Jorge Hounie on the occasion of his 75th birthday anniversary.

1 Introduction

Let $\Omega \subset \mathbb{C}^n$ be a bounded, strictly pseudoconvex domain with C^∞-smooth boundary $\partial\Omega$. Two important geometric objects associated to Ω are the (unique) complete Kähler–Einstein metric g_{KE} with Ricci curvature -1 and the Bergman metric g_B, which is also complete and Kähler. The boundary behaviors of the two metrics are intimately related to the CR geometry of the boundary $\partial\Omega$. The two notions of flatness that we shall consider in this chapter concern the vanishing of an obstruction to C^∞-smoothness up to the boundary of a log potential of the respective metrics. In what follows, the abbreviated terminology "flat CR hypersurface" refers to a strictly pseudoconvex CR hypersurface that is flat in either of these two senses.

In Sect. 2, we first recall the two notions of Bergman logarithmic flatness and obstruction flatness and collect some intriguing open questions. After that, we recall some known results about the two types of such flat CR hypersurfaces. We also establish some new results along the way in our discussion. The 3-dimensional flat CR hypersurfaces are much better understood than higher-dimensional ones. In Sect. 2.4, we discuss some of the known results in the 3-dimensional case. For example, every 3-dimensional Bergman logarithmically flat hypersurface is spherical, which means that their CR structures are locally equivalent. This is

The first author was supported in part by the NSF grant DMS-2154368. The second author was supported in part by the NSF grant DMS-2045104.

P. Ebenfelt (✉) · M. Xiao
Department of Mathematics, University of California at San Diego, La Jolla, CA, USA
e-mail: pebenfelt@ucsd.edu; m3xiao@ucsd.edu

S. Berhanu et al. (eds.), *Geometric Analysis of PDEs and Several Complex Variables*, Latin American Mathematics Series – UFSCar subseries,
https://doi.org/10.1007/978-3-031-69702-9_6

not true for obstruction flat ones, but *compact* 3-dimensional obstruction flat hypersurfaces are expected to be spherical. The latter is sometimes referred to as the Strong Ramadanov Conjecture and is still open.

In Sect. 3, we survey and discuss existence results for flat CR hypersurfaces, noncompact and compact, that are nonspherical in higher dimensions. It remains, however, an interesting question to determine "how many" Bergman logarithmically and obstruction flat (and nonspherical) CR hypersurfaces there are with distinct local CR structures. The question makes sense as the two notions of flatness only depend on the local CR geometry of the hypersurface. A *countably* infinite family of compact such hypersurfaces was constructed in [37] for each real dimension $2n+1$ with $n \geq 2$. The most significant new results of this chapter are probably the ones in Sect. 4. Among other results, we construct an *uncountably* infinite family of flat (in both senses) CR hypersurfaces $\{M_\alpha\}$ in $\mathbb{C}^{n+1}$, for each $n \geq 2$, such that each M_α is homogeneous and has transverse symmetry (but is noncompact) and any pair M_α, M_β with $\alpha \neq \beta$ has distinct local CR structures.

2 Background

2.1 The Kähler–Einstein Metric

Let $\Omega \subset \mathbb{C}^n$ be as in Sect. 1. The (unique) complete Kähler–Einstein metric g_{KE} on Ω with Ricci curvature -1 is given by

$$(g_{KE})_{\alpha\bar{\beta}} = (\log u^{-(n+1)})_{z_\alpha \overline{z_\beta}}, \tag{1}$$

where u is the unique solution in Ω of the Dirichlet problem

$$\begin{cases} J(u) := (-1)^n \det \begin{pmatrix} u & u_{\overline{z_\beta}} \\ u_{z_\alpha} & u_{z_\alpha \overline{z_\beta}} \end{pmatrix} = 1 & \text{in } \Omega, \\ u = 0 & \text{on } \partial\Omega. \end{cases} \tag{2}$$

The uniqueness of a solution u was established by Fefferman [22], who also showed that there are approximate solutions $\rho \in C^\infty(\overline{\Omega})$ to this Dirichlet problem. More precisely, he showed that there are smooth functions ρ in $\overline{\Omega}$ such that $J(\rho) = 1 + O(\rho^{n+1})$, $\rho > 0$ in Ω and $\rho = 0$ on $\partial\Omega$, and any such approximate solution ρ is unique modulo $O(\rho^{n+2})$. Such approximate solutions are called *Fefferman defining functions.*

The existence of an actual solution $u \in C^\infty(\Omega) \cap C^{n+3/2-\epsilon}(\overline{\Omega})$ to (2) was proved by Cheng–Yau [8]. The boundary regularity of u was improved to $C^{n+2-\epsilon}(\overline{\Omega})$ by Lee–Melrose [29], who demonstrated that u has the asymptotic expansion

$$u \sim \rho \sum_{k=0}^{\infty} \eta_k (\rho^{n+1} \log \rho)^k, \quad \eta_k \in C^\infty(\overline{\Omega}), \tag{3}$$

where ρ is a Fefferman defining function for Ω. Graham [23, 24] showed that the coefficients η_k, for $k \geq 1$, are *local* CR invariants of the boundary $\partial\Omega$ modulo $O(\rho^{n+1})$. In particular, the function $\eta_1|_{\partial\Omega}$ is a local CR invariant of $\partial\Omega$, and Graham (*op. cit.*) also proved that, in fact, $\eta_1|_{\partial\Omega} = 0$ if and only if the Cheng–Yau solution u is $C^\infty(\overline{\Omega})$ (which implies that $\eta_k = 0$ in $\overline{\Omega}$ for all $k \geq 1$). The local CR invariant $\eta_1|_{\partial\Omega}$ is called the *obstruction function* and will be denoted by $\mathcal{O}$. We shall say that $\partial\Omega$ is *obstruction flat* if $\mathcal{O} = 0$. The prototype of an obstruction flat hypersurface is the unit sphere or, more generally, any spherical hypersurface, i.e., a hypersurface that is locally CR diffeomorphic to a piece of the unit sphere.

We observe that since the obstruction function $\mathcal{O}$ is determined by the local CR geometry of the strictly pseudoconvex hypersurface $\partial\Omega$, we may define this invariant on a strictly pseudoconvex hypersurface X in a complex manifold in an obvious way. The notion of obstruction flatness then also carries over to strictly pseudoconvex hypersurfaces X in a complex manifold.

2.2 The Bergman Metric

The Bergman metric in Ω is given by

$$(g_B)_{\alpha\bar{\beta}} = (\log K)_{z_\alpha \overline{z_\beta}}, \tag{4}$$

where $K = K(z, \bar{z})$ denotes the Bergman kernel of Ω (the reproducing kernel of the subspace of holomorphic functions in $L^2(\Omega)$) on the diagonal. Fefferman [21] showed that K has an asymptotic expansion

$$K = \frac{\phi}{\rho^{n+1}} + \psi \log \rho, \quad \phi, \psi \in C^\infty(\overline{\Omega}), \tag{5}$$

where we take ρ to be a Fefferman defining function of Ω. The coefficients ϕ mod $O(\rho^{n+1})$ and ψ mod $O(\rho^\infty)$ are determined by the *local* CR geometry of $\partial\Omega$; see [21, 2, 26]. It is further known that $\phi = n!/\pi^n + O(\rho^2)$. We say that $\partial\Omega$ is *Bergman logarithmically flat* if the log coefficient ψ vanishes to infinite order (i.e., $\psi = 0$ mod $O(\rho^\infty)$). By the localization property of the Bergman kernel (cf. [21, 5, 19, 27]), Bergman logarithmic flatness only depends on the local CR geometry of the boundary, and we can define Bergman logarithmic flatness for any germ of strongly pseudoconvex CR hypersurface M in $\mathbb{C}^m$. We say M is Bergman logarithmically flat if, for some (and therefore every, by the localization property of the Bergman kernel) smoothly bounded strongly pseudoconvex domain G with $M \subseteq \partial G$, the coefficient function ψ in the Fefferman expansion (5) of K_G vanishes to the infinite order along M. Furthermore, a CR hypersurface Σ (in a complex manifold) is called Bergman logarithmically flat if it is so as a germ of CR hypersurface at every $p \in \Sigma$. The prototype of a Bergman logarithmically flat hypersurface is the unit sphere or any spherical hypersurface.

In order to observe an analogy with the construction of the Kähler–Einstein metric above, we introduce the function

$$v = K^{-\frac{1}{n+1}}, \tag{6}$$

which is $C^\infty(\Omega) \cap C^{n+2-\epsilon}(\overline{\Omega})$ and $v = 0$ on $\partial\Omega$. In fact, we have an asymptotic expansion of the same type as that for the Cheng–Yau solution u,

$$v \sim \rho \sum_{k=0}^{\infty} \xi_k (\rho^{n+1} \log \rho)^k, \quad \xi_k \in C^\infty(\overline{\Omega}), \tag{7}$$

where we may take, near the boundary $\partial\Omega$,

$$\xi_k = c_k \left(\frac{\psi}{\phi}\right)^k \phi^{-\frac{1}{n+1}} \tag{8}$$

for some combinatorial constants $c_k \neq 0$ with $c_0 = 1$. We note that in this case C^∞-smoothness up to $\partial\Omega$ is obstructed by the coefficient ψ of the log term in Fefferman's asymptotic expansion of K, as ξ_k, for $k \geq 1$, vanishes to infinite order at the boundary if and only if ψ does (cf. [23]). Thus, the function v is $C^\infty(\overline{\Omega})$ if and only if ψ vanishes to infinite order at $\partial\Omega$. Moreover, in analogy with the Kähler–Einstein metric above, we can express the Bergman metric as

$$(g_B)_{\alpha\bar{\beta}} = (\log v^{-(n+1)})_{z_\alpha \overline{z_\beta}}. \tag{9}$$

While the Cheng–Yau solution u and the function v associated with the Bergman kernel have similar asymptotic expansions at $\partial\Omega$, a major difference between the two functions is that u solves the partial differential equation $J(u) = 1$, whereas v does not (in any apparent way) come from a differential equation. In fact, we have the following result, which is a slight sharpening of a key step in Huang–Xiao's resolution [28] of the Cheng Conjecture.

Proposition 2.1 *Let $\Omega \subset \mathbb{C}^n$, with $n \geq 3$, be a bounded, strictly pseudoconvex domain with C^∞-smooth boundary $\partial\Omega$, and let v be the function given by* (6). *Then,*

$$J(v) = \frac{\pi^n}{n!} + O(\rho^2) \tag{10}$$

and the following are equivalent:

(i) It holds that

$$J(v) = \frac{\pi^n}{n!} + o(\rho^2). \tag{11}$$

(ii) It holds that

$$J(v) = \frac{\pi^n}{n!} + O(\rho^{n+2}). \tag{12}$$

(iii) $\partial\Omega$ is spherical.

Remark 2.2 The implication (i) $\Longrightarrow$ (iii) is contained in [34], but in terms of the so-called Bergman function instead of $J(v)$; see the remark at the end of this subsection.

Proof of Proposition 2.1 First, we recall the following result that follows from Fefferman's calculation on pp. 399–400 in [22]:

Lemma 2.3 *Let χ and η be C^2-smooth functions in a neighborhood of $\partial\Omega$ such that $\chi = 0$ on $\partial\Omega$. Then, for any $s \geq 2$, $v = \chi + \eta\chi^s$ satisfies*

$$J(v) = \left(1 + (n+2-s)s\eta\chi^{s-1}\right) J(\chi) + O(\chi^s). \tag{13}$$

By repeating this calculation, *mutatis mutandi*, with $v = \chi + \eta\chi^s \log\chi$ instead of $v = \chi + \eta\chi^s$, we also obtain the following:

Lemma 2.4 *Let χ and η be C^2-smooth functions in a neighborhood of $\partial\Omega$ such that $\chi = 0$ on $\partial\Omega$. Then, for any $s \geq 3$, $v = \chi + \eta\chi^s \log\chi$ satisfies*

$$J(v) = \left(1 + (n+2-s)\eta\chi^{s-1} s \log\chi + (n+2-2s)\eta\chi^{s-1})\right) J(\chi) + O(\chi^s \log\chi). \tag{14}$$

We shall now apply Lemma 2.4 to the function v given by (6). We observe that, by the asymptotic expansion (7), there is a function $\tilde{\eta} \in C^{n+1-\epsilon}$ near $\partial\Omega$ such that $v = \rho\xi_0 + \tilde{\eta}\rho^{n+2}\log\rho$. We note at any point $p \in \partial\Omega$, and it holds that $\rho(p) = 0$, $\xi_0(p) = c := (\frac{n!}{\pi^n})^{-\frac{1}{n+1}}$ and $\tilde{\eta}(p) = a$ for some $a \in \mathbb{R}$. We want to rewrite this in the form $v = \chi + \eta\chi^{n+2}\log\chi$, where $\chi = \rho\xi$ for some $\xi \in C^{n+1-\epsilon}$ near $\partial\Omega$. By comparing the two expressions, we note that this can be done if we can solve the two equations

$$\begin{aligned} &\xi + \eta\xi^{n+2}\rho^{n+1}\log\xi = \xi_0 \\ &\eta\xi^{n+2} = \tilde{\eta}, \end{aligned} \tag{15}$$

for $\xi = \xi(\rho, \xi_0, \tilde{\eta})$ and $\eta = \eta(\rho, \xi_0, \tilde{\eta})$ near the point $(\rho, \xi_0, \tilde{\eta}; \xi, \eta) = (0, c, a; c, \frac{a}{c^{n+2}})$. One can readily verify the system (15) is nondegenerate in (ξ, η) at this point. Therefore a straightforward application of the Implicit Function Theorem guarantees that this is the case. Moreover, the functions $\xi = \xi(\rho, \xi_0, \tilde{\eta})$ and $\eta = \eta(\rho, \xi_0, \tilde{\eta})$ are smooth (in fact, real analytic) in the arguments $(\rho, \xi_0, \tilde{\eta})$; thus ξ, η becomes $C^{n+1-\epsilon}$ functions near $\partial\Omega$ after the composition. Moreover, by the first equation in (15), we see $\xi = \xi_0 + O(\rho^{n+1})$. We now apply Lemma 2.4 to the function $v = \chi + \eta\chi^{n+2}\log\chi$ with $\chi = \rho\xi$ as above. Lemma 2.4 implies that

$$J(v) = J(\rho\xi) + O(\rho^{n+1}). \tag{16}$$

Since $\rho\xi = \rho\xi_0 + O(\rho^{n+2})$, Lemma 2.3 with $s = n+2$ then implies $J(\rho\xi) = J(\rho\xi_0) + O(\rho^{n+2})$. Consequently,

$$J(v) = J(\rho\xi_0) + O(\rho^{n+1}). \tag{17}$$

If we introduce the function

$$P_2 = \frac{\phi - n!/\pi^n}{\rho^2},$$

then we can write $\phi = n!/\pi^n + P_2\rho^2$, which implies that $\rho\xi_0 = \rho\phi^{-\frac{1}{n+1}} = \chi + \eta\chi^3$, where

$$\chi = \left(\frac{\pi^n}{n!}\right)^{1/(n+1)}\rho$$

and where

$$\eta|_{\partial\Omega} = -\frac{1}{n+1}\left(\frac{\pi^n}{n!}\right)^{(n-1)/(n+1)} P_2|_{\partial\Omega}. \tag{18}$$

If we apply Lemma 2.3 with $v = \rho\xi_0 = \chi + \eta\chi^3$, then we conclude that

$$J(\rho\xi_0) = (1 + 3(n-1)\eta\chi^2)J(\chi) + O(\rho^3) = (1 + 3(n-1)\eta\chi^2)\frac{\pi^n}{n!} + O(\rho^3), \tag{19}$$

since $J(\chi) = \pi^n/n!J(\rho) = \pi^n/n! + O(\rho^{n+1})$ and $n \geq 3$. Here we have used the fact that ρ is a Fefferman defining function. The conclusion (10) follows from (17) and (19).

It remains to establish the equivalence of the statements (i)–(iii). We first prove (i) $\Longrightarrow$ (iii). Equations (17) and (19) together imply that (11) holds if and only if $\eta|_{\partial\Omega} = 0$, which by (18) is equivalent to $P_2|_{\partial\Omega} = 0$. On the other hand, $P_2|_{\partial\Omega}$ is known to be a nonzero constant multiple of $\|S\|^2$, where $S = S_{\alpha\bar\beta\nu\bar\mu}$ is the Cartan–Chern–Moser CR curvature tensor of $\partial\Omega$ and $\|S\|^2 = S_{\alpha\bar\beta\nu\bar\mu}S^{\alpha\bar\beta\nu\bar\mu}$ (cf. [10, 23, 28]). Since being spherical can be characterized by the vanishing of $\|S\|^2$, the implication (i) $\Longrightarrow$ (iii) of the proposition is proved.

To prove (iii) $\Longrightarrow$ (ii), we recall that ϕ mod $O(\rho^{n+1})$ and ψ (mod $O(\rho^\infty)$) are determined by the local CR geometry of $\partial\Omega$. Thus, if $\partial\Omega$ is spherical, then

$\phi = n!/\pi^n + O(\rho^{n+1})$ and $\psi = 0$ to infinite order at $\partial\Omega$. Moreover, the obstruction function $\mathcal{O} = 0$, which implies that the Cheng–Yau solution u is $C^\infty(\overline{\Omega})$. Thus, the Cheng–Yau solution u is a Fefferman defining function. Taking $\rho = u$ in (5), by (6), we get

$$v = u\phi^{-\frac{1}{n+1}} + O(u^\infty) = u\left(\frac{\pi^n}{n!}\right)^{\frac{1}{n+1}} + O(u^{n+2}).$$

The conclusion (12) now follows from Lemma 2.3 with $s = n + 2$.

To complete the proof of the equivalence of (i)–(iii) and, hence, that of the proposition, we simply note that (ii) obviously $\Longrightarrow$ (i). □

We end this subsection with some remarks that are somewhat tangential to the main theme of this chapter. We recall that, for $n \geq 3$, a boundary point $p \in \partial\Omega$ is said to *CR umbilical* if the CR curvature tensor $S_{\alpha\bar{\beta}\nu\bar{\mu}}$ vanishes at that point. Thus, being spherical is equivalent to being CR umbilical in an open neighborhood. We observe that the proof of Proposition 2.1 readily yields the following pointwise result.

Proposition 2.5 *Let $\Omega \subset \mathbb{C}^n$, with $n \geq 3$, be a bounded, strictly pseudoconvex domain with C^∞-smooth boundary $\partial\Omega$ and let v be the function given by* (6). *Then, a point $p \in \partial\Omega$ is CR umbilical if and only if*

$$\lim_{z\to p} \rho(z)^{-2}\left(J\,(v(z)) - \frac{\pi^n}{n!}\right) = 0. \tag{20}$$

For our second remark, we observe that if (11) holds, then the Bergman metric is *asymptotically Kähler–Einstein* (with Ricci equal to -1) in the sense that $Ric(g_B) = -g_B + o(1)$, i.e., every component of the tensor $Ric(g_B) + g_B$ is $o(1)$ as $z \to \partial\Omega$. The converse can also be verified. Indeed, a pointwise result along the lines of Proposition 2.5 can be established using the same ideas as in the proof of Proposition 2.1.

Proposition 2.6 *Let $\Omega \subset \mathbb{C}^n$, with $n \geq 3$, be a bounded, strictly pseudoconvex domain with C^∞-smooth boundary $\partial\Omega$. Then, a point $p \in \partial\Omega$ is CR umbilical if and only if*

$$\lim_{z\to p} \left(Ric(g_B)(z) + g_B(z)\right) = 0. \tag{21}$$

Proof It follows easily from the identity $Ric(g) = -(\log G)_{z_\alpha\overline{z_\beta}}$, where $G = \det g$, and the well-known formula

$$\det[(-\log f)_{z_\alpha\overline{z_\beta}}] = \frac{J(f)}{f^{n+1}} \tag{22}$$

that the components of the tensor $Ric(g_B)(z) + g_B(z)$ can be expressed as

$$(Ric(g_B)(z) + g_B(z))_{\alpha\bar{\beta}} = -(\log J(v))_{z_\alpha \overline{z_\beta}}. \tag{23}$$

It follows from (17) and (19) in the proof of Proposition 2.1 that

$$J(v) = (1 + 3(n-1)\eta\chi^2)\frac{\pi^n}{n!} + O(\rho^3), \tag{24}$$

where χ and η are as in that proof. It follows that there is a C^2-smooth function $\tilde{P}_2$ such that

$$\log J(v) = \rho^2 \tilde{P}_2. \tag{25}$$

Moreover, in view of (18) and the arguments in the proof of Proposition 2.1, it holds that $\tilde{P}_2|_{\partial\Omega}$ is a nonzero constant times the norm squared of the CR curvature, $\|S\|^2$. Thus, (23) implies that

$$\lim_{z \to p} (Ric(g_B)(z) + g_B(z))_{\alpha\bar{\beta}} = c\rho_{z_\alpha}(z)\rho_{\overline{z_\beta}}(p)\|S(p)\|^2, \quad \forall\, 1 \le \alpha, \beta \le n, \tag{26}$$

for some nonzero constant c. The proposition then follows from the above identity. □

As our final remark, we note that by using the identity (22), it is not difficult to show that

$$J(v) = \frac{1}{(n+1)^n}\frac{G}{K},$$

where $G = \det g_B$ as above. The function $B = G/K$ is sometimes called the *Bergman function*. This is the function referred to in Remark 2.2.

2.3 The Ramadanov Conjecture and an Analogous Question for Obstruction Flat Boundaries

We first recall the Ramadanov Conjecture [35]:

Ramadanov Conjecture *If $\Omega \subset \mathbb{C}^n$, $n \ge 2$, is a bounded, strictly pseudoconvex domain with C^∞-smooth boundary $\partial\Omega$ and if $\partial\Omega$ is Bergman logarithmically flat, then $\partial\Omega$ is spherical.*

The Ramadanov Conjecture has been established in $\mathbb{C}^2$ by the work of Burns, Graham, and Boutet de Monvel [24, 4]. Indeed, in this case it suffices that $\psi =$

$O(\rho^2)$ for $\partial\Omega$ to be spherical. The Ramadanov Conjecture is still open in $\mathbb{C}^n$ with $n \geq 3$, but it is known to fail if we instead consider domains Ω in more general complex manifolds [20] or Stein spaces [17] as long as the complex dimension n of the space is at least 3.

We observe that the Ramadanov Conjecture can be formulated in terms of the flatness of the first log coefficient ξ_1 in the asymptotic expansion of the function v introduced in (6), since ξ_1 is flat if and only if ψ is flat. As observed above, flatness of ψ is also equivalent to C^∞-smoothness up to the boundary of v.

One may ask the same question in the context of the Kähler–Einstein metric. Asking for C^∞-smoothness up to the boundary of the Cheng–Yau solution is, as mentioned above, equivalent to obstruction flatness. Thus, the analogue of the Ramadanov Conjecture for the Cheng–Yau solution would be:

Obstruction Flatness Conjecture *If $\Omega \subset \mathbb{C}^n$, $n \geq 2$, is a bounded, strictly pseudoconvex domain with C^∞-smooth boundary $\partial\Omega$ and if $\partial\Omega$ is obstruction flat, then $\partial\Omega$ is spherical.*

The Obstruction Flatness Conjecture is open in all dimensions $n \geq 2$, although there is ample evidence for its correctness when $n = 2$. For example, the first author and Curry have shown that in $\mathbb{C}^2$ the conjecture holds if one assumes in addition that $\partial\Omega$ has an infinitesimal CR symmetry [15, 13] or, more generally, an approximate infinitesimal automorphism [11]. We recall that $\partial\Omega$ has an infinitesimal CR symmetry if and only if there exists a holomorphic vector field near $\partial\Omega$ whose real part is tangent along $\partial\Omega$. Possessing an *approximate* infinitesimal automorphism is equivalent to the existence of a holomorphic vector field such that its real part is, roughly speaking, more tangent than normal along $\partial\Omega$; the reader is referred to [11] for the precise definition.

2.4 The Two-Dimensional Case

For domains $\Omega \subseteq \mathbb{C}^2$ (or more generally in a complex surface), there is a precise and direct relation between the two notions of flatness. Graham [24] has shown that in this case, the obstruction function $\mathcal{O}$ equals a nonzero constant times the restriction to $\partial\Omega$ of the log term coefficient ψ of the Bergman kernel K. Thus, Bergman logarithmic flatness implies obstruction flatness. We recall that in $\mathbb{C}^2$ (or in a complex surface), the Ramadanov Conjecture holds and, in fact, a stronger statement holds: If $\psi = O(\rho^2)$ along the boundary $\partial\Omega$, then $\partial\Omega$ is spherical. The Obstruction Flatness Conjecture can then be reformulated as asserting the stronger statement that $\psi = O(\rho)$ implies that $\partial\Omega$ is spherical. Hence, in $\mathbb{C}^2$ the Obstruction Flatness Conjecture has sometimes been called the *Strong Ramadanov Conjecture*.

For domains $\Omega \subset \mathbb{C}^2$, a *local* version of the Ramadanov Conjecture holds. Indeed, if $\psi = O(\rho^2)$ along an open piece U of the boundary $\partial\Omega$, then U is spherical. This follows from the proof of this conjecture in [24, 4]. The corresponding local version of the Obstruction Flatness Conjecture (a.k.a. the Strong Ramadanov Conjecture) is false, however, as was first demonstrated by Graham [23]. In fact, the following, which is a consequence of Theorem 3.2 in [16] and Chern–Moser's normal forms theory [9], holds true:

Theorem 2.7 *There are uncountably infinite families of obstruction flat strictly pseudoconvex hypersurfaces in $\mathbb{C}^2$ that are pairwise locally CR inequivalent.*

Proof Recall from [9, pp. 246–247] that a real analytic hypersurface X in $\mathbb{C}^2$ with a non-umbilical point $p_0 \in X$ can be put in a special Chern–Moser normal form at $p_0 = (0,0)$ as follows:

$$\operatorname{Im} w = |z|^2 + 2\operatorname{Re} z^4\bar{z}^2\left(1 + A^0_{5\bar{2}}z + iA^1_{4\bar{2}}\operatorname{Re} w\right) + O(8), \tag{27}$$

where $A^0_{5\bar{2}} \in \mathbb{C}$, $A^1_{4\bar{2}} \in \mathbb{R}$ are such that $(A^0_{5\bar{2}})^2$ and $A^1_{4\bar{2}}$ are invariants, and $O(8)$ denotes terms that are of (unweighted) order at least 8 in $(z, \bar{z}, \operatorname{Re} w)$. By the proof of Theorem 3.2 in [16], we may pose a Cauchy problem for the totally characteristic partial differential operator K from [23], associated with obstruction flat hypersurfaces, with the data

$$\rho = \operatorname{Im} w - |z|^2 - 2\operatorname{Re} z^4\bar{z}^2\left(1 + A^0_{5\bar{2}}z + iA^1_{4\bar{2}}\operatorname{Re} w\right) \tag{28}$$

on the real hyperplane $\operatorname{Re} z = 0$, and produce, via the Cauchy–Kowalevski Theorem, a unique obstruction flat hypersurface whose Chern–Moser normal form at 0 is of the special form (27). Clearly, using this construction, the existence of families of obstruction flat hypersurfaces as in the statement of Theorem 2.7 follows. □

It was also shown in [16] that there are nonspherical obstruction flat hypersurfaces in $\mathbb{C}^2$ with transverse infinitesimal CR symmetries, which implies that the local versions of the positive results on the Obstruction Flatness Conjecture in [15, 13, 11] do not hold.

2.5 Flat Rigidity of the Unit Sphere

While both the Ramadanov Conjecture in $\mathbb{C}^n$, for $n \geq 3$, and the Obstruction Flatness Conjecture, for all $n \geq 2$, are open, they are known to hold for small deformations of the unit sphere. To describe this more precisely, we let $Z_1, \ldots, Z_n$ denote a local frame, near some point $p \in S^{2n-1}$, for the $(1,0)$-vector fields that define the standard CR structure on the unit sphere $S^{2n-1} = \partial\mathbb{B}^n$ in $\mathbb{C}^n$. Any abstract smooth CR structure on the sphere S^{2n-1} whose underlying contact structure is isotopic to the standard one is CR diffeomorphic to a deformation of the standard structure given, near p, by the vector fields

$$Z_\alpha^{(\mu)} = Z_\alpha + \mu_\alpha{}^{\bar\beta} Z_{\bar\beta}, \tag{29}$$

where $\mu = \mu_\alpha{}^{\bar\beta}$ is a smooth deformation tensor with L^∞-norm < 1; and we are using the summation convention. The deformation tensor $\mu = \mu_\alpha{}^{\bar\beta}$, defined locally in this way near p, defines a global deformation tensor, i.e., a global section of $(T^{1,0}S^{2n-1})^* \otimes T^{1,0}S^{2n-1}$. Since the underlying contact structure for these deformations is fixed (to be the standard one on S^{2n-1}), the deformation tensor μ completely determines the CR structure, and we shall denote the deformed structure by (S^{2n-1}, μ).

We shall be concerned with CR structures that are close to the standard structure $(S^{2n-1}, 0)$. More precisely, we shall require the C^k-norm of μ, $\|\mu\|_k$ for a sufficiently large k (e.g., $k \geq 6$ suffices), to be small; in particular, the CR structure (S^{2n-1}, μ) will remain strictly pseudoconvex. We observe a fundamental difference between the lowest dimensional case, $n = 2$, and $n \geq 3$. In the former case, most deformations (S^3, μ) are not CR diffeomorphic to a CR structure on S^3 obtained by deforming the unit sphere inside $\mathbb{C}^2$ (indeed, they are not even embeddable locally) and the ones that form a proper Fréchet submanifold inside the space of all deformations; the reader is referred to [7, 30, 12] for a discussion. We point out that even when the CR structure (S^3, μ) is not locally embeddable, the CR invariant $\mathcal{O}$ can still be defined, and, hence, the notion of obstruction flatness carries over to this situation. In this case, however, the connection with an ambient complete Kähler–Einstein metric is no longer valid, although a connection to a different ambient metric problem does exist; the reader is referred, e.g., to [14] for a discussion. The notion of Bergman logarithmically flat makes less sense for CR structures that are not locally embeddable.

In the case $n \geq 3$, it follows from Boutet de Monvel's CR embedding theorem [3] and Lempert's stability theorem [30] that *any* deformed structure (S^{2n-1}, μ) is CR diffeomorphic to one obtained by deforming the unit sphere inside $\mathbb{C}^n$.

By combining the results of Curry and the first author [13, 11, 11] (for the obstruction flat statement) and the resolution of the Ramadanov Conjecture (for the Bergman logarithmically flat statement) in the case $n = 2$ and that of Hirachi [25] in the case $n \geq 3$, one obtains the following:

Theorem 2.8 *Let (S^{2n-1}, μ), for $n \geq 2$, be a sufficiently small deformation of the standard CR structure $(S^{2n-1}, 0)$ of the unit sphere. If (S^{2n-1}, μ) is obstruction flat or Bergman logarithmically flat (in the locally embeddable case), then (S^{2n-1}, μ) is CR diffeomorphic to $(S^{2n-1}, 0)$.*

While Theorem 2.8 offers evidence of the validity of the Ramadanov Conjecture and the Obstruction Flatness Conjecture for domains in $\mathbb{C}^n$, we note that the conjectures can also be posed for smoothly bounded (relatively compact) strictly pseudoconvex domains Ω in more general complex manifolds and, in this context, both conjectures would fail when $n \geq 3$. The remainder of this chapter will focus on investigating this phenomenon more closely in the special context of disk bundles over Kähler manifolds.

3 Flat Hypersurfaces in Higher Dimension

In this section, we shall investigate the local CR geometry of compact, obstruction flat, strictly pseudoconvex hypersurfaces in higher dimensions. While the Obstruction Flatness Conjecture is still open in two dimensions, i.e., for 3-dimensional CR manifolds (embeddable in complex manifolds or not), there is ample evidence for its validity. If true, then the local CR geometry of 3-dimensional, compact, obstruction flat, strictly pseudoconvex hypersurfaces is trivial; they are all spherical. As mentioned above, however, both the Ramadanov and Obstruction Flatness Conjecture fail for boundaries of domains in complex manifolds of higher dimension. Hence, the next challenge is to understand the (local) CR equivalence classes of compact, obstruction flat, strictly pseudoconvex hypersurfaces. As a first step in this direction, we shall consider a special class of such, namely, unit circle bundles over Kähler manifolds.

3.1 The Setup

Let M be a Kähler manifold of dimension m, endowed with a Kähler metric g. The associated Kähler form will be denoted by ω. Let $L \to M$ be a negative holomorphic line bundle equipped with a Hermitian metric h "quantizing" (M, g); i.e., the metric g is induced by the curvature of $(L, h) \to M$ via

$$\omega = \frac{\sqrt{-1}}{2\pi}\partial\bar{\partial}\log h.$$

We shall denote the dimension of the complex manifold L by $n = m + 1$. The Ricci curvature $Ric(g)$ induces an endomorphism $Ric(g)^\sharp : TM \to TM$, given in a local coordinate chart $z = (z_1, \ldots, z_m)$ by

$$X^\alpha \mapsto R_\gamma{}^\alpha X^\gamma,$$

where

$$X = X^\gamma \partial/\partial z_\gamma, \qquad R_{\alpha\bar{\beta}} = -\frac{\partial^2}{\partial z_\alpha \overline{z_\beta}}\log G, \qquad G = \det g,$$

and where we raise and lower indices by the metric; e.g., $R_\alpha{}^\beta = g^{\beta\bar\mu} R_{\alpha\bar\mu}$ with $g^{-1} = g^{\alpha\bar\beta}$ denoting the inverse of the matrix $g = g_{\alpha\bar\beta}$.

We shall denote by $D(M, h)$ and $\Sigma(M, h)$ the unit disk bundle and the unit circle bundle, respectively, i.e.,

$$\begin{aligned} D(M,h) &= \{v \in L\colon |v|_h < 1\}, \\ \Sigma(M,h) &= \{v \in L\colon |v|_h = 1\}. \end{aligned} \tag{30}$$

When M and $(L, h) \to M$ are fixed, we shall abbreviate $D(M, h)$ and $\Sigma(M, h)$ by D and Σ, respectively. It is a well-known result of Grauert that Σ is a smooth strictly pseudoconvex hypersurface in L. If M is compact, then Σ is compact and comprises the boundary ∂D. If M is noncompact, then in general Σ is merely a proper subset of ∂D, and ∂D may be neither smooth nor strictly pseudoconvex.

3.2 Sufficient Conditions for Obstruction Flatness

The endomorphism $Ric(g)^\sharp$ has, pointwise, m real eigenvalues $\lambda_1 \le \cdots \le \lambda_m$ that form real-valued functions on M. We have the following sufficient condition for obstruction flatness from the authors' joint work with H. Xu [18]:

Theorem 3.1 *Let $(L, h) \to M$ be a negative Hermitian holomorphic line bundle over a Kähler manifold (M, g) such that the curvature of h induces the metric g as in Sect. 3.1. If the eigenvalues of the Ricci endomorphism $Ric(g)^\sharp$ are constant, then $\Sigma(M, h)$ is obstruction flat.*

Remark 3.2 The prototype of a situation (but not the only one) where the Ricci eigenvalues are constant is where (M, g) is Kähler–Einstein or, more generally, a product of Kähler–Einstein manifolds. If, for example, $M = M_1 \times \ldots \times M_k$ is a product of Kähler–Einstein manifolds (M_i, g^i) such that $Ric(g^i) = \mu_i g^i$, then the Ricci eigenvalues are $\mu_1, \ldots, \mu_k$ with each eigenvalue having multiplicity = dim M_i. Another prototype is when (M, g) is a homogeneous Kähler manifold.

3.3 Examples of Bergman Logarithmic Flatness

Recall the only Bergman logarithmically flat hypersurfaces in $\mathbb{C}^2$ are the spherical ones, by the work of Burns, Graham, and Boutet de Monvel [23, 4]. In higher dimensions, there exist noncompact Bergman logarithmically flat hypersurfaces in $\mathbb{C}^n$ that are not spherical. There are also compact examples realized as circle bundles $\Sigma(M, h)$ that arise from a negative line bundle (L, h) quantizing a compact Kähler manifold (M, g) as in (30). The first such examples were constructed by

Engliš and Zhang [20]. They showed that the corresponding circle bundle must be Bergman logarithmically flat if (M, g) is an irreducible Hermitian symmetric space of compact type. Furthermore, this circle bundle is spherical precisely when M is the projective space (note this fact also follows from the combined work of Webster [36] and Bryant [6], cf. Theorem 4.1). Later Loi–Mossa–Zuddas [32] proved that $\Sigma(M, h)$ is Bergman logarithmically flat if (M, g) is a compact simply connected homogeneous Kähler–Einstein manifold of classical type.

It is then natural to ask about the case when (M, g) is a Hermitian symmetric space of noncompact type, i.e., a bounded symmetric domain. We pause to recall some preliminaries on bounded symmetric domains. A Hermitian symmetric space of noncompact type can be realized as a convex and circular bounded domain containing 0 in some complex Euclidean space, via the Harish–Chandra realization, equipped with the (possibly normalized) Bergman metric. Such realizations will be called bounded symmetric domains. A bounded symmetric domain is called irreducible if it cannot be written as the product of two of such domains. We next recall the notion of generic norms. Let Ω_0 be an irreducible bounded symmetric domain and denote by $K(z, \overline{z})$ its Bergman kernel. Write V_{Ω_0} for the volume of Ω_0 in the Euclidean measure, and γ for the genus of Ω_0 (which is a positive integer associated to Ω_0, cf. [33]). Then $(V_{\Omega_0} K(z, \overline{z}))^{-\frac{1}{\gamma}}$ gives a Hermitian polynomial, denoted by $N(z, \overline{z})$. The polynomial N is called the generic norm of Ω_0.

Now let $\Omega = \Omega_1 \times \cdots \times \Omega_l$ be a bounded symmetric domain in $\mathbb{C}^n_z$, where each Ω_i, $1 \leq i \leq l$, is an irreducible bounded symmetric domain in $\mathbb{C}^{n_i}_{z_i}$ with $n = \sum_{i=1}^{l} n_i$. Denote by $N_i(z_i, \overline{z_i})$, $1 \leq i \leq l$, the generic norm of Ω_i. Next let $L = \Omega \times \mathbb{C}$ be the trivial line bundle over Ω. Equip L with the Hermitian metric h, where

$$h(z, \overline{z}) = \prod_{i=1}^{l} \left(N_i(z_i, \overline{z_i})\right)^{-k_i}. \tag{31}$$

Here every k_i is a positive real number. Note the negative of the Chern class $-c_1(L, h) = \partial\overline{\partial} \log h$ induces a complete homogeneous Kähler metric ω_Ω on Ω. Note ω_Ω is also a product of Kähler–Einstein metrics on Ω_i's. Moreover, the disk and circle bundles of (L, h) are given by the following, respectively:

$$D(\Omega, h) := \left\{ (z, \xi) \in \Omega \times \mathbb{C} : |\xi|^2 < \prod_{i=1}^{l} \left(N_i(z_i, \overline{z_i})\right)^{k_i} \right\} \subseteq \mathbb{C}^{n+1}; \tag{32}$$

$$\Sigma(\Omega, h) := \left\{ (z, \xi) \in \Omega \times \mathbb{C} : |\xi|^2 = \prod_{i=1}^{l} \left(N_i(z_i, \overline{z_i})\right)^{k_i} \right\} \subseteq \mathbb{C}^{n+1}. \tag{33}$$

The Bergman kernel $K_D(z, \xi, \overline{z}, \overline{\xi})$ of $D(\Omega, h)$ was computed by Ahn–Park [1]. Write

$$r(z, \xi, \overline{z}, \overline{\xi}) = 1 - \frac{|\xi|^2}{\prod_{i=1}^{l} \left(N_i(z_i, \overline{z_i})\right)^{k_i}}.$$

Note r is a local defining function of $\Sigma(\Omega, h)$. By Theorem 2.5 in [1], we see $r^{n+1} K_D$ extends smoothly across $\Sigma(\Omega, h)$. Let $D' \subset D(\Omega, h)$ be a small smoothly bounded strongly pseudoconvex domain that shares an open piece of boundary $M \subset \Sigma(\Omega, h)$ with $D(\Omega, h)$. Then writing $K_{D'}$ for the Bergman kernel of D', by the localization property of the Bergman kernel (see, e.g., Theorem 4.2 in [19]), $r^{n+1} K_{D'}$ extends smoothly across M. By inspecting the Fefferman expansion of $K_{D'}$, we conclude that M is Bergman logarithmically flat; since M can be arbitrarily chosen, so is $\Sigma(\Omega, h)$. Thus we have observed the following fact.

Proposition 3.3 *Let $\Omega = \Omega_1 \times \cdots \times \Omega_l$ be a bounded symmetric domain in $\mathbb{C}^n_z$, where each Ω_i, $1 \leq i \leq l$, is an irreducible bounded symmetric domain in $\mathbb{C}^{n_i}_{z_i}$ with $n = \sum_{i=1}^{l} n_i$. Denote by $N_i(z_i, \overline{z_i})$, $1 \leq i \leq l$, the generic norm of Ω_i. If all k_i, $1 \leq i \leq l$, are positive real numbers, then $\Sigma(\Omega, h)$ in (33) is Bergman logarithmically flat.*

4 CR Equivalence Classes of Flat Hypersurfaces

Although Theorem 3.1 allows for a wide range of compact (or noncompact, of course) locally isometrically inequivalent Kähler manifolds (M, g) as base manifolds, it is not at all clear that the corresponding unit sphere bundles $\Sigma(M, h)$ are locally CR inequivalent. To elucidate the issue, and also characterize those (M, g) in the context of Theorem 3.1 that give rise to spherical unit circle bundles $\Sigma(M, h)$, we state the following result, which is a consequence of the work of Webster [36] and Bryant [6] combined with Theorem 3.1:

Theorem 4.1 *Let $(L, h) \to M$ be a negative Hermitian holomorphic line bundle over a Kähler manifold (M, g) such that the curvature of h induces the metric g as in Sect. 3.1. Assume that the eigenvalues of the Ricci endomorphism $Ric(g)^\sharp$ are constant. Then, $\Sigma(M, h)$ is obstruction flat. Furthermore, $\Sigma(M, h)$ is spherical if and only if (M, g) is locally holomorphically isometric to one of the following:*

(i) $(\mathbb{B}^n, \mu g_{-1})$ *for some* $\mu \in \mathbb{R}^+$
(ii) $(\mathbb{CP}^n, \mu g_1)$ *for some* $\mu \in \mathbb{R}^+$
(iii) $(\mathbb{C}^n, g_0)$
(iv) $(\mathbb{B}^l \times \mathbb{CP}^{n-l}, \mu g_{-1} \times \mu g_1)$ *for some* $1 \leq l \leq n-1$ *and some* $\mu \in \mathbb{R}^+$

where g_c denotes the canonical complete metric of constant holomorphic curvature c on the corresponding space form.

We note that the base manifolds within each category (i)–(iv) are conformally isometric, whereas two manifolds from different categories are locally and conformally isometrically inequivalent. Nevertheless, their unit sphere bundles are all spherical and, hence, locally CR equivalent. We shall see in the following two subsections that this spherical case is in a sense special. In particular, we shall consider more general products of base manifolds with constant holomorphic sectional curvature metrics and provide conditions that guarantee that the unit circle bundles are locally CR equivalent if and only if the base manifolds are conformally isometrically equivalent.

4.1 Rigidity of Unit Circle Bundles Over Bounded Symmetric Domains

As a special case of Theorem 3.1, if a negative line bundle (L, h) quantizes a homogeneous Kähler manifold (M, g), then the circle bundle Σ is obstruction flat. In general, homogeneous manifolds are more tractable than the broader class of Kähler manifolds with constant Ricci eigenvalues. Among homogeneous Kähler manifolds, one can further distinguish the Hermitian symmetric spaces. Recently, the second author studied the CR geometry of the corresponding circle bundle when the base manifold (M, g) is a Hermitian symmetric space of noncompact type, i.e., a bounded symmetric domain. A rigidity phenomenon was discovered for local CR mappings between such circle bundles. Using this rigidity result, one can construct countably infinite families of compact, obstruction flat CR manifolds that are pairwise locally CR inequivalent (see Theorem 4.3).

Following the notions in Sect. 3.3, let $\Omega = \Omega_1 \times \cdots \times \Omega_l$ be a bounded symmetric domain in $\mathbb{C}^n_z$, where each Ω_i, $1 \le i \le l$, is an irreducible bounded symmetric domain in $\mathbb{C}^{n_i}_{z_i}$ with $n = \sum_{i=1}^{l} n_i$. Denote by $N_i(z_i, \overline{z_i})$, $1 \le i \le l$, the generic norm of Ω_i. Let $L = \Omega \times \mathbb{C}$ be the trivial line bundle over Ω. Equip L with the Hermitian metric h as defined in (31), where every k_i, there is a positive real number. Note the negative of the Chern class $-c_1(L, h) = \partial\overline{\partial} \log h$ of (L, h) induces a complete homogeneous Kähler metric ω_Ω on Ω, which is a product of Kähler–Einstein metrics on Ω_i's. Let $D(\Omega, h)$ and $\Sigma(\Omega, h)$ be as defined in (32) and (33), respectively. From Proposition 3.3, we see $\Sigma(\Omega, h)$ is Bergman logarithmically flat; from Theorem 3.1 (see also Remark 3.2), we know $\Sigma(\Omega, h)$ is obstruction flat as well.

We also consider a similar set of disk and circle bundles as above. Let $\widetilde{\Omega} = \widetilde{\Omega}_1 \times \cdots \times \widetilde{\Omega}_m$ be a bounded symmetric domain in $\mathbb{C}^n$, where each $\widetilde{\Omega}_j$, $1 \le j \le m$, is an irreducible bounded symmetric domain in $\mathbb{C}^{\widetilde{n}_j}_{w_j}$ with $n = \sum_{j=1}^{m} \widetilde{n}_j$. Denote by $\widetilde{N}_j(w_j, \overline{w_j})$, $1 \le j \le m$, the generic norm of $\widetilde{\Omega}_j$. Next let $\widetilde{L} = \widetilde{\Omega} \times \mathbb{C}$ be the trivial line bundle over $\widetilde{\Omega}$, and equip $\widetilde{L}$ with the Hermitian metric $\widetilde{h}$, where

$$\widetilde{h}(w, \bar{w}) = \prod_{i=1}^{m} \left(\widetilde{N}_i(w_i, \overline{w_i})\right)^{-\widetilde{k}_i}, \quad \widetilde{k}_i \in \mathbb{R}, \widetilde{k}_i > 0.$$

As before, the negative of the Chern class $-c_1(\widetilde{L}, \widetilde{h}) = \partial\bar{\partial}\log\widetilde{h}$ of $(\widetilde{L}, \widetilde{h})$ induces a complete homogeneous Kähler metric $\omega_{\widetilde{\Omega}}$ on $\widetilde{\Omega}$. The disk bundle of $(\widetilde{L}, \widetilde{h})$ is given by

$$D(\widetilde{\Omega}, \widetilde{h}) := \{(w, \eta) \in \Omega \times \mathbb{C} : |\eta|^2 < \prod_{i=1}^{t} \left(\widetilde{N}_i(w_i, \overline{w_i})\right)^{\widetilde{k}_i}\} \subseteq \mathbb{C}^{n+1}.$$

The corresponding circle bundle $\Sigma(\widetilde{\Omega}, \widetilde{h})$ is defined similarly by replacing the inequality with an equation. In [37], the second author proved the following rigidity result.

Theorem 4.2 *Let all k_i's and $\widetilde{k_i}$'s in the above be positive integers, and assume at least one k_i and at least one $\widetilde{k_i}$ equal to 1. Let F be a nonconstant continuous CR map from an open connected piece of $\Sigma(\Omega, h)$ to $\Sigma(\widetilde{\Omega}, \widetilde{h})$. Then the following statements hold:*

(1) The map F extends to a rational biholomorphism from $D(\Omega, h)$ to $D(\widetilde{\Omega}, \widetilde{h})$.
(2) The two bounded symmetric domains (Ω, ω_Ω) and $(\widetilde{\Omega}, \omega_{\widetilde{\Omega}})$ are holomorphically isometric.

Note $\Sigma(\Omega, h)$ is not compact in $\mathbb{C}^{n+1}$, and its (compact) closure in $\mathbb{C}^{n+1}$ is not a smooth hypersurface in general. On the other hand, we may consider the circle bundles over compact quotients of bounded symmetric domains. Let $X = X_1 \times \cdots \times X_l$ be a compact quotient of $\Omega = \Omega_1 \times \cdots \times \Omega_l$, where $X_i, 1 \le i \le l$, is a compact manifold covered by Ω_i. Note X_i naturally inherits the Kähler–Einstein metric from Ω_i. Write $L_i, 1 \le i \le l$, for the anti-canonical line bundle of X_i, which locally is the same as that of Ω_i. Let $q = (q_1, \cdots, q_l)$ be a tuple of positive integers and consider the line bundle

$$L_q := L_1^{q_1} \otimes \ldots \otimes L_l^{q_l}.$$

It is routine to verify that the circle bundle of L_q is locally CR equivalent to the hypersurface defined by

$$|\xi|^2 = \prod_{i=1}^{l} \left(N_i(z_i, \overline{z_i})\right)^{\gamma_i q_i}.$$

Here γ_i is the genus of Ω_i. By picking $q_i = \frac{k_i \prod_{j=1}^{l} \gamma_j}{\gamma_i}, 1 \le i \le l$ for some positive integers k_i's, and taking an appropriate root of ξ locally, we further see the above hypersurface is locally CR equivalent to the one defined in (33). Using this observation and Theorem 4.2, we can prove the following result (Theorem 1.14 in [37]). See [37] for the detailed proof.

Theorem 4.3 *Let $n \geq 2$. There exists a countably infinite family $\mathcal{F}$ of compact real analytic CR hypersurfaces (realized as circle bundles over complex manifolds) of real dimension $2n+1$ such that the following hold:*

(1) Every CR hypersurface $M \in \mathcal{F}$ is obstruction flat and Bergman logarithmically flat. Moreover, every $M \in \mathcal{F}$ is locally homogeneous and has transverse symmetry.
(2) For every pair of (distinct) CR hypersurfaces $M_1, M_2 \in \mathcal{F}$, any open pieces $U \subseteq M_1$ and $V \subseteq M_2$ are not CR diffeomorphic.

4.2 CR Curvature Eigenvalues of Unit Circle Bundles over Products of Space Forms

Consider two Kähler manifolds (M, g) and (M', g') quantized by two negative line bundles (L, h) and (L', h'), respectively, as in Sect. 3.1. We observe that if M and M' are locally (holomorphically) conformally isometric, then their corresponding circle bundles $\Sigma(M, h)$ and $\Sigma(M', h')$ are locally CR equivalent. To see that, assume f is a local holomorphic map at $p_0 \in M$ from M to M' and preserves the metric: $f^*(g') = \mu g$ up to a conformal constant $\mu > 0$. Note in some local charts at $p_0 \in M$ and $q_0 := f(p_0) \in M'$, the circle bundles $\Sigma(M, h)$ and $\Sigma(M', h')$ can be written as follows, respectively:

$$|\xi|^2 h(z, \overline{z}) = 1; \; |\eta|^2 h'(w, \overline{w}) = 1.$$

Here z and w denote the local coordinates of M and M', while ξ, η denote the local fiber coordinates of L, L', respectively. By the conformal isometry assumption, we have

$$\partial\overline{\partial} \log h'(f, \overline{f}) = \mu \partial\overline{\partial} \log h(z, \overline{z}).$$

This yields that $h'(f, \overline{f}) = h^{\mu} |e^{\phi}|^2$ for some local holomorphic function ϕ at p_0. It follows that we have a local CR diffeomorphism from $\Sigma(M, h)$ and $\Sigma(M', h')$ over p_0 and q_0 given by

$$(z, \xi) \to (f(z), e^{-\phi} \xi^{\mu}). \tag{34}$$

It is natural to ask whether the converse of the above observation holds. By Theorem 4.1, the converse fails in general, even for space forms and their products. In the following, we shall show the converse still holds for products of three or more space forms under some mild conditions: If $\Sigma(M, h)$ and $\Sigma(M', h')$ are locally CR diffeomorphic, then (M, g) and (M', g') are locally conformally isometric.

For that, we shall consider unit circle bundles over a base manifold of the form

$$M = M_{c_1} \times \ldots \times M_{c_p},$$

where each M_{c_i} is a complex space form (always taken to be simply connected) of dimension m_i equipped with a complete metric g^i of constant holomorphic sectional curvature c_i and where M is equipped with the product metric $g = g^1 \times \ldots \times g^p$. Let (M', g') be another such product of space forms. We note that for the two products of space forms (M, g) and (M', g'), being locally conformally isometric is the same as being (globally) conformally isometric. The latter means there is a biholomorphism from M and M' that preserves the metric: $f^*(g') = \mu g$ up to a conformal constant $\mu > 0$. We also observe the following easy fact, whose proof will be omitted.

Lemma 4.4 *Let (M, g) and (M', g') be two products of space forms as above, where $M = M_{c_1} \times \ldots \times M_{c_p}$ and $M' = M_{c'_1} \times \ldots \times M_{c'_q}$. Write the two tuples $c = (c_1, \cdots, c_p)$ and $c' = (c'_1, \cdots, c'_q)$. Assume $p = q$ and, after a permutation of $M'_{c_i}s$ and a scaling of c by a factor $a > 0 : c \to ac$, we have* $\dim M_{c_i} = \dim M_{c'_i}$ *for all i and $c = c'$. Then (M, g) and (M', g') are conformally isometric.*

We also assume that there is a Hermitian holomorphic line bundle $(L, h) \to M$ inducing the metric g as in Sect. 3.1; we can, e.g., take $L = L^1 \otimes \ldots \otimes L^p$ with metric $h = h^1 \ldots h^p$, where each $(L^i, h^i) \to M_{c_i}$ is as in Sect. 3.1. (Note the existence of such a line bundle poses an extra condition on those $c_i > 0$.) The unit circle bundle $\Sigma(M, h)$ gives rise to a strictly pseudoconvex hypersurface in L which is obstruction flat by Theorem 3.1. The CR dimension of $\Sigma(M, h)$ is $m = n - 1$. Moreover, since M is a simply connected homogeneous Kähler manifold, $\Sigma(M, h)$ is a homogeneous CR hypersurface.

We shall now assume that $n = \dim L = m + 1 = m_1 + \ldots + m_p + 1 \geq 3$; in other words, the dimension m of M is at least 2. A fundamental CR invariant of $\Sigma = \Sigma(M, h)$ is then the Chern–Moser CR curvature tensor $S_\alpha{}^\beta{}_{\nu\bar{\mu}}$, which can be defined using either Moser's normal form [9], Cartan–Chern's parabolic geometry construction, or Webster's pseudohermitian geometry [36]; the reader is referred to these papers for details of the construction. We lower the index β using the Levi form to obtain a trace-free $(2, 2)$-tensor $S_{\alpha\bar{\beta}\nu\bar{\mu}}$ with Hermitian curvature symmetries and with the property that the CR manifold is spherical if and only if the CR curvature vanishes in a neighborhood. When defined relative to a pseudohermitian structure, i.e., with a given choice of contact form θ, the CR curvature tensor $S_\alpha{}^\beta{}_{\nu\bar{\mu}}$ remains invariant under changes of contact form $\tilde{\theta} = e^u\theta$, whereas the Levi form $g_{\alpha\bar{\beta}}$ changes by $\tilde{g}_{\alpha\bar{\beta}} = e^u g_{\alpha\bar{\beta}}$. Using the Levi form to raise and lower indices, we obtain an endomorphism

$$S^\sharp \colon T^{1,0}\Sigma \odot T^{1,0}\Sigma \to T^{1,0}\Sigma \odot T^{1,0}\Sigma,$$

where $\odot$ denotes the symmetric tensor product, given in a local frame $Z_1, \ldots, Z_m$ for $T^{1,0}\Sigma$ and with a choice of θ by

$$X \odot Y \mapsto S_{\gamma}{}^{\alpha}{}_{\nu}{}^{\beta} X^{\gamma} Y^{\nu},$$

where $X = X^{\alpha} Z_{\alpha}$, $Y = Y^{\beta} Z_{\beta} \in T^{1,0}\Sigma$. This endomorphism is a weighted CR invariant in the sense that it is a pseudohermitian invariant, which transforms under changes of contact form $\tilde{\theta} = e^{u}\theta$ by $\tilde{S}^{\sharp} = e^{-u} S^{\sharp}$, as is easily verified. In particular, if we denote by $\lambda_1, \ldots, \lambda_m$ the eigenvalues (repeated according to multiplicity) of $S^{\sharp}$ at a point $p \in \Sigma$ relative to some contact form θ, then the m-tuple $(\lambda_1, \ldots, \lambda_m)$ is a CR invariant modulo permutations and scaling $(\lambda_1, \ldots, \lambda_m) \mapsto a(\lambda_1, \ldots, \lambda_m)$ for $a > 0$. We shall make use of this observation to study the CR equivalence classes of $\Sigma = \Sigma(M, h)$ for $M = M_{c_1} \times \ldots \times M_{c_p}$ as above.

Now, let $M = M_{c_1} \times \ldots \times M_{c_p}$, take $\rho = |\xi|^2 h(z, \bar{z}) - 1$ as a local defining function for $\Sigma = \Sigma(M, h)$ in some local chart and trivialization, and choose the contact form $\theta = i\bar{\partial} \log |\xi|^2 h(z, \bar{z})$. Then (see [36]), the Levi form of Σ is S^1-invariant and coincides with the Kähler metric g; the pseudohermitian curvature of Σ is S^1-invariant and coincides with the curvature tensor $R_{\alpha}{}^{\beta}{}_{\nu\bar{\mu}}$ of M; the CR curvature tensor $S_{\alpha}{}^{\beta}{}_{\nu\bar{\mu}}$ is S^1-invariant as well, coincides with the Bochner tensor of M, and is obtained by the following formula:

$$\begin{aligned} S_{\alpha\bar{\beta}\nu\bar{\mu}} = R_{\alpha\bar{\beta}\nu\bar{\mu}} - \frac{1}{m+2}(R_{\alpha\bar{\beta}} g_{\nu\bar{\mu}} + R_{\alpha\bar{\mu}} g_{\nu\bar{\beta}} + R_{\nu\bar{\beta}} g_{\alpha\bar{\mu}} + R_{\nu\bar{\mu}} g_{\alpha\bar{\beta}}) \\ + \frac{R}{(m+1)(m+2)}(g_{\alpha\bar{\beta}} g_{\nu\bar{\mu}} + g_{\nu\bar{\beta}} g_{\alpha\bar{\mu}}), \end{aligned} \tag{35}$$

where $R_{\alpha\bar{\beta}} = R_{\gamma}{}^{\gamma}{}_{\alpha\bar{\beta}}$ is the Ricci tensor and $R = R_{\gamma}{}^{\gamma}$ is the scalar curvature.

Next, recall that if (M, g) is a Kähler manifold of constant holomorphic sectional curvature c (and not the product manifold for a moment), then the curvature tensor is given by

$$R_{\alpha\bar{\beta}\nu\bar{\mu}} = c(g_{\alpha\bar{\beta}} g_{\nu\bar{\mu}} + g_{\nu\bar{\beta}} g_{\alpha\bar{\mu}}).$$

Thus, for our product manifold $M = M_{c_1} \times \ldots \times M_{c_p}$, the curvature tensor can be written

$$R_{\alpha\bar{\beta}\nu\bar{\mu}} = \sum_{i=1}^{p} c_i \left(\hat{g}^i_{\alpha\bar{\beta}} \hat{g}^i_{\nu\bar{\mu}} + \hat{g}^i_{\nu\bar{\beta}} \hat{g}^i_{\alpha\bar{\mu}} \right), \tag{36}$$

where the tensors $\hat{g}^i_{\alpha\bar{\beta}}$ are defined as follows. For each $i = 1, \ldots, p$, let $Z^i_{\alpha^i}$, $\alpha^i = 1, \ldots, m_i$, be a local frame for M_{c_i} and Z_{α}, $\alpha = 1, \ldots, m$, the corresponding local frame for M such that $(\pi_i)_* Z_{\alpha} = Z^i_{\alpha-(m_1+\ldots+m_{i-1})}$ when $\alpha \in \{m_1 + \ldots + m_{i-1} + 1, \ldots, m_1 + \ldots + m_i\}$ and $= 0$ otherwise; here, π_i denotes the projection $M \to M_{c_i}$.

We define

$$\hat{g}^i_{\alpha\bar{\beta}} = g^i((\pi_i)_* Z_\alpha, (\pi_i)_* Z_\beta).$$

One readily checks that

$$R_{\alpha\bar{\beta}} = \sum_{i=1}^{p} c_i(m_i+1)\hat{g}^i_{\alpha\bar{\beta}}, \qquad R = \sum_{i=1}^{p} c_i m_i(m_i+1). \tag{37}$$

By an abuse (but great simplification) of notation, we shall denote by Z_α a collection of $(1,0)$-vector fields on Σ such that the projection $\pi\colon \Sigma \to M$ identifies the various curvature tensors on Σ with those on M as explained above.

To investigate the eigenvalues of the CR curvature endomorphism $S^\sharp$, we note that the collection of local sections

$$\zeta^{ij}_{\alpha^i\beta^j} = Z_\alpha \odot Z_\beta, \qquad i \le j = 1,\ldots,p, \quad \alpha^i = 1,\ldots,m_i, \quad \beta^j = 1,\ldots,m_j, \tag{38}$$

where $\alpha = m_1 + \ldots + m_i + \alpha^i$ and $\beta = m_1 + \ldots + m_j + \beta^j$, forms a frame for $T^{1,0}\Sigma \odot T^{1,0}\Sigma$. It is straightforward to check, using (35) and (37), that the $\zeta^{ij}_{\alpha^i\beta^j}$ are also eigenvectors for $S^\sharp$. Moreover, the eigenvalues are readily computed and seen to be constant as functions of $p \in \Sigma$. The details of these calculations are left to the diligent reader. We summarize these observations in the following:

Proposition 4.5 *Let $M = M_{c_1} \times \ldots \times M_{c_p}$, $(L,h) \to M$, and $\Sigma = \Sigma(M,h)$, be as described above. Then, $\zeta^{ij}_{\alpha^i\beta^j}$ given by* (38) *form a complete system of eigenvectors for the CR curvature operator $S^\sharp$. The eigenvalues λ_{ij} and their multiplicities d_{ij}, for which $\zeta^{ij}_{\alpha^i\beta^j}$, $\alpha^i = 1,\ldots,m_i$, $\beta^j = 1,\ldots,m_j$, are the eigenvectors, are constant as functions of $p \in \Sigma$ and given by*

$$\lambda_{ii} = \frac{2}{(m+1)(m+2)}\sum_{l=1}^{p} m_l(m_l+1)c_l + \frac{2}{m+2}(m-2m_i)c_i, \tag{39}$$

$$d_{ii} = \frac{m_i(m_i+1)}{2},$$

and for $i < j$,

$$\lambda_{ij} = \frac{2}{(m+1)(m+2)}\sum_{l=1}^{p} m_l(m_l+1)c_l - \frac{2}{m+2}\big((m_i+1)c_i + (m_j+1)c_j\big), \tag{40}$$

$$d_{ij} = m_i m_j.$$

Remark 4.6 We observe that it may happen that there is an eigenvalue λ such that $\lambda = \lambda_{ij}$ for multiple different pairs (i, j). In this case, the multiplicity of λ is then (of course) the sum of the corresponding individual multiplicities d_{ij}.

Remark 4.7 If there is only one factor (i.e., M itself has constant holomorphic sectional curvature), then there is only one eigenvalue and that is $\lambda = 0$; in other words, the CR curvature is identically zero, implying that Σ is spherical. If there are two factors, one checks readily that all eigenvalues are multiples of $c_1 + c_2$. In particular, if $c_1 + c_2 = 0$, then the CR curvature is identically zero and Σ is again spherical. This is consistent with Theorem 4.1 above.

We temporarily fix p and $m_1, \cdots, m_p$, and as above let $m = \sum_{i=1}^{p} m_i$. Denote by $\Lambda\colon \mathbb{R}^p \to \mathbb{R}^p$ the linear map whose ith component Λ_i, $1 \le i \le p$, is given by

$$\Lambda_i(c_1, \ldots, c_p) = \frac{2}{(m+1)(m+2)} \sum_{l=1}^{p} m_l(m_l+1)c_l + \frac{2}{m+2}(m-2m_i)c_i. \tag{41}$$

As noted in Remark 4.7, when $p = 1$, the rank of Λ is zero; when $p = 2$, the rank is either zero or one. In neither case is Λ invertible. However, for $p \ge 3$, we have the following:

Lemma 4.8 *Fix $p \ge 3$ and $m_1, \cdots, m_p \ge 1$, and let $m = \sum_{i=1}^{p} m_i$. Assume that $m - 2m_i \ge 0$ for $i = 1, \ldots, p$. Then, the linear map $\Lambda\colon \mathbb{R}^p \to \mathbb{R}^p$ given by* (41) *is invertible. Consequently, if $\Lambda(c) = a\Lambda(c')$ for some $c, c' \in \mathbb{R}^p$ and $a \in \mathbb{R}$, then $c = ac'$.*

Proof Let us denote by

$$\alpha_i = \frac{2m_i(m_i+1)}{(m+1)(m+2)}, \qquad \beta_i = \frac{2(m-2m_i)}{m+2}.$$

We observe that the algebraic structure of Λ noted above, we may assume (after a permutation of the components if necessary) that $\beta_1 \ge \beta_2 \ge \ldots \ge \beta_p \ge 0$. Moreover, only the last (smallest) number β_p can equal 0 (since $p \ge 3$). A simple row reduction scheme now shows that

$$\det \Lambda = \beta_1 \ldots \beta_{p-1} \left(\alpha_p + \beta_p \left(1 + \sum_{l=1}^{p-1} \frac{\alpha_l}{\beta_l} \right) \right). \tag{42}$$

The conclusion easily follows, since $\alpha_i > 0$ for all $i = 1, \ldots, p$ and $\beta_i > 0$ for $i \le p - 1$. □

Let (M, g) be a product of space forms and $\Sigma = \Sigma(M, h)$ is the unit circle bundle in $(L, h) \to (M, g)$ as above. We will show under some conditions, we can recover (M, g), up to conformal isometry, from the CR curvature of Σ. The strategy can be roughly summarized as follows: First under some mild conditions, the value of p can be recovered, and the subcollection of the p eigenvalues λ_{ii}, for $i = 1, \ldots, p$, can be identified among the full collection of eigenvalues, relative to some contact form θ; and the λ'_{ii}s are mutually distinct. From this we can then recover the dimensions of factor M'_{c_i}s from the multiplicities d_{ii} of λ_{ii}. Finally with the help of Lemma 4.8, the set of the holomorphic sectional curvatures $\{c_1, \ldots, c_p\}$ can be uniquely determined modulo scaling $a \to \{ac_1, \ldots, ac_p\}$ for $a > 0$.

The first technical problem in this context, which is alluded to in Remark 4.6, occurs when the CR curvature endomorphism $S^\sharp$ has an eigenvalue λ that corresponds to multiple λ'_{ij}s. In this case, one would need to properly split the eigenvalue λ and its multiplicity into individual λ_{ij}s and d_{ij}s, which a priori could potentially be done in multiple ways. To address this issue, we introduce the full eigenvalue map $\hat{\Lambda}\colon \mathbb{R}^p \to \mathbb{R}^{p(p+1)/2}$, given in components $\hat{\Lambda}_{ij}$ for $i \le j$ by the right-hand sides of (39) and (40); we are using here pairs (i, j) with $i \le j$ to index the components of a vector $\xi \in \mathbb{R}^{p(p+1)/2}$; we note that $\hat{\Lambda}_{ii} = \Lambda_i$. The following is easy to check, and we omit its proof.

Lemma 4.9 *Write $\mathcal{S}$ for the set of $c \in \mathbb{R}^p$, $p \ge 3$, for which either $\hat{\Lambda}_{ii}(c) = \hat{\Lambda}_{i'j'}(c)$ for some i and $i' \ne j'$, or $\hat{\Lambda}_{ii}(c) = \hat{\Lambda}_{i'i'}(c)$ for some $i \ne i'$. Then $\mathcal{S}$ is a finite union of $(p-1)$-dimensional subspaces of $\mathbb{R}^p$.*

The scheme described above to recover the holomorphic sectional curvatures c_i and the dimensions m_i of the M_i in the product $M = M_{c_1} \times \ldots \times M_{c_p}$, provided that we can identify the subcollection of eigenvalues λ_{ii}, now works for $c = (c_1, \ldots, c_p) \in \mathbb{R}^p \setminus \mathcal{S}$.

We proceed by identifying two situations where this scheme to recover the conformal isometry class of the base manifold can be executed. The first result is the following:

Proposition 4.10 *Fix $m \ge 3$. Let $\mathcal{F}_1$ be the family of unit circle bundles $\Sigma(M, h)$ of CR dimension m, which arises from a negative line bundle (L, h) quantizing a Kähler manifold (M, g) satisfying the following conditions:*

(i) *(M, g) is a product of space forms: $M = M_{c_1} \times \ldots \times M_{c_p}$, for some $p = p(M) \ge 3$ and M'_{c_i}s, depending on M. Besides, their dimensions satisfy $m - 2m_i \ge 0$, for $i = 1, \ldots p$.*
(ii) *$c_i \ge 0$ (or $c_i \le 0$), for all $i = 1, \ldots p$.*
(iii) *$c = (c_1, \ldots, c_p) \notin \mathcal{S}$, where $\mathcal{S}$ is given by Lemma 4.9.*

Then $\Sigma(M, h)$, $\Sigma(M', h') \in \mathcal{F}_1$ are locally CR equivalent if and only if (M, g) and (M', g') are conformally isometric.

Proof By the discussion at the beginning of this subsection, it suffices to prove the "only if" part. For that, we first give the algorithm of recovering M, up to conformal isometry, from the eigenvalues of the CR curvature operator $S^\sharp$ of $\Sigma(M, h)$. Note that conditions (ii) and (iii) guarantee that the p largest (or smallest) eigenvalues of $S^\sharp$ correspond to the λ_{ii} for $i = 1, \ldots, p$, and the λ'_{ii}s are mutually distinct. Without loss of generality, we assume they are the largest ones. Furthermore, by rearranging the factors M'_{c_i}s in M, we can assume $\lambda_{11} > \cdots > \lambda_{pp}$.

Next pick the largest eigenvalue of $S^\sharp$. Call it λ_1 and write k_1 for its multiplicity. Then $\lambda_1 = \lambda_{11}$, and by Proposition 4.5, k_1 can be uniquely written as $k_1 = \frac{m_1(m_1+1)}{2}$. Note m_1 is the dimension of M_{c_1}. Then pick the second largest eigenvalue λ_2 of $S^\sharp$, which must equal to λ_{22}, and write k_2 for its multiplicity. Again by Proposition 4.5, k_2 can be uniquely written as $k_2 = \frac{m_2(m_2+1)}{2}$, and m_2 is the dimension of M_{c_2}. Keep doing this until the m_i's add up to m. Note the number of m_i's is then precisely p. In this way, we recover both p and m_i's. Once they are determined, we can define the linear map $\Lambda\colon \mathbb{R}^p \to \mathbb{R}^p$ as in (41). Finally, since we know the values of $\lambda_{11} > \cdots > \lambda_{pp}$ (relative to some contact form θ), by Lemma 4.8, we can recover the values of c_i's modulo positive factor scaling.

If $\Sigma(M, h), \Sigma(M', h') \in \mathcal{F}_1$ are locally CR equivalent, then their CR curvature operators have the same set of eigenvalues, up to a positive factor scaling. Moreover, if we sort both sets of eigenvalues in descending order, the corresponding multiplicities are equal. Write $M' = M_{c_1'} \times \ldots \times M_{c_q'}$. Repeating the above algorithm to (M', g'), we see M and M' satisfy the conditions in Lemma 4.4. Hence, they must be conformally isometric. □

Remark 4.11 The proof of Proposition 4.10 does not provide much information on the structure of the local CR diffeomorphism between $\Sigma(M, h), \Sigma(M', h')$ (if exists). We stress that, unlike Theorem 4.2, one cannot expect the rationality and rigidity result to hold for the local CR mappings in Proposition 4.10, as the local CR diffeomorphism may take the form as in (34). In general, this map is not rational and does not extend to a biholomorphism between the corresponding disk bundles.

Our second result along these lines allows mixed signs of the holomorphic sectional curvatures, but at the expense of fixing the dimensions of the factors:

Proposition 4.12 *Fix $p \geq 3$ and $m = pm_0$ for some $m_0 \geq 2$. Consider the family $\mathcal{F}_1$ of unit circle bundles $\Sigma(M, h)$, where $M = M_{c_1} \times \ldots \times M_{c_p}$, $(L, h) \to M$, and $\Sigma(M, h)$ are as in Proposition 4.5 and where $m_i = m_0$ for $i = 1, \ldots p$ and $c = (c_1, \ldots, c_p) \notin \mathcal{S}$. Then $\Sigma(M, h), \Sigma(M', h') \in \mathcal{F}_1$ are locally CR equivalent if and only if (M, g) and (M', g') are conformally isometric.*

Proof In this case, we identify the eigenvalues λ_{ii} by considering the multiplicities of the eigenvalues of $S^\sharp$. Let λ be an eigenvalue of $S^\sharp$. If λ corresponds to some λ_{ii}, then it corresponds to precisely one λ_{ij} (as $c \notin \mathcal{S}$), and thus its multiplicity equals to $\frac{m_0(m_0+1)}{2}$. If λ corresponds to some λ_{ij} with $i < j$, then it cannot correspond to any λ_{ii}, but it may correspond to multiple λ_{ij}'s with $i < j$: $\lambda_{i_1 j_1}, \cdots, \lambda_{i_k j_k}$ for some $k \geq 1$. Then its multiplicity equals to km_0^2. Since $m_0 \geq 2$, we have $km_0^2 > m_0(m_0+1)/2$. Thus the multiplicity $\frac{m_0(m_0+1)}{2}$ identifies the λ_{ii}. The proof is concluded as above and the details are left to the reader. □

Using the above results, we can construct some families of obstruction flat CR manifolds that are pairwise locally CR inequivalent. To construct examples of compact types, in the settings of Propositions 4.10 and 4.12, we may use as the building blocks (M_{c_i}, g^i) of $(\mathbb{P}^{m_i}, \mu_i g_1)$, and compact quotients of $(\mathbb{B}^{m_i}, \mu_i g_{-1})$

and $(\mathbb{C}^{m_i}, g_0)$. Observe that since we insist that the metrics are induced by holomorphic line bundles $(L_i, h_i) \to M_{c_i}$, we are restricted to, e.g., taking L_i to be integral powers of the (anti-)canonical line bundle, which means that we obtain holomorphic sectional curvatures of the form $c_i = \pm 1/k_i$ (or 0) for integers k_i. These families are necessarily only countably infinite. Since this compact case was already covered by Theorem 4.3, we will not elaborate it here. It remains an interesting question to ask whether there exists an uncountably infinite family of compact hypersurfaces with the properties in Theorem 4.3. In the remainder of the section, we shall concentrate on the construction of noncompact obstruction flat CR manifolds.

Theorem 4.13 *Let $n \geq 2$. There exists a family $\mathcal{F}$, parameterized by*

$$\mathcal{J} = \{(\lambda_2, \cdots, \lambda_n) : 1 > \lambda_2 > \cdots > \lambda_n > 0\},$$

of (noncompact) real analytic CR hypersurfaces in $\mathbb{C}^{n+1}$ such that the following hold:

(1) *Every CR hypersurface $M \in \mathcal{F}$ is obstruction flat and Bergman logarithmically flat. Moreover, every $M \in \mathcal{F}$ is homogeneous and has transverse symmetry.*
(2) *For every pair of (distinct) CR hypersurfaces $M_1, M_2 \in \mathcal{F}$, any open pieces $U \subseteq M_1$ and $V \subseteq M_2$ are not CR diffeomorphic.*

Remark 4.14 We recall that in $\mathbb{C}^2$, every Bergman logarithmically flat CR hypersurface must be spherical; and there exist (noncompact) real analytic nonspherical CR hypersurfaces that are all obstruction flat and transversally symmetric (see [23, 24, 16]).

Proof of Theorem 4.13 We will utilize a simple and useful case of Proposition 4.10 where all $m_i = 1$ and $c_i < 0$ (so that $m = p$). In this case, each M_i is a unit disk equipped with the (normalized) Poincaré metric. We first fix various notations and make some observations. Fix $n \geq 2$ and λ be a $(n-1)$ tuple of real numbers $\lambda = (\lambda_2, \cdots, \lambda_n) \in \mathcal{J}$, where $\mathcal{J} = \{(\lambda_2, \cdots, \lambda_n) : 1 > \lambda_2 > \cdots > \lambda_n > 0\}$. □

Consider the homogeneous CR hypersurface

$$\Sigma_\lambda = \Big\{(z, \xi) = (z_1, \cdots, z_n, \xi) \in \Delta^n \times \mathbb{C} : |\xi|^2 = (1 - |z_1|^2) \prod_{j=2}^{n} (1 - |z_j|^2)^{\lambda_j} \Big\} \subset \mathbb{C}^{n+1}.$$

We first note:

Lemma 4.15 *Every $\Sigma_\lambda \subset \mathbb{C}^{n+1}, \lambda \in \mathcal{J}$, is a homogeneous, obstruction flat and Bergman logarithmically flat, strongly pseudoconvex real hypersurface with transverse symmetry.*

Proof Write $h(z,\overline{z}) = (1-|z_1|^2)^{-1}\prod_{j=2}^{n}(1-|z_j|^2)^{-\lambda_j}$. Note with M being the n-dimensional polydisk Δ^n, we have Σ_λ is indeed the circle bundle $\Sigma(M,h)$ of the Hermitian line bundle $(M\times\mathbb{C},h)$. Consequently, Σ_λ has transverse symmetry. By earlier discussions of this subsection, Σ_λ is homogeneous and obstruction flat. Furthermore, note the polydisk is a special case of bounded symmetric domains and the generic norm of a disk is $1-|z_j|^2$. Then Σ_λ is a special case of $\Sigma(\Omega,h)$, which is defined in (33). By Proposition 3.3, Σ_λ is Bergman logarithmically flat. □

We now prove Theorem 4.13. For that, we first consider the case $n\geq 3$.

Lemma 4.16 *Let $n\geq 3$. Then $\mathcal{F} := \{\Sigma_\lambda\subset\mathbb{C}^{n+1}|\lambda\in\mathcal{J}\}$ gives a family of CR hypersurfaces in $\mathbb{C}^{n+1}$ with mutually distinct local CR structure. More precisely, Σ_λ and $\Sigma_{\lambda'}$ in $\mathcal{F}$ are locally CR equivalent if and only if $\lambda=\lambda'$.*

Proof It suffices to prove for the "only if" part. For that, we shall apply Proposition 4.10 for the special case where all $m_i=1$ and $c_i<0$ (so that $m=p$), and $p=n\geq 3$. For $c<0$, denote by Δ_c the unit disk equipped with the normalized Poincaré metric of constant Gaussian curvature c. Write M for the product of the normalized Poincaré disks $M=\Delta_{-2}\times\Delta_{-\frac{2}{\lambda_2}}\times\cdots\Delta_{-\frac{2}{\lambda_n}}, \lambda\in\mathcal{J}$. As already observed, the metric on M is induced by the Kähler form $\partial\overline{\partial}\log h$, where as above, $h(z,\overline{z}) = (1-|z_1|^2)^{-1}\prod_{j=2}^{n}(1-|z_j|^2)^{-\lambda_j}$. Moreover, Σ_λ is the circle bundle $\Sigma(M,h)$. We also note that conditions (i) and (ii) of Proposition 4.10 are satisfied in this setting. Next, let the set $\mathcal{S}$ be as in condition (iii) there, which is a finite union of $(n-1)$-dimensional subspaces of $\mathbb{R}^n$, and write

$$\mathcal{T} := \left\{(\lambda_2,\cdots,\lambda_n)\in\mathcal{J}\ \middle|\ c := \left(-2,-\frac{2}{\lambda_2},\cdots,-\frac{2}{\lambda_n}\right)\in\mathcal{S}\right\}.$$

Assume Σ_λ and $\Sigma_{\lambda'}$, where $\lambda,\lambda'\in\mathcal{J}\setminus\mathcal{T}$, are locally CR equivalent. Write the two tuples

$$c := (c_1,\cdots,c_n) = \left(-2,-\frac{2}{\lambda_2},\cdots,-\frac{2}{\lambda_n}\right),\qquad\text{and}$$

$$c' := (c'_1,\cdots,c'_n) = \left(-2,-\frac{2}{\lambda'_2},\cdots,-\frac{2}{\lambda'_n}\right).$$

Then it follows from (the proof of) Proposition 4.10 that, after a permutation of the factors if necessary, c and c' are proportional by a constant $a>0$. But c and c' are already in the decreasing order, and the largest component for both tuples is -2. We thus must have $a=1$, and $\lambda_i=\lambda'_i$ for all i, i.e., $\lambda=\lambda'$.

Finally to finish the proof of the lemma, it suffices to prove that $\mathcal{T}=\emptyset$. We recall $\mathcal{S}$ is defined in Lemma 4.9 in terms of $\hat{A}_{ij}$ for $i\leq j$, where the latter is given by the right-hand sides of (39) and (40). By these expressions, as all $m_i=1$ and $c_i<0$, we have $\hat{A}_{ii}(c)<\hat{A}_{i'j'}(c)$ for any i and $i'\neq j'$. Moreover, for $\lambda\in\mathcal{J}$, the corresponding c satisfies $c_i\neq c_j$ for $i\neq j$. This yields that $\hat{A}_{ii}(c)\neq\hat{A}_{jj}(c)$ for $i\neq j$. Consequently, $\mathcal{T}=\emptyset$. □

It is clear that Theorem 4.13 follows from Lemmas 4.15 and 4.16 when $n \geq 3$. It remains to prove it for the case $n = 2$. In this case, $\Sigma_\lambda \subset \mathbb{C}^3$ is defined by

$$|\xi|^2 = (1 - |z_1|^2)(1 - |z_2|^2)^\lambda, (z_1, z_2) \in \Delta^2, \lambda \in (0, 1).$$

By taking a local logarithm of the above equation and applying a simple change of coordinates, we see Σ_λ is locally CR diffeomorphic to the hypersurface

$$H_\lambda := \{(z_1, z_2, \xi) \in \Delta^2 \times \mathbb{C} : \mathrm{Im}\xi = \log(1 - |z_1|^2) + \lambda \log(1 - |z_2|^2)\}, \lambda \in (0, 1).$$

The latter hypersurface was studied by Loboda in [31, Theorem 1], where it was proved that for $\lambda, \lambda' \in (0, 1)$, H_λ and $H_{\lambda'}$ are locally CR equivalent if and only if $\lambda = \lambda'$. Consequently, we have the following lemma which corresponds to $n = 2$.

Lemma 4.17 *Write $\mathcal{G} := \{\Sigma_\lambda \subset \mathbb{C}^3 : \lambda \in (0, 1)\}$. Then $\mathcal{G}$ is a family of CR hypersurfaces with mutually distinct local CR structure. More precisely, Σ_λ and $\Sigma_{\lambda'}$ in $\mathcal{G}$ are locally CR equivalent if and only if $\lambda = \lambda'$.*

We finally combine Lemmas 4.15 and 4.17 to establish Theorem 4.13 for $n = 2$.

□

References

1. Heungju Ahn and Jong-Do Park. The explicit forms and zeros of the Bergman kernel function for Hartogs type domains. *Journal of Functional Analysis*, 262(8):3518–3547, 2012.
2. Toby N. Bailey, Michael G. Eastwood, and C. Robin Graham. Invariant theory for conformal and CR geometry. *Ann. of Math. (2)*, 139(3):491–552, 1994.
3. L. Boutet de Monvel. Intégration des équations de Cauchy-Riemann induites formelles. In *Séminaire Goulaouic-Lions-Schwartz 1974–1975; Équations aux derivées partielles linéaires et non linéaires*, pages Exp. No. 9, 14. 1975.
4. L. Boutet de Monvel. Le noyau de Bergman en dimension 2. In *Séminaire sur les Équations aux Dérivées Partielles 1987–1988*, pages Exp. No. XXII, 13. École Polytech., Palaiseau, 1988.
5. L. Boutet de Monvel and J. Sjöstrand. Sur la singularité des noyaux de Bergman et de Szegő. In *Journées: Équations aux Dérivées Partielles de Rennes (1975)*, pages 123–164. Astérisque, No. 34–35. Soc. Math. France, Paris, 1976.
6. Robert L. Bryant. Bochner-Kähler metrics. *J. Amer. Math. Soc.*, 14(3):623–715, 2001.
7. Daniel M. Burns and Charles L. Epstein. Embeddability for three-dimensional CR-manifolds. *J. Amer. Math. Soc.*, 3(4):809–841, 1990.
8. Shiu Yuen Cheng and Shing Tung Yau. On the existence of a complete Kähler metric on noncompact complex manifolds and the regularity of Fefferman's equation. *Comm. Pure Appl. Math.*, 33(4):507–544, 1980.
9. Shiing-Shen Chern and J. K. Moser. Real hypersurfaces in complex manifolds. *Acta Math.*, 133:219–271, 1974.
10. H. J. Christoffers. The Bergman kernel and Monge–Ampere approximations. *ProQuest LLC, Ann Arbor, Ph.D. Thesis, The University of Chicago, Chicago*, 1981.
11. Sean N. Curry and Peter Ebenfelt. Bounded strictly pseudoconvex domains in $\mathbb{C}^2$ with obstruction flat boundary II. *Adv. Math.*, 352:611–631, 2019.
12. Sean N. Curry and Peter Ebenfelt. Deformations and embeddings of three-dimensional strictly pseudoconvex CR manifolds. *Math. Ann.*, to appear, https://arxiv.org/abs/2007.00547, 2020.

13. Sean N. Curry and Peter Ebenfelt. Bounded strictly pseudoconvex domains in $\mathbb{C}^2$ with obstruction flat boundary. *Amer. J. Math.*, 143(1):265–306, 2021.
14. Sean N. Curry and Peter Ebenfelt. Obstruction flat rigidity of the CR 3-sphere. *J. Reine Angew. Math.*, 781:105–126, 2021.
15. Peter Ebenfelt. The log term in the Bergman and Szego kernels in strictly pseudoconvex domains in $\mathbb{C}^2$. *Doc. Math.*, 23:1659–1676, 2018.
16. Peter Ebenfelt, Ming Xiao, and Hang Xu. On the analytic and geometric aspects of obstruction flatness. *preprint, arXiv:2212.04034*, 2022.
17. Peter Ebenfelt, Ming Xiao, and Hang Xu. On the classification of normal stein spaces and finite ball quotients with Bergman-Einstein metrics. *Int. Math. Res. Not. IMRN*, (19):15240–15270, 2022.
18. Peter Ebenfelt, Ming Xiao, and Hang Xu. Kähler-Einstein metrics and obstruction flatness of circle bundles. *Journal de Mathématiques Pures et Appliquées*, 177:368–414, 2023.
19. Miroslav Engliš. Pseudolocal estimates for $\overline{\partial}$ on general pseudoconvex domains. *Indiana Univ. Math. J.*, 50(4):1593–1607, 2001.
20. Miroslav Engliš and Genkai Zhang. Ramadanov conjecture and line bundles over compact Hermitian symmetric spaces. *Math. Z.*, 264(4):901–912, 2010.
21. Charles L. Fefferman. The Bergman kernel and biholomorphic mappings of pseudoconvex domains. *Invent. Math.*, 26:1–65, 1974.
22. Charles L. Fefferman. Monge-Ampère equations, the Bergman kernel, and geometry of pseudoconvex domains. *Ann. of Math. (2)*, 103(2):395–416, 1976.
23. C. Robin Graham. Higher asymptotics of the complex Monge-Ampère equation. *Compositio Math.*, 64(2):133–155, 1987.
24. C. Robin Graham. Scalar boundary invariants and the Bergman kernel. In *Complex analysis, II (College Park, Md., 1985–86)*, volume 1276 of *Lecture Notes in Math.*, pages 108–135. Springer, Berlin, 1987.
25. Kengo Hirachi. Personal communication. *Announced in talk at 'Symmetry and Geometric Structures' at IMPAN, Warsaw, November 12–18, 2017.*
26. Kengo Hirachi. Construction of boundary invariants and the logarithmic singularity of the Bergman kernel. *Ann. of Math. (2)*, 151(1):151–191, 2000.
27. Xiaojun Huang and Xiaoshan Li. Bergman-Einstein metric on a Stein space with a strongly pseudoconvex boundary. *arXiv:2008.03645v1*, 2020.
28. Xiaojun Huang and Ming Xiao. Bergman-Einstein metrics, a generalization of Kerner's theorem and Stein spaces with spherical boundaries. *Journal für die reine und angewandte Mathematik (Crelles Journal)*, 2021(770):183–203, 2021.
29. John M. Lee and Richard Melrose. Boundary behaviour of the complex Monge-Ampère equation. *Acta Math.*, 148:159–192, 1982.
30. László Lempert. Embeddings of three-dimensional Cauchy-Riemann manifolds. *Math. Ann.*, 300(1):1–15, 1994.
31. A V Loboda. Homogeneous strictly pseudoconvex hypersurfaces in $\mathbb{C}^3$ with two-dimensional isotropy groups. *Sbornik: Mathematics*, 192(12):1741, Dec 2001.
32. Andrea Loi, Roberto Mossa, and Fabio Zuddas. The log-term of the Bergman kernel of the disc bundle over a homogeneous Hodge manifold. *Ann. Global Anal. Geom.*, 51(1):35–51, 2017.
33. Ottmar Loos. Bounded symmetric domains and jordan pairs. 1977. Math. Lectures, University of California, Irvine.
34. A. H. Martin. Boundary asymptotics for convex and strongly pseudoconvex domains. *Ph.D. Thesis, The University of California Riverside*, 2021.
35. I. P. Ramadanov. A characterization of the balls in $\mathbf{C}^n$ by means of the Bergman kernel. *C. R. Acad. Bulgare Sci.*, 34(7):927–929, 1981.
36. S. M. Webster. Pseudo-Hermitian structures on a real hypersurface. *J. Differential Geom.*, 13(1):25–41, 1978.
37. Ming Xiao. A new Poincaré type rigidity phenomenon with applications. *preprint, arXiv:2309.13768*, 2023.

Analyticity in a Dispersive Camassa–Holm Equation with Cubic Nonlinearities

A. Alexandrou Himonas and Gerson Petronilho

Dedicated to Professor Jorge Hounie in celebration of his 75th birthday

2020 Mathematics Subject Classification Primary: 35Q55, 35G31, 35G16, 37K10

1 Introduction and Results

In this chapter we consider a Camassa–Holm equation having cubic nonlinearities and being modified (regularized) by a third-order dispersion, denoted by mCH3, and study its Cauchy problem, which is

$$\begin{cases} \partial_t u + \partial_x^3 u + \frac{1}{3}\partial_x(u^3) + (1-\partial_x^2)^{-1}\partial_x\left[\frac{5}{3}u^3 + u(\partial_x u)^2\right] = 0, \\ u(0,x) = \varphi(x), \;\; t \in \mathbb{R}, \;\; x \in \mathbb{R}. \end{cases} \tag{1}$$

First, we prove that it is locally well-posed for data in analytic Gevrey spaces $G^{\delta,s}(\mathbb{R})$, $s > 1/2$, and then, utilizing an almost conservation law, we show that the local solutions $u(x,t)$ extend as analytic functions in x for all time t, and derive a lower bound for their radius of spatial analyticity.

A. A. Himonas (✉)
Department of Mathematics, University of Notre Dame, Notre Dame, IN, USA
e-mail: himonas.1@nd.edu

G. Petronilho
Departamento de Matemática, Universidade Federal de São Carlos, São Carlos, SP, Brazil
e-mail: gersonp@ufscar.br

S. Berhanu et al. (eds.), *Geometric Analysis of PDEs and Several Complex Variables*, Latin American Mathematics Series – UFSCar subseries,
https://doi.org/10.1007/978-3-031-69702-9_7

The motivation for studying mCH3 comes from the following generalized Camassa–Holm equation:

$$\partial_t u+\frac{1}{p+1}\partial_x(u^{p+1})+(1-\partial_x^2)^{-1}\partial_x\Big[\frac{p(p+3)}{2(p+1)}u^{p+1}+\frac{p}{2}u^{p-1}(\partial_x u)^2\Big]=0, \quad (2)$$

which was considered by Hakkaev and Kirchev in [18], where p is a positive integer. There, they proved that the Cauchy problem for Eq. (2) is well-posed in Sobolev spaces $H^s(\mathbb{R})$ for $s > 3/2$, via the method of the pseudoparabolic regularization used earlier by Bona and Smith [4] in the case of the Korteweg–de Vries (KdV) equation, and also by Li and Olver [38] in the case of the Camassa–Holm (CH) equation. We note that for $p = 1$ Eq. (2) is the well-known CH equation (having quadratic nonlinearities), and for $p = 2$ Eq. (2) is our Eq. (1) without the dispersive term $\partial_x^3 u$. Furthermore, we note that the well-posedness of the generalized CH Eq. (2) can be established via a mollification of the Burgers term $\partial_x(u^{p+1})$. For this method applied to Camassa–Holm type equations, we refer the reader to [21, 19] and the references therein. Finally, we note that Eq. (2) has solitary wave solutions, whose stability was studied in [18].

The almost conservation law, which we mentioned earlier and which will be the key ingredient for studying the evolution of the radius of analyticity of solution to the Cauchy problem (1), is based on the fact that solutions $u(x,t)$ of the mCH3 equation conserve their H^1-norm, that is,

$$\frac{d}{dt}\int_{\mathbb{R}}[u^2+(\partial_x u)^2]dx=0. \quad (3)$$

To see this conservation law for mCH3 (which is inherited by Eq. (2)) we multiply mCH3 by $(1-\partial_x^2)$ and write it in the local form

$$\partial_t u-\partial_t\partial_x^2 u=-\partial_x^3 u-\partial_x^5 u-2\partial_x\big[u^3\big]+\partial_x\Big[uu_x^2+u^2u_{xx}\Big]=0. \quad (4)$$

Now, using relation (4) and integration by parts, we get

$$\begin{aligned}
&\frac{1}{2}\frac{d}{dt}\int\Big(u^2+(\partial_x u)^2\Big)dx\\
&=\int u\cdot\Big(\partial_t u-\partial_x^2\partial_t u\Big)dx\\
&=\int u\cdot\Big(-\partial_x^3 u+\partial_x^5 u-2\partial_x\big[u^3\big]+\partial_x\big[uu_x^2+u^2u_{xx}\big]\Big)dx\\
&=\int\Big(-\frac{1}{2}\partial_x(u^4)-\Big[\frac{u^2}{2}\Big]_x\cdot u_x^2-u_xu^2u_{xx}\Big)dx=0. \quad (5)
\end{aligned}$$

Next, we recall the definitions needed for stating our main results precisely. We begin with the data spaces, which are the analytic Gevrey spaces $G^{\delta,s}(\mathbb{R})$ that are defined by the norm (see [15])

$$\|\varphi\|_{G^{\delta,s}}^2 = \int_{\mathbb{R}} e^{2\delta|\xi|}(1+|\xi|)^{2s}|\widehat{\varphi}(\xi)|^2 d\xi, \tag{6}$$

where $\delta > 0$, $s \in \mathbb{R}$, and $\widehat{\varphi}(\xi) \doteq \int_{\mathbb{R}} e^{-ix\xi}\varphi(x)dx$ is the Fourier transform on $\mathbb{R}$. We note that, by the Paley–Wiener theorem, a function belongs in $G^{\delta,s}(\mathbb{R})$ if and only if it can be extended as a holomorphic function of $z = x + iy$ in the strip $|y| < \delta$ around the real axis and whose x-Sobolev norms are uniformly bounded in y [29]. For this reason δ is called **radius of spatial analyticity**. Also, we note that when $\delta = 0$, then we obtain the norm in the Sobolev spaces on the line $H^s(\mathbb{R})$. Also, we define the solution spaces. These are hybrids between the analytic Gevrey spaces and the Bourgain spaces. More precisely, for $\delta \geq 0$, $s \in \mathbb{R}$, and $b \in \mathbb{R}$, the spaces $X^{\delta,s,b}(\mathbb{R}^2)$ are the Banach spaces defined by the norm

$$\|u\|_{X^{\delta,s,b}(\mathbb{R}^2)}^2 = \|u\|_{\delta,s,b}^2 = \int_{\mathbb{R}}\int_{\mathbb{R}} e^{2\delta|\xi|}(1+|\xi|)^{2s}(1+|\tau-\xi^3|)^{2b}|\widehat{u}(\xi,\tau)|^2 d\xi d\tau, \tag{7}$$

where $\widehat{u}(\xi,\tau) \doteq \int_{\mathbb{R}}\int_{\mathbb{R}} e^{-i(x\xi+t\tau)}u(x,t)\,dxdt$ is the space-time Fourier transform. For $\delta = 0$, $X^{\delta,s,b}(\mathbb{R}^2)$ coincides with the space $X^{s,b}(\mathbb{R}^2)$ introduced by Bourgain [5, 6], and Kenig, Ponce and Vega [31]. The norm of $X^{s,b}(\mathbb{R}^2)$ is explicitly given by

$$\|u\|_{X^{s,b}(\mathbb{R}^2)}^2 = \int_{\mathbb{R}}\int_{\mathbb{R}} (1+|\xi|)^{2s}(1+|\tau-\xi^3|)^{2b}|\widehat{u}(\tau,\xi)|^2 d\tau d\xi. \tag{8}$$

Also, $X_T^{\delta,s,b}$ denotes the restricted analytic Bourgain space defined by

$$X_T^{\delta,s,b} = \{u;\, u(x,t) = v(x,t) \;\text{ on }\; \mathbb{R}\times(-T,T) \text{ with } v \in X^{\delta,s,b}(\mathbb{R}^2)\}. \tag{9}$$

endowed with the norm

$$\begin{aligned}\|u\|_{X_T^{\delta,s,b}} &= \|u\|_{X_T^{\delta,s,b}(\mathbb{R}^2)} \\ &= \inf_{v\in X^{\delta,s,b}}\{\|v\|_{\delta,s,b};\; v(x,t) = u(x,t) \;\text{ on }\; \mathbb{R}\times(-T,T)\}, \quad T > 0.\end{aligned} \tag{10}$$

We note that all the well-known properties of the standard $X^{s,b}(\mathbb{R}^2)$ spaces carry over to the Gevrey-modified spaces by making use of the substitution $u \to e^{\delta|D_x|}u$.

Finally, to keep the labeling of the constants appearing in our estimates simple, in the sequel we shall use the letter C to denote a constant, which may change a finite number of times.

With the above terminology in place, we can state our first main result as follows.

Theorem 1.1 (Local Well-Posedness in Analytic Gevrey Spaces) *Let $s > \frac{1}{2}$ and $\delta > 0$. Then for any initial data $\varphi \in G^{\delta,s}(\mathbb{R})$, there exists a positive time $T_{\delta,s} \doteq T_{\delta,s,\varphi}$, depending only on δ, s and $\|\varphi\|_{G^{\delta,s}}$, and a unique solution u of Cauchy problem (1) on the interval $[-T_{\delta,s}, T_{\delta,s}]$ such that $u \in X^{\delta,s,b}_{T_{\delta,s}} \hookrightarrow C([-T_{\delta,s}, T_{\delta,s}], G^{\delta,s}(\mathbb{R}))$. Furthermore, the solution u satisfies the bound*

$$\|u(t)\|_{G^{\delta,s}(\mathbb{R})} \leq C\|u\|_{X^{\delta,s,b}(\mathbb{R}^2)} \leq 2C\|\varphi\|_{G^{\delta,s}(\mathbb{R})}, \quad |t| \leq T_{\delta,s} \tag{11}$$

with the lifespan $T_{\delta,s}$ satisfying the estimate

$$T_{\delta,s} = \frac{1}{\left(2 + 48C^5 2^{\epsilon}\|\varphi\|^2_{G^{\delta,s}(\mathbb{R})}\right)^{\frac{2}{\varepsilon}}}, \tag{12}$$

for some $b \in (\frac{1}{2}, 1)$ and some $\varepsilon \in (0, \frac{1}{2})$. Finally, the solution depends Lip-continuously on the data φ.

The proof of Theorem 1.1 is based on two trilinear estimates. The first estimates the mKdV nonlinearity $\partial_x(u^3)$, and the second estimates the nonlocal Camassa–Holm nonlinearity $(1 - \partial_x^2)^{-1}\partial_x[u(\partial_x u)^2]$, both in the standard Bourgain spaces norm (8). Then, these trilinear estimates are expressed in their analytic Gevrey form and used to establish a fixed point (solution) of the iteration map of the mCH3 Cauchy problem in a ball of the spaces $X^{\delta,s,b}(\mathbb{R}^2)$. We note that the analogous to Theorem 1.1 result for the dispersive regularization of the classical Camassa–Holm equation was established in [25].

While there is no $G^{\delta_0,1}$ conservation law, we can prove an almost conservation law for the mCH3 equation in analytic Gevrey spaces, thanks to the H^1 conservation law (3) and the local well-posedness Theorem 1.1. This is our second main result and reads as follows.

Theorem 1.2 (Almost Conservation Law) *Given $0 < \delta_0 \leq 1$ and initial data $u_0 \in G^{\delta_0,1}(\mathbb{R})$, let $u \in X^{\delta,1,b}_{T_{\delta,1}} \hookrightarrow C([0, T_{\delta,1}], G^{\delta,1}(\mathbb{R}))$ be the local solution obtained in Theorem 1.1 for $0 < \delta < \delta_0$. Then, we have*

$$\sup_{|t| \leq \rho^*} \|u(t)\|^2_{G^{\delta,1}} \leq \|u(0)\|^2_{G^{\delta,1}} + C\frac{\delta^{\theta}}{(\delta_0 - \delta)^5}\|u\|^4_{X^{\delta_0,1,b}_{T_{\delta_0,1}}}, \tag{13}$$

where $0 < \rho^* < T_{\delta_0,1}$ and $0 < \theta \leq 1$.

Remark 1.1 It follows from Theorem 1.1 with $s = 1$ and $\delta = \delta_0$ that

$$\|u(t)\|_{G^{\delta_0,1}(\mathbb{R})} \leq C\|u\|_{X^{\delta_0,1,b}_{T_{\delta_0,1}}(\mathbb{R}^2)} \leq 2C\|u(0)\|_{G^{\delta_0,1}(\mathbb{R})}, \quad |t| \leq T_{\delta_0,1}.$$

Therefore it follows from Theorem 1.2 that

$$\sup_{|t|\le\rho^*} \|u(t)\|^2_{G^{\delta,1}} \le \|u(0)\|^2_{G^{\delta,1}} + C\frac{\delta^\theta}{(\delta_0-\delta)^5}\|u(0)\|^4_{G^{\delta_0,1}(\mathbb{R})}. \tag{14}$$

Our third main result is about the evolution of the radius of spatial analyticity $\delta(t)$ of mCH3 solutions $u(t)$. Using our Theorem 1.1, we get a local solution, and then applying our almost conservation law, we extend this solution step by step for all times while at the same time obtaining a lower bound for its radius of spatial analyticity. More precisely, we have the following result.

Theorem 1.3 (Lower Bound for Radius of Spatial Analyticity) *If* $0 < \delta_0 \le 1$, $s > \frac{1}{2}$, *and* $\varphi \in G^{\delta_0,s}(\mathbb{R})$, *then the solution* u *obtained in Theorem 1.1 extends globally in time, and for any* $T > 0$, *we have*

$$u \in C([-T,T], G^{\delta(T),s}(\mathbb{R})), \quad \text{with } \delta(T) = \min\left\{\delta_0, \frac{c_0}{Te^{cT}}\right\}, \tag{15}$$

where c_0 *and* c *are positive constants depending on* φ, δ_0, *and* s.

Theorem 1.3 constitutes a significant progress in the study of the evolution of the radius of spatial analyticity for dispersive Camassa–Holm equations, although the lower bound (15) proved here may not be optimal. Obtaining better lower bounds for the mCH3 equation is an interesting open problem. Improving the almost conservation law obtained here seems to be a natural direction.

The study of dispersive Camassa–Holm equations was initiated in [22, 23], where well-posedness of the Cauchy problem for a family of such equations in Sobolev space was proved. The idea was to view Camassa–Holm equations as limits of their dispersive regularization for which one could utilize the Bourgain spaces theory and multilinear estimates developed in the 1990s for their solving (see [5, 6, 9, 10, 27, 31, 32, 39, 43]). Then, going to the limit, one could obtain the well-posedness of the original equation. The guiding model for this is the KdV regularization $u_t+\varepsilon^2u_{xxx}-6uu_x=0$ of the Burgers equation $u_t-6uu_x=0$, by Lax and Levermore [33, 34, 35, 36]. Concerning local dispersive equations, like KdV, the first lower bound for their spatial analyticity radius was obtained in Bona, Grujić and Kalisch [3], and in Grujić and Kalisch [17]. Some of these estimates were improved in Selberg and Silva [42] via bilinear and almost conservation laws, a method which we adopt here for the case of nonlocal equations with cubic nonlinearities. For related results on local and nonlocal equations, we refer the reader to [1, 2, 14, 24, 20, 40], and the references therein. Finally, we note that Camassa–Holm type equations arise in the theory of water waves and can be thought as toy models of the Euler equations. Also, they arise in the theory of integrable equations and systems. For some results on these, we refer to [7, 8, 11, 12, 13, 41, 16, 37].

The rest of this chapter is organized as follows. In Sect. 2, using two trilinear estimates in analytic Bourgain spaces, we prove the local well-posedness, that is, Theorem 1.1. In Sect. 3, using Theorem 1.1 and the H^1 conservation law, we prove the almost conservation law in $G^{\delta,1}$ described in Theorem 1.2. Finally, in Sect. 4, using this law, we extend the local solution for all time step by step, using an induction argument, and derive the lower bound for its radius of analyticity stated in Theorem 1.3.

In the sequel, we shall use the same letter C or c to represent a constant which may change a finite number of times.

2 Well-Posedness in Analytic Gevrey Spaces

We begin with the linear estimates for the Cauchy problem

$$\begin{cases} u_t + u_{xxx} = F(t,x) \\ u(0,x) = \varphi(x). \end{cases} \tag{16}$$

By using Duhamel's formula, we may write the solution as follows:

$$u(t,x) = W(t)\varphi(x) - \int_0^t W(t-t')F(t',x)dt', \tag{17}$$

where $W(t) = e^{-tL}$, L is the linear operator $L = \partial_x^3$ and

$$W(t)\varphi(x) = \int_{\mathbb{R}} e^{i\{x\xi - \xi^3 t\}}\widehat{\varphi}(\xi)d\xi. \tag{18}$$

Furthermore, looking for solutions local in time, say they exist for $|t| \leq \frac{1}{2}$, we multiply the right-hand side of (17) by a cut-off function $\psi \in C_0^\infty(-1,1)$ such that

$$\psi(t) = \begin{cases} 1, & |t| \leq \frac{1}{2}, \\ 0, & |t| > 1, \end{cases}$$

and for $0 < T < 1$, we set $\psi_T(t) = \psi(t/T)$. Thus, we obtain the following localized form of the integral equation (17):

$$\Gamma(u)(x,t) = \psi(t)W(t)\varphi(x) - \psi(t)\int_0^t W(t-t')F(t',x)\,dt', \tag{19}$$

which coincides with integral equation (17) for $|t| \leq 1/2$. For more details about this approach we suggest the reader [25].

We begin with the following result, whose proof follows the lines of the proof of Lemma 3.1 in [30].

Lemma 2.1 *For all $s \in \mathbb{R}$, $\delta > 0$ and $b > \frac{1}{2}$, we have*

$$\|\psi(t)W(t)\varphi(x)\|_{X^{\delta,s,b}(\mathbb{R}^2)} \le C\|\varphi\|_{G^{\delta,s}(\mathbb{R})},$$

where $C > 0$ depends on ψ.

We also need the following results.

Lemma 2.2 *For all $s \in \mathbb{R}$, $\delta > 0$, $b \in (1/2, 1)$, we have*

$$\left\|\psi(t)\int_0^t W(t-t')F(t')dt'\right\|_{X^{\delta,s,b}(\mathbb{R}^2)} \le C\|F\|_{X^{\delta,s,b-1}(\mathbb{R}^2)},$$

where the constant C depends on ψ.

Lemma 2.3 *Let $\varphi \in \mathcal{S}(\mathbb{R})$ be a Schwartz function in time. If $-\frac{1}{2} < b_1 < b_1' < \frac{1}{2}$, then for any $T \in (0, 1)$ and $\delta > 0$ we have*

$$\|\psi_T(t)u\|_{X^{\delta,s,b_1}(\mathbb{R}^2)} \le C(\varphi, b_1, b_1')T^{b_1'-b_1}\|u\|_{X^{\delta,s,b_1'}(\mathbb{R}^2)}.$$

The proofs of Lemmas 2.2 and 2.3 for $\delta = 0$ can be found in [25] and in Lemma 2.11 of [43], respectively.

These inequalities clearly remain valid for $\delta > 0$. Note that if for a real-valued function u we define the real-valued function U

$$U(x,t) = e^{\delta|D_x|}u(x,t) \quad or \quad \widehat{U}(\xi,\tau) = e^{\delta|\xi|}\widehat{u}(\xi,\tau), \tag{20}$$

then we have the following norm relations:

$$\|U\|_{s,b} = \|e^{\delta|D_x|}u(x,t)\|_{s,b} = \|u\|_{\delta,s,b} \tag{21}$$

and

$$\|U(t)\|_{H^s} = \|e^{\delta|D_x|}u(t)\|_{H^s} = \|u(t)\|_{G^{\delta,s}}. \tag{22}$$

The next result provides the essential trilinear estimates needed for the proof of Theorem 1.1. We start with the following $mKdV$ trilinear estimate that is proved in [26].

Theorem 2.1 (mKdV Trilinear Estimates) *If $s > \frac{1}{4}$, then there exist ν and ν' with $1/2 < \nu' \le \nu < 1$ such that the following trilinear estimates hold:*

$$\|\partial_x(fgh)\|_{X^{s,\nu-1}} \le c_{s,\nu}\|f\|_{X^{s,\nu'}}\|g\|_{X^{s,\nu'}}\|h\|_{X^{s,\nu'}}, \quad f, g, h \in X^{s,\nu'}(\mathbb{R}^2). \tag{23}$$

In fact, ν and ν′ depend on s and can be chosen as follows:

$$\frac{1}{2} < \nu' \leq \nu \leq \frac{1}{2} + \beta_1, \tag{24}$$

where

$$\beta_1 = \beta_1(s) \doteq \min\left\{\frac{1}{6}\left(s - \frac{1}{4}\right), \frac{1}{8}\right\}. \tag{25}$$

Remark 2.1 Noting that

$$\partial_x \widehat{(f_1 f_2 f_3)}(\tau, \xi) = \xi \left(\int_{\mathbb{R}^4} \widehat{f_1}(\tau - \tau_1, \xi - \xi_1)\widehat{f_2}(\tau_1 - \tau_2, \xi_1 - \xi_2)\widehat{f_3}(\tau_2, \xi_2) d\tau_1 d\tau_2 d\xi_1 d\xi_2 \right),$$

and setting

$$g_i(\tau, \xi) := (1 + |\xi|)^s (1 + |\tau - \xi^3|)^{\nu'} \widehat{f_i}(\tau, \xi), \ \ i = 1, 2, 3,$$

the estimate of Theorem 2.1 can be rewritten as

$$\begin{aligned}
\|\partial_x(f_1 f_2 f_3)\|_{X^{s,\nu-1}(\mathbb{R}^2)} &= \left\| \frac{\xi(1 + |\xi|)^s}{(1 + |\tau - \xi^3|)^{1-\nu}} \widehat{f_1 f_2 f_3}(\tau, \xi) \right\|_{L^2_{\tau,\xi}(\mathbb{R}^2)} \\
&= \left\| \frac{\xi(1 + |\xi|)^s}{(1 + |\tau - \xi^3|)^{1-\nu}} \right. \\
&\quad \times \left. \int_{\mathbb{R}^4} \widehat{f_1}(\tau - \tau_1, \xi - \xi_1)\widehat{f_2}(\tau_1 - \tau_2, \xi_1 - \xi_2)\widehat{f_3}(\tau_2, \xi_2) d\xi_1 d\xi_2 \tau_1 d\tau_2 \right\|_{L^2_{\tau,\xi}(\mathbb{R}^2)} \\
&= \left\| \frac{\xi(1 + |\xi|)^s}{(1 + |\tau - \xi^3|)^{1-\nu}} \int_{\mathbb{R}^4} \frac{g_1(\tau - \tau_1, \xi - \xi_1)}{(1 + |\xi - \xi_1|)^s (1 + |(\tau - \tau_1) - (\xi - \xi_1)^3|)^{\nu'}} \right. \\
&\quad \times \frac{g_2(\tau_1 - \tau_2, \xi_1 - \xi_2)}{(1 + |\xi_1 - \xi_2|)^s (1 + |(\tau_1 - \tau_2) - (\xi_1 - \xi_2)^3)|)^{\nu'}} \\
&\quad \times \left. \frac{g_3(\tau_2, \xi_2)}{(1 + |\xi_2|)^s (1 + |\tau_2 - \xi_2^3|)^{\nu'}} d\xi_1 d\xi_2 d\tau_1 d\tau_2 \right\|_{L^2_{\tau,\xi}(\mathbb{R}^2)} \\
&\leq C \|g_1\|_{L^2_{\tau,\xi}(\mathbb{R}^2)} \|g_2\|_{L^2_{\tau,\xi}(\mathbb{R}^2)} \|g_3\|_{L^2_{\tau,\xi}(\mathbb{R}^2)}.
\end{aligned} \tag{26}$$

Corollary 2.1 *If $s > \frac{1}{4}$, then there exist ν and ν′ with $1/2 < \nu' \leq \nu < 1$ such that the following trilinear estimates hold:*

$$\|\partial_x(v_1v_2v_3)\|_{X^{\delta,s,\nu-1}(\mathbb{R}^2)} \leq C\|v_1\|_{X^{\delta,s,\nu'}(\mathbb{R}^2)}\|v_2\|_{X^{\delta,s,\nu'}(\mathbb{R}^2)}\|v_3\|_{X^{\delta,s,\nu'}(\mathbb{R}^2)}$$

for some constant $C > 0$. In fact, ν and ν' depend on s and can be chosen as in (24) *and* (25).

Proof This estimate can be restated in the form of Remark 2.1 modified by the factor $e^{\delta|\xi|}$ in the first line of (26), and in the other lines, we can replace, using the triangle inequality, $e^{\delta|\xi|}$ by $e^{\delta|\xi-\xi_1|}e^{\delta|\xi_1-\xi_2|}e^{\delta|\xi_2|}$. More precisely, setting

$$f_i(\tau,\xi) \doteq e^{\delta|\xi|}(1+|\xi|)^s(1+|\tau-\xi^3|)^{\nu'}\widehat{v_i}(\tau,\xi),\ i=1,2,3, \tag{27}$$

the estimate of Corollary 2.1 can be rewritten as

$$\begin{aligned}\|\partial_x(v_1v_2v_3)\|_{X^{\delta,s,\nu-1}(\mathbb{R}^2)} &= \left\|\frac{e^{\delta|\xi|}\xi(1+|\xi|)^s}{(1+|\tau-h(\xi)|)^{1-\nu}}\widehat{v_1v_2v_3}(\tau,\xi)\right\|_{L^2_{\tau,\xi}(\mathbb{R}^2)}\\ &\leq \left\|\frac{\xi(1+|\xi|)^s}{(1+|\tau-h(\xi)|)^{1-\nu}}\int_{\mathbb{R}^4} e^{\delta|\xi-\xi_1|}\widehat{v_1}(\tau-\tau_1,\xi-\xi_1)\right.\\ &\quad\left.\times\, e^{\delta|\xi_1-\xi_2|}\widehat{v_2}(\tau_1-\tau_2,\xi_1-\xi_2)e^{\delta|\xi_2|}\widehat{v_3}(\tau_2,\xi_2)d\xi_1 d\xi_2\tau_1 d\tau_2\right\|_{L^2_{\tau,\xi}(\mathbb{R}^2)}.\end{aligned} \tag{28}$$

By using (27), we can write

$$\begin{aligned}\|\partial_x(v_1v_2v_3)\|_{X^{\delta,s,\nu-1}(\mathbb{R}^2)} &= \left\|\frac{\xi(1+|\xi|)^s}{(1+|\tau-h(\xi)|)^{1-\nu}}\right.\\ &\quad\times\int_{\mathbb{R}^4}\frac{f_1(\tau-\tau_1,\xi-\xi_1)}{(1+|\xi-\xi_1|)^s(1+|(\tau-\tau_1)-(\xi-\xi_1)^3|)^{\nu'}}\\ &\quad\times\frac{f_2(\tau_1-\tau_2,\xi_1-\xi_2)}{(1+|\xi_1-\xi_2|)^s(1+|(\tau_1-\tau_2)-(\xi_1-\xi_2)^3|)^{\nu'}}\\ &\quad\left.\times\frac{f_3(\tau_2,\xi_2)}{(1+|\xi_2|)^s(1+|\tau_2-\xi_2^3|)^{\nu'}}d\xi_1 d\xi_2 d\tau_1 d\tau_2\right\|_{L^2_{\tau,\xi}(\mathbb{R}^2)}\\ &\leq C\|f_1\|_{L^2_{\tau,\xi}(\mathbb{R}^2)}\|f_2\|_{L^2_{\tau,\xi}(\mathbb{R}^2)}\|f_3\|_{L^2_{\tau,\xi}(\mathbb{R}^2)} = C||v_1||_{\delta,s,\nu'}||v_2||_{\delta,s,\nu'}||v_3||_{\delta,s,\nu'}.\end{aligned} \tag{29}$$

□

The second trilinear estimates are less standard and are proved in [26].

Theorem 2.2 (Nonlocal Trilinear Estimate) *If $s > \frac{1}{2}$, then there exist β and β' with $\frac{1}{2} < \beta' \leq \beta < 1$, such that the following trilinear estimate holds*

$$\begin{aligned}&\|(1-\partial_x^2)^{-1}\partial_x[f(\partial_x g)(\partial_x h)]\|_{X^{s,\beta-1}}\\ &\quad\leq c_{s,\beta}\|f\|_{X^{s,\beta'}}\|g\|_{X^{s,\beta'}}\|h\|_{X^{s,\beta'}},\quad f,g,h\in X^{s,\beta'}(\mathbb{R}^2),\end{aligned} \tag{30}$$

for some constant $C_{s,\beta} > 0$. *In fact,* β *and* β' *depend on* s *and can be chosen as follows:*

$$\frac{1}{2} < \beta' \leq \beta \leq \frac{1}{2} + \beta_2, \tag{31}$$

where

$$\beta_2 = \beta_2(s) \doteq \min\left\{\frac{1}{6}\left(s - \frac{1}{2}\right), \frac{1}{8}\right\}. \tag{32}$$

Following the lines of Remark 2.1 and the proof of Corollary 2.1, one can prove the following:

Corollary 2.2 *Let* $\delta > 0$, $s > \frac{1}{2}$, *and then there exist* β *and* β' *with* $\frac{1}{2} < \beta' \leq \beta < 1$, *such that the following trilinear estimate holds*

$$\|\partial_x(1-\partial_x^2)^{-1}[f(\partial_x g)(\partial_x h)\|_{X^{\delta,s,\beta-1}(\mathbb{R}^2)}$$
$$\leq C_{s,\beta}\|f\|_{X^{\delta,s,\beta'}(\mathbb{R}^2)}\|g\|_{X^{\delta,s,\beta'}(\mathbb{R}^2)}\|h\|_{X^{\delta,s,\beta'}(\mathbb{R}^2)}, \quad f, g, h \in X^{\delta,s,\beta'}(\mathbb{R}^2)$$

for some constant $C_{s,\beta} > 0$. *In fact,* β *and* β' *depend on* s *and can be chosen as in* (31) *and in* (32).

For the proof of local well-posedness in the analytic Gevrey spaces, we will use the standard Banach contraction principle for functions $u \in X^{\delta,s,b}(\mathbb{R}^2)$ in a given closed ball B. For simplicity we can write the mCH3 equation as

$$\partial_t u + \partial_x^3 u + G(u) = 0, \tag{33}$$

where $G_u(t) = \frac{1}{3}\partial_x(u^3) + F(u)$ with $F(u) = (1-\partial_x^2)^{-1}\partial_x\left[\frac{5}{3}u^3 + u(\partial_x u)^2\right)\right]$.

Next, for fixed $\delta > 0$, $s > \frac{1}{2}$, and $u_0 \in G^{\delta,s}(\mathbb{R})$, we consider the integral operator

$$\Gamma(u)(t,x) := \psi(t)W(t)\varphi(x) + \psi(t)\int_0^t W(t-t')G_u(t')dt'. \tag{34}$$

Furthermore, in order to deal with arbitrary size data for which we can prove existence of solution with small lifespan T, we shall assume that

$$0 < T < 1/2, \tag{35}$$

and we will replace the integral equation (34) with the following localization:

$$\Gamma_T(u)(x,t) = \psi(t)W(t)\varphi(x) + \psi_T(t)\int_0^t W(t-t')\psi_{2T}(t')G_u(t')dt' \doteq \Gamma_{T,u_0}(u), \tag{36}$$

where $\psi_T(t) = \psi(t/T)$. Notice that for $|t| \leq T$, the fixed point of the iteration map (36) is the solution to the mCH3 ivp (1), since $\Gamma_T(u) = \Gamma(u)$ if $|t| \leq T$.

For $s \geq \frac{1}{2}$, $b \in (\frac{1}{2}, 1)$, and $T \in (0, \frac{1}{2})$, it follows from Lemmas 2.1 and 2.2 that

$$\|\Gamma_T(u)\|_{X^{\delta,s,b}(\mathbb{R}^2)} \leq C\|\varphi\|_{G^{\delta,s}(\mathbb{R})} + C\|\psi_{2T}(t)G_u(t)\|_{X^{\delta,s,b-1}(\mathbb{R}^2)}. \tag{37}$$

Now, we proceed by keeping in mind that our constants may change a finite number of times, although we use the same symbol for denoting them. In order to estimate the term $\|\psi_{2T}(t)G_u(t)\|_{X^{\delta,s,b-1}(\mathbb{R}^2)}$, we will use Lemma 2.3 with $b_1 = b-1$, $b_1' = b-1+\epsilon$, with $0 < \epsilon < \frac{1}{2}$, and therefore we have $b_1' - b_1 = \epsilon$. Applying this lemma, we get

$$\|\psi_{2T}(t)G_u(t)\|_{X^{\delta,s,b-1}(\mathbb{R}^2)} \leq C(2T)^{\epsilon}\|G_u(t)\|_{X^{\delta,s,b_1'}(\mathbb{R}^2)}. \tag{38}$$

Thus, from (37) and (38), we have

$$\|\Gamma_T(u)\|_{X^{\delta,s,b}(\mathbb{R}^2)} \leq C\|\varphi\|_{G^{\delta,s}(\mathbb{R})} + C^2 2^{\epsilon} T^{\epsilon}\|G_u(t)\|_{X^{\delta,s,b_1'}(\mathbb{R}^2)}. \tag{39}$$

In order to analyze $\|G_u(t)\|_{X^{\delta,s,b_1'}(\mathbb{R}^2)}$, we will analyze separately the terms $\frac{1}{3}\|\partial_x u^3\|_{X^{\delta,s,b_1'}(\mathbb{R}^2)}$, $\|(1-\partial_x^2)^{-1}\partial_x(\frac{5}{3}u^3)\|_{X^{\delta,s,b_1'}(\mathbb{R}^2)}$ and $\|(1-\partial_x^2)^{-1}\partial_x \left[u(u_x)^2\right]\|_{X^{\delta,s,b_1'}(\mathbb{R}^2)}$.

To analyze the first one, we apply Corollary 2.1 with $v_1 = v_2 = v_3 = u$, $\nu - 1 = b_1'$, $\nu' = b$, and we get

$$\frac{1}{3}\|\partial_x u^3\|_{X^{\delta,s,b_1'}(\mathbb{R}^2)} \leq C\|u\|^3_{X^{\delta,s,b}(\mathbb{R}^2)}. \tag{40}$$

Note that our choice of the parameter is good for this corollary. In fact, since $b \in (\frac{1}{2}, 1)$, we have $\frac{1}{2} < b = \nu'$. Also, since $\nu = b_1' + 1 = (b-1+\epsilon)+1 = b+\epsilon$, we conclude that $\nu' < \nu$. By choosing $0 < \epsilon = \frac{1-b}{2}$, we obtain $b+\epsilon = b + \frac{1-b}{2} = b + \frac{1}{2} - \frac{b}{2} = \frac{b}{2} + \frac{1}{2} = \frac{b+1}{2} < \frac{1+1}{2} = 1$.

Now we will analyze the term $F(u) \doteq F_1(u) + F_2(u)$ where $F_1(u) = (1-\partial_x^2)^{-1}\partial_x(\frac{5}{3}u^3)$, $F_2(u) = (1-\partial_x^2)^{-1}\partial_x[u(u_x)^2]$. Working as above, we have

$$\begin{aligned}\|F_1(u)\|_{X^{\delta,s,b_1'}(\mathbb{R}^2)} &= \left\|\left(1-\partial_x^2\right)^{-1}\partial_x\left(\frac{5}{3}u^3\right)\right\|_{X^{\delta,s,b_1'}(\mathbb{R}^2)} \leq \left\|\partial_x\left(\frac{5}{3}u^3\right)\right\|_{X^{\delta,s-2,b_1'}(\mathbb{R}^2)} \\ &\leq \left\|\partial_x\left(\frac{5}{3}u^3\right)\right\|_{X^{\delta,s,b_1'}(\mathbb{R}^2)} \leq C\|u\|^3_{X^{\delta,s,b}(\mathbb{R}^2)}.\end{aligned} \tag{41}$$

To analyze the term $F_2(u)$, we apply Corollary 2.2 with $f = g = h = u$, $\beta - 1 = b_1'$, $\beta' = b$, and we get

$$\|F_2(u)\|_{X^{\delta,s,b_1'}(\mathbb{R}^2)} = \|(1-\partial_x^2)^{-1}\partial_x[u(u_x)^2]\|_{X^{\delta,s,b_1'}(\mathbb{R}^2)} \leq C\|u\|^3_{X^{\delta,s,b}}. \quad (42)$$

Finally, combining estimates (39), (40), (41), and (42), we obtain that

$$\begin{aligned}\|\Gamma_T(u)\|_{X^{\delta,s,b}(\mathbb{R}^2)} &\leq C\|\varphi\|_{G^{\delta,s}(\mathbb{R})} + C^2 2^\epsilon T^\epsilon \Big(3C\|u\|^3_{X^{\delta,s,b}(\mathbb{R}^2)}\Big) \\ &= C\|\varphi\|_{G^{\delta,s}(\mathbb{R})} + 3C^3 2^\epsilon T^\epsilon \|u\|^3_{X^{\delta,s,b}(\mathbb{R}^2)}. \quad (43)\end{aligned}$$

Now, we let $\mathcal{B}$ be the closed ball given by $\mathcal{B} = \{u \in X^{\delta,s,b}(\mathbb{R}^2) : \|u\|_{X^{\delta,s,b}(\mathbb{R}^2)} \leq 2CT^{-\frac{1}{6}\epsilon}\|\varphi\|_{G^{\delta,s}(\mathbb{R})}\}$, where C comes from (43). For $u \in \mathcal{B}$, from (43), we obtain

$$\begin{aligned}\|\Gamma_T(u)\|_{X^{\delta,s,b}(\mathbb{R}^2)} &\leq C\|\varphi\|_{G^{\delta,s}(\mathbb{R})} + 3C^3 2^\epsilon T^\epsilon (2C)^3 T^{-\frac{1}{2}\epsilon}\|\varphi\|^3_{G^{\delta,s}(\mathbb{R})}. \\ &= C\|\varphi\|_{G^{\delta,s}(\mathbb{R})} + 24C^6 2^\epsilon T^{\frac{1}{2}\epsilon}\|\varphi\|^3_{G^{\delta,s}(\mathbb{R})}. \quad (44)\end{aligned}$$

Thus, if we fix T such that

$$T^{\frac{1}{2}\epsilon} \leq \frac{1}{2 + 48C^5 2^\epsilon \|\varphi\|^2_{G^{\delta,s}(\mathbb{R})}}, \quad (45)$$

which implies that

$$2^\epsilon 24C^5 T^{\frac{1}{2}\epsilon}\|\varphi\|^2_{G^{\delta,s}(\mathbb{R})} \leq \frac{2^\epsilon 24C^5\|\varphi\|^2_{G^{\delta,s}(\mathbb{R})}}{2 + 48C^5 2^\epsilon \|\varphi|^2_{G^{\delta,s}(\mathbb{R})}} \leq \frac{2^\epsilon 24C^5\|\varphi\|^2_{G^{\delta,s}(\mathbb{R})}}{48C^5 2^\epsilon \|\varphi|^2_{G^{\delta,s}(\mathbb{R})}} = \frac{1}{2},$$

then we have that

$$\|\Gamma_T(u)\|_{X^{\delta,s,b}(\mathbb{R}^2)} \leq C\|\varphi\|_{G^{\delta,s}(\mathbb{R})} + \frac{1}{2}C\|\varphi\|_{G^{\delta,s}(\mathbb{R})} < 2C\|\varphi\|_{G^{\delta,s}(\mathbb{R})}, \quad (46)$$

and since $0 < T < 1$, we obtain

$$\|\Gamma_T(u)\|_{X^{\delta,s,b}(\mathbb{R}^2)} < 2CT^{-\frac{1}{6}\epsilon}\|\varphi\|_{G^{\delta,s}(\mathbb{R})}, \quad (47)$$

and therefore we have $\Gamma_T(\mathcal{B}) \subset \mathcal{B}$, which completes the onto proof.

Next we prove that Γ_T is a contraction on $\mathcal{B}$. For this we observe that

$$\Gamma_T(u) - \Gamma_T(v) = -\psi(t)\int_0^t W(t-t')\psi_{2T}(t')\Big[G_u - G_v\Big](t')dt'. \quad (48)$$

And, applying Lemmas 2.2 and 2.3 in the same way as in the previous calculations, we obtain

$$\begin{aligned}
&\|\Gamma_T(u)-\Gamma_T(v)\|_{X^{\delta,s,b}(\mathbb{R}^2)}\\
&=\left\|\psi(t)\int_0^t W(t-t')\psi_{2T}(t')\Big[G_u-G_v\Big](t')dt'\right\|_{X^{\delta,s,b}(\mathbb{R}^2)}\\
&\le C\left\|\psi_{2T}(t')\Big[G_u-G_v\Big](t')\right\|_{X^{\delta,s,b-1}(\mathbb{R}^2)}\le CC(2T)^{\varepsilon}\left\|\Big[G_u-Gv\Big](t')\right\|_{X^{\delta,s,b_1'}(\mathbb{R}^2)}
\end{aligned}\tag{49}$$

$$\begin{aligned}
&=C^2 2^{\varepsilon}T^{\varepsilon}\left\|\frac{1}{3}\partial_x(u^3-v^3)+(1-\partial_x^2)^{-1}\partial_x\right.\\
&\qquad\left.\times\Big[\frac{5}{3}(u^3-v^3)+u(\partial_x u)^2-v(\partial_x v)^2\Big]\right\|_{X^{\delta,s,b_1'}(\mathbb{R}^2)}\\
&\le C^2 2^{\varepsilon}T^{\varepsilon}\Bigg(\left\|\frac{1}{3}\partial_x(u^3-v^3)\right\|_{X^{\delta,s,b_1'}(\mathbb{R}^2)}+\left\|(1-\partial_x^2)^{-1}\partial_x\Big[\frac{5}{3}(u^3-v^3)\Big]\right\|_{X^{\delta,s,b_1'}(\mathbb{R}^2)}\\
&\qquad+\left\|(1-\partial_x^2)^{-1}\partial_x\Big[u(\partial_x u)^2-v(\partial_x v)^2\Big]\right\|_{X^{\delta,s,b_1'}(\mathbb{R}^2)}\Bigg)\doteq C^2 2^{\varepsilon}T^{\varepsilon}\Big(I+II+III\Big).
\end{aligned}$$

For the term $I=\|\frac{1}{3}\partial_x(u^3-v^3)\|_{X^{\delta,s,b_1'}(\mathbb{R}^2)}$ using the identity $u^3-v^3=(u^2+uv+v^2)(u-v)$, we get

$$\begin{aligned}
I\le\ &\frac{1}{3}\left\|\partial_x\big(u^2(u-v)\right\|_{X^{\delta,s,b_1'}(\mathbb{R}^2)}+\frac{1}{3}\left\|\partial_x\big(uv(u-v)\big)\right\|_{X^{\delta,s,b_1'}(\mathbb{R}^2)}\\
&+\frac{1}{3}\|\partial_x\big(v^2(u-v)\big)\|_{X^{\delta,s,b_1'}(\mathbb{R}^2)}.
\end{aligned}$$

Then, by Corollary 2.1, from the last inequality we obtain

$$I\le C\Big[\|u\|^2_{X^{\delta,s,b}(\mathbb{R}^2)}+\|u\|_{X^{\delta,s,b}(\mathbb{R}^2)}\|v\|_{X^{\delta,s,b}(\mathbb{R}^2)}+\|v\|^2_{X^{\delta,s,b}(\mathbb{R}^2)}\Big]\|u-v\|_{X^{\delta,s,b}(\mathbb{R}^2)}.\tag{50}$$

Similarly working, for term II we have

$$\begin{aligned}
II&=\left\|(1-\partial_x^2)^{-1}\partial_x\Big[\frac{5}{3}(u^3-v^3)\Big]\right\|_{X^{\delta,s,b_1'}(\mathbb{R}^2)}\le\left\|\partial_x\Big[\frac{5}{3}(u^3-v^3)\Big]\right\|_{X^{\delta,s,b_1'}(\mathbb{R}^2)}\\
&\le C\Big[\|u\|^2_{X^{\delta,s,b}(\mathbb{R}^2)}+\|u\|_{X^{\delta,s,b}(\mathbb{R}^2)}\|v\|_{X^{\delta,s,b}(\mathbb{R}^2)}+\|v\|^2_{X^{\delta,s,b}(\mathbb{R}^2)}\Big]\|u-v\|_{X^{\delta,s,b}(\mathbb{R}^2)}.
\end{aligned}\tag{51}$$

For term III, using the identity

$$u(\partial_x u)^2 - v(\partial_x v)^2 = (u - v)(\partial_x u)^2 + v\partial_x(u - v)\partial_x(u + v)$$

and Corollary 2.2 we estimate as follows:

$$\begin{aligned}
III &\le \left\|(1-\partial_x^2)^{-1}\partial_x\left[(u-v)(\partial_x u)^2\right]\right\|_{X^{\delta,s,b'_1}(\mathbb{R}^2)} \\
&\quad + \left\|(1-\partial_x^2)^{-1}\partial_x\left[v\partial_x(u-v)\partial_x(u+v)\right]\right\|_{X^{\delta,s,b'_1}(\mathbb{R}^2)} \\
&\le C\|u\|^2_{X^{\delta,s,b}(\mathbb{R}^2)}\|(u-v)\|_{X^{\delta,s,b}(\mathbb{R}^2)} \\
&\quad + C\|v\|_{X^{\delta,s,b}(\mathbb{R}^2)}\|(u-v)\|_{X^{\delta,s,b}(\mathbb{R}^2)}\|u+v\|_{X^{\delta,s,b}(\mathbb{R}^2)} \\
&\le C\left[\|u\|^2_{X^{\delta,s,b}(\mathbb{R}^2)} + \|v\|_{X^{\delta,s,b}(\mathbb{R}^2)}\|u\|_{X^{\delta,s,b}(\mathbb{R}^2)} + \|v\|^2_{X^{\delta,s,b}(\mathbb{R}^2)}\right] \\
&\quad \times \|(u-v)\|_{X^{\delta,s,b}(\mathbb{R}^2)}. \qquad (52)
\end{aligned}$$

Now by using (49), (50), (51), and (52), we have

$$\begin{aligned}
&\|\Gamma(u) - \Gamma(v)\|_{X^{\delta,s,b}(\mathbb{R}^2)} \le C^2 2^\varepsilon T^\varepsilon\Big[C\Big(\|u\|^2_{X^{\delta,s,b}(\mathbb{R}^2)} + \|u\|_{X^{\delta,s,b}(\mathbb{R}^2)}\|v\|_{X^{\delta,s,b}(\mathbb{R}^2)} + \\
&\quad + \|v\|^2_{X^{\delta,s,b}(\mathbb{R}^2)}\Big)\|u-v\|_{X^{\delta,s,b}(\mathbb{R}^2)} + C\Big(\|u\|^2_{X^{\delta,s,b}(\mathbb{R}^2)} + \|u\|_{X^{\delta,s,b}(\mathbb{R}^2)}\|v\|_{X^{\delta,s,b}(\mathbb{R}^2)} \\
&\quad + \|v\|^2_{X^{\delta,s,b}(\mathbb{R}^2)}\Big)\|u-v\|_{X^{\delta,s,b}(\mathbb{R}^2)} + C\Big(\|u\|^2_{X^{\delta,s,b}(\mathbb{R}^2)} + \|v\|_{X^{\delta,s,b}(\mathbb{R}^2)}\|u\|_{X^{\delta,s,b}(\mathbb{R}^2)} \\
&\quad + \|v\|^2_{X^{\delta,s,b}(\mathbb{R}^2)}\Big)\|(u-v)\|_{X^{\delta,s,b}(\mathbb{R}^2)}\Big] \qquad (53) \\
&= 3C^3 2^\varepsilon T^\varepsilon\Big(\|u\|^2_{X^{\delta,s,b}(\mathbb{R}^2)} + \|v\|^2_{X^{\delta,s,b}(\mathbb{R}^2)} + \|u\|_{X^{\delta,s,b}(\mathbb{R}^2)}\|v\|_{X^{\delta,s,b}(\mathbb{R}^2)}\Big) \\
&\quad \times \|u-v\|_{X^{\delta,s,b}(\mathbb{R}^2)}.
\end{aligned}$$

It follows from the last inequality that for $u, v \in \mathcal{B}$ we have

$$\begin{aligned}
&\|\Gamma_T(u) - \Gamma_T(v)\|_{X^{\delta,s,b}(\mathbb{R}^2)} \\
&\quad \le 3C^3 2^\varepsilon T^\varepsilon\Big[2^2C^2T^{-\frac{1}{3}\varepsilon}\|\varphi\|^2_{G^{\delta,s}(\mathbb{R})} + 2^2C^2T^{-\frac{1}{3}\varepsilon}\|\varphi\|^2_{G^{\delta,s}(\mathbb{R})} \\
&\qquad +(2CT^{-\frac{1}{6}\varepsilon}\|\varphi\|_{G^{\delta,s}(\mathbb{R})})^2\Big]\|u-v\|_{X^{\delta,s,b}(\mathbb{R}^2)} \\
&\quad = 3C^3 2^\varepsilon T^\varepsilon\Big[3\times 2^2C^2T^{-\frac{1}{3}\epsilon}\|\varphi\|^2_{G^{\delta,s}(\mathbb{R})}\Big]\|u-v\|_{X^{\delta,s,b}(\mathbb{R}^2)} \qquad (54) \\
&\quad = C^5 2^\epsilon 36 T^{\frac{2}{3}\varepsilon}\|\varphi\|^2_{G^{\delta,s}(\mathbb{R})}\|u-v\|_{X^{\delta,s,b}(\mathbb{R}^2)}.
\end{aligned}$$

By using the fact that $0 < T < 1$, we conclude that

$$36 \cdot 2^{\varepsilon} C^5 T^{\frac{2}{3}\varepsilon} \|\varphi\|^2_{G^{\delta,s}(\mathbb{R})} \leq \frac{3}{4} 48 \cdot 2^{\varepsilon} C^5 T^{\frac{1}{2}\varepsilon} \|\varphi\|^2_{G^{\delta,s}(\mathbb{R})}.$$

Now due to (45) it follows from the last inequality that

$$36 \cdot 2^{\varepsilon} C^5 T^{\frac{2}{3}\varepsilon} \|\varphi\|^2_{G^{\delta,s}(\mathbb{R})} \leq \frac{3}{4}. \tag{55}$$

Finally it follows from (54) and from (55) that

$$\|\Gamma_T(u) - \Gamma_T(v)\|_{X^{\delta,s,b}(\mathbb{R}^2)} \leq \frac{3}{4} \|u - v\|_{X^{\delta,s,b}(\mathbb{R}^2)}, \quad \text{for } u, v \in \mathcal{B}, \tag{56}$$

and hence, Γ_T is a contraction in the Banach space $\mathcal{B}$. Thus, Γ_T has a unique fixed point $u \in \mathcal{B}$. The function u solves the initial-value problem (1) in the space $X^{\delta,s,b}(\mathbb{R}^2)$, where $b \in (\frac{1}{2}, 1)$, with lifespan $|t| \leq T_{\delta,s} \doteq T$ with T given in (45), by our cut-off function ψ_T.

The second estimate in (11) follows from the following elementary result.

Lemma 2.4 *Let $s \in \mathbb{R}$ and $b > 1/2$. Then for $u \in X^{\delta,s,b}(\mathbb{R}^2)$ we have*

$$|u|_{C(\mathbb{R}),\delta,s} \doteq \sup_{t \in \mathbb{R}} \|u(t, \cdot)\|_{G^{\delta,s}(\mathbb{R})} \leq c\|u\|_{X^{\delta,s,b}(\mathbb{R}^2)}, \tag{57}$$

where $c > 0$ depends only on b.

It follows from Lemma 2.4, with $T_{\delta,s} = T$, that we have proved the existence of a solution to our Cauchy problem which belongs to the space $X_T^{\delta,s,b} \hookrightarrow C([-T_{\delta,s}, T_{\delta,s}], G^{\delta,s}(\mathbb{R}))$, for $s > \frac{1}{2}$, $\delta > 0$ if $u_0(x) \in G^{\delta,s}(\mathbb{R})$.

Lip Continuity in Initial Data Their proof is similar to the one presented in [25] for the dispersive regularization of a higher-order Camassa–Holm equation with quadratic nonlinearities.

3 Proof of the Almost Conservation Law

The main ingredients in our proof are the H^1 conservation law, and the trilinear estimates stated earlier, which hold for $s > 1/2$. We begin with the H^1 conservation law. Using the operator (analytic multiplier) $e^{\delta|D_x|}$, for a real-valued function $u(t, x)$, we define $U(x, t)$ by

$$U(x, t) = e^{\delta|D_x|} u(x, t) \text{ or } \widehat{U}(\xi, \tau) = e^{\delta|\xi|} \widehat{u}(\xi, \tau).$$

Next we notice that the function U is also real. Now we apply the operator $e^{\delta|D_x|}$ to the mCH3 Eq. (1), and we get the following forced mCH3 equation:

$$\partial_t U - \partial_t \partial_x^2 U + \partial_x^3 U - \partial_x^5 U + 2\partial_x U^3 - \frac{1}{3}\partial_x^3(U^3) + \partial_x[U(\partial_x U)^2] = H, \tag{58}$$

where

$$H = \frac{1}{3}E + \frac{8}{3}F + G \tag{59}$$

with

$$E = \partial_x(1-\partial_x^2)\Big[e^{\delta|D_x|}u \cdot e^{\delta|D_x|}u \cdot e^{\delta|D_x|}u - e^{\delta|D_x|}(u\cdot u\cdot u)\Big], \tag{60}$$

$$F = \partial_x\Big[e^{\delta|D_x|}u \cdot e^{\delta|D_x|}u \cdot e^{\delta|D_x|}u - e^{\delta|D_x|}(u\cdot u\cdot u)\Big], \tag{61}$$

and

$$G = \partial_x\Big[e^{\delta|D_x|}u \cdot e^{\delta|D_x|}u_x \cdot e^{\delta|D_x|}u_x - e^{\delta|D_x|}(u\cdot u_x\cdot u_x)\Big]. \tag{62}$$

Doing for U similar H^1 conservation computations like the ones done in (5), we get

$$\frac{1}{2}\frac{d}{dt}||U(t)||_{H^1}^2 = \int_{\mathbb{R}} (UH)(t,x)dx. \tag{63}$$

Integrating in time for $t \in [0, \rho]$, we get

$$\|u(t)\|_{G^{\delta,1}}^2 \le \|u(0)\|_{G^{\delta,1}}^2 + 2\int_0^\rho \int_{\mathbb{R}} (UH)(t,x)\,dt\,dx, \tag{64}$$

which gives the inequality

$$||u(t)||_{G^{\delta,1}}^2 \le ||u(0)||_{G^{\delta,1}}^2 + 2\Big|\int_{\mathbb{R}^2} \chi_{[0,\rho]}(t)\cdot(UH)(t,x)\,dtdx\Big|, \tag{65}$$

where $0 < \rho < T_{\delta_{0,1}}$.

Next, we shall estimate the integral in (65). For this estimation and the ones to follow in the rest of this chapter, we recall that we are using the letter C after an inequality to denote a constant, which may change a finite number of times. Now, using the decomposition (59) of H and writing

$$\Big|\int_{\mathbb{R}^2} \chi_{[0,\rho]}(t)\,(UH)(t,x)\,dtdx\Big| \le C(I + II + III), \tag{66}$$

where

$$I \doteq \Big| \int_{\mathbb{R}^2} \chi_{[0,\rho]}(t)\,(UE)(t,x)\,dtdx \Big|, \qquad II \doteq \Big| \int_{\mathbb{R}^2} \chi_{[0,\rho]}(t)\,(UF)(t,x)\,dtdx \Big|, \tag{67}$$

and

$$III \doteq \Big| \int_{\mathbb{R}^2} \chi_{[0,\rho]}(t)\,(UG)(t,x)\,dtdx \Big|, \tag{68}$$

we see that it suffices to estimate each one of these terms separately.

Estimation of (II) Now we will estimate

$$II \doteq \Big| \int_{\mathbb{R}^2} \chi_{[0,\rho]}(t)\,(UF)(t,x)\,dtdx \Big|, \tag{69}$$

where F is given by

$$F = \partial_x \Big[e^{\delta|D_x|}u \cdot e^{\delta|D_x|}u \cdot e^{\delta|D_x|}u - e^{\delta|D_x|}(u \cdot u \cdot u) \Big].$$

Multiplying the integrant by $(1+|\tau-\xi^3|)^{1-b}(1+|\tau-\xi^3|)^{b-1}$ and applying first Parseval's identity and then Cauchy–Schwarz inequality, we obtain

$$\begin{aligned}
II &= \Big| \int_{\mathbb{R}^2} \widehat{\chi_{[0,\rho]}(t)U} \cdot \widehat{\chi_{[0,\rho]}(t)F}\, d\xi d\tau \Big| \\
&= \Big| \int_{\mathbb{R}^2} \widehat{\chi_{[0,\rho]}(t)U}(1+|\tau-\xi^3|)^{1-b} \cdot (1+|\tau-\xi^3|)^{b-1}\widehat{\chi_{[0,\rho]}(t)F}\, d\xi d\tau \Big| \\
&\le \Big(\int_{\mathbb{R}^2} |(\widehat{\chi_{[0,\rho]}U}(\xi,\tau)|^2 (1+|\tau-\xi^3|)^{2(1-b)} d\xi d\tau \Big)^{1/2} \\
&\quad \times \Big(\int_{\mathbb{R}^2} (1+|\tau-\xi^3|)^{2(b-1)} |\widehat{\chi_{[0,\rho]}F}(\xi,\tau)|^2 d\xi d\tau \Big)^{1/2} \\
&= \|\chi_{[0,\rho]}U\|_{X^{0,1-b}(\mathbb{R}^2)} \|\chi_{[0,\rho]}F\|_{0,b-1}.
\end{aligned} \tag{70}$$

Now, we need the following result in Selberg and Silva [42].

Lemma 3.1 ([42]) *Let $\delta \ge 0$, $s \in \mathbb{R}$, $-1/2 < \gamma < 1/2$, and $\rho > 0$. Then, for any time interval $I \subset [-\rho,\rho]$, we have*

$$\|\chi_I u\|_{\delta,s,\gamma} \le c_2 \|u\|_{X_\rho^{\delta,s,\gamma}}, \tag{71}$$

where χ_I is the characteristic of I and c_2 depends only on γ.

Noticing that for $\frac{1}{2} < b < 1$ we obtain $-\frac{1}{2} < 1 - b < \frac{1}{2}$ and $-\frac{1}{2} < b - 1 < \frac{1}{2}$. Thus, using estimate (71) one time with $\gamma = 1 - b$ and the other with $\gamma = b - 1$, from (70), we get

$$II \leq c_2 \|U\|_{X_\rho^{0,1-b}} \|F\|_{X_\rho^{0,b-1}},$$

where $X_\rho^{s,b}(\mathbb{R}^2)$ denotes the restriction to the time interval $[0, \rho]$. Also, since $1-b < \frac{1}{2} < b$, we have $\|U\|_{X_\rho^{0,1-b}} \leq \|U\|_{X_\rho^{0,b}} = \|u\|_{X_\rho^{\delta,0,b}}$, and so we arrive at the estimate

$$II \leq c_2 ||u||_{X_\rho^{\delta,0,b}} ||F||_{X_\rho^{0,b-1}} \leq c_2 ||u||_{X_\rho^{\delta,1,b}} ||F||_{X_\rho^{0,b-1}}. \tag{72}$$

Now, from (72), we see that in order to prove the desired estimate for term (II), we need to estimate $||F||_{X_\rho^{0,b-1}}$. In fact, it suffices to estimate F in the whole space $X^{0,b-1}(\mathbb{R}^2)$. Then, we obtain the desired result by restricting in the time slab $\mathbb{R} \times [0, \rho]$. By the definition we have

$$||F||^2_{X^{0,b-1}} \doteq \int_{\mathbb{R}^2} (1 + |\tau - \xi^3|)^{2(b-1)} |\widehat{F}(\xi, \tau)|^2 d\xi d\tau, \tag{73}$$

and by the properties of the Fourier transform, we have

$$\begin{aligned} \widehat{F}(\xi, \tau) &= i\xi \left[\widehat{e^{\delta|D_x|}u} * \widehat{e^{\delta|D_x|}u} * \widehat{e^{\delta|D_x|}u} - e^{\delta|\xi|} \widehat{u} * \widehat{u} * \widehat{u} \right] \\ &= i\xi \int_{\mathbb{R}^4} \left(e^{\delta|\xi_1|} e^{\delta|\xi_2|} e^{\delta|\xi-\xi_1-\xi_2|} - e^{\delta|\xi|} \right) \widehat{u}(\tau_1, \xi_1) \widehat{u}(\tau_2, \xi_2) \\ &\quad \times \widehat{u}(\tau - \tau_1 - \tau_2, \xi - \xi_1 - \xi_2) d\tau_1 d\tau_2 d\xi_1 d\xi_2. \end{aligned} \tag{74}$$

Also, we need the following elementary result [40] for estimating the combination of exponentials appearing in the last relation.

Lemma 3.2 *For $\delta > 0$, $\theta \in [0, 1]$, and $\xi, \xi_1, \xi_2 \in \mathbb{R}$, we have*

$$e^{\delta|\xi_1|} e^{\delta|\xi_2|} e^{\delta|\xi-\xi_1-\xi_2|} - e^{\delta|\xi|} \leq \left[4\delta \frac{(1 + |\xi - \xi_1 - \xi_2|)(1 + |\xi_1|)(1 + \xi_2|)}{1 + |\xi|} \right]^\theta e^{\delta|\xi_1|} e^{\delta|\xi_2|} e^{\delta|\xi-\xi_1-\xi_2|}.$$

Estimation of $\|F\|_{X^{0,b-1}}$ Our strategy is to use the trilinear estimates directly or with some modifications. We begin by forming the norm $\|F\|_{X^{0,b-1}}$. From (74) and Lemma 3.2 we have that

$$\begin{aligned} &|\widehat{F}(\tau, \xi)| \\ &\leq |\xi| \int_{\mathbb{R}^4} \left[4\delta \frac{(1 + |\xi - \xi_1 - \xi_2|)(1 + |\xi_1|)(1 + \xi_2|)}{1 + |\xi|} \right]^\theta e^{\delta|\xi_1|} e^{\delta|\xi_2|} e^{\delta|\xi-\xi_1-\xi_2|} \end{aligned}$$

$$\times |\widehat{u}(\tau_1,\xi_1)| \cdot |\widehat{u}(\tau_2,\xi_2)| \cdot |\widehat{u}(\tau-\tau_1-\tau_2,\xi-\xi_1-\xi_2)| d\tau_1 d\tau_2 d\xi_1 d\xi_2. \tag{75}$$

Therefore,

$$\|F\|_{X^{0,b-1}} = \left\|(1+|\tau-\xi^3|)^{b-1}\widehat{F}(\tau,\xi)\right\|_{L^2_{\tau,\xi}} = \left\|\frac{1}{(1+|\tau-\xi^3|)^{1-b}}\widehat{F}(\tau,\xi)\right\|_{L^2_{\tau,\xi}}$$

$$\leq 4^\theta\delta^\theta\Big[\int_{\mathbb{R}^2}\frac{1}{(1+|\tau-\xi^3|)^{2(1-b)}}|\xi|^2$$

$$\times\Big(\int_{\mathbb{R}^4}\frac{(1+|\xi-\xi_1-\xi_2|)^\theta(1+|\xi_1|)^\theta(1+\xi_2|)^\theta}{(1+|\xi|)^\theta}e^{\delta|\xi_1|}e^{\delta|\xi_2|}e^{\delta|\xi-\xi_1-\xi_2|}$$

$$\times |\widehat{u}(\tau_1,\xi_1)| \cdot |\widehat{u}(\tau_2,\xi_2)| \cdot |\widehat{u}(\tau-\tau_1-\tau_2,\xi-\xi_1-\xi_2)|$$

$$\times\, d\tau_1 d\tau_2 d\xi_1 d\xi_2\Big)^2 d\tau d\xi\Big]^{\frac{1}{2}}.$$

Taking the term $(1+|\xi|)^\theta$ outside the $d\tau_1 d\tau_2 d\xi_1 d\xi_2$ integral and bounding it with $(1+|\xi|)^{2(1-\theta)}$, we get

$$\|F\|_{X^{0,b-1}} \leq 4^\theta\delta^\theta\Big[\int_{\mathbb{R}^2}\frac{1}{(1+|\tau-\xi^3|)^{2(1-b)}}|\xi|^2(1+|\xi|)^{2(1-\theta)}$$

$$\Big(\int_{\mathbb{R}^4}(1+|\xi-\xi_1-\xi_2|)^\theta(1+|\xi_1|)^\theta(1+\xi_2|)^\theta e^{\delta|\xi_1|}e^{\delta|\xi_2|}e^{\delta|\xi-\xi_1-\xi_2|}$$

$$\times|\widehat{u}(\tau_1,\xi_1)| \cdot |\widehat{u}(\tau_2,\xi_2)| \cdot |\widehat{u}(\tau-\tau_1-\tau_2,\xi-\xi_1-\xi_2)| d\tau_1 d\tau_2 d\xi_1 d\xi_2\Big)^2 d\tau d\xi\Big]^{\frac{1}{2}}$$

$$= 4^\theta\delta^\theta\Big\|\frac{|\xi|}{(1+|\tau-\xi^3|)^{1-b}}(1+|\xi|)^{1-\theta}$$

$$\times\int_{\mathbb{R}^4}(1+|\xi-\xi_1-\xi_2|)^\theta(1+|\xi_1|)^\theta(1+\xi_2|)^\theta e^{\delta|\xi_1|}e^{\delta|\xi_2|}e^{\delta|\xi-\xi_1-\xi_2|}$$

$$\times |\widehat{u}(\tau_1,\xi_1)| \cdot |\widehat{u}(\tau_2,\xi_2)| \cdot |\widehat{u}(\tau-\tau_1-\tau_2,\xi-\xi_1-\xi_2)| d\tau_1 d\tau_2 d\xi_1 d\xi_2\Big\|_{L^2_{\tau,\xi}}. \tag{76}$$

Remark In this last inequality the usual idea is to set $-\theta = s$ and use the L^2 formulation of trilinear estimates, which in our situation they hold for $s > 1/2$, and therefore we get $\theta < -\frac{1}{2}$, which is not good since $0 < \theta \leq 1$.

Modification 1 This forces us to introduce some modification. For this we use the inequality

$$(1+|\xi|)^{1-\theta} \leq (1+|\xi_1|)^{1-\theta}(1+|\xi_2|)^{1-\theta}(1+|\xi-\xi_1-\xi_2|)^{1-\theta},\ 0 \leq \theta \leq 1. \tag{77}$$

Using (77), from (76), we obtain

$$\|F\|_{X^{0,b-1}} \leq 4^\theta \delta^\theta \Big\| \frac{|\xi|}{(1+|\tau-\xi^3|)^{1-b}} \int_{\mathbb{R}^4} (1+|\xi-\xi_1-\xi_2|)(1+|\xi_1|)(1+\xi_2|)$$
$$\times e^{\delta|\xi_1|} e^{\delta|\xi_2|} e^{\delta|\xi-\xi_1-\xi_2|}$$
$$\times |\widehat{u}(\tau_1,\xi_1)| \cdot |\widehat{u}(\tau_2,\xi_2)| \cdot |\widehat{u}(\tau-\tau_1-\tau_2,\xi-\xi_1-\xi_2)| d\tau_1 d\tau_2 d\xi_1 d\xi_2 \Big\|_{L^2_{\tau,\xi}}. \tag{78}$$

Modification 2 Our next modification (new idea) is to choose a parameter κ such that

$$\kappa > 3/2, \quad \text{(or better } k-1 > 1/4, \text{ i.e. } k > 5/4) \tag{79}$$

whose meaning and utility will be seeing later. Then, we define the following combination of u

$$\tilde{f}_\kappa(\tau,\xi) = (1+|\tau-\xi^3|)^b (1+|\xi|)^\kappa e^{\delta|\xi|} |\widehat{u}(\tau,\xi)| \tag{80}$$

and we obtain

$$e^{\delta|\xi|}\widehat{u}(\tau,\xi) = \frac{\tilde{f}_\kappa(\tau,\xi)}{(1+|\tau-\xi^3|)^b(1+|\xi|)^\kappa}. \tag{81}$$

Now it follows from (78) and (81) that

$$\|F\|_{X^{0,b-1}} \leq 4^\theta \delta^\theta \Big\| \frac{|\xi|}{(1+|\tau-\xi^3|)^{1-b}} \int_{\mathbb{R}^4} (1+|\xi-\xi_1-\xi_2|)(1+|\xi_1|)(1+\xi_2|)$$
$$\times \frac{\tilde{f}_\kappa(\tau_1,\xi_1)}{(1+|\tau_1-\xi_1^3|)^b(1+|\xi_1|)^\kappa}$$
$$\times \frac{\tilde{f}_\kappa(\tau_2,\xi_2)}{(1+|\tau_2-\xi_2^3|)^b(1+|\xi_2|)^\kappa}$$
$$\times \frac{\tilde{f}_\kappa(\tau-\tau_1-\tau_2,\xi-\xi_1-\xi_2)}{(1+|(\tau-\tau_1-\tau_2)-(\xi-\xi_1-\xi_2)^3|)^b(1+|\xi-\xi_1-\xi_2|)^\kappa}$$
$$\times d\tau_1 d\tau_2 d\xi_1 d\xi_2 \Big\|_{L^2_{\tau,\xi}}$$
$$= 4^\theta \delta^\theta \Big\| \frac{|\xi|}{(1+|\tau-\xi^3|)^{1-b}} \int_{\mathbb{R}^4} \frac{\tilde{f}_\kappa(\tau_1,\xi_1)}{(1+|\tau_1-\xi_1^3|)^b(1+|\xi_1|)^{\kappa-1}}$$

$$\times \frac{\tilde{f}_\kappa(\tau_2, \xi_2)}{(1+|\tau_2-\xi_2^3|)^b(1+|\xi_2|)^{\kappa-1}}$$
$$\times \frac{\tilde{f}_\kappa(\tau-\tau_1-\tau_2, \xi-\xi_1-\xi_2)}{(1+|(\tau-\tau_1-\tau_2)-(\xi-\xi_1-\xi_2)^3|)^b(1+|\xi-\xi_1-\xi_2|)^{\kappa-1}}$$
$$\times\, d\tau_1 d\tau_2 d\xi_1 d\xi_2 \Big\|_{L^2_{\tau,\xi}}.$$

To make the above expression to look like the L^2-norm in the LHS of the mKdV trilinear estimates (see (3.5) in [26]), we introduce the term $(1+|\xi|)^{\kappa-1} > 1$ in the numerator of the function inside the norm, and we get

$$\|F\|_{X^{0,b-1}} \le 4^\theta \delta^\theta \Big\| \frac{|\xi|(1+|\xi|)^{\kappa-1}}{(1+|\tau-\xi^3|)^{(1-b)}} \int_{\mathbb{R}^4} \frac{\tilde{f}_\kappa(\tau_1,\xi_1)}{(1+|\tau_1-\xi_1^3|)^b(1+|\xi_1|)^{\kappa-1}}$$
$$\cdot \frac{\tilde{f}_\kappa(\tau_2, \xi_2)}{(1+|\tau_2-\xi_2^3|)^b(1+|\xi_2|)^{\kappa-1}}$$
$$\times \frac{\tilde{f}_\kappa(\tau-\tau_1-\tau_2, \xi-\xi_1-\xi_2)}{(1+|(\tau-\tau_1-\tau_2)-(\xi-\xi_1-\xi_2)^3|)^b(1+|\xi-\xi_1-\xi_2|)^{\kappa-1}}$$
$$\times\, d\tau_1 d\tau_2 d\xi_1 d\xi_2 \Big\|_{L^2_{\tau,\xi}}. \tag{82}$$

Now, while the s-exponents look "good," we have a problem with the combination $\tilde{f}_\kappa$ defined in (80). The term $(1+|\xi|)^\kappa$ must be replaced by $(1+|\xi|)^{\kappa-1}$ as formulation (3.5) in [26] indicates. To take care of this problem, we introduce an auxiliary function $f_{\kappa-1} = f_{\kappa-1,\delta_0}$

$$f_{\kappa-1}(\tau,\xi) = (1+|\tau-\xi^3|)^b(1+|\xi|)^{\kappa-1}e^{\delta_0|\xi|}|\widehat{u}(\tau,\xi)|, \tag{83}$$

where $0<\delta<\delta_0$.

To compare $\tilde{f}_\kappa(\tau,\xi)$and $f_{\kappa-1}(\tau,\xi)$, we use the following elementary lemma.

Lemma 3.3 *For* $0<\delta<\delta_0$ *we have*

$$\frac{\tilde{f}_\kappa(\tau,\xi)}{f_{\kappa-1}(\tau,\xi)} = \frac{1+|\xi|}{e^{(\delta_0-\delta)|\xi|}} \le e^{\delta_0-\delta-1}\cdot\frac{1}{\delta_0-\delta} < e^{\delta_0-1}\frac{1}{\delta_0-\delta} = \frac{C}{\delta_0-\delta}. \tag{84}$$

Proof of Lemma 3.3 Setting $t \doteq |\xi|$ and $a \doteq \delta_0-\delta$, we have

$$\frac{\tilde{f}_\kappa(\tau,\xi)}{f_{\kappa-1}(\tau,\xi)} = \frac{1+|\xi|}{e^{(\delta_0-\delta)|\xi|}} = \frac{1+t}{e^{at}} \doteq g(t), \quad t\ge 0,\ 0<a<1. \tag{85}$$

Since $(1+t)e^{-at} \longrightarrow 0$ when $t \longrightarrow +\infty$ to find the maximum of $g(t)$ over $[0, \infty)$, we compute its critical points. Since $g'(t) = (1 - a - at)e^{-at}$, we see that it has a single critical point at $t = (1 - a)/a$. Furthermore, since the second derivative is equal to $g''(t) = (-2a + a^2 + a^2 t)e^{-at}$ and its value at the critical point is equal to

$$g''\Big(\frac{1-a}{a}\Big) = \Big(-2a + a^2 + a^2 \cdot \frac{1-a}{a}\Big)e^{-1+a} = -ae^{-1+a} < 0,$$

which is negative, we conclude that $g(t)$ has global maximum at $t = (1-a)/a$ with maximum value equal to

$$g\Big(\frac{1-a}{a}\Big) = e^{-1+a} \cdot \Big(1 + \frac{1-a}{a}\Big) = e^{-1+a} \cdot \frac{1}{a} = e^{\delta_0 - \delta - 1} \cdot \frac{1}{\delta_0 - \delta} < e^{\delta_0 - 1} \cdot \frac{1}{\delta_0 - \delta} = \frac{C}{\delta_0 - \delta}.$$

This completes the proof of Lemma 3.3. □

Next we try to obtain formulation (3.5) in [26] of the mKdV trilinear estimates. Replacing $\tilde{f}_\kappa(\tau, \xi)$ by $f_{\kappa-1}(\tau, \xi)$ in inequality (82) and taking into account Lemma 3.3, we obtain

$$\begin{aligned} \|F\|_{X^{0,b-1}} \le C4^\theta \frac{\delta^\theta}{(\delta_0 - \delta)^3} \Bigg\| & \frac{|\xi|(1 + |\xi|)^{\kappa-1}}{(1 + |\tau - \xi^3|)^{1-b}} \\ \times \int_{\mathbb{R}^4} & \frac{f_{\kappa-1}(\tau_1, \xi_1)}{(1 + |\tau_1 - \xi_1^3|)^b (1 + |\xi_1|)^{\kappa-1}} \frac{f_{\kappa-1}(\tau_2, \xi_2)}{(1 + |\tau_2 - \xi_2^3|)^b (1 + |\xi_2|)^{\kappa-1}} \\ \times & \frac{f_{\kappa-1}(\tau - \tau_1 - \tau_2, \xi - \xi_1 - \xi_2)}{(1 + |(\tau - \tau_1 - \tau_2) - (\xi - \xi_1 - \xi_2)^3|)^b (1 + |\xi - \xi_1 - \xi_2|)^{\kappa-1}} \\ \times & \, d\tau_1 d\tau_2 d\xi_1 d\xi_2 \Bigg\|_{L^2_{\tau,\xi}}, \end{aligned}$$

which is the LHS in [26] formulation of the trilinear estimates with $s = \kappa - 1$, $\nu = \nu' = b$, where $\frac{1}{2} < b = \nu' = \nu < 1$. (Also, it follows from Remark 2.1, which is the L^2 version of the mKdV trilinear estimate stated in Theorem 2.1, with $s = \kappa - 1$, $\nu = \nu' = b$, where $\frac{1}{2} < b = \nu' = \nu < 1$.) Therefore, LHS is bounded by the RHS, that is,

$$\|F\|_{X^{0,b-1}} \le C4^\theta \frac{\delta^\theta}{(\delta_0 - \delta)^3} \|f_{\kappa-1}\|^3_{L^2_{\tau,\xi}}. \tag{86}$$

Now we notice that

$$\|f_{\kappa-1}\|^2_{L^2_{\tau,\xi}} = \int_{\mathbb{R}^2} |f_{\kappa-1}(\tau, \xi)|^2 d\tau d\xi$$

$$= \int_{\mathbb{R}^2} (1+|\xi|)^{2(\kappa-1)}(1+|\tau-\xi^3|)^{2b} e^{2\delta_0|\xi|} |\widehat{u}(\tau,\xi)|^2 d\tau d\xi$$
$$= \|u\|^2_{X^{\delta_0,\kappa-1,b}}. \tag{87}$$

Since $s = \kappa - 1$ and in order to use Remark 4.1 we need $s > \frac{1}{2}$, therefore we conclude that $\kappa > \frac{3}{2}$. Thus, in order to get $s = 1$ we choose $\kappa = 2$ and we obtain

$$\|F\|_{X^{0,b-1}} \le C4^\theta \frac{\delta^\theta}{(\delta_0-\delta)^3} \|f_{\kappa-1}\|^3_{L^2_{\tau,\xi}} \le C4^\theta \frac{\delta^\theta}{(\delta_0-\delta)^3} \|u\|^3_{X^{\delta_0,1,b}}. \tag{88}$$

Summing up what we have proved, we can conclude that, for $0 \le \theta \le 1$, we have

$$\|F\|_{X^{0,b-1}} \le C4^\theta \frac{\delta^\theta}{(\delta_0-\delta)^3} \|u\|^3_{\delta_0,1,b}. \tag{89}$$

Finally, it follows from (72) and (89) that

$$II \le C4^\theta \frac{\delta^\theta}{(\delta_0-\delta)^3} \|u\|^4_{\delta_0,1,b}. \tag{90}$$

Estimation of I Since E is expressed in terms of F as follows:

$$E = (1-\partial_x^2)F, \quad F = \partial_x\Big[e^{\delta|D_x|}u \cdot e^{\delta|D_x|}u \cdot e^{\delta|D_x|}u - e^{\delta|D_x|}(u\cdot u\cdot u)\Big], \tag{91}$$

utilizing the identity (see [25]),

$$\int_{\mathbb{R}} VW dx = \int_{\mathbb{R}} V(1-\partial_x^2)^{-1} W dx + \int_{\mathbb{R}} V_x (1-\partial_x^2)^{-1} W_x dx, \tag{92}$$

with $V = U$ and $W = (1-\partial_x^2)F$, we have

$$\begin{aligned}\int_{\mathbb{R}} U(1-\partial_x^2)F\,dxdt &= \int_{\mathbb{R}} U(1-\partial_x^2)^{-1}(1-\partial_x^2)F dx \\ &\quad + \int_{\mathbb{R}} U_x(1-\partial_x^2)^{-1}(1-\partial_x^2)F_x dx \\ &= \int_{\mathbb{R}} UF dx + \int_{\mathbb{R}} U_x F_x dx.\end{aligned} \tag{93}$$

Thus we can write I as

$$I \doteq \left| \int_{\mathbb{R}^2} \chi_{[0,\rho]}(t)\, U(1-\partial_x^2)F\,dxdt \right| \le A + B, \tag{94}$$

where

$$A \doteq \left| \int_{\mathbb{R}^2} \chi_{[0,\rho]}(t)\, U F \, dxdt \right| \quad \text{and} \quad B \doteq \left| \int_{\mathbb{R}^2} \chi_{[0,\rho]}(t)\, U \partial_x^2 F \, dxdt \right|.$$

Now we notice that $A = II$, which we have just finished its estimation and obtained

$$A = II \leq C4^{\theta} \frac{\delta^{\theta}}{(\delta_0 - \delta)^3} \|u\|^4_{X_{\rho}^{\delta_0,1,b}}, \tag{95}$$

where C depends on θ.

For estimating B, working like in the case of inequality (72), we obtain the corresponding inequality

$$\begin{aligned} B \doteq \left| \int_{\mathbb{R}^2} \chi_{[0,\rho]}(t)\, U_x F_x \, dxdt \right| &\leq C||U_x||_{X_{\rho}^{0,1-b}} ||F_x|||_{X_{\rho}^{0,b-1}} \\ &\leq C||U_x||_{X_{\rho}^{0,b}} ||F_x|||_{X_{\rho}^{0,b-1}} \leq C||U||_{X_{\rho}^{1,b}} ||F_x|||_{X_{\rho}^{0,b-1}} = C||u||_{X_{\rho}^{\delta,1,b}} ||F_x|||_{X_{\rho}^{0,b-1}}. \end{aligned} \tag{96}$$

Furthermore, working similarly to estimating F in term (II), we derive the following estimates for F_x

$$\|F_x\|_{X^{0,b-1}} \leq C4^{\theta} \frac{\delta^{\theta}}{(\delta_0 - \delta)^4} \|u\|^3_{X^{\delta_0,1,b}}. \tag{97}$$

Thus, from (96) and (97), we get

$$B \leq C4^{\theta} \frac{\delta^{\theta}}{(\delta_0 - \delta)^4} \|u\|^4_{\delta_0,1,b},$$

which combined with estimate (95) gives the inequality

$$I \leq C \frac{\delta^{\theta}}{(\delta_0 - \delta)^3} \|u\|^4_{X_{\rho}^{\delta_0,1,b}} + \frac{\delta^{\theta}}{(\delta_0 - \delta)^4} \|u\|^4_{\delta_0,1,b}. \tag{98}$$

Estimation of III This estimation is analogous to the case of the term (II) after making appropriate modification. The estimate obtained is

$$III \leq C \frac{\delta^{\theta}}{(\delta_0 - \delta)^5} \|u\|^4_{\delta_0,1,b}. \tag{99}$$

Finally, putting together estimates (90), (98), and (99), we obtain the desired almost approximation law (13) and complete the proof of Theorem 1.2. □

4 Lower Bound for the Radius of Spatial Analyticity

We begin by observing that mCH3 is invariant under the change of variables $(t, x) \mapsto (-t, -x)$. Thus, we may assume that $t \geq 0$. Furthermore, we shall prove the lower bound first in the case $s = 1$ by taking advantage of the H^1 almost conservation law. Then, we will reduce all other cases to this one.

The Case $s = 1$ For given data $u(0) \in G^{\delta_0,1}(\mathbb{R})$, applying Theorem 1.1, we get a unique solution $u \in C([0, T_{\delta_0,1}], G^{\delta_0,1}(\mathbb{R}))$ to the Cauchy problem (1) satisfying the size estimate

$$\|u(t)\|_{G^{\delta_0,1}} \leq 2C\|u(0)\|_{G^{\delta_0,1}}, \quad |t| \leq T_{\delta_0,1}, \tag{100}$$

where $T_{\delta_0,1}$ satisfies the estimate (12), that is,

$$T_{\delta_0,1} \doteq T_0 \doteq \frac{1}{(2 + c_0\|u(0)\|^2_{G^{\delta_0,1}})^a}, \quad \text{for} \quad c_0 = 48C^5 2^\epsilon > 0 \text{ and } a = \frac{2}{\epsilon} > 1, \tag{101}$$

for some $b \in (\frac{1}{2}, 1)$ and some $\varepsilon \in (0, \frac{1}{2})$.

Next, we define the maximal lifespan by

$$T^* \doteq \sup\{T_{\delta_0,1} : u \in C([0, T_{\delta_0,1}]; G^{\delta_0,1}(\mathbb{R})) \text{ solves Cauchy problem (1) and satisfies (100)}\}, \tag{102}$$

and distinguish two possible cases. The first case is $T^* = \infty$, which means that $u \in C([0, \infty); G^{\delta_0,1}(\mathbb{R}))$, and thus we have persistence of the uniform radius of spatial analyticity of $u(t)$ for all time. That is $\delta_0(T) = \delta_0$, which proves Theorem 1.3 in this case.

The second case is when T^* is finite. Then $u \in C([0, T^*); G^{\delta_0,1}(\mathbb{R}))$ and

$$\infty > T^* \geq T_0 = T_{\delta_0,1} \doteq \frac{1}{(2 + c_0||u(0)||^2_{G^{\delta_0,1}})^a} > \frac{1}{(2 + 2c_0||u(0)||_{G^{\delta_0,1}})^a} \doteq \rho^*. \tag{103}$$

Given $\rho*$ defined above, for any given $T > T^*$, there is a positive integer $n = n(T)$ such that

$$n\rho^* \leq T < (n+1)\rho^*. \tag{104}$$

Fix $T \geq T^*$. We will show that, for $\delta > 0$ sufficiently small,

$$\|u(t)\|^2_{\delta,1} \leq 2\|u(0)\|^2_{\delta_0,1}, \quad \text{for,} \quad t \in [0, T]. \tag{105}$$

To prove this we will use repeatedly Theorems 1.1 and 1.2 with the time step ρ^*.

Claim If for given $\delta > 0$, $0 < \delta_0 \leq 1$ fixed such that if the conditions

$$0 < \delta < \delta_0 \quad \text{and} \quad (n+1)C(2C)^{4(n+1)} \frac{\delta^\theta}{(\delta_0 - \delta)^5} ||u(0)||^2_{G^{\delta_0,1}} \leq 1 \tag{106}$$

hold, then for each $k \in \{1, 2, \ldots, n+1\}$ we have that

$$\sup_{t \in [0, k\rho^*]} ||u(t)||^2_{G^{\delta,1}} \leq \|u(0)\|^2_{G^{\delta,1}} + kC(2C)^{4k} \frac{\delta^\theta}{(\delta_0 - \delta)^5} \|u(0)\|^4_{G^{\delta_0,1}} \tag{107}$$

and

$$\sup_{t \in [0, k\rho^*]} ||u(t)||^2_{G^{\delta,1}} \leq 2||u(0)||^2_{G^{\delta_0,1}}, \tag{108}$$

for $k \in [1, \ldots, n+1]$, where $n \in \mathbb{N}$ is chosen so that $T \in [n\rho^*, (n+1)\rho^*]$.

Proof of Claim In the first step, we apply Theorem 1.2 over the interval $[0, \rho^*]$, and (14) gives

$$\begin{aligned}\sup_{t \in [0, \rho^*]} ||u(t)||^2_{G^{\delta,1}} &\leq ||u(0)||^2_{G^{\delta,1}} + C \frac{\delta^\theta}{(\delta_0 - \delta)^5} ||u(0)||^4_{G^{\delta_0,1}} \\ &\leq ||u(0)||^2_{G^{\delta,1}} + C(2C)^4 \frac{\delta^\theta}{(\delta_0 - \delta)^5} ||u(0)||^4_{G^{\delta_0,1}}.\end{aligned} \tag{109}$$

This verifies (107) for $k = 1$, and now, (108) follows from (109) and using (106), i.e.,
$C(2C)^4 \frac{\delta^\theta}{(\delta_0 - \delta)^5} ||u(0)||^2_{G^{\delta_0,1}} \leq 1$ and the fact that $||u(0)||^2_{G^{\delta,1}} \leq ||u(0)||^2_{G^{\delta_0,1}}$. More precisely, we have

$$\begin{aligned}\sup_{t \in [0, \rho^*]} ||u(t)||^2_{G^{\delta,1}} &\leq ||u(0)||^2_{G^{\delta,1}} + C(2C)^4 \frac{\delta^\theta}{(\delta_0 - \delta)^5} ||u(0)||^2_{G^{\delta_0,1}} \cdot ||u(0)||^2_{G^{\delta_0,1}} \\ &\leq 2||u(0)||^2_{G^{\delta_0,1}}.\end{aligned}$$

Next, assuming that (107) and (108) hold for some $k \in \{1, \ldots, n\}$, we will prove that they hold for $k+1$.

First we extend Theorem 1.2 and (14) for $[k\rho^*, (k+1)\rho^*]$, i.e., it follows from (14), where we replace $t = 0$ by $t = k\rho^*$, and we have

$$\sup_{t \in [k\rho^*, (k+1)\rho^*]} \|u(t)\|^2_{G^{\delta,1}} \leq \|u(k\rho^*)\|^2_{G^{\delta,1}} + C \frac{\delta^\theta}{(\delta_0 - \delta)^5} \|u(k\rho^*)\|^4_{G^{\delta_0,1}(\mathbb{R})}. \tag{110}$$

We now use Theorem 1.1 replacing δ by δ_0 in the intervals $[0, \rho^*]$, $[\rho^*, 2\rho^*]$, … $[(k-1)\rho^*, k\rho^*]$.

We start with the interval $[0, \rho^*]$, and we obtain

$$\|u(t)\|_{G^{\delta_0,1}(\mathbb{R})} \leq 2C\|u(0)\|_{G^{\delta_0,1}(\mathbb{R})}, \ t \in [0, \rho^*] \\ \left(\text{thus } \|u(\rho^*)\|_{G^{\delta_0,1}(\mathbb{R})} \leq 2C\|u(0)\|_{G^{\delta_0,1}(\mathbb{R})}\right). \tag{111}$$

In the interval $[\rho^*, 2\rho^*]$, we get

$$\|u(t)\|_{G^{\delta_0,1}(\mathbb{R})} \leq 2C\|u(\rho^*)\|_{G^{\delta_0,1}(\mathbb{R})}, \ t \in [\rho^*, 2\rho^*] \\ \left(\text{thus } \|u(2\rho^*)\|_{G^{\delta_0,1}(\mathbb{R})} \leq (2C)^2\|u(0)\|_{G^{\delta_0,1}(\mathbb{R})}\right). \tag{112}$$

Now in the interval $[2\rho^*, 3\rho^*]$, we get

$$\|u(t)\|_{G^{\delta_0,1}(\mathbb{R})} \leq 2C\|u(2\rho^*)\|_{G^{\delta_0,1}(\mathbb{R})}, \ t \in [2\rho^*, 3\rho^*] \\ \left(\text{thus } \|u(3\rho^*)\|_{G^{\delta_0,1}(\mathbb{R})} \leq (2C)^3\|u(0)\|_{G^{\delta_0,1}(\mathbb{R})}\right). \tag{113}$$

Proceeding this way, we conclude that in the interval $[(k-1)\rho^*, k\rho^*]$ we have

$$\|u(k\rho^*)\|_{G^{\delta_0,1}(\mathbb{R})} \leq (2C)^k\|u(0)\|_{G^{\delta_0,1}(\mathbb{R})}, \quad t \in [(k-1)\rho^*, k\rho^*], \tag{114}$$

and therefore we have

$$\|u(k\rho^*)\|^4_{G^{\delta_0,1}(\mathbb{R})} \leq (2C)^{4k}\|u(0)\|^4_{G^{\delta_0,1}(\mathbb{R})}. \tag{115}$$

Thus, from (110) and (115) we conclude that

$$\sup_{t\in[k\rho^*,(k+1)\rho^*]} \|u(t)\|^2_{G^{\delta,1}} \leq \|u(k\rho^*)\|^2_{G^{\delta,1}} + C(2C)^{4k}\frac{\delta^\theta}{(\delta_0-\delta)^5}\|u(0)\|^4_{G^{\delta_0,1}(\mathbb{R})}. \tag{116}$$

By using (107) for k we have

$$\|u(k\rho^*)\|^2_{G^{\delta,1}(\mathbb{R})} \leq \|u(0)\|^2_{G^{\delta,1}(\mathbb{R})} + kC(2C)^{4k}\frac{\delta^\theta}{(\delta_0-\delta)^5}\|u(0)\|^4_{G^{\delta_0,1}}. \tag{117}$$

Replacing $\|u(k\rho^*)\|^2_{G^{\delta,1}}$ in (116) by (117), we obtain the following inequality:

$$\begin{aligned}
&\sup_{t\in[k\rho^*,(k+1)\rho^*]} \|u(t)\|^2_{G^{\delta,1}} \\
&\quad \le \left(\|u(0)\|^2_{G^{\delta,1}(\mathbb{R})} + kC(2C)^{4k}\frac{\delta^\theta}{(\delta_0-\delta)^5}\|u(0)\|^4_{G^{\delta_0,1}}\right) \\
&\qquad + C(2C)^{4k}\frac{\delta^\theta}{(\delta_0-\delta)^5}\|u(0)\|^4_{G^{\delta_0,1}(\mathbb{R})} \\
&\quad \le \|u(0)\|^2_{G^{\delta,1}(\mathbb{R})} + (k+1)C(2C)^{4k}\frac{\delta^\theta}{(\delta_0-\delta)^5}\|u(0)\|^4_{G^{\delta_0,1}} \\
&\quad \le \|u(0)\|^2_{G^{\delta,1}(\mathbb{R})} + (k+1)C(2C)^{4(k+1)}\frac{\delta^\theta}{(\delta_0-\delta)^5}\|u(0)\|^4_{G^{\delta_0,1}},
\end{aligned} \tag{118}$$

verifying (107) with k replaced by $k+1$.

To get (108) with k replaced by $k+1$, it is then enough to have

$$(k+1)C(2C)^{4(k+1)}\frac{\delta^\theta}{(\delta_0-\delta)^5}\|u(0)\|^4_{G^{\delta_0,1}(\mathbb{R})} \le \|u(0)\|^2_{G^{\delta_0,1}(\mathbb{R})},$$

but this holds by (106) since $k+1 \le n+1$. The proof of the Claim is now complete.

End of Claim's Proof for $s=1$ Applying estimate (108) with $k=n+1$, we get

$$\sup_{t\in[0,(n+1)\rho^*]} ||u(t)||^2_{G^{\delta,1}} \le 2||u(0)||^2_{G^{\delta_0,1}}, \tag{119}$$

if δ-size conditions (106) hold. In other words we have proved the following implication:

$$\begin{aligned}
&0<\delta<\delta_0\le 1 \text{ and } (n+1)C\,(2C)^{4(n+1)}\frac{\delta^\theta}{(\delta_\theta-\delta)^5}||u(0)||^2_{G^{\delta_0,1}} \le 1 \\
&\qquad \Rightarrow u\in C([0,(n+1)\rho^*], G^{\delta,1}(\mathbb{R})).
\end{aligned} \tag{120}$$

Thus we can conclude that

$$\frac{\delta^\theta}{(\delta_\theta-\delta)^5} \le \frac{1}{(n+1)C\,(2C)^{4(n+1)}||u(0)||^2_{G^{\delta_0,1}}}. \tag{121}$$

Since $0 < \delta < \delta_0 \leq 1$, we have that $(\delta_0 - \delta)^5 < 1$ and therefore $\frac{1}{(\delta_0-\delta)^5} > 1$. This implies that

$$\delta^\theta < \frac{\delta^\theta}{(\delta_0 - \delta)^5} \leq \frac{1}{(n+1)C\,(2C)^{4(n+1)}||u(0)||^2_{G^{\delta_0,1}}}. \tag{122}$$

Also, since $n\rho^* \leq T < (n+1)\rho^*$ and $T > \rho^*$, we have that $n + 1 \leq \frac{T}{\rho^*} + 1 < \frac{T}{\rho^*} + \frac{T}{\rho^*} = \frac{2T}{\rho^*}$. Therefore, the δ-size condition (122) holds if we chose δ to satisfy the condition

$$\delta^\theta \leq \frac{\rho^*}{2TC(2C)^{\frac{8T}{\rho^*}}\|u(0)\|^2_{\delta_0,1}} = \frac{1}{T(2C)^{\frac{8T}{\rho^*}}}\frac{\rho^*}{2C\|u(0)\|^2_{\delta_0,1}}. \tag{123}$$

Now, defining $C_0 \doteq (2C)^{\frac{8}{\rho^*}} > 1$ and $C_1 \doteq \frac{\rho^*}{2C\|u(0)\|^2_{\delta_0,1}}$, we conclude that

$$\delta^\theta \leq \frac{1}{TC_0^T}C_1 \quad \text{or} \quad \delta \leq T^{-\frac{1}{\theta}}C_0^{-\frac{T}{\theta}}C_1^{\frac{1}{\theta}},$$

and therefore for $\theta = 1$ we get

$$\delta \leq \frac{C_1}{TC_0^T} = \frac{C_1}{Te^{(\ln C_0)T}}. \tag{124}$$

This proves Theorem 1.3 when $s = 1$.

The General Case For $s > \frac{1}{2}$ and $s \neq 1$ we use the following embedding relation, (see [28]), $G^{\delta,s}(\mathbb{R}) \subset G^{\delta',s'}(\mathbb{R})$ for all $0 < \delta' < \delta$ and $s, s' \in \mathbb{R}$, i.e., $||\varphi||_{G^{\delta',s'}} \leq C_{\delta,\delta',s,s'}||\varphi||_{G^{\delta,s}}$. It follows from this embedding that $u_0 \in G^{\delta_0,s}(\mathbb{R}) \subset G^{\frac{\delta_0}{2},1}(\mathbb{R})$. It follows from the case $s = 1$, already proved, that there is a $T_0 > 0$ such that $u \in C([0, T_0], G^{\delta_0/2,1}(\mathbb{R}))$ and $u \in C\Big([0, T], G^{\frac{2C_1}{Te^{(ln(C_0))T}},1}(\mathbb{R})\Big)$ for $T \geq T_0$, where $C_1 > 0$ depends on u_0, δ_0. Applying again the above embedding, we now conclude that $u \in C\Big([0, T], G^{\frac{C_1}{Te^{(ln(C_0))T}},s}(\mathbb{R})\Big)$, for $T \geq T_0$, and this proves Theorem 1.3 with $\theta = 1$. □

Acknowledgments The first author was partially supported by a grant from the Simons Foundation (#524469 to Alex Himonas). The second author was partially supported by Conselho Nacional de Desenvolvimento Científico e Tecnológico-CNPq, Grant 305868/2019-5, and São Paulo Research Foundation (FAPESP), Grant 2018/14316-3. The authors would like to thank the referee for helpful suggestions.

References

1. R.F. Barostichi, A. Himonas and G. Petronilho, *Autonomous Ovsyannikov theorem and applications to nonlocal evolution equations and systems*, J. Funct. Anal., **270** (2016) 330–358.
2. R. Barostichi, R.O. Figueira and A. Himonas, *The modified KdV equation with higher dispersion in Sobolev and analytic spaces on the line*. J. Evol. Equ. 21 (2021), no. 2, 2213–2237.
3. J. L. Bona, Z. Grujić, and H. Kalisch, *Algebraic lower bounds for the uniform radius of spatial analyticity for the generalized KdV equation*, Ann. Inst. H. Poincaré Anal. Non Linéaire, **22** (6) (2005), 783–797.
4. J. Bona and R. Smith, *The initial-value problem for the Korteweg-de Vries equation*. Phil. Trans. Roy. Soc. London Ser. A **278**(1287):555–601.
5. J. Bourgain, *Fourier transform restriction phenomena for certain lattice subsets and applications to nonlinear evolution equations. Part 2: KdV equation*, Geom. Funct. Anal. **3** (1993), 209–262.
6. J. Bourgain, *Fourier transform restriction phenomena for certain lattice subsets and applications to nonlinear evolution equations. Part 1: Schrödinger equation*, Geom. Funct. Anal. **3** (2) (1993), 107–156.
7. A. Bressan and A. Constantin, *Global conservative solutions of the Camassa-Holm equation*. Arch. Ration. Mech. Anal., **183** (2007), 215–239.
8. R. Camassa and D. Holm, *An integrable shallow water equation with peaked solitons*, Phys. Rev. Lett., **71** (1993), 1661-1664.
9. J. Colliander, M. Keel, G. Staffilani, H. Takaoka, and T. Tao, *Sharp global well-posedness for KdV and modified KdV on* $\mathbb{R}$ *and* $\mathbb{T}$, J. Amer. Math. Soc., **16** (3) (2003), 705–749.
10. J. Colliander, M. Keel, G. Staffilani, H. Takaoka, and T. Tao, *Multilinear estimates for periodic KdV equations, and applications*. J. Funct. Anal. **211** (1) (2004), 173–218. ch cl cs-2000 d
11. A. Constantin and D. Lannes, *The hydrodynamical relevance of the Camassa-Holm and Degasperis-Procesi equations*, Arch. Ration. Mech. Anal., **192** (2009), 165–186.
12. A. Constantin and W. A. Strauss, *Stability of peakons*, Comm. Pure Appl. Math., 53 (2000), 603–610.
13. R. Danchin, *A few remarks on the Camassa-Holm equation*, Differential Integral Equations, **14** (8) (2001), 953–988.
14. R. Figueira, A. Himonas and F. Yan, *A higher dispersion KdV equation on the line*. Nonlinear Anal. **199** (2020), 112055, 1–38.
15. C. Foias, R. Temam, *Gevrey class regularity for the solutions of the Navier-Stokes equations*, J. Funct. Anal., **87** (1989), 359–369.
16. A. Fokas, B. Fuchssteiner, *Symplectic structures, their Bäcklund transformations and hereditary symmetries*, Phys. D, **4** (1) (1981/1982), 47–66.
17. Z. Grujić and H. Kalisch, *Local well-posedness of the generalized Korteweg-de Vries equation in spaces of analytic functions*, Differential and Integral Equations, **15**, No 11, 1325–1334, (2002).
18. S. Hakkaev and K. Kirchev, *Local well-posedness and orbital stability of solitary wave solutions for the generalized Camassa-Holm equation*. Comm. Partial Differential Equations **30** (2005), no. 4-6, 761–781.
19. A. Himonas and C. Holliman, *The Cauchy problem for a generalized Camassa-Holm equation*. Adv. Differential Equations **19** (2014), no. 1-2, 161–200.
20. A. Himonas, H. Kalisch and S. Selberg, *On persistence of spatial analyticity for the dispersion-generalized periodic KdV equation*, Nonlinear Anal. Real World Appl, **30** (2017), 35–48.
21. A. Himonas and C. Kenig, *Non-uniform dependence on initial data for the CH equation on the line*, Differential Integral Equations, **22** (2009), 201–224.
22. A. Himonas and G. Misiołek, *Global well-posedness of the Cauchy problem for a shallow water equation on the circle*, J. Differential Equations, **161** (2000), 479–495.

23. A. Himonas and G. Misiołek, *The Cauchy problem for a shallow water type equation*, Comm. Partial Differential Equations, **23** (1-2) (1998), 123–139.
24. A. Himonas, G. Misiołek, *Analyticity of the Cauchy problem for an integrable evolution equation*, Math. Ann., **327** (3) (2003), 575–584.
25. A. Himonas and G. Petronilho, *A G^1 almost conservation law for mCH and evolution of its radius of spatial analyticity*, Discrete and Continuous Dynamical Systems, **41**, No. 5, (2021), 2031–2050.
26. A. Himonas and F. Yan, *On Well-Posedness of nonlocal evolution equations*, to appear in Vietnam Journal of Mathematics, 2023.
27. T. Kato, *On the Cauchy problem for the (generalized) Korteweg-de Vries equation*, Advances in Mathematics Supplementary Studies, Studies in Applied Math., **8** (1983), 93–128.
28. T. Kato and K. Masuda, *Nonlinear Evolution Equations and Analyticity I*, Ann. Inst. H. Poincaré Anal. Non Linéaire, **3** (6) (1986), 455–467.
29. Y. Katznelson, *An introduction to Harmonic Analysis*, corrected ed., Dover Publications, Inc., New York, (1976).
30. C. E. Kenig, G. Ponce and L. Vega, *The Cauchy problem for the Korteweg-de Vries equation in Sobolev spaces of negative indices*, Duke Math. J., **71**, 1–21, (1993).
31. C. Kenig, G. Ponce and L. Vega, *A bilinear estimate with applications to the KdV equation*, J. Amer. Math. Soc., **9** (2) (1996), 573–603.
32. C. Kenig, G. Ponce and L. Vega, *Well-posedness and scattering results for the generalized Korteweg-de Vries equation via the contraction principle*, Comm. Pure Appl. Math., **46** (4) (1993), 527–620.
33. P. Lax, D. Levermore, *The zero dispersion limit for the Korteweg-de Vries (KdV) equation.* Proc. Natl. Acad. Sci. U.S.A. **76** (1979), no. 8, 3602–3606.
34. P. Lax, D. Levermore, *The small dispersion limit of the Korteweg-de Vries equation.* I, Commun. Pure Appl. Math. **36** (1983), no. 3, 253–290.
35. P. Lax, D. Levermore, *The small dispersion limit of the Korteweg-de Vries equation.* II, Commun. Pure Appl. Math. **36** (1983), no. 5, 571–593.
36. P. Lax, D. Levermore, *The small dispersion limit of the Korteweg-de Vries equation.* III, Commun. Pure Appl. Math. **36** (1983), no. 6, 809–829.
37. J. Lenells, *Traveling wave solutions of the Camassa-Holm equation*, J. Differential Equations, **217** (2) (2005), 393–430.
38. Y. Li and P. Olver, *Well-posedness and blow-up solutions for an integrable nonlinearly dispersive model wave equation*, J. Differential Equations, **162** (2000), 27–63.
39. F. Linares and G. Ponce, *Introduction to nonlinear dispersive equations*, Universitext Springer, New York, 2009.
40. G. Petronilho and P. Leal da Silva, *On the radius of spatial analyticity for the modified Kawahara equation on the line*, Mathematische Nachrichten, **292**, (2019), 2032–2047.
41. G. Rodriguez-Blanco, *On the Cauchy problem for the Camassa-Holm equation*, Nonlinear Anal., **46** (2001), 309–327.
42. S. Selberg and D. O. Silva, *Lower Bounds on the Radius of a Spatial Analyticity for the KdV Equation*, Ann. Henri Poincaré, **18** (2016), 1009–1023.
43. T. Tao, *Nonlinear Dispersive Equations-Local and Global Analysis*, Published for the Conference Board of the Mathematical Sciences, Washington, DC; by the American Mathematical Society, Providence, RI (2006).

Distributions with Decay and Restriction Problems

Gustavo Hoepfner and Andrew Raich

To the memory of Emilie Hounie

2020 Mathematics Subject Classification 35A18, 35A27, 26E10, 42B10

1 Introduction

Our motivating problem is the following: Given a function space $\mathfrak{X} \subset \mathcal{S}'(\mathbb{R}^d)$, determine necessary and sufficient conditions on $\hat{\psi} \in \mathcal{S}'(\mathbb{R}^d)$ to ensure $\psi \in \mathfrak{X}$, that is, to understand when $T \in \mathcal{S}'(\mathbb{R}^d)$ satisfies

$$T = \hat{\psi} \quad \text{for some} \quad \psi \in \mathfrak{X}. \tag{1}$$

This question leads us to define and explore microglobal wavefront sets defined in terms of the Fourier transform. Additionally, we apply this new technology to the classical restriction problem and prove a new type of restriction phenomenon that we call the *restriction problem with moments*. Essentially, the restriction problem with moments is the restriction problem applied to a measure (as in the classical case)

This work was partially supported by CNPq (308826/2018-3), FAPESP (2018/14316-3 and 2019/04995-3), and Simons Foundation (707123, ASR)

G. Hoepfner
Departamento de Matemática, Universidade Federal de São Carlos, São Carlos, SP, Brazil
e-mail: ghoepfner@ufscar.br

A. Raich (✉)
Department of Mathematical Sciences, SCEN 309, 1 University of Arkansas, Fayetteville, AR, USA
e-mail: araich@uark.edu

S. Berhanu et al. (eds.), *Geometric Analysis of PDEs and Several Complex Variables*, Latin American Mathematics Series – UFSCar subseries,
https://doi.org/10.1007/978-3-031-69702-9_8

and moments of the measure. Moments of the measure correspond to derivatives of the Fourier transform, so we are, in effect, showing that the restriction problem can hold for certain tempered distributions. This is a new phenomenon for an old problem.

The origin of our investigation is the paper of Boggess and Raich [BR13] in which they find conditions on the Fourier transform of a function f that guarantees that f exhibits a specific type of exponential decay. This led to the development of the global L^q Gevrey and global L^q Denjoy–Carleman function classes [AHR17, HR19a]. Our hope was to use these function classes to explore global properties of partial differential operators, but we discovered that the Fourier transforms of such functions can be highly nonsmooth (e.g., a measure supported on a Salem set). To overcome this difficulty, we substituted the FBI transform for the Fourier transform and developed microglobal tools to analyze operators [HR19b]. The FBI transform was introduced by Bros and Iagolnitzer [BI73] to study local analyticity but was quickly recognized as a powerful tool to investigate microlocal (hypo) regularity in a broad array of function classes, including (real) analytic, Gevrey, Denjoy–Carleman, and C^∞ (see [BCT83, BH12, Bon77, Chr97, HM18, Hör90, Sjö82]). It has even been used to study the local and microlocal regularity of CR functions as well as solutions of more general vector fields and some nonlinear partial differential equations (see [AB12, AH10, AH15, Asa95, BT82, BP09, Ber10, BH12, Ber10, BX15, EG03, HT92, HM18] and references therein). Like the Fourier transform, the FBI transform interchanges smoothness with decay. One particularly useful feature of the FBI transform is that one can apply it on a (much) wider class of objects than the Fourier transform.

Despite the success of the FBI transform as a problem solving tool, there is (at least) one inherent weakness with it—the near impossibility of actually computing it and its inverse. Many objects are naturally analyzed with Fourier transforms, and often the Fourier transform of a given differential operator has an accessible form, and it is therefore natural to gravitate toward the Fourier transform, despite its well-documented weaknesses. Therefore, the positive results that we have had developing a microglobal analysis for the FBI transform encouraged us to develop a Fourier transform-centered microglobal analysis based on a theory of distributions with decay.

The Fourier transform interchanges smoothness with decay so function spaces defined in terms of smoothness estimates, e.g., Sobolev spaces, global L^q Denjoy–Carleman spaces, etc., ought to have characterizations in terms of the decay of the Fourier transform. One complicating factor is that integrability conditions at ∞ are decay conditions, and hence the Fourier transform of an L^1 function is smoother than that of an L^q function, $q > 1$. Additionally, Fourier transforms of L^q functions, regardless of their smoothness, need not be functions at all [HR19a]. In fact, our viewpoint of restriction theorems is that Fourier transforms of L^q functions, $1 \leq q < 2$, have sufficient additional smoothness to allow them restrict to the support of a measure supported on an appropriate lower-dimensional subset.

The notion that we introduce in this chapter is that of a distribution with decay. The latter sections of this chapter are devoted to applications and provide motivation for defining distributions with decay. First, in Sect. 3, we define Fourier microglobal regularity and a Fourier microglobal wavefront. We then characterize distributions with decay in terms of the Fourier microglobal wavefront set as well as comparing them to the FBI microglobal wavefront set defined in [HR19b]. The advantage of establishing a microglobal analysis in terms of the Fourier transform is that the FBI transform is difficult to compute, and for objects that are tempered distributions, the Fourier transform is more natural.

We demonstrate the usefulness of these ideas in Sect. 4. We apply the distribution with decay machinery to measures supported on lower-dimensional subsets of $\mathbb{R}^d$, specifically, the Salem measure and the surface area measure of the sphere. By proving that the Fourier transforms of these measures are global L^q-Gevrey functions, we can show that these measures, as well as all of their moments, satisfy the L^p-L^2 restriction problem for a range of p that we compute. We call this the *L^p-L^2 restriction problem with moments*. It is a generalization of and a stronger result than the classical L^p-L^2 restriction problem. Additionally, by keeping track of the decay of the derivatives of the transform and the fact that the transform originates from a compactly supported measure, we are able to establish the range $1 \le p < \frac{2(d+1)}{d+3}$ for the sphere and $1 \le p < \frac{2(2-2\alpha+\beta)}{4(1-\alpha)+\beta}$ for Salem's measure.

2 General Function Classes

In order to motivate Definition 2.5, we provide several illuminating examples.

Example 2.1 (**L^q Functions with Polynomial Decay of Order** k) Let $f \in L^q(\mathbb{R}^d)$, $1 < q < \infty$, so that

$$\|x^\alpha f\|_{L^q(\mathbb{R}^d)} \le C, \qquad \text{for all } |\alpha| \le k. \tag{2}$$

We call such a function an *L^q function with polynomial decay of order k*. By Hölder's inequality and duality, (2) is equivalent to

$$|\langle x^\alpha f, \varphi\rangle| \le C\|\varphi\|_{L^p(\mathbb{R}^d)}$$

for all $\varphi \in \mathcal{S}(\mathbb{R}^d)$.

Example 2.2 (Fourier Transforms of Functions in a Sobolev Space) Let $1 \le q \le \infty$, $k \ge 1$, and $f \in W^{k,q}(\mathbb{R}^d)$. If $|\alpha| \le k$ and $\varphi \in \mathcal{S}(\mathbb{R}^d)$, then

$$|\langle \xi^\alpha \hat{f}, \varphi\rangle| = |\langle D^\alpha f, \hat{\varphi}\rangle| \le \|f\|_{W^{k,q}(\mathbb{R}^d)}\|\hat{\varphi}\|_{L^p(\mathbb{R}^d)},$$

where $\frac{1}{q} + \frac{1}{p} = 1$. In this case, $\hat{f}$ is a distribution with decay, but it is fundamentally different than an L^q function with decay because there is no reason to believe that $\hat{f} \in L^p(\mathbb{R}^d)$ (or even a function), unless, of course, $1 \le q \le 2$. Additionally, the decay is only apparent when measured against $\hat{\varphi}$. In this case, we say that $\hat{f}$ is a *distribution with polynomial decay of order k with respect to the Fourier transform.*

Example 2.3 (Fourier Transforms of Global L^q-Gevrey Functions) Let $1 \leq q \leq \infty$. We say that f is a *global L^q-Gevrey function* and write $f \in \mathcal{G}^{q,s}(\mathbb{R}^d)$ (see (10) below for a more general framework for these function spaces) if there exist constants $C, A > 0$ so that $\|D^\alpha f\|_{L^q(\mathbb{R}^d)} \leq CA^{|\alpha|}|\alpha|^{|\alpha|s}$. If $\varphi \in \mathcal{S}(\mathbb{R}^d)$, then

$$|\langle \xi^\alpha \hat{f}, \varphi \rangle| \leq CA^{|\alpha|}|\alpha|^{|\alpha|s}\|\hat{\varphi}\|_{L^p(\mathbb{R}^d)}.$$

This shows that $\hat{f}$ is also a distribution with decay with respect to the Fourier transform, and we now introduce a notation to quantify it.

2.1 Semi-norms and Function Spaces

In order to capture the widest array of function spaces and tempered distributions, we define our spaces using the general classes of semi-norms that define the Schwartz class.

One of the most important features of the Fourier transform is its exchange of smoothness for decay and vice versa. This feature underlies all Paley–Wiener Theorems. Exactly how to capture this interplay is challenging, and we focus on using classes of semi-norms. One of the flexible features of $\mathcal{S}(\mathbb{R}^d)$ is that it can be defined using L^p norms for any $1 \leq p \leq \infty$, either on the function side or on the Fourier transform side. In [BR13, AHR17], this type of duality was used to capture exponential decay in terms of inequalities on the Fourier transform.

We notate a *parameter set* $\mathfrak{S}$ by

$$\mathfrak{S} = \{(\alpha, \beta, q) : \alpha, \beta \in \mathbb{N}_0^d, 1 \leq q \leq \infty\},$$

and for simplicity, we assume that q is fixed over the parameter set. We impose the additional requirement that if $(\alpha, \beta, q) \in \mathfrak{S}$, then so is (α', β', q) for all $\alpha' \subset \alpha$ and $\beta' \subset \beta$ where $\alpha' \subset \alpha$ means that $\alpha'_j \leq \alpha_j$ for all $1 \leq j \leq d$. Given a parameter set $\mathfrak{S}$, we also have an associated *growth set*

$$\mathfrak{G} = \{a_\mathfrak{p} : \mathfrak{p} \in \mathfrak{S}\} \subset (0, \infty)$$

and call the pair $(\mathfrak{S}, \mathfrak{G})$ a *parameter pair*. To each element $(\mathfrak{s}, a_\mathfrak{s}) \in (\mathfrak{S}, \mathfrak{G})$, define the semi-norm

$$\rho(f; \mathfrak{s}; a_\mathfrak{s}) = \rho_{\mathfrak{s}, a_\mathfrak{s}}(f) = a_\mathfrak{s}\|x^\beta D^\alpha f\|_{L^q(\mathbb{R}^d)}$$

for $f \in \mathcal{S}(\mathbb{R}^d)$.

We are not exploring the most general setup for the semi-norm and parameter classes. We could have defined our norms on $\Omega \subset \mathbb{R}^d$ and also included compact sets $K \subset \Omega$ with $\mathfrak{S} = \{(\alpha, q, \Omega, K) : \alpha, \beta \in \mathbb{N}_0^d, 1 \le q \le \infty, \Omega \subset \mathbb{R}^d\}$ and $\rho(f; \mathfrak{s}; a_\mathfrak{s}) = a_\mathfrak{s} \|x^\beta D^\alpha f\|_{L^q(K)}$. This would have allowed us to recover the classical Denjoy–Carleman spaces, i.e., with $q = \infty$, but we do not pursue these spaces in this chapter because it is known that these spaces possess a characterization via the Fourier transform, see [HM18].

Our interest is in function spaces $\mathfrak{X} \subset \mathcal{S}'(\mathbb{R}^d)$ that are defined by a finite or countable collection of semi-norms $\rho_{\mathfrak{s},a_\mathfrak{s}}$ or an increasing/decreasing union of such spaces.

For concreteness, we are going to concentrate on the cases when $\mathfrak{X}$ is defined by a finite collection of semi-norms (e.g., Examples 2.1 and 2.2), a global L^q Gevrey space [AHR17] (e.g., Example 2.3), or a global L^q Denjoy–Carleman space as in [HR19b] (see (10)).

If $\mathfrak{X}$ is defined by a finite index set $\mathfrak{S}$,

$$\mathfrak{X} = \{f \in \mathcal{S}'(\mathbb{R}^d) : \rho(f; \alpha, \beta, q; C) < \infty, \text{ for all } (\alpha, \beta, q) \in \mathfrak{S} \text{ and some } C > 0\}, \tag{3}$$

and the growth set $\mathfrak{G}$ plays no role.

For function spaces $\mathfrak{X}$ defined by an infinite collection of semi-norms $\mathfrak{S}$, we require that there exist increasing sequences of positive numbers $M = (M_j)_{j \in \mathbb{N}_0}$ and $M' = (M'_j)_{j' \in \mathbb{N}_0}$ so that

$$j! \le \min\{M_j, M'_j\} \qquad \forall j \in \mathbb{N}_0, \tag{4}$$

$$\mathfrak{X}_A = \left\{ f \in \mathcal{S}'(\mathbb{R}^d) : \left\{ \frac{\|x^\beta D^\alpha f\|_{L^q(\mathbb{R}^d)}}{A^{|\alpha|+|\beta|} M_{|\alpha|} M'_{|\beta|}} : (\alpha, \beta, q) \in \mathfrak{S} \right\} \in \ell^q \right\} \tag{5}$$

and

$$\begin{aligned} \mathfrak{X} &= \bigcup_{A>0} \mathfrak{X}_A \\ &= \Big\{ f \in \mathcal{S}'(\mathbb{R}^d) : \text{ there exist } C, A > 0 \text{ so that} \\ &\qquad \|x^\beta D^\alpha f\|_{L^q(\mathbb{R}^d)} \le C A^{|\alpha|+|\beta|} M_{|\alpha|} M'_{|\beta|} \|f\|_{L^q(\mathbb{R}^d)}, \ (\alpha, \beta, q, j) \in \mathfrak{S} \Big\}. \end{aligned} \tag{6}$$

Remark 2.4 Since the number of terms with $\alpha' \subset \alpha$ and $\beta' \subset \beta$ grows geometrically in $|\alpha|$ and $|\beta|$, we can estimate $\|D^\beta \{x^\alpha f\}\|_{L^q(\mathbb{R}^d)}$ by increasing the geometric constant A.

2.2 Distributions with Decay

Definition 2.5 Let $T \in \mathcal{S}'(\mathbb{R}^d)$ and $(\mathfrak{P}, \mathfrak{G})$ be a parameter pair. We say that:

- T is a *tempered distribution with* $(\mathfrak{P}, \mathfrak{G})$ *decay* if for all $\varphi \in \mathcal{S}(\mathbb{R}^d)$ it holds that

$$\left\{\frac{|\langle x^\beta D^\alpha T, \varphi\rangle|}{c_{\mathfrak{p}}\|\varphi\|_{L^p(\mathbb{R}^d)}} : (\alpha, \beta, q) = \mathfrak{p} \in \mathfrak{P},\ c_{\mathfrak{p}} \in \mathfrak{G}\right\} \in \ell^q. \tag{7}$$

- T is a *tempered distribution with* $(\mathfrak{P}, \mathfrak{G})$ *decay in the direction* $x \neq 0$ if there exists an open cone Γ containing x so that (7) holds for all $\varphi \in \mathcal{S}(\mathbb{R}^d)$ with $\operatorname{supp} \varphi \subset \Gamma$.
- T is a *tempered distribution with* $(\mathfrak{P}, \mathfrak{G})$ *decay at* $x = 0$ if (7) holds with Γ replaced by $B(0, 1)$.
- T is a *tempered distribution with* $(\mathfrak{P}, \mathfrak{G})$ *decay with respect to the Fourier transform* if

$$\left\{\frac{|\langle x^\beta D^\alpha T, \varphi\rangle|}{c_{\mathfrak{p}}\|\hat{\varphi}\|_{L^q(\mathbb{R}^d)}} : (\alpha, \beta, q) = \mathfrak{p} \in \mathfrak{P},\ c_{\mathfrak{p}} \in \mathfrak{G}\right\} \in \ell^q. \tag{8}$$

- T is a *tempered distribution with* $(\mathfrak{P}, \mathfrak{G})$ *decay in the direction* $x \neq 0$ *(resp.,* $x = 0$*)) with respect to the Fourier transform* if there exists an open cone Γ containing x (resp., $B(0, 1)$) so that (8) holds for all $\varphi \in \mathcal{S}(\mathbb{R}^d)$ with $\operatorname{supp} \varphi \subset \Gamma$ (resp., $\operatorname{supp} \varphi \subset B(0, 1)$).

To motivate the correct notion of microglobal regularity, observe that if $f \in \mathcal{G}^{q,s}(\mathbb{R}^d)$, then

$$|\langle \xi^\alpha \hat{f}, \check{\varphi}\rangle| = |\langle D^\alpha f, \varphi\rangle| \leq C A^{|\alpha|} |\alpha|^{|\alpha| s} \|\varphi\|_{L^p(\mathbb{R}^d)}.$$

Given a parameter set $\mathfrak{P}$, we define the *dual parameter set* $\hat{\mathfrak{P}}$ by $\hat{\mathfrak{P}} = \{(\alpha, \beta, q) \in \mathbb{N}_0 \times \mathbb{N}_0 \times [1, \infty] : (\beta, \alpha, p) \in \mathfrak{P}\}$, where $\frac{1}{p} + \frac{1}{q} = 1$.

Definition 2.6 Let $T \in \mathcal{S}'(\mathbb{R}^d)$, $\xi \in \mathbb{R}^d$, and $(\mathfrak{P}, \mathfrak{G})$ be a parameter pair. We say that T is *Fourier microglobally regular in the direction* $\xi \in \mathbb{R}^d$ *with respect to the parameter pair* $(\mathfrak{P}, \mathfrak{G})$ (or simply Fourier $(\mathfrak{P}, \mathfrak{G})$-microglobal regular in the direction $\xi \in \mathbb{R}^d$), if $\hat{T}$ is a tempered distribution with $(\hat{\mathfrak{P}}, \mathfrak{G})$-decay in the direction ξ (possibly with respect to the Fourier transform).

We define the *Fourier wavefront set of* T *with respect to the parameter pair* $(\mathfrak{P}, \mathfrak{G})$ or the Fourier $(\mathfrak{P}, \mathfrak{G})$-wavefront set, denoted by $\mathcal{WF}_{\mathfrak{P},\mathfrak{G}}(T)$, by

$$\mathcal{WF}_{\mathfrak{P},\mathfrak{G}}(T) = \{\xi \in \mathbb{R}^d : T \text{ is not Fourier } (\mathfrak{P}, \mathfrak{G}) - \text{microglobal regular in the direction } \xi\}.$$

If $\mathfrak{X}$ is a function space defined by the parameter pair $(\mathfrak{P}, \mathfrak{G})$, that is,

$$\mathfrak{X} = \{f \in L^q(\mathbb{R}^d) : \rho(f; \mathfrak{s}; c_{\mathfrak{s}}) < \infty \text{ for all } \mathfrak{s} \in \mathfrak{P}\},$$

then we will call the $(\mathfrak{P}, \mathfrak{G})$-wavefront set by $\mathfrak{X}$-*wavefront set* and denote it by $\mathcal{WF}_{\mathfrak{X}}$. Similarly, we say that a distribution is $\mathfrak{X}$ microglobal regular in the direction ξ.

Theorem 2.7 *Let* $T \in \mathcal{S}'(\mathbb{R}^d)$, $1 < q < \infty$, *and* $\mathfrak{X} \subset \mathcal{S}'(\mathbb{R}^d)$ *be defined by the parameter pair* $(\mathfrak{S}, \mathfrak{G})$. *There exists* $\psi \in \mathfrak{X}$ *so that* $\hat{T} = \psi$ *if and only if the* $\mathcal{WF}_{\mathfrak{X}}(T) = \emptyset$.

Proof The forward direction follows from Remark 2.4 and the inequality

$$\begin{aligned}\left|\left(x^\alpha D^\beta T, \varphi\right)\right| &= \left|\left(x^\alpha D^\beta \hat{\psi}, \varphi\right)\right| = \left|\left(D^\alpha \{x^\beta \psi\}, \hat{\varphi}\right)\right| \\ &\leq \sum_{\beta' \subset \alpha \cap \beta} \binom{\alpha}{\beta'} \frac{\beta!}{(\beta-\beta')!} \left(x^{\beta-\beta'} D^{\alpha-\beta'} \psi, \hat{\varphi}\right).\end{aligned}$$

Now suppose that $T \in \mathcal{S}'(\mathbb{R}^d)$ is a tempered distribution with $(\hat{\mathfrak{S}}, \mathfrak{G})$-decay given by (8). We know that ξ^α is a smooth, slowly increasing function, and we use the argument of the forward direction to establish

$$\left|\left\langle x^\beta D^\alpha \hat{T}, \varphi\right\rangle\right| = \left|\left\langle D^\beta \{\xi^\alpha T\}, \hat{\varphi}\right\rangle\right| \leq c_{\alpha,\beta,q} \|\varphi\|_{L^p(\mathbb{R}^d)}$$

since $\hat{\hat{\varphi}}(x) = \varphi(-x)$. Since $1 < p < \infty$, we know that $\mathcal{S}(\mathbb{R}^d)$ is dense in $L^p(\mathbb{R}^d)$ and that $L^p(\mathbb{R}^d)' = L^q(\mathbb{R}^d)$. Consequently,

$$\varphi \mapsto \left\langle x^\beta D^\alpha \hat{T}, \varphi\right\rangle$$

extends to a bounded linear operator on $L^p(\mathbb{R}^d)$. Therefore, there exists $\psi_{\alpha,\beta} \in L^q(\mathbb{R}^d)$ so that

$$\left\langle x^\beta D^\alpha \hat{T}, \varphi\right\rangle = \int_{\mathbb{R}^d} \psi_{\alpha,\beta}(\xi)\varphi(\xi)\, dV(\xi).$$

We claim that

$$\psi_{\alpha,\beta}(\xi) = x^\beta D^\alpha \psi_{0,0}(\xi).$$

If the claim holds, then

$$\|x^\beta D^\alpha \psi_{0,0}\|_{L^q(\mathbb{R}^d)} \leq C_{|\alpha|,|\beta|},$$

for all $(\alpha, \beta, q) \in \mathfrak{S}$. This would mean $\psi_{0,0} \in \mathfrak{X}$. To prove the claim we first assume $\beta = 0$ and observe that $\psi_{0,0} \in L^q(\mathbb{R}^d)$, and hence for any $\varphi \in \mathcal{S}(\mathbb{R}^d)$,

$$\langle D^\alpha \psi_{0,0}, \varphi \rangle = (-1)^{|\alpha|} \langle \psi_{0,0}, D^\alpha \varphi \rangle = (-1)^{|\alpha|} \langle \hat{T}, D^\alpha \varphi \rangle = \langle D^\alpha \hat{T}, \varphi \rangle = \langle \psi_{\alpha,0}, \varphi \rangle.$$

Again by density, it must be the case that $D^\alpha \psi_{0,0} = \psi_{\alpha,0}$. The proof that $\psi_{\alpha,\beta} = x^\beta \psi_{\alpha,0}$ is similar. □

Remark 2.8

(1) The forward direction of this proof holds for $1 \leq q \leq \infty$. It is only the reverse direction that eliminates $q = 1$ and ∞ from the theorem.
(2) If $1 \leq q \leq 2$, $\frac{1}{q} + \frac{1}{p} = 1$, and $\mathfrak{X} = \mathcal{G}^{q,s}(\mathbb{R}^d)$, then the Hausdorff–Young inequality implies that

$$|\langle \xi^\alpha T, \varphi \rangle| \leq C A^{|\alpha|} |\alpha|^{|\alpha| s} \|\hat{\varphi}\|_{L^p(\mathbb{R}^d)} \leq C A^{|\alpha|} |\alpha|^{|\alpha| s} \|\varphi\|_{L^q(\mathbb{R}^d)},$$

which, of course, means that T agrees with a function that decays exponentially in the L^p sense (appropriately defined, see also [BR13, AHR17]).

Theorem 2.9 *Let $T \in \mathcal{S}'(\mathbb{R}^d)$ and $\mathfrak{X} \subset \mathcal{S}'(\mathbb{R}^d)$ be a function space defined by the parameter pair $(\mathfrak{S}, \mathfrak{G})$. Then T is given by integration against a function $\psi \in \mathfrak{X}$ if and only if T is Fourier $\mathfrak{X}$-microglobally regular with respect to $(\mathfrak{S}, \mathfrak{G})$ for all directions $\xi \in \mathbb{R}^d$.*

Proof The forward direction follows immediately from Theorem 2.7. Therefore, we may assume that T is Fourier $\mathfrak{X}$-microglobally regular in every direction $\xi \in \mathbb{R}^d$. Let $\chi_1, \ldots, \chi_N$ be a smooth partition of unity on $\mathbb{S}^{d-1}$ so that $\operatorname{supp} \chi_j \subset \Gamma_j$ where Γ_j is a cone given by Definition 2.6 corresponding to some $\xi_j \in \mathbb{R}^d \setminus \{0\}$. We extend χ_j homogeneously of degree 0 to $\mathbb{R}^d \setminus \{0\}$. We also let $\chi \in C_c^\infty(\mathbb{R}^d)$ so that $\chi \equiv 1$ on $B(0, 1/2)$ and $\operatorname{supp} \chi \subset B(0, 1)$. Let $\varphi \in \mathcal{S}(\mathbb{R}^d)$ and write

$$\check{\varphi} = \chi \check{\varphi} + (1 - \chi) \sum_{k=1}^{N} \chi_k \check{\varphi}. \tag{9}$$

Let $(\alpha, \beta, q) = \mathfrak{s} \in \mathfrak{S}$ so that $(\beta, \alpha, p) \in \widehat{\mathfrak{S}}$. Since $(1 - \chi)\chi_k \check{\varphi} \in \mathcal{S}(\mathbb{R}^d)$ satisfies $\operatorname{supp}(1 - \chi)\chi_k \check{\varphi} \subset \Gamma_k$, we estimate

$$\begin{aligned} |\langle x^\beta D^\alpha T, \mathcal{F}\{(1 - \chi)\chi_k \check{\varphi}\} \rangle| &= |\langle D^\beta \{\xi^\alpha \hat{T}\}, (1 - \chi)\chi_k \check{\varphi} \rangle| \\ &\leq c_{\mathfrak{s}} \|\mathcal{F}\{(1 - \chi)\chi_k \check{\varphi}\}\|_{L^p(\mathbb{R}^d)} \end{aligned}$$

for the appropriate $C_{|\alpha|,|\beta|}$, depending on the definition of $\mathfrak{X}$. Moreover, $(1 - \chi)\chi_k \in L^\infty$ is a Mikhlin multiplier, so its Fourier transform is the convolution kernel of an

L^p-bounded operator, $1 < p < \infty$. Hence

$$\|\mathcal{F}\{(1-\chi)\chi_k\check{\varphi}\}\|_{L^p(\mathbb{R}^d)} = \|\widehat{(1-\chi)\chi_k} * \varphi\|_{L^p(\mathbb{R}^d)} \leq C\|\varphi\|_{L^p(\mathbb{R}^d)}.$$

The argument for decay at 0 follows from the same argument with χ replacing by $(1-\chi)\chi_k$. The theorem now follows from the decomposition of φ given in (9). □

3 A Special Case: The Space of Global L^q Denjoy–Carleman Functions

If $M = (M_j)_{j\in\mathbb{N}_0}$ is an appropriately defined increasing sequence, then the global L^q Denjoy–Carleman space

$$\mathcal{E}^{q,M}(\Omega) = \bigcup_{A>0} \mathcal{E}^{q,M}_A(\Omega),$$

where

$$\mathcal{E}^{q,M}_A(\Omega) = \big\{ f \in C^\infty(\mathbb{R}^d) : \text{there is a constant } C > 0 \text{ so that} \\ \|D^\alpha f\|_{L^q(\Omega)} \leq C A^{|\alpha|} M_{|\alpha|} \text{ for all multi-indices } \alpha \big\}. \tag{10}$$

We will not need to know the requirements on M_j apart from that it is increasing and satisfies (4) (see [HR19b]). The most important example of the global L^q Denjoy–Carleman spaces is the global L^q-Gevrey spaces $\mathcal{G}^{q,s}(\mathbb{R}^d)$ in which case $M_j = j^{js}$ [AHR17, HR19a].

Proposition 3.1 *Let* $1 \leq q \leq \infty$ *and* $\mathfrak{X}$ *be a function space contained by* $W^{1,q}(\mathbb{R}^d)$. *There exists* $T \in \mathcal{S}'(\mathbb{R}^d)$ *so that* T *is* $\mathfrak{X}$*-microglobally regular for all* $\xi \in \mathbb{R}^d \setminus \{0\}$ *but not* $\mathfrak{X}$*-microglobally regular at* $\{0\}$.

Remark 3.2 Examples for $\mathfrak{X}$ include $W^{k,q}(\mathbb{R}^d)$, $1 \leq k \leq \infty$, and $\mathcal{E}^{q,M}(\mathbb{R}^d)$.

Proof Let $T \in \mathcal{S}'(\mathbb{R}^d)$ be defined by

$$\langle T, \varphi \rangle = \int_{\mathbb{R}^d} x_1 \varphi(x)\, dV(x).$$

Then

$$\hat{T} = i \frac{\partial \delta_0}{\partial \xi_1}.$$

Since $\operatorname{supp} \hat{T} = \{0\}$, it follows immediately that T is $\mathfrak{X}$-microglobally regular for all $\xi \in \mathbb{R}^d \setminus \{0\}$. However, T is clearly not $\mathfrak{X}$-microglobally regular at $\{0\}$ because

$$\langle T, \varphi \rangle = \langle T, \hat{\check{\varphi}} \rangle = \langle \hat{T}, \check{\varphi} \rangle = i \Big\langle \frac{\partial \delta_0}{\partial \xi_1}, \check{\varphi} \Big\rangle = -i \frac{\partial \check{\varphi}(0)}{\partial \xi_1}$$

is certainly bounded by neither $\|\check{\varphi}\|_{L^p(\mathbb{R}^d)}$ nor $\|\varphi\|_{L^p(\mathbb{R}^d)}$ for any $1 \leq p \leq \infty$. □

For the global L^q Denjoy–Carleman spaces, we now have two notions of the microglobal wavefront set—the Fourier $\mathcal{E}^{q,M}$-microglobal wavefront set and the FBI $\mathcal{E}^{q,M}$-microglobal wavefront set. The former is defined on $\mathcal{S}'(\mathbb{R}^d)$ and the latter on $\mathcal{E}^{q,M}(\mathbb{R}^d)'$. It is a natural question to know whether or not these notions agree for distributions in $\mathcal{S}'(\mathbb{R}^d) \cap \mathcal{E}^{q,M}(\mathbb{R}^d)'$. First, though, we recall the FBI transform and FBI $\mathcal{E}^{q,M}$-microglobal regularity [HR19b] but need to introduce one piece of notation. For $\xi \in \mathbb{R}^d$, define

$$\langle \xi \rangle = \sqrt{1 + \xi \cdot \xi}$$

and the form ω and the function $\alpha(x, \xi)$ by

$$\begin{aligned} \omega &= dx_1 \wedge \cdots \wedge dx_d \wedge d(\xi_1 + ix_1\langle \xi \rangle) \wedge \cdots d(\xi_d + ix_d\langle \xi \rangle) \\ &:= \alpha(x, \xi) dx_1 \wedge \cdots \wedge dx_d \wedge d\xi_1 \wedge \cdots \wedge d\xi_d. \end{aligned}$$

The function α is a sum of terms of the form

$$\alpha_\beta(x, \xi) = c_\beta x^\beta \Big(\frac{\xi}{\langle \xi \rangle}\Big)^\beta,$$

where $\beta \in \{0, 1\}^d$.

Finally, if $M = (M_j)_{j \in \mathbb{N}}$ is an appropriately defined increasing sequence, then its associated function $M(t)$ is given by

$$M(t) = \sup_{p \in \mathbb{N}} \log \frac{t^p}{M_p}.$$

Recall that $\mathcal{G}^{q,\frac{1}{2}}(\mathbb{R}^d)$ is the set of global L^q Gevrey functions of order $\frac{1}{2}$.

Definition 3.3 Let $\mathcal{E}^{q,M}(\mathbb{R}^d)$ be a global L^q Denjoy–Carleman function class. We say that $\mathcal{E}^{q,M}(\mathbb{R}^d)$ *completely contains* $\mathcal{G}^{q,\frac{1}{2}}(\mathbb{R}^d)$ if $\mathcal{E}^{q,M}_A(\mathbb{R}^d) \supset \mathcal{G}^{q,\frac{1}{2}}(\mathbb{R}^d)$ for every $A > 0$.

Given a global L^q Denjoy–Carleman class $\mathcal{E}^{q,M}(\mathbb{R}^d)$ that completely contains $\mathcal{G}^{q,\frac{1}{2}}(\mathbb{R}^d)$ and a distribution $u \in \mathcal{E}^{q,M}(\mathbb{R}^d)'$, define the *FBI transform* of u by

$$\mathfrak{F}u(x, \xi) = \big\langle u, e^{i(x-\cdot)\cdot\xi - \langle \xi \rangle (x-\cdot)^2} \alpha(x - \cdot, \xi) \big\rangle.$$

The function $\mathfrak{F}u$ is well defined for $u \in \mathcal{E}_A^{q,M}(\mathbb{R}^d)'$ since the exponential function $e^{-a|x|^2} \in \mathcal{G}^{q,\frac{1}{2}}(\Omega)$ [AHR17, Section 4.1].

Definition 3.4 Let $u \in \mathcal{E}^{q,M}(\mathbb{R}^d)'$ and $\xi^0 \in \mathbb{R}^d \setminus \{0\}$. We say that u is *FBI* $\mathcal{E}^{q,M}$*-microglobally regular at* $\mathbb{R}^d \times \{\xi^0\}$ (or simply ξ^0) if there exist a conic neighborhood Γ_0 of ξ^0 in $\mathbb{R}^d \setminus \{0\}$ and constants $c, C > 0$ such that for each $q \le r \le \infty$,

$$\|D_x^J \mathfrak{F}u(x,\xi)\|_{L^r(\mathbb{R}^d)} \le C A_0^{|J|} M_{|J|} e^{-\frac{1}{c}M(a|\xi|)}, \quad \forall \xi \in \Gamma_0. \tag{11}$$

We define the *FBI* $\mathcal{E}^{q,M}$*-wavefront set of* u as the complement of the set of the directions ξ in which u is FBI $\mathcal{E}^{q,M}$-microglobal regular, that is,

$$WF_{\mathcal{E}^{q,M}}(u) := \{\xi \in \mathbb{R}^d : u \text{ is not } \mathcal{E}^{q,M}\text{-microglobal regular at } \xi\}.$$

Proposition 3.5 *There exists* $T \in \mathcal{S}'(\mathbb{R}^d) \cap \mathcal{E}^{1,M}(\mathbb{R}^d)'$ *so that*

$$\mathcal{WF}_{\mathcal{E}^{1,M}}(T) \neq WF_{\mathcal{E}^{1,M}}(T).$$

Proof Let T be the distribution that is given by integration against 1. To compute $\mathcal{WF}_{\mathcal{E}^{1,M}}(T)$, we observe that if $\hat{\varphi} \in \mathcal{S}(\mathbb{R}^d)$ is supported in an open cone, then the fact that $\hat{T} = \delta_0$ is supported at $\{0\}$ forces

$$|\langle D^\alpha T, \varphi\rangle| = |\langle \delta_0, \xi^\alpha \check{\varphi}\rangle| = 0.$$

Hence, $\xi \notin \mathcal{WF}_{\mathcal{E}^{1,M}}(T)$ for all $\xi \neq 0$. On the other hand, for $\varphi \in \mathcal{S}(\mathbb{R}^d)$ with $\operatorname{supp}\hat{\varphi} \subset B(0,1)$, we observe that

$$\langle T, \varphi\rangle = \int_{\mathbb{R}^d} \varphi(x)\, dV(x) = \hat{\varphi}(0)$$

and

$$\hat{\varphi} \mapsto \hat{\varphi}(0)$$

does not extend to a bounded linear operator on $L^1(\mathbb{R}^d)$. Hence $0 \in \mathcal{WF}_{\mathcal{E}^{1,M}}(T)$.

We now compute $WF_{\mathcal{E}^{1,M}}(T)$. Note, though, that

$$f \mapsto \int_{\mathbb{R}^d} f\, dx$$

is a bounded linear operator $\mathcal{E}^{1,M}(\mathbb{R}^d)$ (but not on $\mathcal{E}^{q,M}(\mathbb{R}^d)$ if $q > 1$). We observe that

$$\mathfrak{F}T(x,\xi) = \int_{\mathbb{R}^d} e^{i(x-y)\cdot\xi - \langle\xi\rangle(x-y)^2} \alpha(x-y,\xi)\, dV(y)$$

$$= (-1)^d \int_{\mathbb{R}^d} e^{iy\cdot\xi - \langle\xi\rangle y^2} \alpha(y,\xi)\, dV(y)$$

$$= (-1)^d \sum_{\beta\in\{0,1\}^d} \left\{ \left(\frac{\xi}{\langle\xi\rangle}\right)^\beta c_\beta \prod_{k=1}^d \int_{\mathbb{R}} e^{iy_k\xi_k - \langle\xi\rangle y_k^2} y_k^{\beta_k}\, dy_k \right\}.$$

Since $\beta_k = 0$ or 1, we easily compute that if $b > 0$, then

$$\int_{\mathbb{R}} e^{iya-by^2}\, dy = \frac{\sqrt{\pi}}{\sqrt{b}} e^{-\frac{a^2}{4b}} \qquad \text{and} \qquad \int_{\mathbb{R}} e^{iya-by^2} y\, dy = -\frac{ia\sqrt{\pi}}{2b^{3/2}} e^{-\frac{a^2}{4b}}.$$

The result of this calculation is that the FBI transform of T is *independent* of x, and hence there is no way the resulting function is in $L^q(\mathbb{R}^d)$ for $1 \le q < \infty$. Hence $WF_{\mathcal{E}^1,M}(T) = \mathbb{R}^d \setminus \{0\}$. □

4 Applications to the Restriction Problem

We follow [Ste93, Chapter VIII.4]. Let $S \subset \mathbb{R}^d$ be a subset supporting a measure $d\sigma$ on S (possibly the induced Lebesgue measure). We say that S has the L^p-L^q *restriction property* if there exists $A_{p,q} > 0$ so that

$$\left(\int_S |\hat{f}(\xi)|^q\, d\sigma(\xi) \right)^{1/q} \le A_{p,q} \|f\|_{L^p(\mathbb{R}^d)}, \quad \text{for all } f \in \mathcal{S}(\mathbb{R}^d).$$

For $\xi \in S$, the restriction operator

$$Rf(\xi) = \int_{\mathbb{R}^d} e^{-ix\cdot\xi} f(x)\, dx.$$

If R^* is the adjoint operator and

$$Uf(x) := R^*Rf(x) = \int_{\mathbb{R}^d} \int_S e^{i\xi(x-y)}\, d\sigma(\xi)\, f(y)\, dy = f * \check{d\sigma}(x),$$

then Stein points out that the following are equivalent [Ste93]:

(i) $R : L^p(\mathbb{R}^d) \to L^2(S, d\mu)$ is bounded (that is, the L^p-L^2 restriction property holds).
(ii) $U : L^p(\mathbb{R}^d) \to L^{p'}(\mathbb{R}^d)$ is bounded where $\frac{1}{p} + \frac{1}{p'} = 1$.

Definition 4.1 Fix $1 \le p < \infty$. Let $\mathfrak{X} \subset \mathcal{S}'(\mathbb{R}^d)$ be a function space defined by a parameter pair $(\mathfrak{P}, \mathfrak{G})$. If T is a tempered distribution so that $f * \hat{T} \in \mathfrak{X}$ for all

$f \in L^p(\mathbb{R}^d)$, then we say that L^p-L^2 *restriction problem with* $\mathfrak{X}$*-moments* holds for T.

We use the terminology *moment* in Definition 4.1 because if $\mathfrak{X}$ is a function space defined by smoothness estimates such as Sobolev spaces or global L^q-Denjoy–Carleman space, then estimating $D^\alpha \widehat{d\sigma}$ gives us information about $x^\alpha \, d\sigma$, the moments of $d\sigma$.

Remark 4.2 If $\mathfrak{X} \subset L^{p'}(\mathbb{R}^d)$, then T satisfies the classical L^p-L^2 restriction problem.

4.1 The Salem Example

In [HR19a], we showed that for each $q > 2$, the measure $d\mu$ constructed by Salem [Sal51] satisfies $\|D^k \widehat{d\mu}\|_{L^q(\mathbb{R})} \leq C$ for some constant C. We proved this by showing $\|D^k \widehat{d\mu}\|_{L^\infty(\mathbb{R})} \leq 1$ and $|D^k \widehat{d\mu}(\xi)| \leq C|\xi|^{-\beta/2}$ for some β that does not depend on k. It is also well known that the measure is supported on a set of Hausdorff dimension α (which can be made arbitrarily close to, but greater than, $2/q$). The latter two facts mean that we can employ Mockenhaupt's adaption of the Tomas–Stein argument to prove the following theorem.

Theorem 4.3 *The* L^p-L^2 *restriction problem with* $\mathcal{G}_1^{p',0}(\mathbb{R})$*-moments holds for the Salem measure when* $1 \leq p < \frac{2(2-2\alpha+\beta)}{4(1-\alpha)+\beta}$.

Proof The proof follows by the argument of the proof of Theorem 4.4 below or by combining (13) with the proof of [Moc00, Theorem 4.1]. □

4.2 Surface Area Measure on the Sphere

For each $m \in \mathbb{N}$, denote the mth Bessel J function by

$$J_m(r) = \frac{1}{2\pi} \int_0^{2\pi} e^{ir\sin\theta} e^{-im\theta} \, d\theta.$$

Theorem 4.4 *Let* $d\sigma$ *be the surface area measure on sphere* S^{d-1}*. Then:*

1.

$$\widehat{d\sigma}(\xi) = 2\pi^{\frac{d}{2}} \frac{J_{\frac{d-2}{2}}(|\xi|)}{|\xi|^{\frac{d-2}{2}}} \in \mathcal{G}^{q,1}(\mathbb{R}^d)$$

if and only if $q > \frac{2d}{d-1}$.

2. *The sphere satisfies the* L^p-L^2 *restriction problem with* $\mathcal{G}^{p',1}(\mathbb{R}^d)$*-moments when* $1 \le p < \frac{2(d+1)}{d+3}$.

Remark 4.5 The Bessel J functions decay for large x *and* oscillate. If we ignored the oscillation and applied Young's inequality to the estimates in Part 1 of Theorem 4.4, then we could only say that the L^p-L^2 restriction problem holds with $\mathcal{G}^{q,1}(\mathbb{R}^d)$-moments for $1 \le p < \frac{4d}{3d+1}$.

The next lemma is the main combinatorial lemma that we use to control the derivatives of $\widehat{d\sigma}$.

Lemma 4.6 *Let* $f_m(r) = \frac{J_m(r)}{r^m}$. *Then*

$$f_m^{(2k)}(r) = \sum_{j=0}^{k} (-1)^{j+k} a_{jk} r^{2j} f_{m+k+j}(r) \tag{12}$$

$$f_m^{(2k+1)}(r) = \sum_{j=0}^{k} (-1)^{j+k+1} b_{jk} r^{2j+1} f_{m+k+j+1}(r)$$

for some positive constants a_{jk}, b_{jk} *that grow no faster than* $CA^k k!$. *Moreover,*

$$a_{kk} = b_{kk} = 1$$

for all k *and the constants satisfy the recursive relationships that for* $k \ge 1$,

$$b_{jk} = \begin{cases} 2(j+1)a_{j+1,k} + a_{jk} & \text{if } 0 \le j \le k-1 \\ 1 & \text{if } j = k \end{cases}$$

and

$$a_{jk} = \begin{cases} b_{0,k-1} & \text{if } j = 0 \\ b_{j,k-1}(2j+1) + b_{j-1,k-1} & \text{if } 1 \le j \le k-1 \\ 1 & \text{if } j = k \end{cases}.$$

Proof To prove the lemma, we recall the well-known properties

$$f_m'(r) = -r f_{m+1}(r) \qquad \text{and} \qquad f_m''(r) = r^2 f_{m+2}(r) - f_{m+1}(r), \qquad \text{for all } m \in \mathbb{N}_0.$$

Moreover, assuming the formula holds for $f^{(2k)}(r)$ and suppressing the variable of f,

$$f_m^{(2k+1)} = (-1)^{k+1} a_{0k} r f_{m+k+1}$$

$$+\sum_{j=1}^{k}\left[(-1)^{j+k}a_{jk}(2j)r^{2j-1}f_{m+k+j}+(-1)^{j+k+1}a_{jk}r^{2j+1}f_{m+k+j+1}\right]$$

$$=\sum_{j=0}^{k-1}(-1)^{j+k+1}\left[2(j+1)a_{j+1,k}+a_{jk}\right]r^{2j+1}f_{m+k+j+1}$$

$$-a_{kk}r^{2k+1}f_{m+2k+1},$$

which yields

$$b_{jk}=2(j+1)a_{j+1,k}+a_{jk},\ 0\leq j\leq k-1,\quad \text{and}\quad b_{kk}=a_{kk}.$$

Similarly, assuming the formula holds for $f_m^{(2k-1)}$, we compute

$$f_m^{(2k)}(r)$$

$$=\sum_{j=0}^{k-1}(-1)^{j+k}b_{j,k-1}r^{2j+1}f_{m+k+j}(r)$$

$$=\sum_{j=0}^{k-1}\left[(-1)^{j+k}b_{j,k-1}(2j+1)r^{2j}f_{m+k+j}+(-1)^{j+k+1}b_{j,k-1}r^{2(j+1)}f_{m+k+j+1}\right]$$

$$=(-1)^k b_{0,k-1}f_{m+k}+\sum_{j=1}^{k-1}(-1)^{j+k}\left[b_{j,k-1}(2j+1)+b_{j-1,k-1}\right]r^{2j}f_{m+k+j}$$

$$+b_{k-1,k-1}r^{2k}f_{m+2k},$$

and we see that

$$a_{0,k}=b_{0,k-1}\quad\text{and}\quad a_{jk}=b_{j,k-1}(2j+1)+b_{j-1,k-1},\ 1\leq j\leq k-1\quad\text{and}$$

$$a_{kk}=b_{k-1,k-1}=1.$$

It is immediate that the coefficients grow at worst like $CA^k k!$ for some constants $A, C>0$. □

Proof of Theorem 4.4 The calculation of $\widehat{d\sigma}(\xi)$ is well known and can be found in [Ste93, Chapter VIII.3]. For large r, we know that

$$J_m(r)=\left(\frac{2}{\pi}\right)^{1/2}r^{-1/2}\cos(r-\pi m/2-\pi/4)+O(r^{-3/2}).$$

We start by computing the range of q for which $f_{\frac{d-2}{2}}(|\xi|) \in L^q(\mathbb{R}^d)$. Since $|\cos r| \geq \frac{1}{2}$ if $|r - k\pi| \leq \frac{\pi}{3}$, for N sufficiently large and $q \geq 1$, it follows that

$$\int_{\mathbb{R}^d} |f_{\frac{d-2}{2}}(|\xi|)|^q \, dV(\xi) \geq \frac{1}{100^q} \sum_{k=N}^{\infty} \int_{|r-k\pi| \leq \frac{\pi}{3}} \frac{1}{r^{q(\frac{1}{2}+\frac{d-2}{2})-(d-1)}} \, dr$$

$$\geq \frac{1}{100^q} \sum_{k=N}^{\infty} \frac{1}{[(k-\frac{1}{3})\pi]^{(d-1)(\frac{q}{2}-1)}}.$$

This sum converges exactly when $(d-1)(\frac{q}{2}-1) > 1$, that is, the sum converges if and only if $q > \frac{2d}{d-1}$. Similarly, for R large,

$$\int_{\mathbb{R}^d} |f_{\frac{d-2}{2}}(|\xi|)|^q \, dV(\xi) \leq C \int_{|\xi| \geq R} \frac{1}{|\xi|^{q(\frac{1}{2}+\frac{d-2}{2})}} \, dV(\xi) = C \int_R^{\infty} \frac{1}{r^{(d-1)(\frac{q}{2}-1)}} \, dr.$$

The range of q for which the latter integral converges is again $q > \frac{2d}{d-1}$.

The terms in (12) with the worst decay in r correspond to $r^{2k} f_{m+2k}$ and $r^{2k+1} f_{m+2k+1}$, both of which satisfy $r^\ell f_{m+\ell} \sim O(r^{-m-1/2})$. This means $f_m^{(j)}(|\xi|) \in L^q(\mathbb{R}^d)$ for the same range of q as f_m. Moreover, since f_m is analytic and even, it follows that for some constants C and A, if $q > 2 + \frac{2}{d-1}$, then

$$\|f_{\frac{d-2}{2}}^{(k)}\|_{L^q(\mathbb{R}^d)} \leq C_q A^k k^k.$$

The mapping $\xi \mapsto f(|\xi|)$ is even in ξ, real analytic in $|\xi|$ (and hence well behaved on $B(0,1)$), and $\xi \mapsto |\xi|$ is easily shown to be a function in $\mathcal{G}^{\infty,1}(B(0,1)^c)$. Thus, Part 1 of Theorem 4.4 now follows from the arguments in [AHR17, Section 5].

Let α be a multi-index and $d\mu_\alpha = x^\alpha \, d\sigma$. $d\mu_\alpha$ is not necessarily a positive measure, but that will not prevent us from using the Tomas–Stein restriction theorem arguments [Tom75, Moc00]. We want to show that there exist constants $C_p, A_p > 0$ so that if T^α is the operator defined by convolution against $D^\alpha \widehat{d\sigma}$, then

$$\|T^\alpha f\|_{L^{p'}(\mathbb{R}^d)} \leq C_p A_p^{|\alpha|} |\alpha|^{|\alpha|}.$$

Let $\{\varphi_k : k = 0, 1, 2, \ldots\} \subset C_c^\infty(\mathbb{R}^d)$ be a partition of unity so that $\operatorname{supp}\varphi_0 \subset B(0,2)$, and for $k \geq 1$, $\operatorname{supp}\varphi_k \subset \{x : 2^{k-1} \leq |x| \leq 2^{k+1}\}$ and satisfies $\varphi_k(x) = \varphi(x/2^k)$ for some $\varphi \in C_c^\infty(\mathbb{R}^d)$. Set $T_k f = (\varphi_k D^\alpha \widehat{d\sigma}) * f$ so that $\sum_{k=0}^\infty T_k = T$.

It follows immediately from the proof of Theorem 4.4 that there exist constants $C, A > 0$ so that

$$\|T_k\|_{L^1 \to L^\infty} \leq \|\varphi_k D^\alpha \widehat{d\sigma}\|_{L^\infty(\mathbb{R}^d)} \leq C A^{|\alpha|} 2^{-k\frac{d-1}{2}}.$$

Next, we bound $\|T_k\|_{L^2\to L^2}$ by Plancherel's Theorem. Using the fact that $\hat{\varphi} \in \mathcal{S}(\mathbb{R}^d)$ and decays rapidly, for some constant C (that can increase line by line but independent of k)

$$\begin{aligned}\|T_k\|_{L^2\to L^2} = & \sup_{\substack{f,g\in\mathcal{S}(\mathbb{R}^d)\\ \|f\|_{L^2}=\|g\|_{L^2}=1}} \left|\left((\hat{\varphi}_k * d\mu_\alpha)\hat{f},\hat{g}\right)\right| \\ &\leq \sup_{x\in\mathbb{R}^n} |\hat{\varphi}_k * d\mu_\alpha(x)| \\ &\leq C2^{kd} \sup_{x\in\mathbb{R}^d} \int_{\mathbb{R}^d} \frac{1}{(1+2^k|x-y|)^d} |d\mu_\alpha|(y) \\ &\leq C2^{kd} \sup_{x\in\mathbb{R}^d} \int_{S^{d-1}} \frac{1}{(1+2^k|x-y|)^d} d\sigma(y).\end{aligned}$$

It is clear that the supremum occurs whenever $x \in S^{d-1}$, and at $x = 1$, in particular. We compute that

$$\begin{aligned}&\int_{S^{d-1}} \frac{1}{(1+2^k|1-y|)^N} d\sigma(y) \\ &\leq \int_{|1-y|<\frac{1}{2^k}} d\sigma(y) + \sum_{j=0}^{k-1}\left\{\int_{\frac{2^j}{2^k}\leq|1-y|<\frac{2^{j+1}}{2^k}} \left(2^k\frac{2^j}{2^k}\right)^{-d} d\sigma(y)\right\} + \frac{C}{2^{kd}} \\ &\leq \frac{C}{2^{k(d-1)}}.\end{aligned} \tag{13}$$

Thus, $\|T_k\|_{L^2\to L^2} \leq C2^k$. We use the Riesz–Thorin interpolation theorem and bound $\|T_k\|_{L^p\to L^{p'}} \leq CA^{|\alpha|\frac{2-p}{p}} 2^{-\frac{k}{p}(2(d+1)-p(d+3))}$. Since $T^\alpha = \sum_{k=0}^\infty T_k$, this sum will converge when $1 \leq p < \frac{2(d+1)}{d+3}$ and the theorem is proved. □

References

[AB12] Z. Adwan and S. Berhanu. On microlocal analyticity and smoothness of solutions of first-order nonlinear PDEs. *Math. Ann.*, 352(1):239–258, 2012.

[AH10] Z. Adwan and G. Hoepfner. Approximate solutions and micro-regularity in the Denjoy-Carleman classes. *J. Differential Equations*, 249(9):2269–2286, 2010.

[AH15] Z. Adwan and G. Hoepfner. Denjoy-Carleman classes: boundary values, approximate solutions and applications. *J. Geom. Anal.*, 25(3):1720–1743, 2015.

[AHR17] Z. Adwan, G. Hoepfner, and A. Raich. Global L^q-Gevrey functions and their applications. *J. Geom. Anal.*, 27(3):1874–1913, 2017.

[Asa95] C. H. Asano. On the C^∞ wave-front set of solutions of first-order nonlinear PDEs. *Proc. Amer. Math. Soc.*, 123(10):3009–3019, 1995.

[BCT83] S. Baouendi, C. Chang, and F. Trèves. Microlocal hypo-analyticity and extension of CR functions. *J. Differential Geom.*, 18(3):331–391, 1983.

[Ber10] S. Berhanu. On involutive systems of first-order nonlinear partial differential equations. In *Complex analysis*, Trends Math., pages 25–50. Birkhäuser/Springer Basel AG, Basel, 2010.

[BH12] S. Berhanu and J. Hounie. A class of FBI transforms. *Comm. Partial Differential Equations*, 37(1):38–57, 2012.

[BI73] J. Bros and D. Iagolnitzer. Causality and local analyticity: mathematical study. *Ann. Inst. H. Poincaré Sect. A. (N.S.)*, 18:147–184, 1973.

[Bon77] J. M. Bony. Équivalence des diverses notions de spectre singulier analytique. In *Séminaire Goulaouic-Schwartz (1976/1977), Équations aux dérivées partielles et analyse fonctionnelle, Exp. No. 3*. Centre Math., École Polytech., Palaiseau, 1977.

[BP09] R. F. Barostichi and G. Petronilho. Gevrey micro-regularity for solutions to first order nonlinear PDE. *J. Differential Equations*, 247(6):1899–1914, 2009.

[BR13] A. Boggess and A. Raich. Heat kernels, smoothness estimates and exponential decay. *J. Fourier Anal. Appl.*, 19:180–224, 2013.

[BT82] M.S. Baouendi and F. Trèves. A microlocal version of Bochner's tube theorem. *Indiana Univ. Math. J*, 31(6):885–895, 1982.

[BX15] S. Berhanu and M. Xiao. On the C^∞ version of the reflection principle for mappings between CR manifolds. *Amer. J. Math*, 137(5):1365–1400, 2015.

[Chr97] M. Christ. Intermediate optimal Gevrey exponents occur. *Comm. Partial Differential Equations*, 22(3-4):359–379, 1997.

[EG03] M.G. Eastwood and R.C. Graham. Edge of the wedge theory in hypo-analytic manifolds. *Comm. Partial Differential Equations*, 28(11-12):2003–2028, 2003.

[HM18] G. Hoepfner and R. Medrado. The FBI transforms and their use in microlocal analysis. *J. Funct. Anal.*, 275(5):1208–1258, 2018.

[Hör90] L. Hörmander. *The analysis of linear partial differential operators. I*. Springer Study Edition. Springer-Verlag, Berlin, Second edition, 1990.

[HR19a] G. Hoepfner and A. Raich. Global L^q Gevrey functions, Paley-Weiner theorems, and the FBI transform. *Indiana Univ. Math. J.*, 68(3):967–1002, 2019.

[HR19b] G. Hoepfner and A. Raich. Microglobal regularity and the global wavefront set. *Math. Z.*, 291(3-4):971–998, 2019.

[HT92] N. Hanges and F. Trèves. On the analyticity of solutions of first-order nonlinear PDE. *Trans. Amer. Math. Soc.*, 331(2):627–638, 1992.

[Moc00] G. Mockenhaupt. Salem sets and restriction properties of Fourier transforms. *Geom. Funct. Anal.*, 10(6):1579–1587, 2000.

[Sal51] R. Salem. On singular monotonic functions whose spectrum has a given Hausdorff dimension. *Ark. Mat.*, 1:353–365, 1951.

[Sjö82] J. Sjöstrand. Singularités analytiques microlocales. In *Astérisque, 95*, volume 95 of *Astérisque*, pages 1–166. Soc. Math. France, Paris, 1982.

[Ste93] E. M. Stein. *Harmonic Analysis: Real-Variable Methods, Orthogonality, and Oscillatory Integrals*. Princeton Mathematical Series; 43. Princeton University Press, Princeton, New Jersey, 1993.

[Tom75] P. Tomas. A restriction theorem for the Fourier transform. *Bull. Amer. Math. Soc.*, 81:477–478, 1975.

Equality of Commutator Type and Levi Form Type for an $(n-2)$-Dimensional Bundle

Xiaojun Huang, Wanke Yin, and Pingsan Yuan

Dedicated to Jorge Hounie on the occasion of his 75th birthday

1 Introduction

Finite type conditions for real hypersurfaces in $\mathbb{C}^n$ with $n \geq 2$ were originated from the investigation of subelliptic estimates of the $\overline{\partial}$-Neumann equation. For any boundary point of a smoothly bounded domain in $\mathbb{C}^2$, Kohn [11] introduced three numerical invariants taking positive integer values or ∞, which are called, respectively, the regular contact type, the commutator type, and the Levi form type. Kohn proved that these invariants are in fact the same. When these invariants are finite at each boundary point and the domain is pseudoconvex, the $\overline{\partial}$-Neumann operator of the domain possesses subelliptic estimates.

Ever since then, much attention has been paid to generalize Kohn's finite type condition to higher dimensions. Kohn [12] defined the ideal of subelliptic multipliers near each boundary point of a pseudoconvex domain, and if the constant function 1 is in these ideals, the boundary point is said to be of finite ideal type. In [4], D'Angelo introduced a fundamental concept now called the D'Angelo singular type in terms of the order of contact with respect to singular complex analytic varieties. Both of these finite type conditions imply the existence of the subelliptic estimates for smoothly bounded pseudoconvex domains. Bloom [3] naturally generalized Kohn's types in $\mathbb{C}^2$ to higher dimensions. More precisely, for

Supported in part by DMS-2247151 and DMS-2000050.
Supported in part by NSFC-12171372.

X. Huang (✉)
Department of Mathematics, Rutgers University, New Brunswick, NJ, USA

W. Yin · P. Yuan
School of Mathematics and Statistics, Wuhan University, Wuhan, Hubei, P.R. China

S. Berhanu et al. (eds.), *Geometric Analysis of PDEs and Several Complex Variables*, Latin American Mathematics Series – UFSCar subseries,
https://doi.org/10.1007/978-3-031-69702-9_9

a smooth real hypersurface $M \subset \mathbb{C}^n$ and $p \in M$ with $n \geq 2$, Bloom defined the regular contact type $a^{(s)}(M, p)$, the commutator type $t^{(s)}(M, p)$, and the Levi type $c^{(s)}(M, p)$ of M at p for each integer $s \in [1, n-1]$. Bloom conjectured in [3] that these three invariants are the same when the hypersurface is pseudoconvex, which is known as the Bloom conjecture.

Bloom–Graham [1] and Bloom [2] proved the Bloom conjecture for $s = n-1$. In [3], Bloom showed that $a^{(1)}(M, p) = c^{(1)}(M, p)$ when $M \subset \mathbb{C}^3$. Recently, Huang–Yin [8] proved that the Bloom conjecture holds for $s = n-2$. This, in particular, gives a complete solution of the Bloom conjecture in the case of complex dimension 3.

Let $s \in [1, n-1]$ be an integer and let B be an s-dimensional subbundle of $T^{(1,0)}M$. We similarly define the commutator type $t(B, p)$ and the Levi type $c(B, p)$ associated to B as follows:

Let $M \subset \mathbb{C}^n$ be a smooth real hypersurface with $p \in M$, and let ρ be a defining function of M near p. Let $X_1, \cdots, X_s$ be the local sections of B near p. Denote by $\mathcal{M}_1(B)$ the $C^\infty(M)$-module spanned by X_j and $\overline{X_j}$ for each j. For $k \geq 1$, we inductively define $\mathcal{M}_k(B)$ to be the $C^\infty(M)$-module spanned by $\mathcal{M}_{k-1}(B)$ and elements of the form $[X, Y]$ with $X \in \mathcal{M}_{k-1}(B)$ and $Y \in \mathcal{M}_1(B)$. We say the commutator type $t(B, p) = m$ if $\langle F, \partial\rho\rangle(p) = 0$ for any $F \in \mathcal{M}_{m-1}(B)$ but $\langle G, \partial\rho\rangle(p) \neq 0$ for a certain $G \in \mathcal{M}_m(B)$. We define the Levi type $c(B, p) = m$ if for any $m-3$ vector fields $F_1, \cdots, F_{m-3}$ of $\mathcal{M}_1(B)$, we have for each j

$$F_{m-3} \cdots F_1 \langle [X_j, \overline{X_j}], \partial\rho\rangle(p) = 0;$$

and for a certain choice of j and $m-2$ vector fields $G_1, \cdots, G_{m-2}$ of $\mathcal{M}_1(B)$, we have

$$G_{m-2} \cdots G_1 \langle [X_j, \overline{X_j}], \partial\rho\rangle(p) \neq 0.$$

We mention that our notion $c(B, p)$ is different from the one defined in Bloom [3]. However, when the hypersurface is pseudoconvex, by the work in [9], both notions are indeed the same.

D'Angelo [5] conjectured that when the real hypersurface is pseudoconvex and $s = 1$, these two types equal to each other. He confirmed the conjecture when one of the types is exactly 4. Huang–Yin conjectured in [9] that the two types should be the same for any $s \in [1, n-1]$. Recently, the conjecture was confirmed by Huang–Yin [10] if one of the types is less than 8. The present paper is to verify this conjecture for $s = n-2$, by the machinery developed in [8]. Our theorem can be stated as follows:

Theorem 1.1 *Let M be a smooth pseudoconvex hypersurface in $\mathbb{C}^n$ with $n \geq 3$ and $p \in M$. For any $(n-2)$-dimensional subbundle B of $T^{(1,0)}M$, it holds that $t(B, p) = c(B, p)$.*

2 Proof of Theorem 1.1

We proceed to the proof of Theorem 1.1 as follows:

Let $(z, w) = (z', z_{n-1}, w) = (z_1, z_2, \ldots, z_{n-1}, w)$ be the coordinates in $\mathbb{C}^n$. After a holomorphic change of coordinates, we may assume that $p = 0$ and the defining function of M takes the following form:

$$\rho = -\mathrm{Im}w + E(z, w, \overline{z}, \overline{w}), \;\; E(z, w, \overline{z}, \overline{w}) = O(|z|^2 + |zw| + |w^2|). \tag{1}$$

For a sufficiently large integer κ, we can apply a change of holomorphic coordinates by replacing w with $w-\phi(z)$ for a certain holomorphic polynomial $\phi(z)$ to eliminate the harmonic terms of degree less than or equal to κ in the formal expansion of E at 0.

For any integer $j \in [1, n-1]$, write

$$L_j = \frac{\partial}{\partial z_j} - \frac{\partial \rho}{\partial z_j} \cdot \left(\frac{\partial \rho}{\partial w}\right)^{-1} \frac{\partial}{\partial w}. \tag{2}$$

Then $\{L_j\}_{j=1}^{n-1}$ forms a basis of the complex tangent vector fields of type $(1, 0)$ along M near 0. Let B be an $(n-2)$-dimensional bundle of $T^{(1,0)}M$. Let $\{X_j\}_{j=1}^{n-2}$ be a local basis of the sections of B near 0. After a holomorphic change of coordinates, we may assume a priori that they take the following form:

$$X_j = L_j + a_j(z, w, \overline{z}, \overline{w})L_{n-1} \text{ with } a_j(z, w, \overline{z}, \overline{w})|_0 = 0 \text{ for each } j. \tag{3}$$

Define

$$l_0^* = \min_{1\le j\le n-2} \{k_j : k_j = \text{vanishing order of } a_j(z', 0, 0, \overline{z'}, 0, 0) \text{ at } 0\}.$$

By [8, Lemma 3.2], if $l_0^* \neq \infty$, after a fixed procedure of holomorphic change of coordinates which does not change $(z_1, \cdots, z_{n-2}, w)$, we obtain

$$a_j^{(l_0^*)}(0, \cdots, 0, z_j, \cdots, z_{n-2}, 0, \cdots, 0) = 0 . \text{for all } 1 \le j \le n-2. \tag{4}$$

Here $a_j^{(l_0^*)}$ is the homogeneous polynomial of degree l^* in its formal expansion at 0. After this normalization, we may assume X_j in (3) takes the new form $L_j + \hat{a}_j(z, w, \overline{z}, \overline{w})L_{n-1}$. If $\hat{a}_j{}^{(l_0^*)}(z', 0, 0, z', 0, 0) \equiv 0$ for all $1 \le j \le n-2$, we define

$$l_0 = \min_{1\le j\le n-2} \{k_j : k_j = \text{vanishing order of } \hat{a}_j(z', 0, 0, \overline{z'}, 0, 0) \text{ at } 0\}.$$

Then $l_0^* < l_0$. We redefine l_0^* to be l_0 and then repeat such a normalization. By an induction, for any fixed positive integer κ, we can assume that either $l_0 > \kappa$ or the normalization in (4) holds for $l_0 \leq \kappa$. When $l_0 \leq \kappa$ for a certain κ, we then define

$$m_0 = \min\{\text{vanishing order of } \frac{\partial^2}{\partial z_j \partial \overline{z_k}} E(z', 0, 0, \overline{z'}, 0, 0)$$
$$\text{at } 0 \text{ for } 1 \leq j, k \leq n-2\} + 2.$$

The proof of our theorem is carried out in three cases, according to the values of l_0 and m_0.

Case I For any sufficiently large κ, the l_0 and m_0 produced through the above procedures are larger than κ.

Then

$$E(z', 0, \overline{z'}, 0) = O(|z'|^{\kappa+1}),\ a_j(z', 0, \overline{z'}, 0) = O(|z'|^{\kappa+1}). \tag{5}$$

Here E and a_j are defined by (1) and (3), respectively. We set the following weight system:

$$\begin{aligned} &\mathrm{wt}(z_j) = \mathrm{wt}(\overline{z_j}) = 1,\ \mathrm{wt}\left(\frac{\partial}{\partial z_j}\right) = \mathrm{wt}\left(\frac{\partial}{\partial \overline{z_j}}\right) = -1 \text{ for } 1 \leq j \leq n-2, \\ &\mathrm{wt}(z_{n-1}) = \mathrm{wt}(\overline{z_{n-1}}) = \kappa, \mathrm{wt}\left(\frac{\partial}{\partial z_{n-1}}\right) = \mathrm{wt}\left(\frac{\partial}{\partial \overline{z_{n-1}}}\right) = -\kappa, \\ &\mathrm{wt}(w) = \mathrm{wt}(\overline{w}) = \kappa,\ \mathrm{wt}\left(\frac{\partial}{\partial w}\right) = \mathrm{wt}\left(\frac{\partial}{\partial \overline{w}}\right) = -\kappa. \end{aligned} \tag{6}$$

Then the weighted degree of $E(z, w, \overline{z}, \overline{w})$ is at least $\kappa + 1$. From (1)–(2) and (5)–(6), it follows that

$$L_j = \frac{\partial}{\partial z_j} + O_{wt}(0) \text{ for } 1 \leq j \leq n-2,\ L_{n-1} = O_{wt}(-\kappa).$$

Here for an integer k, we have denoted by $O_{wt}(k)$ a vector field of weighted degree at least k. Combining this with (3) and (5), we obtain $X_j = \frac{\partial}{\partial z_j} + O_{wt}(0)$. Hence for any $Y_1, \cdots, Y_\kappa \in \mathcal{M}_1(B)$, $\partial\overline{\partial}\rho(Y_2, \overline{Y_1}) = \partial\overline{\partial}E(Y_2, \overline{Y_1})$, and thus its weighted degree is at least $\kappa - 1$. Since $X_j = \frac{\partial}{\partial z_j} + O_{wt}(0)$, the weighted degree of $[Y_2, \overline{Y_1}]$ is at least -1. Thus both

$$Y_\kappa \cdots Y_3 \partial\overline{\partial}\rho(Y_2, \overline{Y_1}) \text{ and } \langle [Y_\kappa, [Y_{\kappa-1}, \cdots, [Y_3, [Y_2, \overline{Y_1}] \cdots], \partial\rho \rangle$$

are smooth functions of weighted degree at least 1. Hence

$$Y_\kappa \cdots Y_3 \partial\overline{\partial}\rho(Y_2, \overline{Y_1})(0) = \langle [Y_\kappa, [Y_{\kappa-1}, \cdots, [Y_3, [Y_2, \overline{Y_1}] \cdots], \partial\rho \rangle(0) = 0. \tag{7}$$

Namely, $t(B,0) \geq \kappa$ and $c(B,0) \geq \kappa$. Since κ can be made arbitrarily large, it follows that $t(B,0) = c(B,0) = +\infty$.

Case II For a sufficiently large κ_0, after a finite number of procedures mentioned above, it holds that $l_0 < \kappa_0$ and $m_0 \leq l_0$.

Denote by $E^{(m_0)}(z', 0, \overline{z'}, 0)$ the sum of homogeneous terms of degree m_0 in the Taylor expansion of $E(z', 0, \overline{z'}, 0)$ at 0. Then it is not identically zero and contains no harmonic terms. We set the following weight system in this case:

$$\begin{aligned}
&\mathrm{wt}(z_j) = \mathrm{wt}(\overline{z_j}) = 1,\ \mathrm{wt}\left(\frac{\partial}{\partial z_j}\right) = \mathrm{wt}\left(\frac{\partial}{\partial \overline{z_j}}\right) = -1, \text{ for } 1 \leq j \leq n-2,\\
&\mathrm{wt}(z_{n-1}) = \mathrm{wt}(\overline{z_{n-1}}) = m_0,\ \ \mathrm{wt}\left(\frac{\partial}{\partial z_{n-1}}\right) = \mathrm{wt}\left(\frac{\partial}{\partial \overline{z_{n-1}}}\right) = -m_0,\\
&\mathrm{wt}\left(\frac{\partial}{\partial w}\right) = \mathrm{wt}\left(\frac{\partial}{\partial \overline{w}}\right) = -m_0,\ \ \mathrm{wt}(w) = \mathrm{wt}(\overline{w}) = m_0.
\end{aligned} \tag{8}$$

Define

$$\rho^{(m_0)} = -\mathrm{Im} w + E^{(m_0)}(z', 0, \overline{z'}, 0),\ L_{j,-1} = \frac{\partial}{\partial z_j} + 2i \frac{\partial}{\partial z_j} E^{(m_0)}(z', 0, \overline{z'}, 0) \frac{\partial}{\partial w}.$$

With respect to the weight system (8), $\rho^{(m_0)}$ ($L_{j,-1}$, respectively) is the sum of weighted homogeneous terms of weighted degree m_0 (-1, respectively) in ρ (L_j, respectively).

Denote by $O_{wt}(k)$ a smooth function or vector field with weighted degree at least k. Then we have

$$\rho = \rho^{(m_0)} + O_{wt}(m_0 + 1),\ \ L_j = L_{j,-1} + O_{wt}(0).$$

Thus for any $1 \leq j \leq m_0 - 2$ and $Y_1, \cdots, Y_j \in \mathcal{M}_1(B)$, the weighted degrees of terms in

$$Y_j \cdots Y_1 \partial\overline{\partial}\rho(X_k, \overline{X_{k'}}) \text{ and } \langle [Y_j, [Y_{j-1}, \cdots, [Y_1, [X_k, \overline{X_{k'}}] \cdots], \partial\rho \rangle$$

are at least $m_0 - j - 2$. Hence both of them are 0 when restricted to the origin for any $1 \leq j \leq m_0 - 3$. When $j = m_0 - 2$, by considering weighted expansion, we have

$$Y_j \cdots Y_1 \partial\overline{\partial}\rho(X_k, \overline{X_{k'}})(0) = Y_{j,-1} \cdots Y_{1,-1} \partial\overline{\partial}\rho^{(m_0)}(X_{k,-1}, \overline{X_{k',-1}})(0) \tag{9}$$

and

$$\begin{aligned}&\langle [Y_j, [Y_{j-1}, \cdots, [Y_1, [X_k, \overline{X_{k'}}] \cdots], \partial\rho\rangle(0)\\ =&\langle [Y_{j,-1}, [Y_{j-1,-1}, \cdots, [Y_{1,-1}, [X_{k,-1}, \overline{X_{k',-1}}] \cdots], \partial\rho^{(m_0)}\rangle(0).\end{aligned} \tag{10}$$

Here $Y_{h,-1}$ is the sum, in the vector field Y_h, of terms with weighted degree -1.

Notice that $\{X_{k,-1}\}_{k=1}^{n-2}$ are $(1,0)$ tangent vector fields of the real hypersurface defined by $\widetilde{M} = \{\rho^{(m_0)} = 0\} \subset \mathbb{C}^{n-1}$. Denoted by B_{-1} the subbundle of $T^{(1,0)}M$ whose sections near 0 are $\{X_{k,-1}\}_{k=1}^{n-2}$. By the result of Bloom–Graham [1, Theorem 2.4] and Bloom [2, Addendum], we have $t(B_{-1}, p) = c(B_{-1}, p)$ for the $(n-2)$-dimensional bundle B_{-1} in $\widetilde{M} \subset \mathbb{C}^{n-1}$. Thus $t(B_{-1}, 0) = c(B_{-1}, 0) = a^{(n-2)}(\widetilde{M}, 0) = m_0$. Thus in (9) and (10), we can choose Y_h for $1 \le h \le m_0$ such that the two expressions are nonzero. This yields that $t(B, 0) = c(B, 0) = m_0$.

Case III For a certain sufficiently large κ_1, after a finite number of procedures mentioned above, it holds that $l_0 < \kappa_1$ and $l_0 \le m_0 - 1 < \infty$.

As in Case II, we define

$$\begin{aligned}&\mathrm{wt}(z_j) = \mathrm{wt}(\overline{z_j}) = 1,\ \mathrm{wt}(z_{n-1}) = \mathrm{wt}(\overline{z_{n-1}}) = l_0 + 1,\\ &\mathrm{wt}\left(\frac{\partial}{\partial z_j}\right) = \mathrm{wt}\left(\frac{\partial}{\partial \overline{z_j}}\right) = -1,\ \mathrm{wt}\left(\frac{\partial}{\partial z_{n-1}}\right) = \mathrm{wt}\left(\frac{\partial}{\partial \overline{z_{n-1}}}\right) = -(l_0+1).\end{aligned} \tag{11}$$

Denote by m_1 the weighted degree of the lowest nonzero term in the weighted expansion of $\rho(z, 0, \overline{z}, 0)$ in terms of the weights given in (11). Then $m_1 \le m_0 < \infty$. Define

$$\mathrm{wt}(w) = \mathrm{wt}(\overline{w}) = m_1,\ \mathrm{wt}\left(\frac{\partial}{\partial w}\right) = \mathrm{wt}\left(\frac{\partial}{\partial \overline{w}}\right) = -m_1.$$

Write

$$\begin{aligned}\rho^{(m_1)} &= -\mathrm{Im}w + E^{(m_1)}(z, 0, \overline{z}, 0),\\ L_{j,-1} &= \frac{\partial}{\partial z_j} + a_j^{(l_0)}\frac{\partial}{\partial z_{n-1}} + 2i\left(\frac{\partial\rho^{(m_1)}}{\partial z_j} + a^{(l_0)}\frac{\partial\rho^{(m_1)}}{\partial z_{n-1}}\right)\frac{\partial}{\partial w}.\end{aligned}$$

Since $l_0 \le m_0 \le a^{(n-2)}(M, 0)$, ρ and $\{X_j\}_{j=1}^{n-2}$ may be normalized as in [8, Theorem 3.5].

Corollary 2.1 *Keep the same notations and definitions we have made in Case III. After a holomorphic change of coordinates and after choosing a suitable basis of B, we have one of the following three normalizations for $\{a_j^{(l_0)}, \rho^{(m_1)}\}_{j=1}^{n-2}$:*

(I) $a_j^{(l_0)}(z_1, \cdots, z_{n-2}, 0, \overline{z_1}, \cdots, \overline{z_{n-2}}, 0)$ *is holomorphic in* $z_1, \cdots, z_{n-2}$ *for each j, and there exits* $j_0 \in [2, n-2]$ *such that* $a_j^{(l_0)}(z_1, \cdots, z_{n-2}, 0, \overline{z_1}, \cdots, \overline{z_{n-2}}, 0) = 0$ *for* $1 \leq j \leq j_0 - 1$, $a_{j_0}^{(l_0)}(0, \cdots, 0, z_{j_0}, \cdots, z_{n-2}, 0, \cdots, 0) = 0$, *but* $a_{j_0}^{(l_0)}(z_1, \cdots, z_{n-2}, 0, \cdots, 0) \not\equiv 0$.

(II) $a_1^{(l_0)}(z_1, 0, \cdots, 0, \overline{z_1}, 0, \cdots, 0) \not\equiv 0$, $a_1^{(l_0)}(z_1, 0, \cdots, 0) = 0$, $\rho^{(m_1)}(z_1, 0, \cdots, 0, z_{n-1}, 0, \overline{z_1}, 0, \cdots, 0, \overline{z_{n-1}}, 0)$ *is not identically zero (and contains no nontrivial holomorphic terms).*

(III) For a certain $j \in [1, n-2]$, $a_j^{(l_0)}(z_1, \cdots, z_{n-2}, 0, \overline{z_1}, \cdots, \overline{z_{n-2}}, 0)$ *is not holomorphic,* $\sum_{j=1}^{n-2} z_j a_j^{(l_0)}(z_1, \cdots, z_{n-2}, 0, \overline{z_1}, \cdots, \overline{z_{n-2}}, 0) = 0$, *and*

$$\rho^{(m_1)}(z_1, \cdots, z_{n-1}, 0, \overline{z_1}, \cdots, \overline{z_{n-1}}, 0)$$

is not identically zero containing no nontrivial holomorphic terms.

In particular, we have

$$\begin{aligned} &\rho = \rho^{(m_1)} + O_{wt}(m_1 + 1), \;\; X_j = X_{j,-1} + O_{wt}(0), \\ &\rho^{(m_1)} \text{ is not identically 0 and contains no harmonic terms.} \end{aligned} \tag{12}$$

By a similar weighted degree estimate as in Case II, for any $1 \leq j \leq m_1 - 1$ and $Y_1, \cdots, Y_j \in \mathcal{M}_1(B)$, we have

$$Y_j \cdots Y_3 \partial\overline{\partial}\rho(Y_2, \overline{Y_1})(0) = 0 \text{ and } \langle [Y_j, [Y_{j-1}, \cdots, [Y_3, [Y_2, \overline{Y_1}] \cdots], \partial\rho\rangle(0) = 0.$$

Also, for any $k, k' \in [1, n-2]$ and $Y_1, \cdots, Y_{m_1-2} \in \mathcal{M}_1(B)$, we have

$$\begin{aligned} &Y_{m_1-2} \cdots Y_1 \partial\overline{\partial}\rho(X_k, \overline{X_{k'}})(0) \\ &\quad = (Y_{m_1-2})_{-1} \cdots (Y_1)_{-1} \partial\overline{\partial}\rho^{(m_1)}((X_k)_{-1}, \overline{(X_{k'})_{-1}})(0) \end{aligned} \tag{13}$$

and

$$\begin{aligned} &\langle [Y_{m_1-2}, [Y_{m_1-3}, \cdots, [Y_1, [X_k, \overline{X_{k'}}] \cdots], \partial\rho\rangle(0) \\ =&\langle [(Y_{m_1-2})_{-1}, [(Y_{m_1-3})_{-1}, \cdots, [(Y_1)_{-1}, [(X_k)_{-1}, \overline{(X_{k'})_{-1}}] \cdots], \partial\rho^{(m_1)}\rangle(0). \end{aligned} \tag{14}$$

Now $\{L_{j,-1}\}_{j=1}^{n-2}$ are sections of an $(n-2)$ -dimensional bundle B_{-1} associated to the hypersurface $M^0 := \{\rho^{(m_1)} = 0\} \subset \mathbb{C}^n$. We next claim that $t(B_{-1}, 0) = c(B_{-1}, 0) = m_1$.

Since $X_{j,-1}$ are weighted homogeneous, by Freeman's version of the Nagano Theorem [7, Theorem 2.1], $\{\mathrm{Re}(X_{j,-1}), \mathrm{Im}(X_{j,-1})\}_{j=1}^{n-2}$, and their Lie brackets will generate a homogeneous real manifold N^0.

Suppose that $t(B_{-1}, 0) > m_1$, and then for any $Y_1, \cdots, Y_{m_1} \in \mathcal{M}_1(B)$,

$$Y_{m_1} \cdots Y_3 \partial\overline{\partial}\rho(Y_2, \overline{Y_1})(0) = Y_{m_1,-1} \cdots Y_{3,-1} \partial\overline{\partial}\rho^{(m_1)}(Y_{2,-1}, \overline{Y_{1,-1}})(0) = 0.$$

On the other hand, for any $j \geq 0$, $Y_{j,-1} \cdots Y_{3,-1} \partial\overline{\partial}\rho^{(m_1)}(Y_{2,-1}, \overline{Y_{1,-1}})(z, w)$ is a weighted homogeneous polynomial of degree $m_1 - j$. Hence it is 0 at the origin for $j \neq m_1$. Thus for any integer $j \geq 1$ and any choice of $Y_j \in \mathcal{M}_1(B)$, we have

$$\langle [Y_{j,-1}, [Y_{j-1,-1}, \cdots, [Y_{3,-1}, [Y_{2,-1}, \overline{Y_{1,-1}}] \cdots], \partial\rho^{(m_1)} \rangle (0) = 0. \tag{15}$$

Hence $\frac{\partial}{\partial u}|_0 \notin T_0 N^0$ with $u = \mathrm{Re}(w)$. Hence the real dimension of N^0 is at most $2(n-1)$. Since our normalization is exactly the same as that in Section 4 of [8], it follows from [8, Lemma 4.3] that the real dimension of N^0 can only be $2n-3$ or $2n-2$. From Step I of [8, Page 84] for the case with dimension $2n-2$ and Step II of [8, Page 82] for the case with dimension $2n-3$, we conclude that $\rho^{(m_1)} \equiv 0$, which contradicts to the second line of (12). Hence $t(B_{-1}, 0) = m_1$. Together with (13), we obtain $t(B, 0) = m_1$.

Next, suppose that $c(B_{-1}, 0) > m_1$, and then for any $j \leq m_1 - 2$, $1 \leq k, k' \leq n-2$, $Y_1, \cdots, Y_j \in \mathcal{M}_1(B)$, we have

$$Y_j \cdots Y_1 \partial\overline{\partial}\rho(X_k, \overline{X_{k'}})(0) = 0. \tag{16}$$

A similar weighted degree argument shows that for any $j \geq 0$, $Z_1, \cdots, Z_j \in \mathcal{M}_1(B_{-1})$, we have

$$Z_j \cdots Z_1 \partial\overline{\partial}\rho(X_k, \overline{X_{k'}})(0) = 0.$$

This gives that $\partial\overline{\partial}\rho(X_j, \overline{X_j}) \equiv 0$ on N^0. Hence for any $q \in N^0$ around 0, we have

$$\mathrm{Re}(L_{-1})(q),\ \mathrm{Im}(L_{-1})(q) \in T_q^N N^0.$$

Here we have set

$$T_q^N N^0 := \{S \in T_q N^0 :\ S = \mathrm{Re}(X),\ X \in T_q^{(1,0)} M^0,\ \partial\overline{\partial}\rho(X, \overline{X})(q) = 0\}.$$

By [6, Proposition 2], $T_z^N N^0 = T_z N^0$. In particular, any $S \in T_0 N^0$ is in $T_0^h(M^0) := \{Y = \mathrm{Re}(X) :\ X \in T_0^{(1,0)} M^0 = \mathbb{C}^{n-1} \times \{0\}\}$. Since $\partial\rho|_0 = \frac{1}{2i} dw|_0$, we have (15), which gives that $t(B_{-1}, 0) > m_1$. We thus reach a contradiction. Hence we obtain $c(B_{-1}, 0) = m_1$. Together with (14), we obtain $c(B, 0) = m_1$. Hence in this case, we again obtain $t(B, 0) = c(B, 0)$.

The proof of Theorem 1.1 is completed.

References

1. T. Bloom and I. Graham, A geometric characterization of points of type m on real submanifolds of $\mathbb{C}^n$, J. Differential Geometry, 1977, 12: 171–182.
2. T. Bloom, Remarks on type conditions for real hypersurfaces in $\mathbb{C}^n$, Proc. Internat. Conf. on Several Complex Variables, Cortona, Italy, 1976–77, Scuola Norm. Sup. Pisa, Pisa, 1978, pp. 14–24.
3. T. Bloom, On the contact between complex manifolds and real hypersurfaces in $\mathbb{C}^3$, Trans. Amer. Math. Soc., 1981, 263 (2): 515–529.
4. J. D'Angelo, Real hypersurfaces, orders of contact, and applications, Ann. of Math. 1982, 115 (3): 615–637.
5. J. D'Angelo, Iterated commutators and derivatives of the Levi form, Complex analysis (University Park, Pa., 1986), 103–110, Lecture Notes in Math., 1268, Springer, Berlin, 1987.
6. K. Diederich and J. Fornaess, Pseudoconvex domains with real analytic boundary, Ann. of Math., 1978, 107 (3): 371–384.
7. M. Freeman, Integration of analytic differential systems with singularities and some applications to real submanifolds of C^n, J. Math. Soc. Japan, 1978, 30: 571–578.
8. X. Huang, W. Yin, Regular multi-types and the Bloom conjecture, J. Math. Pures Appl., 2021, 146: 69–98.
9. X. Huang, W. Yin, Regular types and order of vanishing along a set of non-integrable vector fields, to appear in *Acta Math Sin, Engl Series*, 2024. (A special issue dedicated to the memory of Professor Zhihua Chen)
10. X. Huang, W. Yin, Commutator type and Levi type of a system of CR vector fields, preprint, 2023.
11. J. Kohn, Boundary behaviour of $\overline{\partial}$ on weakly pseudo-convex manifolds of dimension two, J. Differential Geometry, 1972, 6: 523–542.
12. J. Kohn, Subellipticity of the $\overline{\partial}$-Neumann problem on pseudo-convex domains: sufficient conditions, Acta Math., 1979, 142 (1–2): 79–122.

Characterization of Real-Analytic Infinitesimal CR Automorphisms for a Class of Hypersurfaces in $\mathbb{C}^4$

Cyril Julien and Francine Meylan

To Jorge Hounie for his 75th birthday

1 Introduction

The structure of the Lie algebra $\mathfrak{g}$ of real-analytic infinitesimal CR automorphisms of a model hypersurface M_0 plays a fundamental role when studying the stability group of a hypersurface of finite type in $\mathbb{C}^{n+1}$ whose model hypersurface is M_0. This study was initiated by Poincaré for the unit sphere in $\mathbb{C}^2$ [17] and pursued by many others; see, for instance [1, 3, 4, 6, 15, 16, 19, 20, 21, 22, 23]. While Chern and Moser deal with the Levi nondegenerate case in [5], that is, when the model hypersurface is a hyperquadric, Kolar, Meylan, and Zaitsev, inspired by the technics of Chern and Moser, deal with the Levi degenerate case in [13], that is, when the model hypersurface is polynomial.

More precisely, Theorem 1.1 in [13] describes explicitly the graded components involved in the grading of $\mathfrak{g}$ but does not indicate anything about the dimension or the vanishing of each graded component regarding the model hypersurface.

Recently, Kim and Kolar gave in [7] a complete description of $\mathfrak{g}$ in the case of a pseudoconvex model $M_0 \subset \mathbb{C}^{n+1}$ which is a sum of squares, which is of the form

$$\operatorname{Im} w = \sum_{j=1}^{k} P_j \overline{P_j},$$

where P_j are homogeneous holomorphic polynomials, under the hypothesis that M_0 is holomorphically nondegenerate (which implies $k \geq n$). In the case $k = n$, it can

C. Julien · F. Meylan (✉)
Department of Mathematics, University of Fribourg, Fribourg, Switzerland
e-mail: cyril.julien@unifr.ch; francine.meylan@unifr.ch

S. Berhanu et al. (eds.), *Geometric Analysis of PDEs and Several Complex Variables*, Latin American Mathematics Series – UFSCar subseries,
https://doi.org/10.1007/978-3-031-69702-9_10

be written using the usual scalar product as

$$\operatorname{Im} w =< P, P >,$$

where $P = (P_1, \ldots, P_n)$. This motivates the following generalization:

Let D be the symmetric matrix with respect to the canonical basis $\{e_1, \ldots, e_n\}$ of $\mathbb{C}^n$ associated with the linear transformation sending e_j to $e_{\sigma(j)}$, where $\sigma \in S_n$ is a permutation that is a product of disjoint 2-cycles. Let $P = (P_1, \ldots, P_n)$ be an n-tuple of homogeneous holomorphic polynomials in the variables $z = (z_1, \ldots, z_n)$ for which ${}^t\bar{P}DP$ is a (real) homogeneous polynomial. We consider the model hypersurface denoted by M_0 in $\mathbb{C}^{n+1}$ given by

$$\operatorname{Im} w ={}^t \bar{P}DP, \ (z, w) \in \mathbb{C}^n \times \mathbb{C}. \tag{1}$$

Assuming that M_0 is holomorphically nondegenerate (which implies that P_j, $j = 1 \ldots, n$, are linearly independent over $\mathbb{C}$), we address the question of characterizing in terms of P the Lie algebra $\mathfrak{g}$ of M_0.

If the matrix D is diagonal, that is, M_0 is pseudoconvex and a sum of squares, Kim and Kolar show that there are exactly three nonvanishing graded components, with only one depending on the polynomial P_j. See Theorem 5.3 in [7].

The case $n = 2$ has been solved in [10] and [7].

In this chapter, we study the case $n = 3$ for nondiagonal matrices D, and we assume that P_1, P_2, and P_3 are monomial. We call this problem the *monomial PQR problem*. This can also be seen as the generalization of the case $n = 1$ given in [9, 12] where P_1 is necessarily a monomial of the form cz^k, and M_0 in $\mathbb{C}^2$ is given by

$$\operatorname{Im} w = |c|^2|z|^{2k}, \ (z, w) \in \mathbb{C} \times \mathbb{C}. \tag{2}$$

Note that M_0 given by (2) is the only Levi degenerate model hypersurface in $\mathbb{C}^2$ that possesses a nonlinear symmetry.

We will be particularly interested in the structure of the rigid symmetries of weighted degree 0, called the rotations. We will see that any such element can be decomposed into a diagonal rotation and a sum of at most two nilpotent rotations. Note that the analogous result for $n = 2$ is that any rotation is decomposed into the sum of a diagonal rotation and at most one nilpotent rotation. Note that if D is diagonal, the only rotations are the pure imaginary diagonal rotations (in Jordan form) [7].

Another phenomenon we will discuss is the structure of the generalized rotations that impact the number of derivatives needed to determine completely an automorphism of a smooth perturbation of M_0. Note that if D is diagonal, there are no such symmetries by Proposition 4.1 and Theorem 5.3 in [7].

We emphasize that the study of the structure of rotations may lead to the understanding (and hence the vanishing) of the graded components of generalized rotations. See Proposition 4.1 in [7] and Sect. 3.

In Sect. 2, we recall some definitions needed for the sequel. In Sect. 3, we collect, restate, and prove some (partially) known results on the graded component of $\mathfrak{g}$ of weight 0. In Sects. 4, 5, and 6, we study the graded components of $\mathfrak{g}$ in the case of the monomial PQR problem. Finally in Sect. 7, we state and prove the main results of the chapter, Theorems 7.1 and 7.2.

2 Preliminaries

In this section, we recall the necessary definitions needed in the sequel regarding smooth hypersurfaces in $\mathbb{C}^N$.

Let $M \subseteq \mathbb{C}^{n+1}$ be a smooth hypersurface and $p \in M$ be a point of *finite type* d (in the sense of Bloom-Graham) [2]. We will consider local holomorphic coordinates (z, w) vanishing at p, where $z = (z_1, z_2, ..., z_n)$ and $z_j = x_j + iy_j$, $w = u + iv$. The hyperplane $\{v = 0\}$ is assumed to be tangent to M at p; hence, M is described near p as the graph of a uniquely determined real-valued function

$$v = \psi(z_1, \ldots, z_n, \bar{z}_1, \ldots, \bar{z}_n, u), \ d\psi(0) = 0. \tag{3}$$

We may assume [2, 8] that

$$\psi(z_1, \ldots, z_n, \bar{z}_1, \ldots, \bar{z}_n, u) = Q(z, \bar{z}) + o(|u| + |z|^d), \tag{4}$$

where $Q(z, \bar{z})$ is a nonzero homogeneous polynomial of degree d with no pluriharmonic terms. Note that removing the pluriharmonic terms requires a change of coordinates in the variable w, which "absorbs" the pluriharmonic terms into w.

Reflecting the choice of variables given by (4), the variables w, u, and v are given weights 1, $\dfrac{\partial}{\partial w}$ weight -1, while the complex tangential variables (z_j, $j = 1, \ldots, n$) are given weights $\dfrac{1}{d}$ and $\dfrac{\partial}{\partial z_j}$, $j = 1, \ldots n$, weights $-\dfrac{1}{d}$.

For this choice of weights, $Q(z, \bar{z})$ is a weighted homogeneous polynomial of weighted degree 1, while $o(|u|+|z|^d)$ contains weighted homogeneous polynomials of weighted degree strictly bigger than 1.

Recall that the derivatives of weighted order κ are those of the form

$$\frac{\partial^{|\alpha|+|\hat{\alpha}|+l}}{\partial z^{\alpha} \partial \bar{z}^{\hat{\alpha}} \partial u^{l}}, \quad \kappa = l + \frac{1}{d} \sum_{j=1}^{n} (\alpha_j + \hat{\alpha}_j).$$

We may then rewrite Eq. (4) as

$$v = \psi(z, \bar{z}, u) = Q(z, \bar{z}) + o(1), \tag{5}$$

where $o(1)$ denotes a smooth function whose derivatives at 0 of weighted order less than or equal to 1 vanish. This motivates the following definition:

Definition 2.1 ([11]) Let M be given by (5). We define its associated model hypersurface by

$$M_0 := \{(z, w) \in \mathbb{C}^{n+1} \mid v = Q(z, \bar{z})\}. \tag{6}$$

We say that M is a smooth perturbation of M_0.

When $Q =^t \bar{P}DP$, the associated model hypersurface is given by (1), and we write

$$M_0 = \{(z, w) \in \mathbb{C}^{n+1} \mid \mathsf{Im}\, w =^t \bar{P}DP\}. \tag{7}$$

Remark 2.2 We may also consider a C^{k+d} perturbation of M_0, that is, with C^{k+d} smooth function $o(1)$.

We denote by $\mathsf{Aut}(M, p)$ the stability group of M, that is, those germs at p of biholomorphisms mapping M into itself and fixing p, and by $\mathsf{hol}(M, p)$ the set of germs of holomorphic vector fields in $\mathbb{C}^{n+1}$ whose real part is tangent to M.

Remark 2.3 If M admits a holomorphic vector field X in $\mathsf{hol}(M, p)$ such that $\mathsf{Im}\, X$ is also tangent, then $\mathsf{hol}(M, p)$ is of infinite dimension [18].

We write

$$\mathsf{hol}(M_0, p) = \bigoplus_{\mu \geq -1} \mathfrak{g}_\mu,$$

where $\mathfrak{g}_\mu$ consists of weighted homogeneous vector fields of weight μ. We recall the following definition.

Definition 2.4 A real-analytic hypersurface $M \subset \mathbb{C}^{n+1}$ is *holomorphically nondegenerate at* $p \in M$ if there is no germ at p of a holomorphic vector field X tangent to M.

3 The Structure of $\mathfrak{g}_0$ and Its Consequences

We start this section with a lemma that has been used several times (see, for instance, Proposition 4.1 in [7]), but whose proof was never given: This lemma deals with the properties of a linear vector field given in Jordan normal form. For the sake of clarity, we give a proof of it.

Without loss of generality, we will assume that the nilpotent part is of the form $\sum_{j\in J_1} z_j \frac{\partial}{\partial z_{j+1}}$, where J_1 is the subset of indices corresponding to the Jordan decomposition of the given vector field. We recall the following definition:

Definition 3.1 A linear vector field in the variables z given in Jordan normal form is real diagonal (respectively, imaginary) if it is diagonal with all the entries real (respectively, imaginary). Without loss of generality, we say that X is real (respectively, imaginary) diagonal if its Jordan normal form is real (respectively, imaginary) diagonal.

Lemma 3.2 *Let Q be a real homogeneous polynomial in $(z, \bar{z})$ without pluriharmonic terms, and let $W = X + Z$ be a holomorphic linear vector field in the variables z given in Jordan normal form with X the diagonal and Z the nilpotent part. Then*

$$W(Q) = Q \text{ if and only if } X(Q) = Q,\ Z(Q) = 0. \tag{8}$$

Moreover, if the model hypersurface given by

$$M_0 := \{(z, w) \in \mathbb{C}^{n+1} \mid v = Q(z, \bar{z})\} \tag{9}$$

is holomorphically nondegenerate, then X is real diagonal and $Z = 0$.

Proof Since X is diagonal, we may consider the spectral decomposition

$$Q = \sum Q_\mu, \quad Q_\mu \in \mathcal{Q}_\mu := \langle z^\alpha \bar{z}^\beta : X(z^\alpha \bar{z}^\beta) = \mu z^\alpha \bar{z}^\beta \rangle.$$

Since $[X, Z] = 0$, we obtain that $Z(\mathcal{Q}_\mu) \subset \mathcal{Q}_\mu$. We put the lexicographic order on monomials $cz^\alpha \bar{z}^\beta$ so that $Z(z^\alpha \bar{z}^\beta)$ is a sum of monomials larger than $cz^\alpha \bar{z}^\beta$. We claim that $Q_\mu = 0$ unless $\mu = 1$. Indeed, assume by contradiction that $Q_\mu \neq 0$ for some $\mu \neq 1$, and consider the minimal nontrivial monomial $cz^\alpha \bar{z}^\beta$ in the expansion of Q_μ with respect to the lexicographic order. Then, since $Z(\mathcal{Q}_\mu) \subset \mathcal{Q}_\mu$, $Z(z^\alpha \bar{z}^\beta)$ is a sum of monomials larger than $cz^\alpha \bar{z}^\beta$ in $\mathcal{Q}_\mu$. Therefore the coefficient of $z^\alpha \bar{z}^\beta$ in $W(Q)$ is equal to μ. Since $W(Q) = Q$, we must have $\mu = 1$ contradicting our assumption. Hence $Q = Q_1$ as claimed, implying $X(Q) = Q$ and hence $Z(Q) = 0$. □

Let $X \in \mathsf{hol}(M_0, p)$ be given by $X = \sum_{j=1}^n F_j(z, w)\partial_{z_j} + G(z, w)\partial_w$. We recall that X is rigid if $F_1, \dots, F_n, G$ are all independent of the variable w [11]. The rigid vector fields of weighted degree 0 contained in $\mathfrak{g}_0$ are called rotations, while $\mathfrak{g}_c$ denotes the graded components of rigid vector fields of weight strictly bigger than 0, that is with nonlinear coefficients; they are called generalized rotations or exotic symmetries. It is known that the weighted degree of an element of $\mathfrak{g}_c$ is strictly less than 1 [13].

Theorem 3.3 *Let M_0 be given by* (6) *and holomorphically nondegenerate. Suppose that $\mathfrak{g}_0$ contains no imaginary diagonal rotations. Then $\mathfrak{g}_1 = 0$.*

Proof Suppose by contradiction that $\mathfrak{g}_1 \neq 0$. Then, using a result of [13], we obtain that $\dim \mathfrak{g}_1 = 1$, with a generator given by

$$Y := \sum \varphi_j(z) w \partial_{z_j} + \frac{1}{2} w^2 \partial_w, \quad 2 \sum \varphi_j(z) \partial_{z_j}(Q) = Q. \tag{10}$$

Let $S := 2 \sum \varphi_j(z) \partial_{z_j}$ be the linear vector field associated with Y. By Lemma 3.2, we may find linear holomorphic coordinates in the z variables such that S is real diagonal. We obtain then that iS is an imaginary diagonal rotation since $iS(Q) = iQ$ and hence a contradiction. □

Since the grading element is in $\mathfrak{g}_0$, we have the following corollary:

Corollary 3.4 *Let M_0 be given by* (6) *and holomorphically nondegenerate. Suppose that $\mathfrak{g}_1 \neq 0$. Then* $\dim \mathfrak{g}_0 > 1$.

The following example shows that the converse is not true, unlike the case $n = 2$, [10].

Example 3.5 Consider the hypersurface in $\mathbb{C}^4$ given by

$$\operatorname{Im} w = \left(z_1^2 z_2 z_3^3 + z_1 z_2^4 z_3\right) \overline{z_1}^2 + z_1{}^2 \left(\overline{z_1}^2 \overline{z_2}\,\overline{z_3}^3 + \overline{z_1}\,\overline{z_2}^4 \overline{z_3}\right) + z_1^2 z_2^2 \overline{z_1}^2 \overline{z_2}^2.$$

The reader can check that X given by

$$X := i\,11 z_1 \frac{\partial}{\partial z_1} + i\,3 z_2 \frac{\partial}{\partial z_2} - i\,z_3 \frac{\partial}{\partial z_3}$$

is an imaginary diagonal rotation. It is easy to check that $\dim \mathfrak{g}_1 = 0$, using the characterization (10).

We now state a theorem that has been proved in [7] when M_0 is pseudoconvex (Proposition 4.1 in [7]). It shows the link between the structure of the rotations and the generalized rotations, using a theorem of [13]. The proof is similar to the proof of Proposition 4.1 in [7].

Theorem 3.6 *Let M_0 be given by* (6) *holomorphically nondegenerate. Suppose that $\mathfrak{g}_1 \neq 0$ and that $\mathfrak{g}_0$ contains no real diagonal rotations. Then $\mathfrak{g}_c = 0$.*

Proof Since $\mathfrak{g}_1 \neq 0$, using a result of [13], we obtain that $\dim \mathfrak{g}_1 = 1$, with a generator given by

$$Y := \sum \varphi_j(z) w \partial_{z_j} + \frac{1}{2} w^2 \partial_w, \quad 2 \sum \varphi_j(z) \partial_{z_j}(Q) = Q. \tag{11}$$

Let $S := 2\sum \varphi_j(z)\partial_{z_j}$ be the linear vector field associated with Y. By Lemma 3.2, we may find linear holomorphic coordinates in the z variables such that S is real diagonal. Without loss of generality, we may then assume that Y is of the form

$$Y = \sum \lambda_j z_j w \partial_{z_j} + \frac{1}{2} w^2 \partial_w, \quad \lambda_j \in \mathbb{R}.$$

Assume by contradiction that there exists a nonzero $X \in \mathfrak{g}_c$ of weight $\mu > 0$. Then again by a result of [13], the Lie bracket $[X, Y] = 0$, the reason being that there is no graded component of weight strictly bigger than one under the assumption that M_0 is holomorphically nondegenerate. Let $E \in \mathfrak{g}_0$ be the grading element given by

$$E = \sum_{j=1}^{n} \frac{1}{d} z_j \partial_{z_j} + w\partial_w. \tag{12}$$

Since $[E, X] = \mu X$, $[\partial_w, X] = 0$ and $[[\partial_w, Y], X] = 0$ by the identity of Jacobi, we obtain that

$$E - [\partial_w, Y] = \sum \left(\frac{1}{d} - \lambda_j\right) z_j \partial_{z_j}$$

is a nonzero real diagonal rotation and hence a contradiction. □

4 The Structure of $\mathfrak{g}_0$ for the Monomial PQR Problem

In this section, we consider the monomial PQR problem, that is, when M_0 is given by

$$M_0 = \{(z, w) \in \mathbb{C}^3 \times \mathbb{C} \mid \operatorname{Im} w = P\bar{Q} + Q\bar{P} + R\bar{R}\}, \tag{13}$$

where P, Q, and R are monomials such that $P\bar{Q} + Q\bar{P} + R\bar{R}$ is of degree d. We show that any rotation of M_0 can be decomposed into a rotation that is diagonal *in these same coordinates* and a sum of at most two nilpotent rotations. We start with the following lemma that gives a characterization of holomorphic nondegeneracy.

Lemma 4.1 *Let M_0 be given by* (13)*, with P, Q, and R not necessarily monomials. Then M_0 is holomorphically nondegenerate if and only if the Jacobian of P, Q, and R is nonzero.*

Proof Let $J(P, Q, R)$ be the Jacobian of P, Q, and R. By contradiction, if $J(P, Q, R) = 0$, solving a system of three equations, one can find a holomorphic vector field $X = \sum_{j=1}^{3} f_j(z)\frac{\partial}{\partial z_j}$ satisfying $X(P) = 0$, $X(Q) = 0$, and $X(R) = 0$, hence the contradiction.

Suppose now that $J(P, Q, R) \neq 0$. Let X be a holomorphic vector field tangent to M_0 given by

$$X = \sum_{j=1}^{3} f_j(z, w)\frac{\partial}{\partial z_j} + g(z, w)\frac{\partial}{\partial w}. \tag{14}$$

This implies $g(z, u) = 0$, and hence

$$g(z, w) = 0, \;\; X(P)\overline{Q} + X(Q)\overline{P} + X(R)\overline{R} = 0, \tag{15}$$

and hence, by assumption,

$$X(P) = 0, \;\; X(Q) = 0, \;\; X(R) = 0.$$

This yields to a Cramer system of three equations whose only solution is the zero solution since $J(P, Q, R) \neq 0$. Hence $X = 0$, that is, M_0 is holomorphically nondegenerate. □

Proposition 4.2 *Let M_0 be given by* (13) *with $P = z_1^{\alpha_1} z_2^{\alpha_2} z_3^{\alpha_3}$, $Q = z_1^{\beta_1} z_2^{\beta_2} z_3^{\beta_3}$, and $R = z_1^{\gamma_1} z_2^{\gamma_2} z_3^{\gamma_3}$. Then M is holomorphically nondegenerate if and only if*

$$\begin{vmatrix} \alpha_1 & \beta_1 & \gamma_1 \\ \alpha_2 & \beta_2 & \gamma_2 \\ \alpha_3 & \beta_3 & \gamma_3 \end{vmatrix} \neq 0.$$

Proof By Lemma 4.1, we have M_0 holomorphically nondegenerate if and only if

$$0 \neq J(P, Q, R) = \begin{vmatrix} P_{z_1} & Q_{z_1} & R_{z_1} \\ P_{z_2} & Q_{z_2} & R_{z_2} \\ P_{z_3} & Q_{z_3} & R_{z_3} \end{vmatrix}$$

$$= \begin{vmatrix} \alpha_1 z_1^{\alpha_1 - 1} z_2^{\alpha_2} z_3^{\alpha_3} & \beta_1 z_1^{\beta_1 - 1} z_2^{\beta_2} z_3^{\beta_3} & \gamma_1 z_1^{\gamma_1 - 1} z_2^{\gamma_2} z_3^{\gamma_3} \\ \alpha_2 z_1^{\alpha_1} z_2^{\alpha_2 - 1} z_3^{\alpha_3} & \beta_2 z_1^{\beta_1} z_2^{\beta_2 - 1} z_3^{\beta_3} & \gamma_2 z_1^{\gamma_1} z_2^{\gamma_2 - 1} z_3^{\gamma_3} \\ \alpha_3 z_1^{\alpha_1} z_2^{\alpha_2} z_3^{\alpha_3 - 1} & \beta_3 z_1^{\beta_1} z_2^{\beta_2} z_3^{\beta_3 - 1} & \gamma_3 z_1^{\gamma_1} z_2^{\gamma_2} z_3^{\gamma_3 - 1} \end{vmatrix}$$

$$= \alpha_1 z_1^{\alpha_1 - 1} z_2^{\alpha_2} z_3^{\alpha_3} \begin{vmatrix} \beta_2 z_1^{\beta_1} z_2^{\beta_2 - 1} z_3^{\beta_3} & \gamma_2 z_1^{\gamma_1} z_2^{\gamma_2 - 1} z_3^{\gamma_3} \\ \beta_3 z_1^{\beta_1} z_2^{\beta_2} z_3^{\beta_3 - 1} & \gamma_3 z_1^{\gamma_1} z_2^{\gamma_2} z_3^{\gamma_3 - 1} \end{vmatrix}$$

$$- \beta_1 z_1^{\beta_1 - 1} z_2^{\beta_2} z_3^{\beta_3} \begin{vmatrix} \alpha_2 z_1^{\alpha_1} z_2^{\alpha_2 - 1} z_3^{\alpha_3} & \gamma_2 z_1^{\gamma_1} z_2^{\gamma_2 - 1} z_3^{\gamma_3} \\ \alpha_3 z_1^{\alpha_1} z_2^{\alpha_2} z_3^{\alpha_3 - 1} & \gamma_3 z_1^{\gamma_1} z_2^{\gamma_2} z_3^{\gamma_3 - 1} \end{vmatrix}$$

$$+\gamma_1 z_1^{\gamma_1-1} z_2^{\gamma_2} z_3^{\gamma_3} \begin{vmatrix} \alpha_2 z_1^{\alpha_1} z_2^{\alpha_2-1} z_3^{\alpha_3} & \beta_2 z_1^{\beta_1} z_2^{\beta_2-1} z_3^{\beta_3} \\ \alpha_3 z_1^{\alpha_1} z_2^{\alpha_2} z_3^{\alpha_3-1} & \beta_3 z_1^{\beta_1} z_2^{\beta_2} z_3^{\beta_3-1} \end{vmatrix}$$

$$= \begin{vmatrix} \alpha_1 & \beta_1 & \gamma_1 \\ \alpha_2 & \beta_2 & \gamma_2 \\ \alpha_3 & \beta_3 & \gamma_3 \end{vmatrix} z_1^{\alpha_1+\beta_1+\gamma_1-1} z_2^{\alpha_2+\beta_2+\gamma_2-1} z_3^{\alpha_3+\beta_3+\gamma_3-1}.$$

□

We may now state the main result of this section.

Theorem 4.3 *Let M_0 be given by* (13) *with P, Q, and R monomials and holomorphically nondegenerate. Then any $X \in \mathfrak{g}_0$ given by*

$$X = \sum_{j,k=1}^{3} a_{jk} z_k \frac{\partial}{\partial z_j}, \quad a_{jk} \in \mathbb{C},$$

may be written as

$$X = D + N,$$

where

$$D = \sum_{j=1}^{3} a_{jj} z_j \frac{\partial}{\partial z_j} \in \mathfrak{g}_0, \tag{16}$$

and

$$N = \sum_{j \neq k, j,k=1}^{3} a_{jk} z_k \frac{\partial}{\partial z_j} \in \mathfrak{g}_0 \tag{17}$$

satisfies the following property:

(1) Either N is a nilpotent rotation
(2) Or there exists a subset of indices J_1 of $J = \{(j,k), j \neq k, j,k = 1,2,3,\}$ such that

$$N_1 := \sum\nolimits_{J_1} a_{jk} z_k \frac{\partial}{\partial z_j} \text{ and } N_2 := \sum\nolimits_{J-J_1} a_{jk} z_k \frac{\partial}{\partial z_j}$$

are nilpotent rotations, with $N = N_1 + N_2$.

Proof We write

$$P = c_P z_1^{\alpha_1} z_2^{\alpha_2} z_3^{\alpha_3} = c_P z^{\alpha}, \ Q = c_Q z_1^{\beta_1} z_2^{\beta_2} z_3^{\beta_3} = c_Q z^{\beta},$$

$$R = c_R z_1^{\gamma_1} z_2^{\gamma_2} z_3^{\gamma_3} = c_R z^{\gamma}, \ c_P, \ c_Q, \ c_R \in \mathbb{C}.$$

We first show that $D \in \mathfrak{g}_0$.

By assumption, we have $\operatorname{Re} X(P\bar{Q} + Q\bar{P} + R\bar{R}) = \operatorname{Re} D(P\bar{Q} + Q\bar{P} + R\bar{R}) + \operatorname{Re} N(P\bar{Q} + Q\bar{P} + R\bar{R}) = 0$.

Observing that the terms given by $\operatorname{Re} N(P\bar{Q} + Q\bar{P} + R\bar{R})$ are

$$\begin{array}{lll}
 & a_{12}c_P\overline{c_Q}\alpha_1 z_1^{\alpha_1-1}z_2^{\alpha_2+1}z_3^{\alpha_3}\overline{z_1}^{\beta_1}\overline{z_2}^{\beta_2}\overline{z_3}^{\beta_3} & +a_{13}c_P\overline{c_Q}\alpha_1 z_1^{\alpha_1-1}z_2^{\alpha_2}z_3^{\alpha_3+1}\overline{z_1}^{\beta_1}\overline{z_2}^{\beta_2}\overline{z_3}^{\beta_3} \\
 & +\overline{a_{12}}c_Q\overline{c_P}\alpha_1 z_1^{\beta_1}z_2^{\beta_2}z_3^{\beta_3}\overline{z_1}^{\alpha_1-1}\overline{z_2}^{\alpha_2+1}\overline{z_3}^{\alpha_3} & +\overline{a_{13}}c_Q\overline{c_P}\alpha_1 z_1^{\beta_1}z_2^{\beta_2}z_3^{\beta_3}\overline{z_1}^{\alpha_1-1}\overline{z_2}^{\alpha_2}\overline{z_3}^{\alpha_3+1} \\
 & +a_{12}c_Q\overline{c_P}\beta_1 z_1^{\beta_1-1}z_2^{\beta_2+1}z_3^{\beta_3}\overline{z_1}^{\alpha_1}\overline{z_2}^{\alpha_2}\overline{z_3}^{\alpha_3} & +a_{13}c_Q\overline{c_P}\beta_1 z_1^{\beta_1-1}z_2^{\beta_2}z_3^{\beta_3+1}\overline{z_1}^{\alpha_1}\overline{z_2}^{\alpha_2}\overline{z_3}^{\alpha_3} \\
 & +\overline{a_{12}}c_P\overline{c_Q}\beta_1 z_1^{\alpha_1}z_2^{\alpha_2}z_3^{\alpha_3}\overline{z_1}^{\beta_1-1}\overline{z_2}^{\beta_2+1}\overline{z_3}^{\beta_3} & +\overline{a_{13}}c_P\overline{c_Q}\beta_1 z_1^{\alpha_1}z_2^{\alpha_2}z_3^{\alpha_3}\overline{z_1}^{\beta_1-1}\overline{z_2}^{\beta_2}\overline{z_3}^{\beta_3+1} \\
 & +a_{12}c_R\overline{c_R}\gamma_1 z_1^{\gamma_1-1}z_2^{\gamma_2+1}z_3^{\gamma_3}\overline{z_1}^{\gamma_1}\overline{z_2}^{\gamma_2}\overline{z_3}^{\gamma_3} & +a_{13}c_R\overline{c_R}\gamma_1 z_1^{\gamma_1-1}z_2^{\gamma_2}z_3^{\gamma_3+1}\overline{z_1}^{\gamma_1}\overline{z_2}^{\gamma_2}\overline{z_3}^{\gamma_3} \\
 & +\overline{a_{12}}c_R\overline{c_R}\gamma_1 z_1^{\gamma_1}z_2^{\gamma_2}z_3^{\gamma_3}\overline{z_1}^{\gamma_1-1}\overline{z_2}^{\gamma_2+1}\overline{z_3}^{\gamma_3} & +\overline{a_{13}}c_R\overline{c_R}\gamma_1 z_1^{\gamma_1}z_2^{\gamma_2}z_3^{\gamma_3}\overline{z_1}^{\gamma_1-1}\overline{z_2}^{\gamma_2}\overline{z_3}^{\gamma_3+1} \\
 & & \\
+a_{21}c_P\overline{c_Q}\alpha_2 z_1^{\alpha_1+1}z_2^{\alpha_2-1}z_3^{\alpha_3}\overline{z_1}^{\beta_1}\overline{z_2}^{\beta_2}\overline{z_3}^{\beta_3} & & +a_{23}c_P\overline{c_Q}\alpha_2 z_1^{\alpha_1}z_2^{\alpha_2-1}z_3^{\alpha_3+1}\overline{z_1}^{\beta_1}\overline{z_2}^{\beta_2}\overline{z_3}^{\beta_3} \\
+\overline{a_{21}}c_Q\overline{c_P}\alpha_2 z_1^{\beta_1}z_2^{\beta_2}z_3^{\beta_3}\overline{z_1}^{\alpha_1+1}\overline{z_2}^{\alpha_2-1}\overline{z_3}^{\alpha_3} & & +\overline{a_{23}}c_Q\overline{c_P}\alpha_2 z_1^{\beta_1}z_2^{\beta_2}z_3^{\beta_3}\overline{z_1}^{\alpha_1}\overline{z_2}^{\alpha_2-1}\overline{z_3}^{\alpha_3+1} \\
+a_{21}c_Q\overline{c_P}\beta_2 z_1^{\beta_1+1}z_2^{\beta_2-1}z_3^{\beta_3}\overline{z_1}^{\alpha_1}\overline{z_2}^{\alpha_2}\overline{z_3}^{\alpha_3} & & +a_{23}c_Q\overline{c_P}\beta_2 z_1^{\beta_1}z_2^{\beta_2-1}z_3^{\beta_3+1}\overline{z_1}^{\alpha_1}\overline{z_2}^{\alpha_2}\overline{z_3}^{\alpha_3} \\
+\overline{a_{21}}c_P\overline{c_Q}\beta_2 z_1^{\alpha_1}z_2^{\alpha_2}z_3^{\alpha_3}\overline{z_1}^{\beta_1+1}\overline{z_2}^{\beta_2-1}\overline{z_3}^{\beta_3} & & +\overline{a_{23}}c_P\overline{c_Q}\beta_2 z_1^{\alpha_1}z_2^{\alpha_2}z_3^{\alpha_3}\overline{z_1}^{\beta_1}\overline{z_2}^{\beta_2-1}\overline{z_3}^{\beta_3+1} \\
+a_{21}c_R\overline{c_R}\gamma_2 z_1^{\gamma_1+1}z_2^{\gamma_2-1}z_3^{\gamma_3}\overline{z_1}^{\gamma_1}\overline{z_2}^{\gamma_2}\overline{z_3}^{\gamma_3} & & +a_{23}c_R\overline{c_R}\gamma_2 z_1^{\gamma_1}z_2^{\gamma_2-1}z_3^{\gamma_3+1}\overline{z_1}^{\gamma_1}\overline{z_2}^{\gamma_2}\overline{z_3}^{\gamma_3} \\
+\overline{a_{21}}c_R\overline{c_R}\gamma_2 z_1^{\gamma_1}z_2^{\gamma_2}z_3^{\gamma_3}\overline{z_1}^{\gamma_1+1}\overline{z_2}^{\gamma_2-1}\overline{z_3}^{\gamma_3} & & +\overline{a_{23}}c_R\overline{c_R}\gamma_2 z_1^{\gamma_1}z_2^{\gamma_2}z_3^{\gamma_3}\overline{z_1}^{\gamma_1}\overline{z_2}^{\gamma_2-1}\overline{z_3}^{\gamma_3+1} \\
 & & \\
+a_{31}c_P\overline{c_Q}\alpha_3 z_1^{\alpha_1+1}z_2^{\alpha_2}z_3^{\alpha_3-1}\overline{z_1}^{\beta_1}\overline{z_2}^{\beta_2}\overline{z_3}^{\beta_3} & +a_{32}c_P\overline{c_Q}\alpha_3 z_1^{\alpha_1}z_2^{\alpha_2+1}z_3^{\alpha_3-1}\overline{z_1}^{\beta_1}\overline{z_2}^{\beta_2}\overline{z_3}^{\beta_3} & \\
+\overline{a_{31}}c_Q\overline{c_P}\alpha_3 z_1^{\beta_1}z_2^{\beta_2}z_3^{\beta_3}\overline{z_1}^{\alpha_1+1}\overline{z_2}^{\alpha_2}\overline{z_3}^{\alpha_3-1} & +\overline{a_{32}}c_Q\overline{c_P}\alpha_3 z_1^{\beta_1}z_2^{\beta_2}z_3^{\beta_3}\overline{z_1}^{\alpha_1}\overline{z_2}^{\alpha_2+1}\overline{z_3}^{\alpha_3-1} & \\
+a_{31}c_Q\overline{c_P}\beta_3 z_1^{\beta_1+1}z_2^{\beta_2}z_3^{\beta_3-1}\overline{z_1}^{\alpha_1}\overline{z_2}^{\alpha_2}\overline{z_3}^{\alpha_3} & +a_{32}c_Q\overline{c_P}\beta_3 z_1^{\beta_1}z_2^{\beta_2+1}z_3^{\beta_3-1}\overline{z_1}^{\alpha_1}\overline{z_2}^{\alpha_2}\overline{z_3}^{\alpha_3} & \\
+\overline{a_{31}}c_P\overline{c_Q}\beta_3 z_1^{\alpha_1}z_2^{\alpha_2}z_3^{\alpha_3}\overline{z_1}^{\beta_1+1}\overline{z_2}^{\beta_2}\overline{z_3}^{\beta_3-1} & +\overline{a_{32}}c_P\overline{c_Q}\beta_3 z_1^{\alpha_1}z_2^{\alpha_2}z_3^{\alpha_3}\overline{z_1}^{\beta_1}\overline{z_2}^{\beta_2+1}\overline{z_3}^{\beta_3-1} & \\
+a_{31}c_R\overline{c_R}\gamma_3 z_1^{\gamma_1+1}z_2^{\gamma_2}z_3^{\gamma_3-1}\overline{z_1}^{\gamma_1}\overline{z_2}^{\gamma_2}\overline{z_3}^{\gamma_3} & +a_{32}c_R\overline{c_R}\gamma_3 z_1^{\gamma_1}z_2^{\gamma_2+1}z_3^{\gamma_3-1}\overline{z_1}^{\gamma_1}\overline{z_2}^{\gamma_2}\overline{z_3}^{\gamma_3} & \\
+\overline{a_{31}}c_R\overline{c_R}\gamma_3 z_1^{\gamma_1}z_2^{\gamma_2}z_3^{\gamma_3}\overline{z_1}^{\gamma_1+1}\overline{z_2}^{\gamma_2}\overline{z_3}^{\gamma_3-1} & +\overline{a_{32}}c_R\overline{c_R}\gamma_3 z_1^{\gamma_1}z_2^{\gamma_2}z_3^{\gamma_3}\overline{z_1}^{\gamma_1}\overline{z_2}^{\gamma_2+1}\overline{z_3}^{\gamma_3-1}, &
\end{array} \tag{18}$$

and using Proposition 4.2, we obtain that the terms of $\operatorname{Re} D(P\bar{Q} + Q\bar{P} + R\bar{R})$, which are of the form $z^{\alpha}\overline{z}^{\beta}, \ z^{\beta}\overline{z}^{\alpha}, \ z^{\gamma}\overline{z}^{\gamma}$, can never cancel out with the terms of (18).

Therefore $\operatorname{Re} D(P\bar{Q} + Q\bar{P} + R\bar{R}) = 0$, that is, $D \in \mathfrak{g}_0$, and hence $N \in \mathfrak{g}_0$. We need the following lemma. □

Lemma 4.4 *Let*

$$N = (a_{12}z_2 + a_{13}z_3)\frac{\partial}{\partial z_1} + (a_{21}z_1 + a_{23}z_3)\frac{\partial}{\partial z_2} + (a_{31}z_1 + a_{32}z_2)\frac{\partial}{\partial z_3} \in \mathfrak{g}_0, \ a_{12} \neq 0.$$

Then

(1) Either

$$N_1 = a_{12}z_2\frac{\partial}{\partial z_1} \in \mathfrak{g}_0,$$

and M_0 *is given by* $\mathsf{Im}\, w = 2\mathsf{Re}\left(c_P\overline{c_Q}z_1z_2^{\alpha_2}z_3^{\alpha_3}\overline{z_2}^{\alpha_2+1}\overline{z_3}^{\alpha_3}\right) + c_R\overline{c_R}$ $z_2^{\gamma_2}z_3^{\gamma_3}\overline{z_2}^{\gamma_2}\overline{z_3}^{\gamma_3}$

(2) Or

$$N_1 = a_{12}z_2\frac{\partial}{\partial z_1} + a_{23}z_3\frac{\partial}{\partial z_2} \in \mathfrak{g}_0,$$

and M_0 *is given by* $\mathsf{Im}\, w = 2\mathsf{Re}\left(c_P\overline{c_Q}z_1z_3^k\overline{z_3}^{k+1}\right) + c_R\overline{c_R}z_2z_3^k\overline{z_2}\overline{z_3}^k$

(3) Or

$$N_1 = a_{12}z_2\frac{\partial}{\partial z_1} + a_{31}z_1\frac{\partial}{\partial z_3} \in \mathfrak{g}_0,$$

and M_0 *is given by* $\mathsf{Im}\, w = 2\mathsf{Re}\left(c_P\overline{c_Q}z_2^kz_3\overline{z_2}^{k+1}\right) + c_R\overline{c_R}z_1z_2^k\overline{z_1}\overline{z_2}^k.$

Proof Suppose that $N_1 = a_{12}z_2\frac{\partial}{\partial z_1} \in \mathfrak{g}_0$. It implies that

$$\begin{aligned}
0 = \;& a_{12}c_P\overline{c_Q}\alpha_1z_1^{\alpha_1-1}z_2^{\alpha_2+1}z_3^{\alpha_3}\overline{z_1}^{\beta_1}\overline{z_2}^{\beta_2}\overline{z_3}^{\beta_3}\\
&+\overline{a_{12}}c_Q\overline{c_P}\alpha_1z_1^{\beta_1}z_2^{\beta_2}z_3^{\beta_3}\overline{z_1}^{\alpha_1-1}\overline{z_2}^{\alpha_2+1}\overline{z_3}^{\alpha_3}\\
&+a_{12}c_Q\overline{c_P}\beta_1z_1^{\beta_1-1}z_2^{\beta_2+1}z_3^{\beta_3}\overline{z_1}^{\alpha_1}\overline{z_2}^{\alpha_2}\overline{z_3}^{\alpha_3}\\
&+\overline{a_{12}}c_P\overline{c_Q}\beta_1z_1^{\alpha_1}z_2^{\alpha_2}z_3^{\alpha_3}\overline{z_1}^{\beta_1-1}\overline{z_2}^{\beta_2+1}\overline{z_3}^{\beta_3}\\
&+a_{12}c_R\overline{c_R}\gamma_1z_1^{\gamma_1-1}z_2^{\gamma_2+1}z_3^{\gamma_3}\overline{z_1}^{\gamma_1}\overline{z_2}^{\gamma_2}\overline{z_3}^{\gamma_3}\\
&+\overline{a_{12}}c_R\overline{c_R}\gamma_1z_1^{\gamma_1}z_2^{\gamma_2}z_3^{\gamma_3}\overline{z_1}^{\gamma_1-1}\overline{z_2}^{\gamma_2+1}\overline{z_3}^{\gamma_3}.
\end{aligned} \tag{19}$$

By Proposition 4.2, there are two possibilities:

(1) Either

$a_{12}c_P\overline{c_Q}\alpha_1z_1^{\alpha_1-1}z_2^{\alpha_2+1}z_3^{\alpha_3}\overline{z_1}^{\beta_1}\overline{z_2}^{\beta_2}\overline{z_3}^{\beta_3}$ matches with

$\overline{a_{12}}c_Q\overline{c_P}\alpha_1z_1^{\beta_1}z_2^{\beta_2}z_3^{\beta_3}\overline{z_1}^{\alpha_1-1}\overline{z_2}^{\alpha_2+1}\overline{z_3}^{\alpha_3}$

(2) Or

$a_{12}c_P\overline{c_Q}\alpha_1z_1^{\alpha_1-1}z_2^{\alpha_2+1}z_3^{\alpha_3}\overline{z_1}^{\beta_1}\overline{z_2}^{\beta_2}\overline{z_3}^{\beta_3}$ matches with

$\overline{a_{12}}c_R\overline{c_R}\gamma_1z_1^{\gamma_1}z_2^{\gamma_2}z_3^{\gamma_3}\overline{z_1}^{\gamma_1-1}\overline{z_2}^{\gamma_2+1}\overline{z_3}^{\gamma_3}.$

Case (1) yields to

$$\begin{aligned} \alpha_1 - 1 &= \beta_1 \\ \alpha_2 + 1 &= \beta_2 \\ \alpha_3 \quad &= \beta_3, \end{aligned}$$

which implies

$$\gamma_1 = 0 \;\; \alpha_1 = 1 \;\; \beta_1 = 0.$$

Hence M_0 is given by

$$\begin{aligned} \operatorname{Im} w = {} & c_P\overline{c_Q} z_1 z_2^{\alpha_2} z_3^{\alpha_3} \overline{z_2}^{\alpha_2+1} \overline{z_3}^{\alpha_3} + c_Q\overline{c_P} z_2^{\alpha_2+1} z_3^{\alpha_3} \overline{z_1}\,\overline{z_2}^{\alpha_2} \overline{z_3}^{\alpha_3} \\ & + c_R\overline{c_R} z_2^{\gamma_2} z_3^{\gamma_3} \overline{z_2}^{\gamma_2} \overline{z_3}^{\gamma_3}. \end{aligned} \tag{20}$$

Case (2) yields to

$$\begin{aligned} \gamma_1 &= \alpha_1 - 1 = \beta_1 + 1 \\ \gamma_2 &= \alpha_2 + 1 = \beta_2 - 1 \\ \gamma_3 &= \quad \alpha_3 \quad = \quad \beta_3. \end{aligned} \tag{21}$$

Using Proposition 4.2, (21) implies that M_0 is holomorphically degenerate: Case (2) does not occur.

For the rest of the proof, we assume that $N_1 = a_{12} z_2 \frac{\partial}{\partial z_1} \notin \mathfrak{g}_0$ and suppose $a_{13} \neq 0$.

We claim that terms containing a_{12} cannot interfere with terms containing a_{13}.

The terms under study are

$$\begin{aligned} & a_{12} c_P \overline{c_Q} \alpha_1 z_1^{\alpha_1-1} z_2^{\alpha_2+1} z_3^{\alpha_3} \overline{z_1}^{\beta_1} \overline{z_2}^{\beta_2} \overline{z_3}^{\beta_3} && + a_{13} c_P \overline{c_Q} \alpha_1 z_1^{\alpha_1-1} z_2^{\alpha_2} z_3^{\alpha_3+1} \overline{z_1}^{\beta_1} \overline{z_2}^{\beta_2} \overline{z_3}^{\beta_3} \\ & + \overline{a_{12}} c_Q \overline{c_P} \alpha_1 z_1^{\beta_1} z_2^{\beta_2} z_3^{\beta_3} \overline{z_1}^{\alpha_1-1} \overline{z_2}^{\alpha_2+1} \overline{z_3}^{\alpha_3} && + \overline{a_{13}} c_Q \overline{c_P} \alpha_1 z_1^{\beta_1} z_2^{\beta_2} z_3^{\beta_3} \overline{z_1}^{\alpha_1-1} \overline{z_2}^{\alpha_2} \overline{z_3}^{\alpha_3+1} \\ & + a_{12} c_Q \overline{c_P} \beta_1 z_1^{\beta_1-1} z_2^{\beta_2+1} z_3^{\beta_3} \overline{z_1}^{\alpha_1} \overline{z_2}^{\alpha_2} \overline{z_3}^{\alpha_3} && + a_{13} c_Q \overline{c_P} \beta_1 z_1^{\beta_1-1} z_2^{\beta_2} z_3^{\beta_3+1} \overline{z_1}^{\alpha_1} \overline{z_2}^{\alpha_2} \overline{z_3}^{\alpha_3} \\ & + \overline{a_{12}} c_P \overline{c_Q} \beta_1 z_1^{\alpha_1} z_2^{\alpha_2} z_3^{\alpha_3} \overline{z_1}^{\beta_1-1} \overline{z_2}^{\beta_2+1} \overline{z_3}^{\beta_3} && + \overline{a_{13}} c_P \overline{c_Q} \beta_1 z_1^{\alpha_1} z_2^{\alpha_2} z_3^{\alpha_3} \overline{z_1}^{\beta_1-1} \overline{z_2}^{\beta_2} \overline{z_3}^{\beta_3+1} \\ & + a_{12} c_R \overline{c_R} \gamma_1 z_1^{\gamma_1-1} z_2^{\gamma_2+1} z_3^{\gamma_3} \overline{z_1}^{\gamma_1} \overline{z_2}^{\gamma_2} \overline{z_3}^{\gamma_3} && + a_{13} c_R \overline{c_R} \gamma_1 z_1^{\gamma_1-1} z_2^{\gamma_2} z_3^{\gamma_3+1} \overline{z_1}^{\gamma_1} \overline{z_2}^{\gamma_2} \overline{z_3}^{\gamma_3} \\ & + \overline{a_{12}} c_R \overline{c_R} \gamma_1 z_1^{\gamma_1} z_2^{\gamma_2} z_3^{\gamma_3} \overline{z_1}^{\gamma_1-1} \overline{z_2}^{\gamma_2+1} \overline{z_3}^{\gamma_3} && + \overline{a_{13}} c_R \overline{c_R} \gamma_1 z_1^{\gamma_1} z_2^{\gamma_2} z_3^{\gamma_3} \overline{z_1}^{\gamma_1-1} \overline{z_2}^{\gamma_2} \overline{z_3}^{\gamma_3+1}. \end{aligned} \tag{22}$$

Without loss of generality, it is enough to study the term $a_{12} c_P \overline{c_Q} \alpha_1 z_1^{\alpha_1-1} z_2^{\alpha_2+1} z_3^{\alpha_3} \overline{z_1}^{\beta_1} \overline{z_2}^{\beta_2} \overline{z_3}^{\beta_3}$. Using Proposition 4.2, we conclude that the only possibility for this term to match with a term containing a_{13} is

$$\begin{aligned} \gamma_1 &= \alpha_1 - 1 = \beta_1 + 1 \\ \gamma_2 &= \alpha_2 + 1 = \quad \beta_2 \\ \gamma_3 &= \quad \alpha_3 \quad = \beta_3 - 1, \end{aligned} \tag{23}$$

with $\gamma_1 = 0$. This is a contradiction.

Similarly, one can show that terms containing a_{12} cannot interfere with terms containing a_{21} or a_{32}. We leave the details to the reader.

Therefore, the terms containing a_{12} can only interfere with terms containing a_{23} and a_{31}.

Suppose that $a_{23} \neq 0$ and that a term containing a_{12} matches with a term a_{23}. We have to study the terms containing a_{12} and a_{23} given by

$$\begin{aligned}
&a_{12}c_P\overline{c_Q}\alpha_1 z_1^{\alpha_1-1}z_2^{\alpha_2+1}z_3^{\alpha_3}\overline{z_1}^{\beta_1}\overline{z_2}^{\beta_2}\overline{z_3}^{\beta_3} \quad +a_{23}c_P\overline{c_Q}\alpha_2 z_1^{\alpha_1}z_2^{\alpha_2-1}z_3^{\alpha_3+1}\overline{z_1}^{\beta_1}\overline{z_2}^{\beta_2}\overline{z_3}^{\beta_3} \\
&+\overline{a_{12}}c_Q\overline{c_P}\alpha_1 z_1^{\beta_1}z_2^{\beta_2}z_3^{\beta_3}\overline{z_1}^{\alpha_1-1}\overline{z_2}^{\alpha_2+1}\overline{z_3}^{\alpha_3} +\overline{a_{23}}c_Q\overline{c_P}\alpha_2 z_1^{\beta_1}z_2^{\beta_2}z_3^{\beta_3}\overline{z_1}^{\alpha_1}\overline{z_2}^{\alpha_2-1}\overline{z_3}^{\alpha_3+1} \\[2ex]
&+a_{12}c_Q\overline{c_P}\beta_1 z_1^{\beta_1-1}z_2^{\beta_2+1}z_3^{\beta_3}\overline{z_1}^{\alpha_1}\overline{z_2}^{\alpha_2}\overline{z_3}^{\alpha_3} \quad +a_{23}c_Q\overline{c_P}\beta_2 z_1^{\beta_1}z_2^{\beta_2-1}z_3^{\beta_3+1}\overline{z_1}^{\alpha_1}\overline{z_2}^{\alpha_2}\overline{z_3}^{\alpha_3} \\
&+\overline{a_{12}}c_P\overline{c_Q}\beta_1 z_1^{\alpha_1}z_2^{\alpha_2}z_3^{\alpha_3}\overline{z_1}^{\beta_1-1}\overline{z_2}^{\beta_2+1}\overline{z_3}^{\beta_3} +\overline{a_{23}}c_P\overline{c_Q}\beta_2 z_1^{\alpha_1}z_2^{\alpha_2}z_3^{\alpha_3}\overline{z_1}^{\beta_1}\overline{z_2}^{\beta_2-1}\overline{z_3}^{\beta_3+1} \\[2ex]
&+a_{12}c_R\overline{c_R}\gamma_1 z_1^{\gamma_1-1}z_2^{\gamma_2+1}z_3^{\gamma_3}\overline{z_1}^{\gamma_1}\overline{z_2}^{\gamma_2}\overline{z_3}^{\gamma_3} \quad +a_{23}c_R\overline{c_R}\gamma_2 z_1^{\gamma_1}z_2^{\gamma_2-1}z_3^{\gamma_3+1}\overline{z_1}^{\gamma_1}\overline{z_2}^{\gamma_2}\overline{z_3}^{\gamma_3} \\
&+\overline{a_{12}}c_R\overline{c_R}\gamma_1 z_1^{\gamma_1}z_2^{\gamma_2}z_3^{\gamma_3}\overline{z_1}^{\gamma_1-1}\overline{z_2}^{\gamma_2+1}\overline{z_3}^{\gamma_3} +\overline{a_{23}}c_R\overline{c_R}\gamma_2 z_1^{\gamma_1}z_2^{\gamma_2}z_3^{\gamma_3}\overline{z_1}^{\gamma_1}\overline{z_2}^{\gamma_2-1}\overline{z_3}^{\gamma_3+1}.
\end{aligned} \tag{24}$$

Using Proposition 4.2 again, one can check that this implies

$$\begin{aligned} \gamma_1 &= \alpha_1 - 1 = \quad \beta_1 \\ \gamma_2 &= \alpha_2 + 1 = \beta_2 + 1 \\ \gamma_3 &= \quad \alpha_3 \quad = \beta_3 - 1, \end{aligned} \tag{25}$$

with $\gamma_1 = 0$ and $\alpha_2 = 0$. Therefore, we obtain

$$\begin{aligned} \gamma_1 &= 0 \quad \alpha_1 = 1 \quad \beta_1 = 0 \\ \gamma_2 &= 1 \quad \alpha_2 = 0 \quad \beta_2 = 0 \\ \gamma_3 &= k \quad \alpha_3 = k \quad \beta_3 = k + 1, \end{aligned} \tag{26}$$

and M_0 is given by

$$\mathsf{Im}\, w = c_P\overline{c_Q}z_1 z_3^k\overline{z_3}^{k+1} + c_Q\overline{c_P}z_3^{k+1}\overline{z_1 z_3}^k + c_R\overline{c_R}z_2 z_3^k\overline{z_2 z_3}^k, \tag{27}$$

with

$$N_1 = a_{12}z_2\frac{\partial}{\partial z_1} + a_{23}z_3\frac{\partial}{\partial z_2} \in \mathfrak{g}_0.$$

The case $a_{31} \neq 0$ is similar, and its proof is left to the reader. □

We now return to the proof of Theorem 4.3.

We notice that N_1 given by Lemma 4.4 is nilpotent in all cases.

Let N be given by (17) with $N \neq 0$. After a possible permutation of the variables, we may assume that $a_{12} \neq 0$. If $N_2 := N - N_1 \neq 0$, we apply again Lemma 4.4 to N_2. We obtain that N_2 is nilpotent, with M_0 given, up to permutations of the variables, by

$$\mathsf{Im}\, w = 2\mathsf{Re}\left(c_P\overline{c_Q}z_1z_2^k\overline{z_2}^{k+1}\right) + c_R\overline{c_R}z_2^kz_3\overline{z_2}^k\overline{z_3}, \tag{28}$$

$$\mathsf{Im}\, w = 2\mathsf{Re}\left(c_P\overline{c_Q}z_1z_3^k\overline{z_2}\overline{z_3}^k\right) + c_R\overline{c_R}z_3^{k+1}\overline{z_3}^{k+1}, \tag{29}$$

or

$$\mathsf{Im}\, w = 2\mathsf{Re}\left(c_P\overline{c_Q}z_1z_2^{\alpha_2}z_3^{\alpha_3}\overline{z_2}^{\alpha_2+1}\overline{z_3}^{\alpha_3}\right) + c_R\overline{c_R}z_2^{\gamma_2}z_3^{\gamma_3}\overline{z_2}^{\gamma_2}\overline{z_3}^{\gamma_3}. \tag{30}$$

□

Remark 4.5 It is not hard to see that $D \in \mathfrak{g}_0$ given by (16) depends on three parameters. Indeed, one can decompose D into $D = D_1 + D_2$, where $D_1 \in \mathfrak{g}_0$ has a real diagonal and $D_2 \in \mathfrak{g}_0$ has a purely imaginary diagonal. Using Proposition 4.2, one obtains that D_1 depends on one real parameter and D_2 on two real parameters.

Remark 4.6 Inspecting the proof of Lemma 4.4, we see that N given by (17) depends on three real parameters if M_0 is given by (28), two real parameters if M_0 is given by (29), and one real parameter if M_0 is given by (30). Otherwise, $N = 0$.

We illustrate the above results with the following examples:

Example 4.7 Let M_0 be given by

$$\mathsf{Im}\, w = z_1z_3\overline{z_3}^2 + z_3^2\overline{z_1}\,\overline{z_3} + z_2z_3\overline{z_2}\,\overline{z_3}. \tag{31}$$

Then N is given by

$$N = \begin{pmatrix} 0 & a+ib & ic \\ 0 & 0 & -a+ib \\ 0 & 0 & 0 \end{pmatrix}, \tag{32}$$

where $a, b, c \in \mathbb{R}$. Here, N is nilpotent.

Example 4.8 Let M_0 be given by

$$\operatorname{Im} w = iz_1z_2\overline{z_1}\,\overline{z_3} - iz_1z_3\overline{z_1}\,\overline{z_2} + z_1^2\overline{z_1}^2. \tag{33}$$

Then N is given by

$$N = \begin{pmatrix} 0 & 0 & 0 \\ 0 & 0 & a \\ 0 & b & 0 \end{pmatrix} = \begin{pmatrix} 0 & 0 & 0 \\ 0 & 0 & a \\ 0 & 0 & 0 \end{pmatrix} + \begin{pmatrix} 0 & 0 & 0 \\ 0 & 0 & 0 \\ 0 & b & 0 \end{pmatrix}, \tag{34}$$

with $a, b \in \mathbb{R}$. Here, N is not nilpotent but is the sum of two nilpotent rotations.

5 The Structure of $\mathfrak{g}_c$ for the PQR Problem

In this section, we study the grading component $\mathfrak{g}_c$, that is, the set of those rigid vector fields of weight strictly bigger than 0 (and strictly less than 1 [13]).

We have the following proposition:

Proposition 5.1 *Let M_0 be given by* (13) *with P, Q, and R not necessarily monomials and holomorphically nondegenerate. Then*

$$0 \leq \dim \mathfrak{g}_c \leq 3. \tag{35}$$

Proof Let $X \in \mathfrak{g}_c$. We have

$$\begin{aligned} 0 = {} & X(P)\overline{Q} + X(Q)\overline{P} + X(R)\overline{R} \\ & + Q\overline{X(P)} + P\overline{X(Q)} + R\overline{X(R)}. \end{aligned}$$

The case $\deg P = \deg Q$ yields to

$$X(P)\overline{Q} + X(Q)\overline{P} + X(R)\overline{R} = 0. \tag{36}$$

Using Lemma 4.1, we get a contradiction.

The case $\deg P < \deg Q$ yields to

$$X(Q) = 0, \;\; X(P)\overline{Q} + X(R)\overline{R} + Q\overline{X(P)} + R\overline{X(R)} = 0, \tag{37}$$

which implies

$$X(P) = 0, \tag{38}$$

or

$$X(P)\overline{Q} + R\overline{X(R)} = 0, \tag{39}$$

or

$$X(P)\overline{Q} + Q\overline{X(P)} = 0 \quad \text{et} \quad X(R) = 0.$$

The first case implies $X(R) = 0$, which is a contradiction. The second case implies

$$\begin{aligned} X(P) &= \alpha R, \\ X(R) &= \beta Q, \quad \text{and} \\ X(Q) &= 0, \end{aligned}$$

with $\beta = -\overline{\alpha}$.

The third case implies

$$\begin{aligned} X(P) &= \alpha Q, \\ X(R) &= 0, \quad \text{and} \\ X(Q) &= 0, \end{aligned}$$

with $\alpha = ic$, $c \in \mathbb{R}$. Using Cramer's rule, we obtain the desired conclusion.

□

The following example shows that $\dim \mathfrak{g}_{\mathfrak{c}} = 3$ is reached.

Example 5.2 Let M_0 be given by

$$\operatorname{Im} w = z_1 z_2 z_3^2 \overline{z_1}^2 \overline{z_2 z_3}^3 + {z_1}^2 z_2 {z_3}^3 \overline{z_1 z_2 z_3}^2 + z_1 z_2 z_3^3 \overline{z_1 z_2 z_3}^3. \tag{40}$$

M_0 is holomorphically nondegenerate since the Jacobian

$$J(P, Q, R) = \begin{vmatrix} P_{z_1} & P_{z_2} & P_{z_3} \\ Q_{z_1} & Q_{z_2} & Q_{z_3} \\ R_{z_1} & R_{z_2} & R_{z_3} \end{vmatrix} = \begin{vmatrix} z_2 z_3^2 & z_1 z_3^2 & 2z_1 z_2 z_3 \\ 2z_1 z_2 {z_3}^3 & {z_1}^2 {z_3}^3 & 3{z_1}^2 z_2 {z_3}^2 \\ z_2 z_3^3 & z_1 z_3^3 & 3z_1 z_2 z_3^2 \end{vmatrix} = -z_1^3 z_2^2 z_3^7 \tag{41}$$

is not vanishing.

Let X be given by $X = \sum_{j=1}^{3} f_j(z) \frac{\partial}{\partial z_j}$. Inspecting the proof of Theorem 5.1, we see that there are two cases to study. Either

$$\begin{aligned} X(P) &= \alpha R, \\ X(R) &= \beta Q, \quad \text{and} \\ X(Q) &= 0 \end{aligned}$$

or

$$
\begin{aligned}
X(P) &= \alpha Q, \\
X(R) &= 0, \quad \text{and} \\
X(Q) &= 0.
\end{aligned}
$$

The first case leads to the following system:

$$
\begin{pmatrix} P_{z_1} & P_{z_2} & P_{z_3} \\ R_{z_1} & R_{z_2} & R_{z_3} \\ Q_{z_1} & Q_{z_2} & Q_{z_3} \end{pmatrix} \begin{pmatrix} f_1 \\ f_2 \\ f_3 \end{pmatrix} = \begin{pmatrix} \alpha R \\ \beta Q \\ 0 \end{pmatrix}. \tag{42}
$$

Using Cramer's rule, we obtain

$$
f_1(z) = \frac{\begin{vmatrix} \alpha z_1 z_2 z_3^3 & z_1 z_3^2 & 2 z_1 z_2 z_3 \\ \beta {z_1}^2 z_2 {z_3}^3 & z_1 z_3^3 & 3 z_1 z_2 z_3^2 \\ 0 & {z_1}^2 {z_3}^3 & 3 {z_1}^2 z_2 {z_3}^2 \end{vmatrix}}{z_1^3 z_2^2 z_3^7} = -\beta {z_1}^2,
$$

$$
f_2(z) = \frac{\begin{vmatrix} z_2 z_3^2 & \alpha z_1 z_2 z_3^3 & 2 z_1 z_2 z_3 \\ z_2 z_3^3 & \beta {z_1}^2 z_2 {z_3}^3 & 3 z_1 z_2 z_3^2 \\ 2 z_1 z_2 {z_3}^3 & 0 & 3 {z_1}^2 z_2 {z_3}^2 \end{vmatrix}}{z_1^3 z_2^2 z_3^7} = 3\alpha z_2 z_3 - \beta z_1 z_2,
$$

$$
f_3(z) = \frac{\begin{vmatrix} z_2 z_3^2 & z_1 z_3^2 & \alpha z_1 z_2 z_3^3 \\ z_2 z_3^3 & z_1 z_3^3 & \beta {z_1}^2 z_2 {z_3}^3 \\ 2 z_1 z_2 {z_3}^3 & {z_1}^2 {z_3}^3 & 0 \end{vmatrix}}{z_1^3 z_2^2 z_3^7} = -\alpha z_3^2 + \beta z_1 z_3,
$$

where $\alpha = a + ib$ and $\beta = -\overline{\alpha} = -a + ib$.

The second case leads to the following solutions:

$$
f_1(z) = 0,
$$

$$
f_2(z) = 3\alpha z_1 z_2 z_3, \quad \text{and}
$$

$$
f_3(z) = -\alpha z_1 z_3^2,
$$

$\alpha = ic,\ c \in \mathbb{R}$. Hence $\dim_{\mathbb{R}} \mathfrak{g}_{\mathfrak{c}} = 3$.

Remark 5.3 It is interesting to notice that $\dim_{\mathbb{R}} \mathfrak{g}_c = 0$ if $n = 1$, $\dim_{\mathbb{R}} \mathfrak{g}_c \leq 1$ if $n = 2$, while here we have an example with $\dim_{\mathbb{R}} \mathfrak{g}_c = 3$.

6 The Structure of $\mathfrak{g}_{-\frac{1}{d}}$, $\mathfrak{g}_{1-\frac{1}{d}}$ and $\mathfrak{g}_1$ for the Monomial PQR Problem

We now study the remaining grading components of weights $-\dfrac{1}{d}$, $1-\dfrac{1}{d}$, and 1. Indeed, by Theorem 1.1 of [13], there is one more left, the grading component of weight -1, $\mathfrak{g}_{-1}$, whose real dimension is 1 and generated by ∂_w.

We start with the following lemma:

Lemma 6.1 *Let M_0 be given by* (13) *with P, Q, and R monomials and holomorphically nondegenerate. If $\mathfrak{g}_{-\frac{1}{d}} \neq \{0\}$, then M_0 is, after a possible permutation of the variables, given by*

$$M_0 = \{(z, w) \in \mathbb{C}^3 \times \mathbb{C} \mid \operatorname{Im} w = z_1 \bar{Q} + Q\bar{z_1} + R\bar{R}\}, \tag{43}$$

with Q and R independent of z_1.

Proof Let $X \in \mathfrak{g}_{-\frac{1}{d}}$ be given by

$$X = a\frac{\partial}{\partial z_1} + b\frac{\partial}{\partial z_2} + c\frac{\partial}{\partial z_3} + g(z)\frac{\partial}{\partial w}. \tag{44}$$

We first assume that $g \neq 0$. It implies that P, Q, or R is linear. Without loss of generality, we may assume that P is nonzero linear since R cannot be linear (M_0 is assumed to be Levi degenerate). By permuting the variables, we may assume that $P = z_1$, and hence $X = a\dfrac{\partial}{\partial z_1} + g(z)\dfrac{\partial}{\partial w}$. This implies

$$\operatorname{Re}(X(Q)\bar{z_1} + X(R)\bar{R}) = 0, \tag{45}$$

which is possible if $X(Q) = 0$, and $X(R) = 0$.

We now assume that $g = 0$. By a holomorphic change of coordinates, we may assume that $X = \frac{\partial}{\partial z_1}$. We obtain

$$\operatorname{Re}(X(P)\bar{Q} + X(Q)\bar{P} + X(R)\bar{R}) = 0. \tag{46}$$

Considering the biggest power in z_1 in P, Q, R, it is not hard to conclude that this case cannot happen. □

We have the following proposition:

Proposition 6.2 *Let M_0 be given by* (13) *with P, Q, and R monomials and holomorphically nondegenerate. Then*

$$\dim \mathfrak{g}_{-\frac{1}{d}} > 0 \quad \Leftrightarrow \quad \dim \mathfrak{g}_{1-\frac{1}{d}} > 0.$$

Proof Using Theorem 1.2 in [14], we only have to show that there exists a holomorphic vector field Y satisfying

$$Y(R) = QR, \quad Y(Q) = Q^2. \tag{47}$$

Since M_0 is holomorphically nondegenerate and monomial, we obtain Y explicitly, using Cramer's rule. □

Lemma 6.3 *Let M_0 be given by* (13) *with P, Q, and R monomials and holomorphically nondegenerate. Then*

$$\dim_{\mathbb{R}} \mathfrak{g}_1 = 1. \tag{48}$$

Proof By Theorem 1.1 in [13], we obtain that $\dim_{\mathbb{R}} \mathfrak{g}_1 \neq 0$, and hence $\dim_{\mathbb{R}} \mathfrak{g}_1 = 1$ also by [13], if and only if there exists a holomorphic linear vector field Y with

$$Y(P\bar{Q} + Q\bar{P} + R\bar{R}) = P\bar{Q} + Q\bar{P} + R\bar{R}.$$

Writing explicitly the above equation, we obtain that Y is diagonal. Using Lemma 4.1 and Proposition 4.2, we solve a Cramer system and get Y explicitly. □

7 Statements and Proofs of the Main Results

Theorem 7.1 *Let M_0 be given by* (13) *with P, Q, and R monomials and holomorphically nondegenerate. Assume that* $\dim \mathfrak{g}_c > 0$. *Then*

$$7 \leq \dim \mathfrak{g} \leq 13. \tag{49}$$

Furthermore, there exists a hypersurface M_0 for which $\dim \mathfrak{g} = 13$.

Proof Using the characterization of the existence of a nilpotent rotation given by (28), (29), and (30) and inspecting the proof of Theorem 5.1, we conclude that under the assumption $\dim \mathfrak{g}_c > 0$, there are no nilpotent rotations. Using Lemma 6.1, Proposition 6.2, Lemma 6.3, and Theorem 5.1, we conclude that $7 \leq \dim \mathfrak{g} \leq 13$.

It is not hard to see that the hypersurface given by

$$\operatorname{Im} w = z_1 \overline{z_2}^{l} + z_2^{l} \overline{z_1} + z_2^{\frac{\ell-1}{2}} z_3 \overline{z_2}^{\frac{\ell-1}{2}} \overline{z_3}$$

admits $\dim \mathfrak{g}_c = 3$.

Indeed, the following vector fields are in $\mathfrak{g}_c$:

(1) $X_1 = aiz_2^\ell \frac{\partial}{\partial z_1}, a \in \mathbb{R}$.

(2) $X_2 = -(a+ib)z_2^{\frac{\ell-1}{2}} z_3 \frac{\partial}{\partial z_1} + (a-ib)z_2^{\frac{\ell+1}{2}} \frac{\partial}{\partial z_3}, \ a, b \in \mathbb{R}$.

Hence, Using Lemma 6.1 and Proposition 6.2, we conclude that $\dim \mathfrak{g} = 13$. □

Theorem 7.2 *Let M_0 be given by* (13) *with P, Q, and R monomials and holomorphically nondegenerate. Assume that* $\dim \mathfrak{g}_c = 0$. *Then*

$$6 \le \dim \mathfrak{g} \le 9. \tag{50}$$

Furthermore, there exists a hypersurface M_0 for which $\dim \mathfrak{g} = 9$.

Proof If $\dim \mathfrak{g}_c = 0$, then by Proposition 6.2, $\dim \mathfrak{g}_{-\frac{1}{d}} = 0$. Therefore, we have $6 \le \dim \mathfrak{g} \le 9$. Using Remark 4.6, we conclude that $\dim \mathfrak{g} = 9$ for the hypersurface given by (28). □

References

1. V.K.Beloshapka, V.V. Ezhov, G. Schmalz, *Holomorphic classification of four-dimensional surfaces in $\mathbb{C}^3$*, Izv. Ross. Akad. Nauk Ser. Mat. **72** (2008), 3–18.
2. Bloom, T., Graham, I., *On "type" conditions for generic real submanifolds of C^n*, Invent. Math. **40** (1977), 217–243.
3. Beals, M., Fefferman, C., Graham R., *Strictly pseudoconvex domains in $\mathbb{C}^n$*, Bull. Amer. Math. Soc. (N.S.) **8** (1983), 125–322.
4. E.Cartan : *Sur la géométrie pseudo-conforme des hypersurfaces de deux variables complexes, II*, Ann. Scoula Norm. Sup. Pisa **1** (1932), p. 333–354.
5. Chern, S. S. Moser, J., *Real hypersurfaces in complex manifolds*, Acta Math. **133** (1974), 219–271.
6. Isaev, A. V., Kruglikov, B., *On the symmetry algebras of 5-dimensional CR-manifolds*, Adv. Math. **322** (2017), 530–564.
7. Kim, S. Y., Kolář, M., *Infinitesimal symmetries of weakly pseudoconvex manifolds*, Math. Z., **300** (2022), no. 3, 2451–2466.
8. Kohn, J. J., *Boundary behaviour of $\bar\partial$ on weakly pseudoconvex manifolds of dimension two*, J. Differential Geom. **6** (1972), 523–542.
9. Kolář, M., *Normal forms for hypersurfaces of finite type in $\mathbb{C}^2$*, Math. Res. Lett., **12** (2005), 523–542.
10. Kolář, M., Meylan, F., Infinitesimal CR automorphisms for a class of polynomial models Arch. Math. (Brno) **53** no. 5 (2017): 255–265
11. Kolář, M., Meylan, F., *Chern-Moser operators and weighted jet determination problems*, Geometric analysis of several complex variables and related topics, 75–88, Contemp. Math. 550, 2011.
12. Kolář, M., Meylan, F. *Infinitesimal CR automorphisms of hypersurfaces of finite type in C2.*, Arch. Math. (Brno) **47** no. 5 (2011): 367–375
13. M. Kolář, F. Meylan, D. Zaitsev, *Chern-Moser operators and polynomial models in CR geometry*, Adv. Math. **263** (2014), 321–356.

14. Kolář, M., Meylan, F., *Nonlinearizable CR Automorphisms for Polynomial Models in* $\mathbb{C}^N$., J. Geom. Anal. **33** no. 3(2023): 106.
15. Kruzhilin, N. G., Loboda, A. V., *Linearization of local automorphisms of pseudoconvex surfaces*, Dokl. Akad. Nauk SSSR, **271** (1983), 280–282.
16. Kruglikov B., *Blow-ups and infinitesimal automorphisms of CR-manifolds* Mathematische Zeitschrift **296** (2020), 1701–1724.
17. Poincaré, H., *Les fonctions analytiques de deux variables et la représentation conforme,* Rend. Circ. Mat. Palermo **23** (1907), 185–220.
18. Stanton, N., *Infinitesimal CR automorphisms of real hypersurfaces*, Amer. J. Math. **118** (1996), 209–233.
19. Tanaka N. I., *On the pseudo-conformal geometry of hypersurfaces of the space of n complex variables*. J. Math. Soc. Japan, **14** (1962), pp. 397–429.
20. Treves, F., *A treasure trove of geometry and analysis: the hyperquadric*, Notices Amer. Math. Soc. **47** (2000),
21. Vitushkin, A.G., *Real analytic hypersurfaces in complex manifolds*, Russ. Math. Surv. **40** (1985), 1–35.
22. Webster, S.M., *On the Moser normal form at a non-umbilic point*, Math. Ann. **233** (1978), 97–102.
23. Yang, P., *Automorphism of tube domains*, Amer. J. Math. **104** (1982), 1005–1024.

On Some Singular Integrals Linked to Solvability and Signal Theory

Nicolas Lerner

This article is dedicated to Professor Jorge Hounie in celebration of his 75th birthday.

1 Miscellaneous Historical Elements

1.1 The Local Solvability Problem, from 1955 to 1981

In his thesis [16], published in 1955 by *Acta Mathematica*, L. Hörmander provided many new results about existence and regularity of solutions of linear differential operators with C^∞ coefficients. In particular, he proved some local solvability results for real principal type operators, that is, differential operators whose principal symbol is real-valued, with a nonvanishing differential (with respect to the momentum variable). It turned out later on that these results could be extended readily to real principal type pseudo-differential operators. So it came as a great surprise when H. Lewy proved in 1957 in [30] that the three-dimensional complex vector field,

$$\mathscr{L}_0 = \partial_{x_1} + i\partial_{x_2} + i(x_1 + ix_2)\partial_{x_3}, \tag{1}$$

was not locally solvable[1] at any point of $\mathbb{R}^3$. Of course, it was not contradicting the aforementioned results by L. Hörmander, since $\mathscr{L}_0$ is a complex-valued vector field,

[1] Let Ω be an open subset of $\mathbb{R}^n$, let P be a properly supported pseudo-differential operator on Ω, and let $x_0 \in \Omega$: We shall say that P is *locally solvable at* x_0 if there exists an open neighborhood U of x_0 such that for every $f \in C^\infty(\Omega)$, there exists $v \in \mathscr{D}'(\Omega)$ such that $Pv = f$ in U. The operator P will be said *locally solvable on* Ω if it is locally solvable at every point of Ω.

N. Lerner (✉)
Institut de Mathématiques de Jussieu, Sorbonne Université (formerly Paris VI), Paris, France
e-mail: nicolas.lerner@imj-prg.fr; nicolas.lerner@sorbonne-universite.fr

S. Berhanu et al. (eds.), *Geometric Analysis of PDEs and Several Complex Variables*, Latin American Mathematics Series – UFSCar subseries,
https://doi.org/10.1007/978-3-031-69702-9_11

but nevertheless the simplicity of that vector field apparently made it ineligible for counterexample matters. Anyhow H. Lewy's article [30] was the starting point of a long quest to understand local solvability properties of principal type operators with complex-valued symbols. A first step in the geometric understanding of local solvability was accomplished by L. Hörmander in 1960 with the paper [17] (a pseudo-differential extension was given in 1966 in [19]) in which he proved that if the principal symbol p of an operator P is such that, at some point (x, ξ) in the cotangent bundle, we have

$$p(x, \xi) = 0, \quad \{\operatorname{Re} p, \operatorname{Im} p\}(x, \xi) > 0, \tag{2}$$

then the operator P is not locally solvable in any neighborhood[2] of x. In 1970, L. Nirenberg and F. Treves [35, 36, 37] after a study of complex vector fields in [34] (see also S. Mizohata's paper [32][3]) formulated a conjecture, asserting the equivalence of local solvability of a principal type operator with a geometric condition on its principal symbol, the so-called *Condition* (Ψ), which reads

$$\operatorname{Im} p \texttt{ does not change sign from} - \texttt{to} + \texttt{along the forward flow of } H_{\operatorname{Re} p}, \tag{3}$$

where $H_{\operatorname{Re} p}$ is the Hamiltonian vector field of $\operatorname{Re} p$. The necessity of Condition (Ψ) for local solvability (of principal type operators) was proven in 1981 by L. Hörmander in [21], following closely some earlier two-dimensional work [33] by R.D. Moyer. If P is a differential operator, it is easy to see that Condition (Ψ) is equivalent to Condition (P), which requires that

$$\operatorname{Im} p \texttt{ does not change sign along the flow of } H_{\operatorname{Re} p}. \tag{4}$$

In their aforementioned articles, L. Nirenberg and F. Treves provided massive evidence that condition (P) should be equivalent to local solvability of principal type differential operators, but in their proof of sufficiency of that condition, they used an analyticity assumption for the principal symbol; that strong restriction was removed by R. Beals and C.L. Fefferman in their 1973 articles [2, 1], proving that in the C^∞ category, Condition (P) is sufficient for local solvability of principal

---

[2] For differential operators, it is enough to assume $p(x, \xi) = 0$, $\{\operatorname{Re} p, \operatorname{Im} p\}(x, \xi) \neq 0$, since the principal symbol of a differential operator is a polynomial in the fibers of the cotangent bundle and the Poisson bracket $\{\operatorname{Re} p, \operatorname{Im} p\}$ is of odd degree (if ξ is not suitable, $-\xi$ will do). Moreover, (2) holds true for any $x \in \mathbb{R}^3$ for the vector field (1), choosing $\xi_3 > 0, \quad \xi_1 = x_2\xi_3, \quad \xi_2 = -x_1\xi_3$.

[3] The first examples issued from the papers [32, 34] are illuminating: For $k \in \mathbb{N}$, the operators

$$D_{x_1} \pm i x_1^{2k} D_{x_2}, \quad D_{x_1} - i x_1^{2k+1} |D_{x_2}| \quad \text{are locally solvable,}$$

$$D_{x_1} \pm i x_1^{2k+1} D_{x_2}, \quad D_{x_1} + i x_1^{2k+1} |D_{x_2}| \quad \text{are not locally solvable.}$$

type differential operators. As a consequence, the papers [21, 2, 1] proved the *Nirenberg–Treves conjecture* for differential operators, namely the equivalence between Condition (Ψ) and local solvability for principal type differential operators.

1.2 Nonhomogeneous Classes of Pseudo-Differential Operators

The perspective developed by R. Beals and C.L. Fefferman in their 1973 articles on local solvability [2, 1] was a revolutionary improvement of the L^2 theory of pseudo-differential operators. Instead of reviewing extensively the content of [2], which provides a proof of the Nirenberg–Treves conjecture for local solvability of principal type operators satisfying condition (P), removing in particular the analyticity requirement of L. Nirenberg and F. Treves' paper [36], we raise an abstract-looking question: Let

$$\{q(t,x,\xi,\Lambda)\}_{\substack{t\in[-1,1]\\ \Lambda\ge 1}} \tag{5}$$

be a smooth family of real-valued semiclassical symbols of order 1, i.e.,

$$\sup_{\substack{(t,x,\xi,\Lambda)\in\\ [-1,1]\times\mathbb{R}^n\times\mathbb{R}^n\times[1,+\infty)}} |(\partial_x^\alpha\partial_\xi^\beta q)(t,x,\xi,\Lambda)|\Lambda^{-1+\frac{|\alpha|+|\beta|}{2}} < +\infty. \tag{6}$$

Remark 1.1 The usual formulation of semiclassical symbols looks a bit different with estimates of the following type: A *standard* semiclassical family of symbols of order $m\in\mathbb{R}$, $\{a(x,\xi,\Lambda)\}_{\Lambda\ge 1}$ is a mapping

$$[1,+\infty)\ni\Lambda\mapsto a(\cdot,\cdot,\Lambda)\in C^\infty(\mathbb{R}^n\times\mathbb{R}^n),$$

such that, for all $(\alpha,\beta)\in\mathbb{N}^n\times\mathbb{N}^n$,

$$\sup_{\substack{(x,\xi,\Lambda)\in\\ \mathbb{R}^n\times\mathbb{R}^n\times[1,+\infty)}} |(\partial_x^\alpha\partial_\xi^\beta a)(x,\xi,\Lambda)|\Lambda^{-m+|\beta|} < +\infty. \tag{7}$$

We note that we may consider

$$\tilde a(x,\xi,\Lambda)=a(\Lambda^{-1/2}x,\Lambda^{1/2}\xi,\Lambda) \tag{8}$$

and get back estimates of type (6) with

$$\sup_{\substack{(x,\xi,\Lambda)\in\\ \mathbb{R}^n\times\mathbb{R}^n\times[1,+\infty)}} |(\partial_x^\alpha\partial_\xi^\beta \tilde a)(x,\xi,\Lambda)|\Lambda^{-m+\frac{|\alpha|+|\beta|}{2}} < +\infty. \tag{9}$$

Moreover, it is trivial to check that the operators $\operatorname{Op}(a)$ and $\operatorname{Op}(\tilde{a})$ are unitarily equivalent, with

$$\operatorname{Op}(a(x,\xi,\Lambda)) = \mathcal{U}_\Lambda^* \operatorname{Op}(\tilde{a}(x,\xi,\Lambda))\mathcal{U}_\Lambda, \quad (\mathcal{U}_\Lambda v)(x) = \Lambda^{-n/4} v(\Lambda^{-1/2}x). \tag{10}$$

As a consequence, there is no significant difference between the two definitions (6) and (7) with $m = 1$, and it is more convenient for the upcoming developments to stick with our definition (6), providing a symmetric role for the variables x and ξ. Also, we have chosen to use a *large* parameter $\Lambda \in [1, +\infty)$ instead of using a small parameter $h \in (0, 1]$ (just set $h = 1/\Lambda$ to recover the standard picture).

If we look for natural estimates to be satisfied by the family (5), we may think to tailor these estimates to the very family under scope. For instance, looking for a weight $\lambda(x, \xi)$ so that we would have

$$|q(t,x,\xi,\Lambda)| \le \lambda(x,\xi), \quad \max\bigl\{|\partial_x q(t,x,\xi,\Lambda)|, |\partial_\xi q(t,x,\xi,\Lambda)|\bigr\} \le \lambda(x,\xi)^{\frac12}, \tag{11}$$

we get the natural formula

$$\lambda(x,\xi,\Lambda) = \sup_{t\in[-1,1]} \Bigl\{|q(t,x,\xi,\Lambda)| + |\partial_x q(t,x,\xi,\Lambda)|^2 + |\partial_\xi q(t,x,\xi,\Lambda)|^2\Bigr\} + 1, \tag{12}$$

where the last term "1" is here to avoid that the weight λ vanishes. Now, of course, the estimates (11) hold true, and we have from (12) that

$$\sup_{(x,\xi,\Lambda)\in\mathbb{R}^n\times\mathbb{R}^n\times[1,+\infty)} \lambda(x,\xi,\Lambda)\Lambda^{-1} < +\infty, \tag{13}$$

which implies that for $|\alpha| + |\beta| \ge 2$, we have

$$|(\partial_x^\alpha \partial_\xi^\beta q)(t,x,\xi,\Lambda)| \le C_{\alpha,\beta}\Lambda^{1-\frac{|\alpha|+|\beta|}{2}} \le \widetilde{C}_{\alpha,\beta}\lambda(x,\xi,\Lambda)^{1-\frac{|\alpha|+|\beta|}{2}}, \tag{14}$$

obtaining conditions somewhat analogous to (6) and producing a class of pseudo-differential operators to which $q(t,\cdot)$ belongs and is essentially the only symbol satisfying (14). It is a technical task to prove that if a symbol $r(x, \xi)$ satisfies

$$\sup_{(x,\xi)\in\mathbb{R}^n\times\mathbb{R}^n} |(\partial_x^\alpha \partial_\xi^\beta r)(x,\xi)|\lambda(x,\xi,\Lambda)^{\frac{|\alpha|+|\beta|}{2}} < +\infty, \tag{15}$$

then $\operatorname{Op}(r)$ is $L^2(\mathbb{R}^n)$-bounded. The main improvement that follows from the estimates (14) and Formula (12) is that considering points (x, ξ) such that $\lambda(x, \xi)$ is "equivalent" to $\sup_{|t|\le 1} |q(t, x, \xi)|$, we can find $t_0 \in [-1, +1]$ such that $\pm q(t_0, \cdot)$

is elliptic of order 1; if now $\lambda(x, \xi)$ is "equivalent" to $\sup_{|t|\le 1} |\nabla_{x,\xi} q(t, x, \xi)|^2$, we find a $t_0 \in [-1, +1]$ where $q(t_0, x, \xi)$ behaves essentially as $\lambda^{1/2}\xi_1$. If $\lambda(x, \xi)$ is "equivalent" to 1, then q is a symbol of order 0. Now let us assume that Condition (P) holds true for the operator $D_t + i\mathrm{Op}(q(t, x, \xi))$, i.e., that we have

$$\forall (t, s) \in [-1, 1]^2, \quad q(t, x, \xi) q(s, x, \xi) \ge 0. \tag{16}$$

- If $q(t_0, x, \xi)$ is elliptic positive (respectively, negative) at t_0 as above, it is then nonnegative (respectively, nonpositive) for all t, so that $q(t, y, \eta)$ is either nonnegative for all t or nonpositive for all t as soon as

$$|y - x|^2 + |\eta - \xi|^2 \le \rho\lambda(x, \xi), \quad \rho \text{ small enough.} \tag{17}$$

- If $\nabla_{x,\xi} q(t_0, x, \xi)$ is elliptic at t_0, then on a typical ball (17), we have essentially

$$q(t_0, y, \eta) = \lambda(x, \xi)^{1/2}\big(\zeta_n - \alpha(z, \zeta')\big), \quad z_j = y_j - x_j, \quad \zeta_j = \eta_j - \xi_j,$$

  and then (16) forces

$$q(t, y, \eta) = \lambda(x, \xi)^{1/2}\big(\zeta_n - \alpha(z, \zeta')\big) a_0(t, y, \eta), \quad a_0 \ge 0, \quad \text{of order } 0.$$

- If $\lambda(x, \xi)$ is "equivalent to 1," then the symbols $q(t, y, \eta)$ are of order 0 on the typical ball (17).

The result of that "trichotomy" is that the main local solvability result of the paper [2] is essentially reduced to the following lemma due to L. Nirenberg and F. Treves in [36] (see also Theorem 26.8.1 in [24]). As given above in (12), λ is a (positive) function of (x, ξ, Λ); however, we can also prove that this function is *slowly varying*, which means that there exist positive constants r, C such that

$$|y - x|^2 + |\eta - \xi|^2 \le r^2\lambda(x, \xi, \Lambda) \Longrightarrow C^{-1} \le \frac{\lambda(x, \xi, \Lambda)}{\lambda(y, \eta, \Lambda)} \le C. \tag{18}$$

Using a partition of unity related to a family of Euclidean balls with centers (x_k, ξ_k) and radius $r\lambda(x_k, \xi_k, \Lambda)^{1/2}$, we are reduced to the following cases.

Hypothesis 1.2 *Let* $\{q(t, x, \xi, \lambda)\}_{t\in[-1,1],\lambda\ge 1}$ *be a smooth family of first-order semiclassical symbols such that* (6) *holds true. We shall note*

$$\Gamma = [-1, +1] \times \mathbb{R}^n \times \mathbb{R}^n \times [1, +\infty). \tag{19}$$

We shall assume that one of the following assumptions is verified:

(1)$_+$ For all $(t, x, \xi, \lambda) \in \Gamma$, *we have* $q(t, x, \xi, \lambda) \ge 0$.
(1)$_-$ For all $(t, x, \xi, \lambda) \in \Gamma$, *we have* $q(t, x, \xi, \lambda) \le 0$.

(2) *For all* $(t, x, \xi, \lambda) \in \Gamma$, *we have* $q(t, x, \xi, \lambda) = a_0(t, x, \xi, \lambda)\lambda^{1/2}\xi_1$, *where* a_0 *is a nonnegative symbol of order 0, i.e., such that* λa_0 *satisfies* (6).
(3) *For all* $(t, x, \xi, \lambda) \in \Gamma$, *we have* $q(t, x, \xi, \lambda) = 0$ *for* $\lambda \geq 2$.

Lemma 1.3 (First Hilbertian Lemma) *Let* J *be an interval of the real line, and let* $\{R(t)\}_{t\in J}$ *be a family of bounded operators on* $\mathbb{H} = L^2(\mathbb{R}^n)$ *such that*

$$\rho_0 = \sup_{t\in J} \|R(t)\|_{\mathcal{B}(\mathbb{H})} < +\infty. \tag{20}$$

We assume that $\{Q(t)\}_{t\in J}$ *is a family of bounded self-adjoint operators such that for some* $\sigma_0 \in \mathbb{R}$, *we have*

$$Q(t) + \sigma_0 \geq 0. \tag{21}$$

Then for $\mu \geq \sigma_0 + \rho_0$, *we have for* $u \in C^1(J, \mathbb{H})$

$$\sup_{t\leq T,\ t\in J} \|e^{-\mu(T-t)}u(t)\|_{\mathbb{H}} \leq \|u(T)\|_{\mathbb{H}} + \int_{J\cap(-\infty,T)} \|e^{-\mu(T-s)}\mathcal{L}u(s)\|_{\mathbb{H}}ds, \tag{22}$$

$$\text{with} \quad \mathcal{L} = D_t + iQ(t) + R(t). \tag{23}$$

N.B *A good way to reformulate Inequality* (22) *is to say that the "nonnegativity assumption"* (21) *for* Q *implies well-posedness of the* backward *Cauchy problem, namely that* $\|u(t)\|_{\mathbb{H}}$ *is controlled by* $\|u(T)\|_{\mathbb{H}}$, *provided* $t \leq T$ *and* $\mathcal{L}u(s)$ *belongs to* $\mathbb{H}$ *for* $s \in [t, T]$. *This can be reformulated and a propagation-of-singularities result: Under the previous assumption on* $\mathcal{L}u(s)$, *if* u *is singular at* t *(say* $\|u(t)\|_{\mathbb{H}} = +\infty$*), then we have also that* u *is singular at* T *for* $t \leq T$*; the singularities are propagating* forward.

Proof We consider for $\mu \geq 0$,

$$\mathcal{L}_\mu = e^{\mu t}\big(D_t + iQ(t) + R(t)\big)e^{-\mu t} = D_t + i\big(Q(t) + \mu\big) + R(t), \tag{24}$$

and we have for $T_1 \leq T_2$ both in J,

$$2\,\mathrm{Re}\langle \mathcal{L}_\mu v, i\mathbf{1}_{[T_1,T_2]}(t)v\rangle_{\mathbb{H}} = \|v(T_1)\|_{\mathbb{H}}^2 - \|v(T_2)\|_{\mathbb{H}}^2 + 2\int_{T_1}^{T_2} \langle\big(Q(t)+\mu\big)v, v\rangle_{\mathbb{H}}dt + 2\,\mathrm{Re}\int_{T_1}^{T_2} \frac{1}{i}\langle R(t)v, v\rangle dt,$$

and this implies that if $\mu \geq \sigma_0$ (using (21)),

$$\|v(T_1)\|_{\mathbb{H}}^2 + 2(\mu - \sigma_0) \int_{T_1}^{T_2} \|v\|_{\mathbb{H}}^2 dt$$
$$\leq \|v(T_2)\|_{\mathbb{H}}^2 + 2 \int_{T_1}^{T_2} \|\mathcal{L}_\mu v\|_{\mathbb{H}} \|v\|_{\mathbb{H}} dt + 2\rho_0 \int_{T_1}^{T_2} \|v\|_{\mathbb{H}}^2 dt,$$

and thus for $\mu \geq \sigma_0 + \rho_0$,

$$\|v(T_1)\|_{\mathbb{H}}^2 + 2(\mu - \sigma_0 - \rho_0) \int_{T_1}^{T_2} \|v\|_{\mathbb{H}}^2 dt$$
$$\leq \|v(T_2)\|_{\mathbb{H}}^2 + 2 \int_{T_1}^{T_2} \|\mathcal{L}_\mu v\|_{\mathbb{H}} \|v\|_{\mathbb{H}} dt = f(T_1, T_2). \tag{25}$$

As a consequence, for $t = T_1 \leq T$ both in J, we find that

$$-\frac{\partial}{\partial t}\{f(t, T)^{1/2}\} = f(t, T)^{-1/2} \|\mathcal{L}_\mu v(t)\|_{\mathbb{H}} \|v(t)\|_{\mathbb{H}} \leq \|\mathcal{L}_\mu v(t)\|_{\mathbb{H}},$$

entailing that $-f(T, T)^{1/2} + f(t, T)^{1/2} \leq \int_t^T \|\mathcal{L}_\mu v(s)\|_{\mathbb{H}} ds$, and thus

$$\|v(t)\|_{\mathbb{H}} \leq \left\{\|v(t)\|_{\mathbb{H}}^2 + 2(\mu - \sigma_0 - \rho_0) \int_t^T \|v\|_{\mathbb{H}}^2 dt\right\}^{1/2}$$
$$\leq f(t, T)^{1/2} \leq \|v(T)\|_{\mathbb{H}} + \int_t^T \|\mathcal{L}_\mu v(s)\|_{\mathbb{H}} ds. \tag{26}$$

We apply then (26) to $v(t) = e^{\mu t} u(t)$, and we get for $u \in C^1(J, \mathbb{H}), t \leq T, \mu \geq \sigma_0 + \rho_0$,

$$\|e^{-\mu(T-t)} u(t)\|_{\mathbb{H}} \leq \|u(T)\|_{\mathbb{H}} + \int_t^T \|e^{-\mu(T-s)} \mathcal{L} u(s)\|_{\mathbb{H}} ds, \tag{27}$$

concluding the proof of the lemma.

Remark 1.4 We have proven in fact that for $\mu \geq 2(\sigma_0 + \mu_0)$, $T \in J$,

$$\sup_{t \in J \cap (-\infty, T]} \left(\|v(t)\|_{\mathbb{H}}^2 + \mu \int_t^T \|v\|_{\mathbb{H}}^2 dt\right)^{1/2}$$
$$\leq \|v(T)\|_{\mathbb{H}} + \int_{J \cap (-\infty, T]} \|\mathcal{L}_\mu v(s)\|_{\mathbb{H}} ds. \tag{28}$$

We may notice that no better estimate holds true for the operator D_t itself. Also Lemma 1.3 yields the inequalities

$$\sup_{t \geq T,\ t \in J} \|e^{-\mu(t-T)} u(t)\|_{\mathbb{H}} \leq \|u(T)\|_{\mathbb{H}} + \int_{J \cap [T, +\infty)} \|e^{-\mu(s-T)} \mathcal{L}u(s)\|_{\mathbb{H}} ds, \tag{29}$$

provided $\mu \geq \sigma_0 + \rho_0$, where σ_0 is such that

$$Q(t) - \sigma_0 \leq 0. \tag{30}$$

Finally, Gårding's inequality allows us to handle the cases $(1)_{\pm}$, (3) in Hypothesis 1.2.

N.B *It is certainly pleonastic (but hopefully not useless) to repeat our previous Nota Bene, mutatis mutandis. A good way to interpret Inequality* (29) *is to say that the "nonpositivity assumption"* (30) *for Q implies well-posedness of the* forward *Cauchy problem, namely that $\|u(t)\|_{\mathbb{H}}$ is controlled by $\|u(T)\|_{\mathbb{H}}$, provided $t \geq T$ and $\mathcal{L}u(s)$ belongs to $\mathbb{H}$ for $s \in [T, t]$. This can be reformulated and a propagation-of-singularities result: Under the previous assumption on $\mathcal{L}u(s)$, if u is singular at t (say $\|u(t)\|_{\mathbb{H}} = +\infty$), then we have also that u is singular at T for $t \geq T$; the singularities are propagating* backward.

The remaining case (2) is more difficult to handle and follows from a subtler Hilbertian Lemma due to L. Nirenberg and F. Treves (Lemma 3.1. in [37], see also L. Hörmander's Theorem 26.8.1 in [24]).

Lemma 1.5 (Second Hilbertian Lemma) *Let $\{R(t)\}_{t \in \mathbb{R}}$ be a family of bounded operators on $\mathbb{H} = L^2(\mathbb{R}^n)$ such that*

$$\rho_0 = \sup_{t \in \mathbb{R}} \|R(t)\|_{\mathcal{B}(\mathbb{H})} < +\infty. \tag{31}$$

Let $\{A_0(t)\}_{t \in \mathbb{R}}$ be a family of bounded nonnegative *self-adjoint operators, and let $Q(t)$ be defined by*

$$Q(t) = \frac{1}{2}\big(A_0(t)B + BA_0(t)\big), \quad B \textit{ self-adjoint}. \tag{32}$$

We assume also that

$$\beta_0 = \sup_{t \in \mathbb{R}} \|A_0(t)\|_{\mathcal{B}(\mathbb{H})} < +\infty, \tag{33}$$

$$\beta_1 = \sup_{t \in \mathbb{R}} \|[B, A_0(t)]\|_{\mathcal{B}(\mathbb{H})} < +\infty, \tag{34}$$

$$\beta_2 = \sup_{t \in \mathbb{R}} \|[B, [B, A_0(t)]]\|_{\mathcal{B}(\mathbb{H})} < +\infty. \tag{35}$$

Then there exists a constant α_0 depending only on $(\beta_j)_{0\le j\le 2}$ such that for any interval J of $\mathbb{R}$ such that

$$(\alpha_0 + 4\rho_0)|J| < 1, \tag{36}$$

we have for $u \in C^1_c(J, \mathbb{H})$,

$$\big(1 - (\alpha_0 + 4\rho_0)|J|\big) \sup_{t\in\mathbb{R}} \|u(t)\|_{\mathbb{H}} \le 4 \int_{\mathbb{R}} \|\mathcal{L}u(t)\|_{\mathbb{H}} dt, \tag{37}$$

with

$$\mathcal{L} = D_t + iQ(t) + R(t), \quad D_t = -i\partial_t. \tag{38}$$

N.B *We note that the assumptions* (33), (34), *and* (35) *are fulfilled whenever $A_0(t)$ is a family of pseudo-differential operators of order 0 and B is a fixed pseudo-differential operator of order 1.*

Proof The idea of the proof is rather simple and inspired by the previous Lemma 1.3 and its nonpositive version: We want to use as a multiplier the sign of the operator $Q(t)$ and more precisely the sign of B, which is actually independent of the time variable t. Let S be the sign of the self-adjoint bounded operator B, and let us consider the spectral projections $P_\pm$, $P_+ + P_- = \mathrm{Id}$,

$$S = P_+ - P_-, \quad P_\pm^* = P_\pm = P_\pm^2, \quad B_\pm = \pm P_\pm B = \pm B P_\pm, \tag{39}$$

$$B = B_+ - B_-, \quad B_\pm \ge 0, \quad P_+P_- = P_-P_+ = 0 = B_+B_- = B_-B_+. \tag{40}$$

We start with the calculation of

$$\begin{aligned} &P_+Q(t) + Q(t)P_+ \\ &\quad = \frac{1}{2}P_+A_0(t)B + \frac{1}{2}BA_0(t)P_+ + \frac{1}{2}B_+A_0(t) + \frac{1}{2}A_0(t)B_+ \\ &\quad = \frac{1}{2}P_+\big[A_0(t), B\big] + \frac{1}{2}\big[B, A_0(t)\big]P_+ + B_+A_0(t) + A_0(t)B_+. \end{aligned} \tag{41}$$

We note then that for J_1, J_2 self-adjoint bounded operators on $\mathbb{H}$ with $J_l = J_l^*$, we have that the commutator $[J_1, J_2]$ is skew-adjoint since

$$[J_1, J_2]^* = \big(J_1J_2 - J_2J_1\big)^* = J_2^*J_1^* - J_1^*J_2^* = J_2J_1 - J_1J_2 = -[J_1, J_2]. \tag{42}$$

Using the notation

$$\operatorname{Re} A = \frac{1}{2}(A + A^*), \qquad A \text{ bounded operator on } \mathbb{H}, \tag{43}$$

we obtain also that for J, K, bounded operators on $\mathbb{H}$ with $J = J^*$ and $K^* = -K$,

$$2\operatorname{Re}(JK) = JK + K^*J^* = [J, K]. \tag{44}$$

Resuming the calculation in (41), we get

$$\begin{aligned} 2\operatorname{Re}(P_+Q(t)) &= B_+^{1/2}B_+^{1/2}A_0(t) + A_0(t)B_+^{1/2}B_+^{1/2} + \frac{1}{2}\big[[B, A_0(t)], P_+\big] \\ &= B_+^{1/2}\big[B_+^{1/2}, A_0(t)\big] + B_+^{1/2}A_0(t)B_+{}^{1/2} + \big[A_0(t), B_+^{1/2}\big]B_+^{1/2} + B_+^{1/2}A_0(t)B_+^{1/2} \\ &\qquad + \frac{1}{2}\big[[B, A_0(t)], P_+\big], \quad \text{so that} \end{aligned}$$

$$2\operatorname{Re}(P_+Q(t)) = \big[B_+^{1/2}, \big[B_+^{1/2}, A_0(t)\big]\big] + 2B_+^{1/2}A_0(t)B_+^{1/2} + \frac{1}{2}\big[[B, A_0(t)], P_+\big]. \tag{45}$$

Similarly, we obtain

$$-2\operatorname{Re}(P_-Q(t)) = \big[B_-^{1/2}, \big[B_-^{1/2}, A_0(t)\big]\big] + 2B_-^{1/2}A_0(t)B_-^{1/2} - \frac{1}{2}\big[[B, A_0(t)], P_-\big]. \tag{46}$$

We use now Lemma 26.8.2 in [24] and the assumptions (33), (34), and (35) to obtain the existence[4] of a constant α_0 depending only on $(\beta_j)_{0\le j\le 2}$ such that

$$\|\big[B_+^{1/2}, \big[B_+^{1/2}, A_0(t)\big]\big]\|_{\mathcal{B}(\mathbb{H})} \le \alpha_0/6, \quad \|\big[B_-^{1/2}, \big[B_-^{1/2}, A_0(t)\big]\big]\|_{\mathcal{B}(\mathbb{H})} \le \alpha_0/6,$$
$$\|\big[[B, A_0(t)], P_\pm\big]\|_{\mathcal{B}(\mathbb{H})} \le \alpha_0/3,$$

[4] Following Lemma 26.8.2 and its proof in [24], we can take

$$\alpha_0 = 3\max\Big(2^6\pi^{-2}\beta_0^{1/4}\beta_1^{1/2}\beta_2^{1/4}, \beta_1\Big). \tag{47}$$

so that we get, adding (45) and (46),

$$2\,\mathrm{Re}\big(SQ(t)\big) + \frac{2}{3}\alpha_0 \geq 2B_+^{1/2}A_0(t)B_+^{1/2} + 2B_-^{1/2}A_0(t)B_-^{1/2}. \tag{48}$$

Let $t_0 \in \mathbb{R}$; we define our multiplier $M(t)$ as the self-adjoint bounded operator given by

$$M(t) = S + \sigma H(t - t_0), \quad \text{where } \sigma \text{ is a positive parameter to be chosen later,} \tag{49}$$

and H is the Heaviside function (172) in Appendix A. We note that we have

$$\|M(t)\|_{\mathcal{B}(\mathbb{H})} \leq 1 + \sigma. \tag{50}$$

Claim 1.6 For $\sigma \leq 1$, we have $2\,\mathrm{Re}\big(M(t)Q(t)\big) \geq -\alpha_0$.

Proof Indeed we have

$$\begin{aligned}
2\,\mathrm{Re}\big(M(t)Q(t)\big) &= 2\,\mathrm{Re}(SQ(t)) + 2\sigma H(t - t_0)Q(t)\\
\text{(thanks to (48))} &\geq 2B_+^{1/2}A_0(t)B_+^{1/2} + 2B_-^{1/2}A_0(t)B_-^{1/2} - \frac{2}{3}\alpha_0\\
&\qquad + \sigma H(t - t_0)(A_0B + BA_0). \qquad (51)
\end{aligned}$$

We note that

$$\begin{aligned}
2Q = A_0B + BA_0 &= 2\,\mathrm{Re}\big(A_0B_+^{1/2}B_+^{1/2} - A_0B_-^{1/2}B_-^{1/2}\big)\\
&= 2\,\mathrm{Re}\big([A_0, B_+^{1/2}]B_+^{1/2} - [A_0, B_-^{1/2}]B_-^{1/2}\big) + 2B_+^{1/2}A_0B_+^{1/2} - 2B_-^{1/2}A_0B_-^{1/2}\\
&= \big[[A_0, B_+^{1/2}], B_+^{1/2}\big] - \big[[A_0, B_-^{1/2}], B_-^{1/2}\big]\big) + 2B_+^{1/2}A_0B_+^{1/2} - 2B_-^{1/2}A_0B_-^{1/2}\\
&\geq -\frac{\alpha_0}{3} - 2B_-^{1/2}A_0B_-^{1/2},
\end{aligned}$$

so that (51) entails

$$\begin{aligned}
2\,\mathrm{Re}\big(M(t)Q(t)\big) &= 2\,\mathrm{Re}(SQ(t)) + 2\sigma H(t - t_0)Q(t)\\
&\geq 2B_+^{1/2}A_0(t)B_+^{1/2} + (2 - 2\sigma)B_-^{1/2}A_0(t)B_-^{1/2} - \frac{2}{3}\alpha_0 - \frac{1}{3}\alpha_0\sigma\\
&\geq -\alpha_0, \quad \text{provided } \sigma \leq 1, \qquad (52)
\end{aligned}$$

proving the claim.

On the other hand, we also have, thanks to (50), (31), and (49),

$$2\,\mathrm{Re}\big(MR(t)\big) \geq -2\rho_0(1 + \sigma), \quad \text{for } \sigma \geq 0, \quad \text{and} \quad \dot{M} = \sigma\delta_{t_0}. \tag{53}$$

As a consequence, we obtain that for $\sigma \in [0, 1]$, thanks to (52), (53),

$$2\,\mathrm{Re}\langle \mathcal{L}u, iMu\rangle_{L^2(\mathbb{R};\mathbb{H})} \ge \sigma|u(t_0)|^2 - \big(\alpha_0 + 2\rho_0(1+\sigma)\big)\int_{\mathbb{R}} |u(t)|^2_{\mathbb{H}}dt,$$

implying that for $\sigma = 1$,

$$4\|\mathcal{L}u\|_{L^1(\mathbb{R};\mathbb{H})}\|u\|_{L^\infty(\mathbb{R};\mathbb{H})} + (\alpha_0 + 4\rho_0)\|u\|^2_{L^2(\mathbb{R};\mathbb{H})} \ge \|u\|^2_{L^\infty(\mathbb{R};\mathbb{H})}, \tag{54}$$

implying

$$4\|\mathcal{L}u\|_{L^1(\mathbb{R};\mathbb{H})}\|u\|_{L^\infty(\mathbb{R};\mathbb{H})} + (\alpha_0 + 4\rho_0)|J|\|u\|^2_{L^\infty(\mathbb{R};\mathbb{H})} \ge \|u\|^2_{L^\infty(\mathbb{R};\mathbb{H})}, \tag{55}$$

concluding the proof of the lemma.

Remark 1.7 The second Hilbertian Lemma 1.5 is much weaker than the first Hilbertian Lemma 1.3. In particular in the second lemma, we need to *shrink the size of the support of* u with Condition (36) to get an estimate, although no such condition was required in the assumptions of Lemma 1.3. As a consequence, the estimate (37) is sufficient for local solvability of the adjoint operator but falls short of providing a propagation estimate as Lemma 1.3 does. These difficulties are linked to semi-global solvability of differential equations under condition (P) and to local solvability with loss of one derivative: We invite the reader to consult Theorem 26.11.2 in L. Hörmander's book [24], a result for which it appears that semi-global solvability at a compact K (satisfying a non-trapping assumption) of the equation (of order m) $Pu = f$, for P satisfying Condition (P), occurs with loss of $1+\varepsilon$ derivatives; indeed for f in the Sobolev space H^s_{loc}, the aforementioned theorem gives a solution u in $H^{s+m-1-\varepsilon}_{\mathrm{loc}}$, for any $\varepsilon > 0$. However in two dimensions, P. Cordaro and J. Hounie's article [9] is proving the optimal result of solvability with one derivative. That optimal semi-global solvability result remains an open problem in three (or more) dimensions.

1.3 Metrics on the Phase Space

We have seen with Formula (12) that it was possible to tailor a class of pseudo-differential operators to a particular symbol or to a family of symbols, and it was described in R. Beals and C.L. Fefferman's articles [1, 2], with momentous applications to local solvability questions: Indeed, starting with a semiclassical symbol q satisfying (6) and (16), the authors were able to construct a new pseudo-differential calculus in which they were reduced to study the few cases occurring in Hypothesis 1.2. The cases $(1)_\pm$ and (3) were easy to tackle (see our Lemma 1.3), and the case (2) could be handled with the spectral argument of L. Nirenberg and F. Treves in [36], a version of which is given in our Lemma 1.5. The nonhomogeneous microlocalization obtained in [1, 2], using a Calderón–Zygmund decomposition, allowed the authors to tackle the general smooth case and to avoid a restrictive analyticity assumption. A. Unterberger's paper [38] gave a more general approach to pseudo-differential calculus. In 1979, L. Hörmander published the paper [20], entitled *The Weyl calculus of pseudo-differential operators*, which was a good synthesis of the previous relevant generalizations of classes of pseudo-differential operators (see also Chapter XVIII in the book [23] by the same author). To define classes of symbols, L. Hörmander is using a Riemannian structure on the phase space $\mathbb{R}^n_x \times \mathbb{R}^n_\xi$ endowed with specific properties. You are given a mapping

$$\mathbb{R}^{2n} \ni X \mapsto g_X, \quad \text{a positive-definite quadratic form on } \mathbb{R}^{2n},$$

such that (we are using a definition used in J.-M. Bony and N. Lerner's article [6], see also [29]):

- *g is slowly varying,* i.e.,

$$\exists C > 0, \exists r > 0, \quad \forall X, Y, T \in \mathbb{R}^{2n},$$
$$g_X(Y - X) \le r^2 \Longrightarrow C^{-1} g_Y(T) \le g_X(T) \le C g_Y(T). \tag{56}$$

- *g satisfies the uncertainty principle,* i.e., using the notation (185) in our Appendix,

$$\forall X \in \mathbb{R}^{2n}, \quad g_X \le g^\sigma_X. \tag{57}$$

- *g is temperate,* i.e., using the notation (186) in our Appendix, we have

$$\exists C > 0, \exists N \ge 0, \quad \forall X, Y, T \in \mathbb{R}^{2n}, T \ne 0,$$
$$\frac{g_X(T)}{g_Y(T)} \le C\big(1 + (g^\sigma_X \wedge g^\sigma_Y)(X - Y)\big)^N. \tag{58}$$

Definition 1.8 Let g be defined as above, such that (56), (57), and (58) hold true. Then we shall say that g is an admissible metric on $\mathbb{R}^{2n}$. We define also the *Inverse Planck Function* λ_g as

$$\lambda_g(X) = \inf_{T \ne 0} \left(\frac{g^\sigma_X(T)}{g_X(T)} \right)^{1/2}. \tag{59}$$

N.B *We note that*

$$g_X \le \lambda_g(X)^{-2} g^\sigma_X, \quad \lambda_g(X) \ge 1. \tag{60}$$

The terminology Inverse Planck Function *for the function* λ_g *is justified by the fact that when a is a semiclassical symbol, e.g., when for* $h \in (0, 1]$,

$$a(x, \xi, h) = a_1(x, h\xi), \quad \textit{where } a_1 \in C_b^\infty(\mathbb{R}^{2n}),$$

where $C_b^\infty(\mathbb{R}^{2n})$ *stands for the smooth functions bounded and all their derivatives, we have*

$$|(\partial_x^\alpha \partial_\xi^\beta a)(x, \xi, h)| \le C_{\alpha,\beta} h^{|\beta|}, \quad \textit{i.e.} \quad a \in \mathcal{S}(1, g), \quad g(y, \eta) = |y|^2 + h^2|\eta|^2,$$

and thus $g^\sigma(y, \eta) = h^{-2}|y|^2 + |\eta|^2, \quad \lambda_g = h^{-1}$.

This metric g on the phase space allows us to define the class of symbols $\mathcal{S}(\lambda_g^\mu, g)$ ($\mu \in \mathbb{R}$), as the set of smooth functions a defined on $\mathbb{R}^{2n}$ such that

$$\forall k \in \mathbb{N}, \qquad \sup_{X,T \in \mathbb{R}^{2n}, T \ne 0} |a^{(k)}(X)T^k| g_X(T)^{-k/2} \lambda_g^{-\mu}(X) < +\infty. \tag{61}$$

The reader will find some examples of metrics on the phase space in Sect. 3.5 of our Appendix. A key result in that theory is the following theorem.

Theorem 1.9 *Let* g *be an admissible metric on* $\mathbb{R}^{2n}$, *and let* $a \in \mathcal{S}(1, g)$. *Then, using the notation* (189) *of section "The Weyl Quantization" in our Appendix, we find that* $\operatorname{Op_w}(a)$ *is a bounded operator on* $L^2(\mathbb{R}^n)$.

Note that the pointwise assumption (57) is rather natural and related to the fact that the localization in the phase space is limited by the uncertainty principle, saying roughly that the area of

$$B_g(X, 1) \cap \texttt{symplectic plane containing } X$$

should be bounded from below by a fixed constant.[5] Also the condition (56) appears as some sort of continuity property which is purely local and allows some convenient partitions of unity, discrete or continuous. The third condition (58) (temperance) is obviously of a global type and is using the affine connection; it appears as a key condition to verify the hypotheses of Cotlar's Lemma.

Theorem 1.10 *Let* μ_1, μ_2 *be real numbers, and let* $a_j \in \mathcal{S}(\lambda_g^{\mu_j}, g)$, $j = 1, 2$. *Then there exist symbols* $r \in \mathcal{S}(\lambda_g^{\mu_1+\mu_2-1}, g)$ *and* $s \in \mathcal{S}(\lambda_g^{\mu_1+\mu_2-2}, g)$ *such that*

$$\operatorname{Op_w}(a_1)\operatorname{Op_w}(a_2) = \operatorname{Op_w}(a_1 a_2) + \operatorname{Op_w}(r), \tag{62}$$

$$\big[\operatorname{Op_w}(a_1), \operatorname{Op_w}(a_2)\big] = \operatorname{Op_w}\left(\frac{1}{2\pi i}\{a_1, a_2\}\right) + \operatorname{Op_w}(s). \tag{63}$$

[5] We use here the notation $B_g(X, r) = \{Y \in \mathbb{R}^{2n}, g_X(Y - X) \le r^2\}$.

A specificity of the Weyl quantization is that the symbol r above is such that

$$r \in \frac{1}{4\pi i}\{a_1, a_2\} + \mathcal{S}\left(\lambda_g^{\mu_1+\mu_2-2}, g\right). \tag{64}$$

J.-M. Bony and J.-Y. Chemin's article [5] provided a complete study of the Sobolev spaces attached to an admissible metric g on $\mathbb{R}^{2n}$ and proved in particular that for $\mu \in \mathbb{R}$,

$$\mathcal{H}(\lambda_g^\mu, g) = \{u \in \mathscr{S}'(\mathbb{R}^n),\ \forall a \in S(\lambda_g^\mu, g), \mathrm{Op_w}(a)u \in L^2(\mathbb{R}^n)\} \tag{65}$$

is a Hilbert space and that the following result holds true.

Theorem 1.11 *Let g be an admissible metric on $\mathbb{R}^{2n}$, let μ be a real number and let a be a symbol in $\mathcal{S}(\lambda_g^\mu, g)$. Then for all $\nu \in \mathbb{R}$, the operator* $\mathrm{Op_w}(a)$ *is bounded from $\mathcal{H}(\lambda_g^\nu, g)$ into $\mathcal{H}(\lambda_g^{\nu-\mu}, g)$.*

The 2010 book [29] by the author contains a survey of non-self-adjoint operators and pseudo-differential methods linked to metrics on the phase space. Some more general classes of pseudo-differential operators were defined in J.-M. Bony and N. Lerner's article [6] in which global temperance was replaced by a more local condition but with a different quantization formula.

1.4 The Local Solvability Problem, from 1988 to 2006

We have not said much about local solvability of principal type *pseudo-differential* equations under Condition (Ψ) (see (3)), although two examples in Footnote 3 on page 226 are actually pseudo-differential. Although proven necessary in 1981 for local solvability of principal type pseudo-differential equations (see [33, 21]), this natural geometric condition was also proven invariant by multiplication by an elliptic factor by L. Nirenberg and F. Treves in [35]; key lemmas for that result were given in two papers by J.-M. Bony [3] and H. Brézis [7]. However, the sufficiency of Condition (Ψ) for local solvability of principal type pseudo-differential equations remained an open problem in 1988, although the conjecture was proven in two dimensions in the author's paper [26], using some very specific features of the two-dimensional situation.

We need at this point to mention the notion of *loss of derivatives* for solving differential equations: Let P be a pseudo-differential operator of order m. Local solvability with loss of δ derivatives for P means that the equation $Pu = f$ with $f \in H^s_{\mathrm{loc}}$ has local solutions $u \in H^{s+m-\delta}_{\mathrm{loc}}$. The loss is 0 iff P is elliptic and the known results for principal type differential operators in any dimension and principal type pseudo-differential operators in two dimensions are that Condition (Ψ) implies local solvability with loss of one derivative in these cases. In particular, for real principal type pseudo-differential equations, such as the wave equation, local solvability occurs with loss of one derivative; more generally, we have seen that under Condition (Ψ), *differential* operators of principal type are locally solvable with loss of one derivative. It appeared then natural to expect that Condition (Ψ) should imply solvability with loss of one derivative as well, and it seems that the research efforts on these local solvability questions were naturally oriented toward proving that this optimal (smallest) loss of one derivative occurred under Condition (Ψ).

However in 1994, the author proved in [27] that some three-dimensional principal type pseudo-differential equations satisfying Condition (Ψ) were not locally solvable with loss of one derivative (see also L. Hörmander's [22]). Obviously, the conventional wisdom on these matters needed some drastic refurbishing, and this was accomplished in 2006 with N. Dencker's [12], a paper in which he proved that Condition (Ψ) implies local solvability with loss of two derivatives; an improvement of that result cutting the loss to $\frac{3}{2}+\varepsilon$ (for any positive ε) derivatives was given by N. Dencker in [11]. The paper [28] by the author showed that the loss could in fact be limited to $\frac{3}{2}$ derivatives, and this is also the result of L. Hörmander's article #23 in the book *Unpublished manuscripts—from 1951 to 2007* [25]. The article [13] by N. Dencker gives a historical perspective on this problem and a tribute to L. Nirenberg. We can define

$$\Sigma_\ell = \{\text{Principal type pseudo-differential operators locally solvable with loss of } \ell \text{ derivatives}\},$$

$$\mathsf{OP}[\Psi] = \{\text{Principal type pseudo-differential operators satisfying Condition}(\Psi)\}.$$

Noting that local solvability (with any finite loss of derivatives) for principal type pseudo-differential operators implies Condition (Ψ), it is proven that

$$\Sigma_1 \subsetneq \mathsf{OP}[\Psi] = \Sigma_{3/2}. \tag{66}$$

Since it seems that the counterexamples of [27] are locally solvable with a loss of $1+\varepsilon$ derivatives for any $\varepsilon > 0$, this would be compatible with

$$\Sigma_1 \subsetneq \mathsf{OP}[\Psi] = \cap_{\varepsilon>0}\Sigma_{1+\varepsilon}, \tag{67}$$

although the equality in (67) is not proven.

2 Singular Integrals Involved in the Counterexamples to Optimal Solvability

2.1 Preliminaries

We consider the unbounded operators on $L^2(\mathbb{R}_x) = \mathbb{H}$,

$$Q_1 = D_x, \quad Q_2 = D_x + V(x) = e^{-2i\pi\phi(x)} D_x e^{2i\pi\phi(x)}, \quad \phi' = V, \tag{68}$$

where ϕ is real-valued and $\phi' \ge 0$. We keep in mind that we are ready to accept that V could be a nonnegative Radon measure. We note that

$$Q_1 \le Q_2, \tag{69}$$

since the potential $V = \phi'$ is nonnegative. Both Q_1, Q_2 are unbounded self-adjoint operators, and we shall denote by $E_{j,+}, E_{j,-}$ the spectral projectors corresponding, respectively, to the positive and nonpositive half-line for the operator Q_j. We also have, with H standing for the Heaviside function (172) in Appendix A,

$$E_{1,+} = H(D_x), \qquad E_{1,-} = \check{H}(D_x), \tag{70}$$

$$E_{2,-} = e^{-2i\pi\phi(x)}\check{H}(D_x)e^{2i\pi\phi(x)}, \quad E_{2,+} = e^{-2i\pi\phi(x)}H(D_x)e^{2i\pi\phi(x)}. \tag{71}$$

Let us consider the evolution equation

$$\mathscr{L}_{\theta,\phi} = D_t + i\big[H(-t-\theta)Q_1 + H(t-\theta)Q_2\big], \tag{72}$$

where θ is a nonnegative parameter.

Notation 2.1 We keep our notation $D_x = \frac{\partial}{2i\pi\partial x}$, but we will denote $D_t = \frac{d}{idt}$.

We are interested in local solvability properties for the adjoint operator $\mathscr{L}^*_{\theta,\phi}$ or "injectivity estimates" for $\mathscr{L}_{\theta,\phi}$.

Claim 2.2 For $v \in C^1_c(\mathbb{R}_t, \mathbb{H})$, we have

$$\|\mathscr{L}_{0,\phi}v\|_{L^2(\mathbb{R},\mathbb{H})} \ge \|D_t v\|_{L^2(\mathbb{R},\mathbb{H})}, \tag{73}$$

so that $\mathscr{L}^*_{0,\phi}$ is locally solvable.

Proof of the Claim Indeed we have with

$$Q_{0,\phi}(t) = H(-t)Q_1 + H(t)Q_2, \quad \text{so that } \dot{Q}_{0,\phi}(t) = V(x)\delta_0(t), \tag{74}$$

which is nonnegative, inducing that

$$\begin{aligned}\|\mathscr{L}_{0,\phi}v\|^2_{L^2(\mathbb{R},\mathbb{H})} &= \|D_t v\|^2_{L^2(\mathbb{R},\mathbb{H})} + \langle \dot{Q}_{0,\phi}v, v\rangle_{L^2(\mathbb{R},\mathbb{H})} + \|Q_{0,\phi}(t)v\|^2_{L^2(\mathbb{R},\mathbb{H})} \\ &\ge \|D_t v\|^2_{L^2(\mathbb{R},\mathbb{H})} \ge T^{-2}\|v\|^2_{L^2(\mathbb{R},\mathbb{H})},\end{aligned}$$

if $\operatorname{diam}(\operatorname{supp} v) \le T$, proving the claim.

Note that the previous estimate would hold true for any operator of type

$$D_t + iQ(t), \quad Q(t) \ \text{ self-adjoint increasing (i.e., } \dot{Q}(t) \ge 0). \tag{75}$$

We shall see that a proper choice of ϕ will somehow prevent some injectivity estimates of type (73) for the operator $\mathscr{L}_{\theta,\phi}$ when $\theta > 0$. Since we have

$$Q_{\theta,\phi}(t) = Q_{0,\phi}(t) \underbrace{\big(H(-t-\theta) + H(t-\theta)\big)}_{\text{scalar} \geq 0},$$

this suggests that an $L^2 - L^2$ injectivity estimate of type

$$C\|\mathscr{L}_{\theta,\phi}v\|_{L^2(\mathbb{R},\mathbb{H})} \geq \|v\|_{L^2(\mathbb{R},\mathbb{H})} \tag{76}$$

will not be a consequence of Condition (Ψ) which is stable by the change

$$D_t + iQ(t) \rightsquigarrow D_t + i\alpha(t)Q(t),$$

with α nonnegative scalar function.

2.2 The Drift

In the case where $\theta > 0$, we are looking for a nontrivial solution in the kernel of the operator $\mathscr{L}_{\theta,\phi}$ given by (72) or at least to a sequence of (non-zero) functions $v_k \in \mathrm{Lip}(\mathbb{R}, \mathbb{H}) \cap L^2(\mathbb{R}, \mathbb{H})$ so that

$$\lim_{k\to+\infty} \frac{\|\mathscr{L}_{\theta,\phi}v_k\|_{L^2(\mathbb{R},\mathbb{H})}}{\|v_k\|_{L^2(\mathbb{R},\mathbb{H})}} = 0.$$

In fact, sticking with the search for an element in the kernel, we would need to have

$$\begin{cases} v(t) = e^{(t+\theta)Q_1}w_{1,\varepsilon}, & \text{for } t \leq -\theta, \quad w_{1,\varepsilon} \in E_{1,+}, \\ v(t) = e^{(t-\theta)Q_2}w_{2,\varepsilon}, & \text{for } t \geq \theta, \quad w_{2,\varepsilon} \in E_{2,-}, \end{cases} \qquad \|w_j\|_{\mathbb{H}} = 1,\ j = 1, 2.$$

To get a path from $w_{1,\varepsilon}$ to $w_{2,\varepsilon}$ on the time interval $[-\theta, \theta]$ would require that $\|w_{1,\varepsilon} - w_{2,\varepsilon}\|_{\mathbb{H}}$ should be small, that is, the distance of the respective unit spheres $\mathbb{S}_{1,+}, \mathbb{S}_{2,-}$ of ran $E_{1,+}$, ran $E_{2,-}$ must be small. This is compatible with the assumption that $Q_1 \leq Q_2$ which forces

$$\mathbb{S}_{1,+} \cap \mathbb{S}_{2,-} = \emptyset. \tag{77}$$

Indeed if $w \in \mathbb{S}_{1,+}$, then $\langle Q_1 w, w\rangle_{\mathbb{H}} > 0$, and thus $\langle Q_2 w, w\rangle_{\mathbb{H}} > 0$ so that $w \notin \mathbb{S}_{2,-}$, but we shall see that does not prevent the spheres to *drifting* close to each other while staying disjoint. Let Q_1, Q_2 be self-adjoint operators on a Hilbert space $\mathbb{H}$. Let us define the spectral projections

$$E_{j,+} = \mathbf{1}_{(0,+\infty)}(Q_j), \quad E_{j,-} = \mathbf{1}_{(-\infty,0]}(Q_j), \quad j = 1, 2, \tag{78}$$

$$\text{and} \quad \alpha(Q_1, Q_2) = \inf\big(\text{Spectrum}(E_{2,+} + E_{1,-})\big). \tag{79}$$

Note that we have for $\|\omega\|_{\mathbb{H}} = 1$,

$$\begin{aligned}
&\|E_{2,-}\omega - E_{1,+}\omega\|^2 \\
&= \|E_{2,-}\omega\|^2 + \|E_{1,+}\omega\|^2 - 2\,\mathrm{Re}\langle \omega - E_{2,+}\omega, \omega - E_{1,-}\omega\rangle \\
&= 1 - \|E_{2,+}\omega\|^2 + 1 - \|E_{1,-}\omega\|^2 \\
&\quad - 2 - 2\,\mathrm{Re}\langle E_{2,+}\omega, E_{1,-}\omega\rangle + 2\,\mathrm{Re}\langle E_{2,+}\omega, \omega\rangle + 2\,\mathrm{Re}\langle E_{1,-}\omega, \omega\rangle \\
&= \|E_{2,+}\omega\|^2 + \|E_{1,-}\omega\|^2 - 2\,\mathrm{Re}\langle E_{2,+}\omega, E_{1,-}\omega\rangle.
\end{aligned} \tag{80}$$

- If $\alpha(Q_1, Q_2) = 0$, then for ω_j with $\|\omega_j\| = 1$, we have $\lim_j\big(E_{2,+}\omega_j + E_{1,-}\omega_j\big) = 0$, and thus, making the dot product with ω_j, we get

$$\lim_j \|E_{2,+}\omega_j\| = \lim_j \|E_{1,-}\omega_j\| = 0, \tag{81}$$

  so that applying (80), we get $\lim_j\big(E_{2,-}\omega_j - E_{1,+}\omega_j\big) = 0$ in $\mathbb{H}$. Since (81) implies $\lim_j \|E_{2,-}\omega_j\| = \lim_j \|E_{1,+}\omega_j\| = 1$, we get that

$$\|\mathbb{S}_{2,-} - \mathbb{S}_{1,+}\| = \inf_{\substack{\omega_2 \in \operatorname{ran} E_{2,-} \\ \omega_1 \in \operatorname{ran} E_{1,+} \\ \|\omega_j\|_{\mathbb{H}} = 1}} \|\omega_2 - \omega_1\|_{\mathbb{H}} = 0. \tag{82}$$

- Conversely, if $\|\mathbb{S}_{2,-} - \mathbb{S}_{1,+}\| = 0$, then we can find two sequences $(\omega_{2,j})$, $(\omega_{1,j})$ in the unit sphere of $\mathbb{H}$ with $\omega_{2,j} \in \operatorname{ran} E_{2,-}$, $\omega_{1,j} \in \operatorname{ran} E_{1,+}$ such that $\lim_j(\omega_{2,j} - \omega_{1,j}) = 0$. We have then

$$\begin{aligned}
(E_{2,+} + E_{1,-})(\omega_{2,j}) &= (E_{2,+} + E_{1,-})(\omega_{2,j} - \omega_{1,j}) + (E_{2,+} + E_{1,-})(\omega_{1,j}) \\
&= (E_{2,+} + E_{1,-})(\omega_{2,j} - \omega_{1,j}) + E_{2,+}(\omega_{1,j}) \\
&= (E_{2,+} + E_{1,-})(\omega_{2,j} - \omega_{1,j}) + E_{2,+}(\omega_{1,j} - \omega_{2,j}),
\end{aligned}$$

  which entails that $\lim_j(E_{2,+} + E_{1,-})(\omega_{2,j}) = 0$ in $\mathbb{H}$, proving that $\alpha(Q_1, Q_2) = 0$.

Definition 2.3 Let Q_1, Q_2 be self-adjoint operators on a Hilbert space $\mathbb{H}$. We define the spectral projections by (78) and $\alpha(Q_1, Q_2)$ by (79). We shall say that the drift of (Q_1, Q_2) is bounded (respectively, infinite) whenever $\alpha(Q_1, Q_2) > 0$ (respectively, $\alpha(Q_1, Q_2) = 0$). We have proven above that the drift of (Q_1, Q_2) is infinite iff $\|\mathbb{S}_{2,-} - \mathbb{S}_{1,+}\| = 0$ (as defined by (82)). We shall define

$$\operatorname{drift}(Q_1, Q_2) = \tan\big\{\arcsin\big[1 - \alpha(Q_1, Q_2)\big]_+\big\}. \tag{83}$$

N.B Note that:

- If $Q_1 \le 0 \le Q_2$, we have $E_{2,+} = E_{1,-} = \mathrm{Id}$, so that $\mathrm{drift}(Q_1, Q_2) = 0$.
- If $E_{2,+} + E_{1,-} \ge \mathrm{Id}$, we get that $\mathrm{drift}(Q_1, Q_2) = 0$.
- In particular, if $[Q_1, Q_2] = 0$ and $Q_1 \le Q_2$, we find that $E_{1,+} - E_{1,-} \le E_{2,+} - E_{2,-}$, so that $E_{2,+} + E_{1,-} \ge E_{1,+} + E_{2,-}$, and thus

$$E_{2,+} + E_{1,-} \ge \frac{1}{2}(E_{2,+} + E_{1,-}) + \frac{1}{2}(E_{1,+} + E_{2,-}) = \mathrm{Id},$$

  so that $\mathrm{drift}(Q_1, Q_2) = 0$.
- If for some $\kappa_1, \kappa_2 \in (0, 1)$, we have $\kappa_1 \ge \inf(E_{2,+} + E_{1,-}) \ge \kappa_2$, we get

$$\tan[\arcsin(1 - \kappa_1)] \le \mathrm{drift}(Q_1, Q_2) \le \tan[\arcsin(1 - \kappa_2)].$$

- Since we have for $\kappa \in (0, 1)$, $\tan[\arcsin(1 - \kappa)] = \frac{1}{\sqrt{2\kappa}} \frac{1-\kappa}{\sqrt{1-\frac{\kappa}{2}}}$,

$$\inf\bigl(E_{2,+} + E_{1,-}\bigr) \le \kappa \Longrightarrow \mathrm{drift}(Q_1, Q_2) \ge \frac{1}{\sqrt{2\kappa}}\bigl(1 + O(\kappa)\bigr), \ \kappa \to 0_+.$$

- If $\inf(E_{2,+} + E_{1,-}) = 0$, we get that $\mathrm{drift}(Q_1, Q_2) = +\infty$.

2.3 The Drift for Pseudo-Differential Operators

Lemma 2.4 *Let Q_1, Q_2 be as in* (68) *with $\phi(x) = \frac{1}{2}H(x)$, where H is the Heaviside function* (172) *in Appendix A. Then we have*

$$E_{2,-} = e^{-i\pi H(x)} \check{H}(D_x) e^{i\pi H(x)}, \quad E_{2,+} = e^{-i\pi H(x)} H(D_x) e^{i\pi H(x)}. \tag{84}$$

For any $\varepsilon > 0$, we can find $w_1 \in \operatorname{ran} E_{1,+}$, $w_2 \in \operatorname{ran} E_{2,-}$, such that

$$\|w_1 - w_2\|_{L^2(\mathbb{R})} \le \varepsilon, \quad \|w_j\|_{L^2(\mathbb{R})} = 1, \, j = 1, 2. \tag{85}$$

Proof We note that $e^{i\pi H(x)} = -\operatorname{sign} x$ and that the Fourier multiplier $\operatorname{sign}(D_x)$ has the distribution kernel given by Lemma A.1, that is, $\frac{1}{i\pi}\mathrm{pv}\frac{1}{y-x}$. A function w_1 (respectively, w_2) will belong to $\operatorname{ran} E_{1,+}$ (respectively, $\operatorname{ran} E_{2,-}$) if

$$w_1(x) = \int e^{2i\pi x\xi} \kappa_1(\xi) d\xi, \quad \operatorname{supp} \kappa_1 \subset \mathbb{R}_+, \tag{86}$$

$$-(\operatorname{sign} x) w_2(x) = \int e^{2i\pi x\eta} \kappa_2(-\eta) d\eta, \quad \operatorname{supp} \kappa_2 \subset \mathbb{R}_+. \tag{87}$$

Moreover, we have

$$\langle w_1, w_2 \rangle_{L^2(\mathbb{R})} = \iiint e^{2i\pi x\xi} e^{2i\pi x\eta} (-\operatorname{sign} x) \kappa_1(\xi) \overline{\kappa_2(\eta)} dx d\xi d\eta$$
$$= \iint \frac{1}{i\pi(\xi + \eta)} \kappa_1(\xi) \overline{\kappa_2(\eta)} d\xi d\eta.$$

Let κ be an $L^2(\mathbb{R})$ function with norm 1 supported in $\mathbb{R}_+$, and let us choose

$$\kappa_1(\xi) = \kappa(\xi), \quad \kappa_2(\xi) = -i\kappa(\xi). \tag{88}$$

We get then

$$\langle w_1, w_2 \rangle_{L^2(\mathbb{R})} = \iint \frac{H(\xi)H(\eta)}{\pi(\xi + \eta)} \kappa(\xi) \overline{\kappa(\eta)} d\xi d\eta \tag{89}$$
$$= \iint_{\mathbb{R}_+^2} \frac{1}{\pi(e^s + e^t)} \kappa(e^s) \overline{\kappa(e^t)} e^{s+t} ds dt$$
$$= \iint_{\mathbb{R}_+^2} \frac{e^{\frac{s+t}{2}}}{\pi(e^s + e^t)} \kappa(e^s) e^{s/2} \overline{\kappa(e^t) e^{t/2}} ds dt$$
$$= \frac{1}{2\pi} \iint_{\mathbb{R}^2} \operatorname{sech}\big(\frac{t-s}{2}\big) \psi(s) \overline{\psi(t)} ds dt,$$

with

$$\psi(t) = \kappa(e^t) e^{t/2} \quad \text{so that} \quad \|\psi\|_{L^2(\mathbb{R})} = \|\kappa\|_{L^2(\mathbb{R}_+)} = 1. \tag{90}$$

Using now the classical formula (see our Appendix (175)),

$$\int e^{2i\pi x\xi} \operatorname{sech} x dx = \pi \operatorname{sech}(\pi^2 \xi), \tag{91}$$

we obtain

$$\langle w_1, w_2 \rangle_{L^2(\mathbb{R})} = \frac{1}{2\pi} \iint \operatorname{sech}\big(\frac{t-s}{2}\big) \psi(s) \overline{\psi(t)} ds dt$$
$$= \frac{1}{\pi} \iint \operatorname{sech}\big(t - s\big) \underbrace{2^{1/2} \psi(2s)}_{\nu(s)} 2^{1/2} \overline{\psi(2t)} ds dt$$
$$= \frac{1}{\pi} \langle \operatorname{sech} * \nu, \nu \rangle = \int \operatorname{sech}(\pi^2 \tau) |\hat{\nu}(\tau)|^2 d\tau, \quad \text{with } \|\nu\|_{L^2(\mathbb{R})} = 1.$$

Since the function $\tau \mapsto \operatorname{sech}(\pi^2 \tau)$ is smooth with range $(0, 1]$, we obtain that the operator of multiplication by this function is bounded self-adjoint with $[0, 1]$ as a (continuous) spectrum. In particular, for any $\varepsilon > 0$, we can find $\nu_\varepsilon \in L^2(\mathbb{R})$ such that $\langle w_1, w_2\rangle_{L^2(\mathbb{R})} \in \mathbb{R}$ and

$$\langle w_1, w_2\rangle_{L^2(\mathbb{R})} \ge 1 - \frac{\varepsilon^2}{2}, \tag{92}$$

where w_1, w_2 satisfy (86), (87), and κ_1, κ_2 verify (88), with κ given by (90) and $\nu(s) = 2^{1/2}\psi(2s)$, $\|\nu\|_{L^2(\mathbb{R})} = 1 = \|w_1\|_{L^2(\mathbb{R})} = \|w_2\|_{L^2(\mathbb{R})} = 1$, so that we have

$$\|w_1 - w_2\|^2_{L^2(\mathbb{R})} = 2 - 2\operatorname{Re}\langle w_1, w_2\rangle_{L^2(\mathbb{R})} \le \varepsilon^2,$$

proving the lemma.

Along our way, we have also proven the following result.

Lemma 2.5 *The Hardy operator $\mathscr{H}$ with kernel*

$$\mathfrak{h}(x, y) = \frac{H(x)H(y)}{\pi(x+y)} \tag{93}$$

is self-adjoint bounded on $L^2(\mathbb{R})$ with spectrum $[0, 1]$. *In particular, its $\mathcal{B}(L^2(\mathbb{R}))$ norm is equal to 1.*

N.B It is obvious that the operator with kernel (93) is self-adjoint and bounded with a $\mathcal{B}(L^2(\mathbb{R}))$-norm ≤ 1. Indeed, with the unitary operator $\mathcal{C}$ defined by (170) in Appendix A, we find that

$$\mathscr{H} = iH\mathcal{C}\operatorname{sign}(D)H, \tag{94}$$

where H stands for the orthogonal projection from $L^2(\mathbb{R})$ onto $L^2(\mathbb{R}_+)$ and this equality readily implies that $\|\mathscr{H}\|_{\mathcal{B}(L^2(\mathbb{R}))} \le 1$. The main point of the above discussion is to show that the upper bound of the spectrum is actually 1.

Remark 2.6 Since the operator $\mathscr{H}$ sends $L^2(\mathbb{R}_+)\oplus L^2(\mathbb{R}_-)$ into $L^2(\mathbb{R}_+)$, it is also a self-adjoint bounded operator on $L^2(\mathbb{R}_+)$ with spectrum $[0, 1]$. In particular, we may find a sequence $(w_j)_{j\ge1}$ of unit vectors in $L^2(\mathbb{R}_+)$ such that $\lim_j(\mathscr{H}w_j - w_j) = 0$.

Claim 2.7 Let us consider

$$Q_1 = D_x, \quad Q_2 = e^{-2i\pi aH(x)}D_x e^{2i\pi aH(x)} = D_x + a\delta_0(x).$$

We have then for $a \in [0, 1]$,

$$\operatorname{drift}(Q_1, Q_2) = |\tan \pi a| = \begin{cases} 0 & \text{if } a = 0, 1, \\ 1 & \text{if } a = \frac14, \frac34, \\ +\infty & \text{if } a = \frac12. \end{cases}$$

Proof of the Claim We have indeed

$$\begin{aligned}E_{2,+}(a)+E_{1,-}&=e^{-2i\pi aH(x)}H(D_x)e^{2i\pi aH(x)}+\check{H}(D_x)\\&=e^{-i\pi aH(x)}\big(e^{-i\pi aH(x)}H(D_x)e^{i\pi aH(x)}+e^{i\pi aH(x)}\check{H}(D_x)e^{-i\pi aH(x)}\big)e^{i\pi aH(x)},\end{aligned}$$

so that $E_{2,+}(a)+E_{1,-}$ is unitarily equivalent to

$$F_{1,2}(a)=e^{-i\pi aH(x)}H(D_x)e^{i\pi aH(x)}+e^{i\pi aH(x)}\check{H}(D_x)e^{-i\pi aH(x)}. \tag{95}$$

We have also that the kernel of the operator $e^{-i\pi aH(x)}H(D_x)e^{i\pi aH(x)}$ is[6]

$$\begin{aligned}&\frac{1}{2}e^{-i\pi aH(x)}\Big(\delta_0(y-x)+\frac{1}{i\pi(y-x)}\Big)e^{i\pi aH(y)}=\frac{1}{2}\delta_0(y-x)+\frac{e^{i\pi a(H(y)-H(x))}}{2i\pi(y-x)}\\&=\frac{1}{2}\delta_0(y-x)+\frac{H(x)H(y)+\check{H}(x)\check{H}(y)}{2i\pi(y-x)}+\frac{H(x)\check{H}(y)e^{-i\pi a}+\check{H}(x)H(y)e^{i\pi a}}{2i\pi(y-x)},\end{aligned}$$

and we have also that the kernel of the operator $e^{i\pi aH(x)}\check{H}(D_x)e^{-i\pi aH(x)}$ is

$$\begin{aligned}&\frac{1}{2}e^{i\pi aH(x)}\Big(\delta_0(x-y)+\frac{1}{i\pi(x-y)}\Big)e^{-i\pi aH(y)}=\frac{1}{2}\delta_0(x-y)+\frac{e^{i\pi a(H(x)-H(y))}}{2i\pi(x-y)}\\&=\frac{1}{2}\delta_0(x-y)+\frac{H(x)H(y)+\check{H}(x)\check{H}(y)}{2i\pi(x-y)}+\frac{H(x)\check{H}(y)e^{i\pi a}+\check{H}(x)H(y)e^{-i\pi a}}{2i\pi(x-y)},\end{aligned}$$

so that

$$\begin{aligned}\texttt{kernel}\big(F_{1,2}(a)-\mathrm{Id}\big)&=\frac{H(x)\check{H}(y)(e^{i\pi a}-e^{-i\pi a})+\check{H}(x)H(y)(e^{-i\pi a}-e^{i\pi a})}{2i\pi(x-y)}\\&=\sin(\pi a)\frac{H(x)\check{H}(y)-\check{H}(x)H(y)}{\pi(x-y)}.\end{aligned}$$

We have thus, with C defined by (170) in Appendix A,

$$F_{1,2}(a)=\mathrm{Id}+\big(\mathscr{H}C+C\mathscr{H}\big)\sin(\pi a), \tag{96}$$

[6] The calculation which follows could appear as meaningless since we multiply discontinuous functions by Radon measures as the Dirac mass or distributions of order 1 as the principal value of $1/x$; however, keeping in mind that the operators $E_{2,+}(a)+E_{1,-}$ and $F_{1,2}(a)$ are obviously bounded on $L^2(\mathbb{R})$, it is enough to calculate their actions on compactly supported smooth functions v such that $0\notin\operatorname{supp}v$. We have then $v=v_++v_-$, $v_\pm$ smooth compactly supported, $\operatorname{supp}v_+$, $\operatorname{supp}\check{v}_-\subset(0,+\infty)$. We get then $e^{i\pi aH(x)}v(x)=e^{i\pi a}v_+(x)+v_-(x)$, and we can carry out the computations without difficulty.

since the kernel of $\mathcal{C}\mathscr{H}$ is $\frac{\check{H}(x)H(y)}{\pi(y-x)}$ and the kernel of $\mathscr{H}\mathcal{C}+\mathcal{C}\mathscr{H}$ is

$$\frac{\check{H}(x)H(y)}{\pi(y-x)}+\frac{\check{H}(y)H(x)}{\pi(x-y)}=\frac{H(x)\check{H}(y)-\check{H}(x)H(y)}{\pi(x-y)}.$$

We note that $0 \le E_{2,+}(a)+E_{1,-} \le 2$, so that we get for $a=1/2$,

$$-\operatorname{Id} \le \mathscr{H}\mathcal{C}+\mathcal{C}\mathscr{H} \le \operatorname{Id}.$$

We claim now that -1 belongs to the spectrum of $\mathscr{H}\mathcal{C}+\mathcal{C}\mathscr{H}$: Indeed we know from Remark 2.6 that there exists a sequence $(w_j)_{j\ge 1}$ of unit vectors in $L^2(\mathbb{R}_+)$ such that $\lim_j(\mathscr{H}w_j-w_j)=0$, and using that $\operatorname{supp} w_j \subset \mathbb{R}_+$, we get $\mathscr{H}\mathcal{C}w_j=0$, and thus

$$(\mathscr{H}\mathcal{C}+\mathcal{C}\mathscr{H})(w_j-\mathcal{C}w_j)=\mathcal{C}\mathscr{H}w_j-\mathscr{H}w_j,$$

entailing that $\lim_j\big\{(\mathscr{H}\mathcal{C}+\mathcal{C}\mathscr{H})(w_j-\mathcal{C}w_j)+(w_j-\mathcal{C}w_j)\big\}=0$, proving that (-1) belongs to the spectrum of $\mathscr{H}\mathcal{C}+\mathcal{C}\mathscr{H}$, since $\|w_j-\mathcal{C}w_j\|=\sqrt{2}$. Using the sequence $w_j+\mathcal{C}w_j$, we see that 1 belongs to the spectrum of $\mathscr{H}\mathcal{C}+\mathcal{C}\mathscr{H}$. As a consequence, we get that

$$\alpha(Q_1,Q_2)=\inf\big(E_{2,+}(a)+E_{1,-}\big)=1-|\sin\pi a|,$$

and

$$\operatorname{drift}(Q_1,Q_2)=\tan\big(\arcsin(|\sin\pi a|\big)=\begin{Bmatrix}\tan\pi a, & \text{for } a\in[0,1/2)\\ +\infty, & \text{for } a=\frac{1}{2}\\ -\tan\pi a, & \text{for } a\in(1/2,1]\end{Bmatrix}=|\tan\pi a|,$$

proving the claim.

Remark 2.8 We can be more explicit for the expressions of $w_{j,\varepsilon}$ in Lemma 2.4. For $\varepsilon>0$, we define

$$\kappa_\varepsilon(\xi)=\Gamma(\varepsilon)^{-1/2}e^{-\frac{\xi}{2}}\xi_+^{\frac{\varepsilon-1}{2}}, \tag{97}$$

and we define $w_{j,\varepsilon}$ by (86), and (87), (88), namely

$$w_{1,\varepsilon}(x)=\int e^{2i\pi x\xi}\kappa_\varepsilon(\xi)d\xi, \qquad w_{2,\varepsilon}(x)=i\operatorname{sign}x\int e^{-2i\pi x\xi}\kappa_\varepsilon(\xi)d\xi. \tag{98}$$

In particular, we have

$$w_{1,\varepsilon}\in\operatorname{ran}E_{1,+}, \qquad w_{2,\varepsilon}\in\operatorname{ran}E_{2,-}. \tag{99}$$

We find $\|w_{j,\varepsilon}\|_{\mathbb{H}} = 1$, and according to (89), we have

$$\begin{aligned}
&\langle w_{1,\varepsilon}, w_{2,\varepsilon}\rangle_{\mathbb{H}} \\
&= \iint \frac{H(\xi)H(\eta)}{\pi(\xi+\eta)} e^{-\frac{\xi+\eta}{2}} (\xi\eta)^{\frac{\varepsilon-1}{2}} d\xi d\eta \Gamma(\varepsilon)^{-1}, \quad (\xi = t+s, \eta = t-s) \\
&= \frac{1}{\pi\Gamma(\varepsilon)} \iint H(t-|s|) \frac{e^{-t}}{2t} (t^2 - s^2)^{\frac{\varepsilon-1}{2}} 2dtds, \quad (s = t\sin\theta) \\
&= \frac{1}{\pi\Gamma(\varepsilon)} \iint H(t) \mathbf{1}_{[-\frac{\pi}{2},\frac{\pi}{2}]}(\theta) \frac{e^{-t}t^{\varepsilon-1}}{2t} (\cos\theta)^{\varepsilon-1} 2t\cos\theta d\theta dt \\
&= \frac{2}{\pi} \int_0^{\pi/2} (\cos\theta)^{\varepsilon} d\theta.
\end{aligned}$$

Lemma A.8 in our Appendix gives

$$\langle w_{1,\varepsilon}, w_{2,\varepsilon}\rangle_{\mathbb{H}} \geq 1 - \varepsilon, \tag{100}$$

$$\text{and thus} \quad \|w_{1,\varepsilon} - w_{2,\varepsilon}\|^2_{\mathbb{H}} \leq 2\varepsilon. \tag{101}$$

With $w_{1,\varepsilon}, w_{2,\varepsilon}$ defined in (98), we note that $\langle w_{2,\varepsilon}, w_{1,\varepsilon}\rangle_{\mathbb{H}} \in [1-\varepsilon, 1)$, and we define

$$\alpha_\varepsilon = \arccos(\langle w_{1,\varepsilon}, w_{2,\varepsilon}\rangle), \quad \text{noting that} \quad 0 < \alpha_\varepsilon \leq 2\sqrt{\varepsilon}, \quad \varepsilon \in (0, 1]. \tag{102}$$

We define now

$$w_{3,\varepsilon} = \frac{w_{2,\varepsilon} - w_{1,\varepsilon}\cos\alpha_\varepsilon}{\sin\alpha_\varepsilon}, \quad \text{so that we get } \|w_{3,\varepsilon}\|_{\mathbb{H}} = 1, \langle w_{3,\varepsilon}, w_{1,\varepsilon}\rangle_{\mathbb{H}} = 0. \tag{103}$$

We set then for $t \in [-\theta, \theta]$,

$$v(t) = w_{1,\varepsilon}\cos\left(\alpha_\varepsilon \frac{t+\theta}{2\theta}\right) + w_{3,\varepsilon}\sin\left(\alpha_\varepsilon \frac{t+\theta}{2\theta}\right), \tag{104}$$

so that we have there

$$v(-\theta) = w_{1,\varepsilon}, \quad v(\theta) = w_{1,\varepsilon}\cos\alpha_\varepsilon + w_3\sin\alpha_\varepsilon = w_{2,\varepsilon}, \tag{105}$$

$$\dot{v}(t) = -w_{1,\varepsilon}\frac{\alpha_\varepsilon}{2\theta}\sin\left(\alpha_\varepsilon \frac{t+\theta}{2\theta}\right) + w_{3,\varepsilon}\frac{\alpha_\varepsilon}{2\theta}\cos\left(\alpha_\varepsilon \frac{t+\theta}{2\theta}\right). \tag{106}$$

We define now the linear operator $M_{0,\theta,\varepsilon}$ on $\mathbb{H}$ by

$$M_{0,\theta,\varepsilon}w_{1,\varepsilon} = \frac{\alpha_\varepsilon}{2\theta}w_{3,\varepsilon}, \qquad M_{0,\theta,\varepsilon}w_{3,\varepsilon} = -\frac{\alpha_\varepsilon}{2\theta}w_{1,\varepsilon},$$

$$M_{0,\theta,\varepsilon}w = 0 \text{ for } w \in (\mathbb{C}w_{1,\varepsilon} \oplus \mathbb{C}w_{3,\varepsilon})^\perp, \tag{107}$$

and we note that $M_{0,\theta,\varepsilon}$ is skew-adjoint since, with $w, \tilde{w} \in (\mathbb{C}w_{1,\varepsilon} \oplus \mathbb{C}w_{3,\varepsilon})^\perp$, we have

$$\begin{aligned}&\langle M^*_{0,\theta,\varepsilon}(\beta_1 w_{1,\varepsilon} + \beta_3 w_{3,\varepsilon} + w), (\gamma_1 w_{1,\varepsilon} + \gamma_3 w_{3,\varepsilon} + \tilde{w})\rangle\\ &\quad= \frac{\alpha_\varepsilon}{2\theta}\langle \beta_1 w_{1,\varepsilon} + \beta_3 w_{3,\varepsilon} + w, \gamma_1 w_{3,\varepsilon} - \gamma_3 w_{1,\varepsilon}\rangle = \frac{\alpha_\varepsilon}{2\theta}(-\beta_1\overline{\gamma_3} + \beta_3\overline{\gamma_1}),\end{aligned}$$

and

$$\begin{aligned}&\langle M_{0,\theta,\varepsilon}(\beta_1 w_{1,\varepsilon} + \beta_3 w_{3,\varepsilon} + w), \gamma_1 w_{1,\varepsilon} + \gamma_3 w_{3,\varepsilon} + \tilde{w}\rangle\\ &\qquad= \frac{\alpha_\varepsilon}{2\theta}\langle \beta_1 w_{3,\varepsilon} - \beta_3 w_{1,\varepsilon}, \gamma_1 w_{1,\varepsilon} + \gamma_3 w_{3,\varepsilon}\rangle = \frac{\alpha_\varepsilon}{2\theta}(\beta_1\overline{\gamma_3} - \beta_3\overline{\gamma_1}),\end{aligned}$$

so that we have indeed $M^*_{0,\theta,\varepsilon} = -M_{0,\theta,\varepsilon}$. With $w \in (\mathbb{C}w_{1,\varepsilon} \oplus \mathbb{C}w_{3,\varepsilon})^\perp$, we have

$$\|M_{0,\theta,\varepsilon}(\beta_1 w_{1,\varepsilon} + \beta_3 w_{3,\varepsilon} + w)\|^2 = \big(\frac{\alpha_\varepsilon}{2\theta}\big)^2(|\beta_1|^2 + |\beta_3|^2),$$

so that, thanks to (102),

$$\|M_{0,\theta,\varepsilon}\|_{\mathcal{B}(\mathbb{H})} = \frac{\alpha_\varepsilon}{2\theta} \le \frac{\sqrt{\varepsilon}}{\theta}. \tag{108}$$

In particular, given a positive θ, we can find $\varepsilon > 0$ such that

$$\|M_{0,\theta,\varepsilon}\|_{\mathcal{B}(\mathbb{H})} \le \varepsilon^{1/4} \quad (\text{just take } \varepsilon \le \theta^4).$$

We have also for $|t| \le \theta$, from (106), (107),

$$M_{0,\theta,\varepsilon}v(t) = \frac{\alpha_\varepsilon}{2\theta}w_{3,\varepsilon}\cos\left(\alpha_\varepsilon\frac{t+\theta}{2\theta}\right) - \frac{\alpha_\varepsilon}{2\theta}w_{1,\varepsilon}\sin\left(\alpha_\varepsilon\frac{t+\theta}{2\theta}\right) = \dot{v}. \tag{109}$$

Definition 2.9 Using (72) and (107), we set

$$\mathscr{M}_{\theta,\phi,\varepsilon} = \mathscr{L}_{\theta,\phi} + \mathbf{1}_{[-\theta,\theta]}(t)iM_{0,\theta,\varepsilon} = D_t + iQ(t) + \mathbf{1}_{[-\theta,\theta]}(t)iM_{0,\theta,\varepsilon}, \tag{110}$$

and with $w_{1,\varepsilon}, w_{2,\varepsilon}, w_{3,\varepsilon}$ given by (98) and (103), we define

$$v_{\theta,\varepsilon}(t) = \begin{cases} e^{(t+\theta)Q_1}w_{1,\varepsilon}, & \text{for } t \le -\theta,\\ w_{1,\varepsilon}\cos\left(\alpha_\varepsilon\frac{t+\theta}{2\theta}\right) + w_{3,\varepsilon}\sin\left(\alpha_\varepsilon\frac{t+\theta}{2\theta}\right), & \text{for } t \in (-\theta,\theta),\\ e^{(t-\theta)Q_2}w_{2,\varepsilon}, & \text{for } t \ge \theta.\end{cases} \tag{111}$$

Theorem 2.10 *For $\varepsilon \in (0,1]$, the function $v_{\theta,\varepsilon}$ given in Definition 2.9 belongs to $Lip(\mathbb{R}, \mathbb{H})$ and is such that*

$$\sup_{t\in\mathbb{R}} \|v_{\theta,\varepsilon}(t)\|_{\mathbb{H}} \le 1, \quad \|v_{\theta,\varepsilon}\|^2_{L^2(\mathbb{R},\mathbb{H})} \ge \|v_{\theta,\varepsilon}\|^2_{L^2([-\theta,\theta],\mathbb{H})} = 2\theta, \tag{112}$$

and with $\mathscr{M}_{\theta,\phi,\varepsilon}$ given in (110), *we have*

$$\mathscr{M}_{\theta,\phi,\varepsilon} v_{\theta,\varepsilon} = 0. \tag{113}$$

We note also that

$$\mathscr{M}_{\theta,\phi,\varepsilon} = D_t + iQ(t) + \left[\mathbf{1}_{[-\theta,\theta]}(t) \text{ bounded selfadjoint operator } iM_{0,\theta,\varepsilon}\right],$$

where $\|M_{0,\theta,\varepsilon}\|_{\mathcal{B}(\mathbb{H})} \le \varepsilon^{1/4}$, provided $\varepsilon \le \theta^4$.

Proof We note first that

$$\begin{aligned} v_{\theta,\varepsilon}((-\theta)_-) &= w_{1,\varepsilon} = v_{\theta,\varepsilon}((-\theta)_+), \\ v_{\theta,\varepsilon}(\theta_-) &= w_{1,\varepsilon}\cos\alpha_\varepsilon + w_{3,\varepsilon}\sin\alpha_\varepsilon = w_{2,\varepsilon} = v_{\theta,\varepsilon}(\theta_+), \end{aligned}$$

proving that v_ε belongs to Lip($\mathbb{R}$, $\mathbb{H}$). Moreover, since we have from (98), (103), and (99)

$$w_{1,\varepsilon} \in E_{1,+}, \quad \|v_{\theta,\varepsilon}([-\theta,\theta])\|_{\mathbb{H}} = 1, \quad w_{2,\varepsilon} \in E_{2,-}, \quad \text{we get (112).}$$

We have also

$$i\mathscr{M}_{\theta,\phi,\varepsilon} v_{\theta,\varepsilon} = \frac{dv_{\theta,\varepsilon}}{dt} - \left\{\begin{matrix} Q_1 v_{\theta,\varepsilon}, & t \le -\theta, \\ M_{0,\theta,\varepsilon} v_{\theta,\varepsilon}, & -\theta \le t \le \theta, \\ Q_2 v_{\theta,\varepsilon}, & t \ge \theta, \end{matrix} \quad \text{thanks to (109) this implies} \right\} = 0,$$

completing the proof of the theorem.

2.4 *The Drift Is Governing the Estimates*

Theorem 2.11 *Let Q_1, Q_2 be two self-adjoint operators on a Hilbert space $\mathbb{H}$ such that the* drift(Q_1, Q_2) *is bounded, i.e., using the notations* (79) *and* (78),

$$\alpha(Q_1, Q_2) = \inf(E_{2,+} + E_{1,-}) = \alpha_0 > 0. \tag{114}$$

Then we have for any measurable nonnegative function a defined on $\mathbb{R}$ *and any* $v \in C_c^1(\mathbb{R}, \mathbb{H})$ *the estimate*

$$\left\| D_t v + ia(t)\big[\check{H}(t)Q_1 + H(t)Q_2\big]v \right\|_{L^1(\mathbb{R},\mathbb{H})} \ge \frac{\alpha_0}{12}\|v\|_{L^\infty(\mathbb{R},\mathbb{H})}. \tag{115}$$

Proof We shall use a multiplier method and define

$$M(t) = -\check{H}(t)E_{1,-} + H(t)E_{2,+}, \tag{116}$$

so that we have

$$\dot{M}(t) = (E_{2,+} + E_{1,-})\delta_0(t) \ge \alpha_0\delta_0(t), \quad \|M(t)\|_{\mathcal{B}(\mathbb{H})} \le 1. \tag{117}$$

We calculate

$$\begin{aligned} &2\,\mathrm{Re}\langle D_t v + ia(t)\big[\check{H}(t)Q_1 + H(t)Q_2\big]v, iM(t)v\rangle_{L^2(\mathbb{R},\mathbb{H})} \\ &= \langle \dot{M}v, v\rangle_{\mathbb{H}} + 2\int_{-\infty}^{0} a(t)\langle Q_{1,-}v(t), v(t)\rangle_{\mathbb{H}}dt \\ &\quad + 2\int_0^{+\infty} a(t)\langle Q_{2,+}v(t), v(t)\rangle_{\mathbb{H}}dt \\ &\ge \alpha_0\|v(0)\|_{\mathbb{H}}^2, \end{aligned} \tag{118}$$

where we have used

$$\begin{gathered} Q_{j,+} = E_{j,+}Q_j = Q_jE_{j,+} \ge 0, \quad Q_{j,-} = -E_{j,-}Q_j = -Q_jE_{j,-} \ge 0, \\ E_{j,\pm} = E_{j,\pm}^* = E_{j,\pm}^2, \quad E_{j,+} + E_{j,-} = \mathrm{Id}, \\ Q_j = Q_{j,+} - Q_{j,-}, \quad Q_{j,+}Q_{j,-} = Q_{j,-}Q_{j,+} = 0. \end{gathered}$$

On the other hand we have

$$\begin{aligned} &2\,\mathrm{Re}\langle D_t v + ia(t)\big[\check{H}(t)Q_1 + H(t)Q_2\big]v, iH(t)\,\mathrm{sign}\,Q_2v\rangle_{L^2(\mathbb{R},\mathbb{H})} \\ &\quad \ge \langle \mathrm{sign}(Q_2)v(0), v(0)\rangle, \end{aligned} \tag{119}$$

and
$$\begin{aligned} &2\,\mathrm{Re}\langle D_t v + ia(t)\big[\check{H}(t)Q_1 + H(t)Q_2\big]v, i\check{H}(t)\,\mathrm{sign}\,Q_1v\rangle_{L^2(\mathbb{R},\mathbb{H})} \\ &\quad \ge -\langle \mathrm{sign}(Q_1)v(0), v(0)\rangle. \end{aligned} \tag{120}$$

As a consequence, adding the estimates (118), (119), and (120), we find that, for $\beta \ge 0$,

$$
\begin{aligned}
&2\operatorname{Re}\langle D_t v + iQ(t)v, i\big[M(t)v + \beta \check{H}(t)\operatorname{sign}(Q_1) + \beta H(t)\operatorname{sign}(Q_2)\big]\rangle_{L^2(\mathbb{R},\mathbb{H})} \\
&\quad \geq \langle \big[E_{2,+} + E_{1,-} - \beta(E_{1,+} - E_{1,-}) + \beta(E_{2,+} - E_{2,-})\big]v(0), v(0)\rangle_{\mathbb{H}} \\
&\qquad\qquad + 2\beta \int_{\mathbb{R}} \langle |Q(t)|v, v(t)\rangle_{\mathbb{H}} dt \\
&\quad = \langle \big[E_{2,+}(1+\beta) + E_{1,-}(1+\beta) - \beta E_{2,-} - \beta E_{1,+}\big]v(0), v(0)\rangle_{\mathbb{H}} \\
&\qquad\qquad + 2\beta \int_{\mathbb{R}} \langle |Q(t)|v, v(t)\rangle_{\mathbb{H}} dt \\
&\quad \geq \langle \big[(1+\beta)\alpha_0 - 2\beta\big]v(0), v(0)\rangle_{\mathbb{H}} + 2\beta \int_{\mathbb{R}} \langle |Q(t)|v, v(t)\rangle_{\mathbb{H}} dt. \qquad (121)
\end{aligned}
$$

We note that we have with M defined in (116)

$$
M_\beta(t) = M(t) + \beta \check{H}(t)\operatorname{sign} Q_1 + \beta H(t)\operatorname{sign} Q_2, \qquad (122)
$$

$$
\|M_\beta(t)\|_{\mathcal{B}(\mathbb{H})} \leq 1 + \beta, \qquad (123)
$$

so that with $\mathcal{L} = D_t + ia(t)\big[\check{H}(t)Q_1 + H(t)Q_2\big]$, we obtain from (121) and (122)

$$
\begin{aligned}
&2\operatorname{Re} \int_{\mathbb{R}} \langle (\mathcal{L}v)(t), iM_\beta(t)v(t)\rangle_{\mathbb{H}} dt \\
&\qquad \geq \big((1+\beta)\alpha_0 - 2\beta\big)\|v(0)\|_{\mathbb{H}}^2 + 2\beta \int_{\mathbb{R}} \langle |Q(t)|v, v(t)\rangle_{\mathbb{H}} dt. \qquad (124)
\end{aligned}
$$

This implies for any $T \in \mathbb{R}$, $\gamma > 0$,

$$
\begin{aligned}
&2\operatorname{Re} \int_{\mathbb{R}} \langle (\mathcal{L}v)(t), iM_\beta(t)v(t) + i\gamma H(t-T)v(t)\rangle_{\mathbb{H}} dt \\
&\quad \geq \big((1+\beta)\alpha_0 - 2\beta\big)\|v(0)\|_{\mathbb{H}}^2 + 2\beta \int_{\mathbb{R}} \langle |Q(t)|v, v(t)\rangle_{\mathbb{H}} dt + \gamma \|v(T)\|_{\mathbb{H}}^2 \\
&\qquad\qquad + 2\gamma \int_{\mathbb{R}} H(t-T)\langle Q(t)v(t), v(t)\rangle_{\mathbb{H}} dt \\
&\geq \overbrace{\big((1+\beta)\alpha_0 - 2\beta\big)}^{\geq 0 \text{ if } \alpha_0 \geq \beta(2-\alpha_0)} \|v(0)\|_{\mathbb{H}}^2 + 2(\beta - \gamma) \int_{\mathbb{R}} \langle |Q(t)|v(t), v(t)\rangle_{\mathbb{H}} dt + \gamma \|v(T)\|_{\mathbb{H}}^2. \qquad (125)
\end{aligned}
$$

Noting that $\alpha_0 \in (0, 2]$, we get, choosing $\gamma = \beta = \alpha_0/(4 - \alpha_0)$, using (123), (125), and $\|M_\beta + \gamma \operatorname{Id}\|_{\mathcal{B}(\mathbb{H})} \leq 1 + \beta + \gamma$, the inequality

$$2\|\mathcal{L}v\|_{L^1(\mathbb{R},\mathbb{H})}(1+2\beta)\|v\|_{L^\infty(\mathbb{R},\mathbb{H})} \ge \beta\|v\|^2_{L^\infty(\mathbb{R},\mathbb{H})},$$

$$\text{yielding} \quad \|\mathcal{L}v\|_{L^1(\mathbb{R},\mathbb{H})} \ge \frac{\frac{\alpha_0}{4-\alpha_0}}{2(1+\frac{2\alpha_0}{4-\alpha_0})} = \frac{\alpha_0}{8+2\alpha_0}\|v\|_{L^\infty(\mathbb{R},\mathbb{H})},$$

which gives the sought result since $\alpha_0 \in (0, 2]$.

Remark 2.12 A consequence of the previous discussion is that for the two-step operators

$$\mathcal{L}_{a;Q_1,Q_2} = D_t + ia(t)\big[\check{H}(t)Q_1 + H(t)Q_2\big], \tag{126}$$

where a is a nonnegative measurable scalar function, either the drift is bounded (see (83)), i.e., (114) holds true, and then we find a nice estimate such as (115), or the drift is infinite, i.e., $\inf(E_{2,+} + E_{1,-}) = 0$, and then Theorem 2.10 shows that you can find a bounded self-adjoint operator N in $\mathcal{B}(\mathbb{H})$ and $\theta > 0$ such that the operator

$$D_t + i\mathbf{1}_{(-\theta,\theta)^c}(t)\big[\check{H}(t)Q_1 + H(t)Q_2\big] + \mathbf{1}_{[-\theta,\theta]}(t)N$$

has a nontrivial kernel. In that sense, the drift is indeed governing the occurrence of an estimate for the operator $\mathcal{L}_{a;Q_1,Q_2}$.

2.5 The Hardy Operator and Singular Integrals Leading to an Infinite Drift

Looking back at the Hardy operator $\mathscr{H}$ defined in Lemma 2.5, we see that the very point of matters related to the infinite drift of the pair $\big(D_x, D_x + \frac{1}{2}\delta_0(x)\big)$ is the fact that 1 belongs to the spectrum of the self-adjoint bounded operator $\mathscr{H}$. We gave two different proofs of this fact: the first one was based on a change to logarithmic coordinates, which proves that $\mathscr{H}$ appears in these coordinates as a Fourier multiplier with symbol $\operatorname{sech}(\pi^2\tau)$ (see Lemma 2.4 and its proof), whereas the second proof was an explicit computation (cf. Remark 2.8) exhibiting 1 as a generalized eigenvalue for $\mathscr{H}$. We see that the kernel

$$\mathfrak{h}(x, y) = \frac{H(x)H(y)}{\pi(x + y)}$$

is *not* the kernel of a Calderón–Zygmund operator (in particular, none of the properties of Definition A.3 holds true, cf. the discussion at the end of Sect. 3.5) but is closely related to the Hilbert operator $\operatorname{sign}(D)$ by Formula (94) which reads as

$$\mathscr{H} = HiC \overbrace{\operatorname{sign}(D)}^{\text{Hilbert transform}} H,$$

where H is the orthogonal projection onto $L^2(\mathbb{R}_+)$ and $\mathcal{C}$ is defined by (170) in Appendix A.[7] We shall keep that in mind before we start our next section in which we shall see that singular integrals of a rather similar type occur.

3 Singular Integrals Involved in Counterexamples to the Flandrin Conjecture

3.1 Preliminaries

We study in this section the operator

$$A_0 = \mathrm{Op_w}(H(x)H(\xi)), \tag{127}$$

where $H = \mathbf{1}_{\mathbb{R}_+}$, that is, the Weyl quantization of the characteristic function of the first quarter of the plane.

Lemma 3.1 *The operator A_0 given by* (127) *is bounded self-adjoint on $L^2(\mathbb{R})$.*

Proof Since the Weyl symbol of A_0 is real-valued, A_0 is formally self-adjoint, and it is enough to prove that A_0 is bounded on $L^2(\mathbb{R})$. The kernel of A_0 is[8]

$$k_0(x,y) = H(x+y)\hat{H}(y-x) = H(x+y)\frac{1}{2}\Big(\delta_0(y-x) + \frac{1}{i\pi}\mathrm{pv}\frac{1}{y-x}\Big), \tag{128}$$

so that for u, v smooth compactly supported such that

$$u = u_+ + u_-,\ \operatorname{supp} u_\pm \subset \mathbb{R}^*_\pm, \qquad v = v_+ + v_-,\ \operatorname{supp} v_\pm \subset \mathbb{R}^*_\pm, \tag{129}$$

we find readily

$$\langle k_0(x,y), u(y)\bar{v}(x)\rangle_{\mathscr{S}'(\mathbb{R}^2),\mathscr{S}(\mathbb{R}^2)} = \frac{1}{2}\iint H(x)u(x)\bar{v}(x)dx$$

$$-\frac{1}{2i\pi}\langle \mathrm{pv}\frac{1}{t}, \int_0^{+\infty} u\Big(s-\frac{t}{2}\Big)\bar{v}\Big(s+\frac{t}{2}\Big)ds\rangle_{\mathscr{S}'(\mathbb{R}_t),\mathscr{S}(\mathbb{R}_t)}$$

$$= \frac{1}{2}\langle Hu, v\rangle_{L^2(\mathbb{R})} + \frac{1}{2}\langle (\operatorname{sign} D)u_+, v_+\rangle_{L^2(\mathbb{R})}$$

$$-\iint \frac{H(x-y)H(x)H(y)}{2i\pi(x+y)}(\mathcal{C}u_-)(y)\bar{v}_+(x)dydx$$

$$+\iint \frac{H(-x+y)H(x)H(y)}{2i\pi(y+x)}u_+(y)\overline{\mathcal{C}v_-}(x)dydx.$$

---

[7] We note that $\mathscr{H}$ is self-adjoint since $\big(i\mathcal{C}\operatorname{sign}(D)\big)^* = -i\operatorname{sign}(D)\mathcal{C} = i\mathcal{C}\operatorname{sign}(D)$.

[8] Note that the tensor product of two distributions always makes sense and $T_1(x+y)T_2(x-y)$ as a distribution on $\mathbb{R}^2_{x,y}$ for T_1, T_2 distributions on $\mathbb{R}$.

Since the Hardy operator given in Lemma 2.5 is bounded in $L^2(\mathbb{R})$ and has the nonnegative kernel (93) in Sect. 2, we infer that the operators with respective kernels

$$\begin{aligned} H(x-y)\mathfrak{h}(x,y) &= \frac{H(x-y)H(x)H(y)}{\pi(y+x)}, \quad \frac{H(y-x)H(x)H(y)}{\pi(y+x)} \\ &= H(y-x)\mathfrak{h}(x,y) \end{aligned}$$

are also bounded on $L^2(\mathbb{R})$ (and have an operator-norm smaller than 1). We define

$$\mathcal{K} \text{ as the operator with kernel } H(x-y)\mathfrak{h}(x,y). \tag{130}$$

We have proven that

$$A_0 = \frac{1}{2}H\big(\mathrm{Id} + \operatorname{sign} D\big)H + \frac{1}{2}\big(i\mathcal{K}\mathcal{C} - i\mathcal{C}\mathcal{K}^*\big), \tag{131}$$

and since H is the orthogonal projection onto $L^2(\mathbb{R}_+)$, $\operatorname{sign} D$ is the Hilbert transform (already proven bounded on $L^2(\mathbb{R})$), $\mathcal{K}$ is bounded on $L^2(\mathbb{R})$, and $\mathcal{C}$ is given by (170) in Appendix A, we see at once from (131) that

$$\begin{aligned} \langle A_0 u, u\rangle_{L^2(\mathbb{R})} &= \|H(D)Hu\|^2_{L^2(\mathbb{R})} + \operatorname{Re}\langle i\mathcal{K}\mathcal{C}\check{H}u, Hu\rangle_{L^2(\mathbb{R})} \\ &\le \|Hu\|^2_{L^2(\mathbb{R})} + \operatorname{Re}\langle i\mathcal{K}\mathcal{C}\check{H}u, Hu\rangle_{L^2(\mathbb{R})} \\ &\le \|Hu\|^2_{L^2(\mathbb{R})} + \|Hu\|_{L^2(\mathbb{R})}\|\check{H}u\|_{L^2(\mathbb{R})}, \end{aligned}$$

proving the lemma and also the estimate $\|A_0\|_{\mathcal{B}(L^2(\mathbb{R}))} \le \frac{1+\sqrt{2}}{2} < 1.20711$, a nontrivial estimate that can be drastically improved.

3.2 Diagonal Terms

We wish to calculate now the kernel of the operator HA_0H, where, as before, H stands for the projection onto $L^2(\mathbb{R}_+)$, that is, the operator of multiplication by $H(x)$. For u as in (129), we define for $t \in \mathbb{R}$

$$\phi_+(t) = u_+(e^t)e^{t/2}, \quad \phi_-(t) = u_-(-e^t)e^{t/2}, \tag{132}$$

so that $\|\phi_+\|_{L^2(\mathbb{R})} = \|u_+\|_{L^2(\mathbb{R}_+)} = \|Hu\|_{L^2(\mathbb{R})}$, $\|\phi_-\|_{L^2(\mathbb{R})} = \|u_-\|_{L^2(\mathbb{R}_-)} = \|\check{H}u\|_{L^2(\mathbb{R})}$, and we find

$$\begin{aligned}\langle HA_0Hu, u\rangle_{L^2(\mathbb{R})} &= \frac{1}{2}\int |u_+(x)|^2dx + \frac{1}{2}\langle \mathrm{pv}(\frac{1}{i\pi(y-x)}), u_+(y)\overline{u_+(x)}\rangle \\ \text{(omitting the principal value)} &= \frac{1}{2}\|\phi_+\|^2_{L^2(\mathbb{R})} + \frac{1}{4i\pi}\iint_{\mathbb{R}^2} \frac{2e^{\frac{s+t}{2}}}{e^t - e^s}\phi_+(t)\overline{\phi_+(s)}dsdt \\ &= \frac{1}{2}\|\phi_+\|^2_{L^2(\mathbb{R})} + \frac{i}{4\pi}\iint_{\mathbb{R}^2} \operatorname{csch}(\frac{s-t}{2})\phi_+(t)\overline{\phi_+(s)}dsdt \\ &= \int |\hat{\phi}_+(\tau)|^2\Big[\frac{1}{2} + \frac{i}{2\pi}\widehat{\operatorname{csch}}(2\tau)\Big]d\tau. \end{aligned} \tag{133}$$

We have also

$$\frac{i}{2\pi}\widehat{\operatorname{csch}}(2\tau) = \frac{i}{2\pi}\int e^{-4i\pi t\tau}\frac{t}{\sinh t}\frac{1}{t}dt = \int e^{-2i\pi t\tau}\frac{t}{4\sinh(t/2)}\frac{i}{\pi t}dt, \tag{134}$$

so that introducing the (even and real-valued) Schwartz function ρ_0 given by

$$\rho_0(\tau) = \int e^{-2i\pi t\tau}\frac{t}{4\sinh(t/2)}dt, \tag{135}$$

we get that

$$\begin{aligned}(\operatorname{sign} * \rho_0)(\tau) &= \int e^{2i\pi t\tau}\widehat{\operatorname{sign}}(t)\widehat{\rho_0}(t)dt = \int e^{2i\pi t\tau}\frac{1}{i\pi t}\frac{t}{4\sinh(t/2)}dt \\ &= \frac{i}{4\pi}\int e^{-2i\pi t\tau}\frac{1}{\sinh(t/2)}dt = \frac{i}{2\pi}\widehat{\operatorname{csch}}(2\tau).\end{aligned} \tag{136}$$

We have also $\frac{d}{d\tau}(\operatorname{sign} * \rho_0) = 2\rho_0(\tau)$ and thus

$$(\operatorname{sign} * \rho_0)(\tau) = (\operatorname{sign} * \rho_0)(+\infty) + \int_{+\infty}^{\tau} 2\rho_0(\tau')d\tau' = \underbrace{\widehat{\rho_0}(0)}_{=1/2} - 2\int_{\tau}^{+\infty}\rho_0(\tau')d\tau',$$

so that (134), (136), and (135) give

$$\frac{1}{2} + \frac{i}{2\pi}\widehat{\operatorname{csch}}(2\tau) = 1 \quad \underbrace{-2\int_{\tau}^{+\infty}\rho_0(\tau')d\tau'}_{=\rho_1(\tau),\ \rho_1 \text{ smooth, bounded, rapidly decreasing at } +\infty}, \tag{137}$$

proving the following:

Claim 3.2 There exists a real-valued function ρ_1 defined [9] on $\mathbb{R}$, smooth, bounded, rapidly decreasing at $+\infty$,[10] such that with u and ϕ_+ given by (129) and (132), we have from (133) and (137)

$$\langle HA_0Hu, u\rangle_{L^2(\mathbb{R})} = \int \big(1+\rho_1(\tau)\big)|\widehat{\phi_+}(\tau)|^2 d\tau. \tag{138}$$

3.3 *Off-Diagonal Terms*

We check now the operator

$$\check{H}A_0H + HA_0\check{H} = 2\,\mathrm{Re}(\check{H}A_0H), \tag{139}$$

keeping in mind that

$$\check{H}A_0\check{H} = 0. \tag{140}$$

Indeed, the latter property follows from the obvious identities

$$\mathcal{C}H = \check{H}\mathcal{C}, \quad H\mathcal{C} = \mathcal{C}\check{H} \tag{141}$$

and the equality (131) which entails

$$-2i\check{H}A_0\check{H} = \check{H}\mathcal{K}\mathcal{C}\check{H} - \check{H}\mathcal{C}\mathcal{K}^*\check{H} = \check{H}\mathcal{K}H\mathcal{C} - \mathcal{C}H\mathcal{K}^*\check{H} = 0,$$

since we have the equalities $\check{H}\mathcal{K} = 0 = \mathcal{K}^*\check{H}$, following from (130) with

$$\mathtt{kernel}(\check{H}\mathcal{K}) = \check{H}(x)H(x-y)\frac{H(x)H(y)}{\pi(y+x)} = 0.$$

We have $2\,\mathrm{Re}(\check{H}A_0H) = \mathrm{Re}(i\check{H}\mathcal{K}\mathcal{C}H - i\check{H}\mathcal{C}\mathcal{K}^*H) = \mathrm{Re}(-i\mathcal{C}H\mathcal{K}^*H)$ and

$$\mathtt{kernel}(H\mathcal{K}^*H) = \frac{H(x)H(y)H(y-x)}{\pi(x+y)},$$

$$\mathtt{kernel}(-i\mathcal{C}H\mathcal{K}^*H) = \frac{H(-x)H(y)H(y+x)}{i\pi(y-x)}.$$

As a consequence, we have

[9] We take $\rho_1(\tau) = -2\int_\tau^{+\infty} \rho_0(\tau')d\tau'$, where ρ_0 is the Schwartz function defined by (135).
[10] We may also note that $\rho_1(\tau) + \rho_1(-\tau) = -1$.

$$
\begin{aligned}
\langle 2\,\mathrm{Re}(\check{H}A_0H)u, u\rangle &= \mathrm{Re}\iint \frac{H(-x)H(y)H(y+x)}{i\pi(y-x)}u(y)\bar{u}(x)dydx \\
&= \mathrm{Re}\iint \frac{H(y+x)}{i\pi(y-x)}u_+(y)\bar{u}_-(x)dydx \\
&= \mathrm{Re}\iint \frac{H(e^t-e^s)e^{\frac{t+s}{2}}}{i\pi(e^t+e^s)}\phi_+(t)\bar{\phi}_-(s)dsdt \\
&= \mathrm{Re}\,\frac{i}{\pi 2}\iint H(t-s)\,\mathrm{sech}(\frac{t-s}{2})\phi_-(s)ds\;\bar{\phi}_+(t)dt \\
&= 2\,\mathrm{Re}\,\frac{i}{4\pi}\langle H\,\mathrm{sech}(\frac{\cdot}{2}) * \phi_-, \phi_+\rangle. \qquad (142)
\end{aligned}
$$

We calculate now

$$
\begin{aligned}
&\frac{i}{4\pi}\int H(t)\,\mathrm{sech}\big(\frac{t}{2}\big)e^{-2i\pi t\tau}dt \\
&= \frac{i}{4\pi}\int_0^{+\infty}\mathrm{sech}\big(\frac{t}{2}\big)\cos(2\pi t\tau)dt + \frac{1}{4\pi}\int_0^{+\infty}\mathrm{sech}\big(\frac{t}{2}\big)\sin(2\pi t\tau)dt \\
&= \underbrace{\frac{i}{8\pi}\int_{-\infty}^{+\infty}\mathrm{sech}\big(\frac{t}{2}\big)\cos(2\pi t\tau)dt}_{=\rho_2(\tau),\ \rho_2\in\mathscr{S}(\mathbb{R})} + \frac{1}{4\pi}\int_0^{+\infty}\mathrm{sech}\big(\frac{t}{2}\big)\sin(2\pi t\tau)dt.
\end{aligned}
$$

Lemma 3.3 *We have*

$$
\int_0^{+\infty}\mathrm{sech}\big(\frac{t}{2}\big)\sin(t\lambda)dt = \frac{1}{\lambda} + O(\lambda^{-3}), \quad |\lambda| \to +\infty.
$$

Proof With $I(\lambda)$ standing for the left-hand side of the above equality, we have

$$
\begin{aligned}
\lambda I(\lambda) &= -\int_0^{+\infty}\mathrm{sech}\big(\frac{t}{2}\big)\frac{d}{dt}\big\{\cos(t\lambda)\big\}dt \text{(integrating by parts repeatedly)}, \\
&= 1 + \frac{1}{2\lambda^2}\int_0^{+\infty}\frac{d^2}{dt^2}\Big\{\mathrm{sech}^2\big(\frac{t}{2}\big)\sinh\big(\frac{t}{2}\big)\Big\}\cos(t\lambda)dt + \frac{1}{4\lambda^2} = 1 + O(\lambda^{-2}),
\end{aligned}
$$

proving the lemma.

As a consequence, we have the following:

Claim 3.4 We define on the real line $G(t) = \frac{i}{4\pi}H(t)\,\mathrm{sech}\big(\frac{t}{2}\big)$, and we have

$$
\hat{G}(\tau) = \frac{i}{4\pi}\int H(t)\,\mathrm{sech}\big(\frac{t}{2}\big)e^{-2i\pi t\tau}dt = \frac{1}{8\pi^2\tau} + O(\tau^{-3}), \quad |\tau| \to +\infty. \qquad (143)
$$

N.B *The function G is discontinuous at* $t = 0$*, bounded on* $\mathbb{R}$*, smooth on* $\mathbb{R}^*$*, rapidly decreasing at infinity, and thus in* $L^1(\mathbb{R})$*; as a consequence, its Fourier transform is a continuous function on* $\mathbb{R}$ *and cannot be rapidly decreasing; otherwise, G would be smooth at 0.*[11]

We have thus from (142)

$$\begin{aligned}\langle 2\,\mathrm{Re}(\check{H}A_0H)u, u\rangle &= 2\,\mathrm{Re}\langle G * \phi_-, \phi_+\rangle = \langle \hat{G}(D)\phi_-, \phi_+\rangle + \langle \phi_+, \hat{G}(D)\phi_-\rangle \\ &= \int \hat{G}(\tau)\widehat{\phi_-}(\tau)\overline{\widehat{\phi_+}(\tau)}d\tau + \int \overline{\hat{G}(\tau)\widehat{\phi_-}(\tau)}\widehat{\phi_+}(\tau)d\tau,\end{aligned} \tag{144}$$

so that, thanks to Claim 3.2, (140), and (144), we get that the operator A_0 appears in logarithmic coordinates as the matrix-valued Fourier multiplier

$$\begin{pmatrix} 1+\rho_1(\tau) & \hat{G}(\tau) \\ \overline{\hat{G}(\tau)} & 0 \end{pmatrix}. \tag{145}$$

The characteristic polynomial of (145) is $P_\tau(\lambda) = \lambda^2 - \lambda\big(1+\rho_1(\tau)\big) - |\hat{G}(\tau)|^2$. We note that

$P_\tau(1) = -\rho_1(\tau) - |\hat{G}(\tau)|^2 < 0,$ for τ positive large enough, thanks to (137) and Claim 3.4,

$P_\tau(0) = -|\hat{G}(\tau)|^2 < 0,$ for τ large enough, thanks to Claim 3.4,

proving that the roots $\lambda_\pm(\tau)$ of P_τ are real-valued such that

$$\lambda_-(\tau) < 0 < 1 < \lambda_+(\tau), \quad \text{for } \tau \text{ positive large enough.}$$

3.4 The Invalidation of Flandrin's Conjecture

Finally, we have proven the following result.

Theorem 3.5 *The operator* A_0 *given by* (127) *is bounded self-adjoint on* $L^2(\mathbb{R})$*, and we have*

$$\inf(A_0) < 0 < 1 < \sup(A_0). \tag{146}$$

[11] The Fourier inversion formula gives $G(t) = \int_{\mathbb{R}} e^{2i\pi t\tau}\hat{G}(\tau)d\tau$. If we had $|\hat{G}(\tau)| \le (1+|\tau|)^{-1-\varepsilon}$ for some positive ε, we would obtain that $\hat{G}$ belongs to $L^1(\mathbb{R})$, and then we would get continuity on $\mathbb{R}$ for G.

Corollary 3.6 *There exists $a > 0$ and $\psi \in \mathscr{S}(\mathbb{R})$ with $L^2(\mathbb{R})$ norm 1 such that*

$$\iint_{[0,a]\times[0,a]} \mathcal{W}(\psi, \psi)(x, \xi)dxd\xi > 1. \tag{147}$$

N.B *This corollary invalidates Flandrin's conjecture, formulated in 1988, which claimed (on page 2178 of [15]) that for any K compact and convex subset of $\mathbb{R}^2$,*

$$\forall u \in L^2(\mathbb{R}), \quad \iint_K \mathcal{W}(u, u)(x, \xi)dxd\xi \le \|u\|^2_{L^2(\mathbb{R})}. \tag{148}$$

A disproof of the Flandrin conjecture was already given in the 2020 article [10] by B. Delourme, T. Duyckaerts, and the author, using a numerical argument based upon the homogeneity of the symbol $H(x)H(\xi)$. Our proof above is "theoretical" and close to the arguments of the paper [39] by J.G. Wood and A.J. Bracken, but we have avoided here formal calculations. P. Flandrin's article [15] tackled the case of the disc in $\mathbb{R}^2$, whereas E.H. Lieb and Y. Ostrover in [31] extended his results to the Euclidean balls of $\mathbb{R}^{2n}$: These two latter papers provided a sharp bound strictly smaller than 1 for the integral of the Wigner distribution on Euclidean balls. Eventually, it turns out that not only the quarter-plane but also any convex subset K with a corner (in the sense of Theorem 1.3 in [10]) fails to satisfy (148).

Proof Theorem 3.5 implies that there exists $u_0 \in L^2(\mathbb{R})$ with $\|u_0\|_{L^2(\mathbb{R})} = 1$ such that $\langle A_0 u_0, u_0\rangle_{L^2(\mathbb{R})} = 1 + 4\beta_0$, $\beta_0 > 0$. Let $\varepsilon > 0$ be given, and let $\psi_\varepsilon \in \mathscr{S}(\mathbb{R})$ such that $\|u_0 - \psi_\varepsilon\|_{L^2(\mathbb{R})} \le \varepsilon$. We have thus

$$\begin{aligned}\langle A_0\psi_\varepsilon, \psi_\varepsilon\rangle_{L^2(\mathbb{R})} &= \langle A_0(\psi_\varepsilon - u_0) + A_0 u_0, \psi_\varepsilon - u_0 + u_0\rangle_{L^2(\mathbb{R})} \\ &= \langle A_0 u_0, u_0\rangle_{L^2(\mathbb{R})} + \langle A_0(\psi_\varepsilon - u_0), \psi_\varepsilon - u_0\rangle_{L^2(\mathbb{R})} \\ &\quad + 2\,\mathrm{Re}\langle A_0(\psi_\varepsilon - u_0), u_0\rangle_{L^2(\mathbb{R})} \\ &\ge 1 + 4\beta_0 - (\varepsilon^2 + 2\varepsilon)\|A_0\|_{\mathcal{B}(L^2(\mathbb{R}))} \ge 1 + 3\beta_0,\end{aligned}$$

provided

$$(\varepsilon^2 + 2\varepsilon)\|A_0\|_{\mathcal{B}(L^2(\mathbb{R}))} \le \beta_0. \tag{149}$$

As a consequence, the Rayleigh quotient satisfies $\frac{\langle A_0\psi_\varepsilon, \psi_\varepsilon\rangle_{L^2(\mathbb{R})}}{\|\psi_\varepsilon\|^2_{L^2(\mathbb{R})}} \ge \frac{1+3\beta_0}{(1+\varepsilon)^2} \ge 1 + 2\beta_0$, provided

$$0 < \varepsilon \le \sqrt{\frac{1 + 3\beta_0}{1 + 2\beta_0}} - 1. \tag{150}$$

We may of course choose $\varepsilon > 0$ such that (149) and (150) are satisfied, and then we find $\Psi \in L^2(\mathbb{R})$ such that

$$\langle A_0\Psi, \Psi\rangle \geq 1 + 2\beta_0, \quad \Psi \in \mathscr{S}(\mathbb{R}), \|\Psi\|_{L^2(\mathbb{R})} = 1,$$

i.e., $\iint_{x\geq 0,\xi\geq 0} \mathcal{W}(\Psi, \Psi)(x, \xi)dxd\xi \geq 1 + 2\beta_0$. By the Lebesgue Dominated Convergence Theorem, since $\mathcal{W}(\Psi, \Psi)$ belongs to $\mathscr{S}(\mathbb{R}^2)$, thus to $L^1(\mathbb{R}^2)$, we have

$$\lim_{a\to+\infty} \iint_{[0,a]^2} \mathcal{W}(\Psi, \Psi)(x, \xi)dxd\xi = \iint_{x\geq 0,\xi\geq 0} \mathcal{W}(\Psi, \Psi)(x, \xi)dx \geq 1 + 2\beta_0.$$

As a result, for a large enough, we find $\iint_{[0,a]^2} \mathcal{W}(\Psi, \Psi)(x, \xi)dxd\xi \geq 1 + \beta_0$, which is the sought result.

3.5 Comments

The main point in the above argument, disproving Flandrin's conjecture, is the fact that the spectrum of the bounded self-adjoint operator

$$\mathrm{Op_w}(H(x)H(\xi))$$

contains an interval $[-\beta_1, 1 + \beta_2]$ with $\beta_1, \beta_2 > 0$. With the quarter-plane

$$\mathcal{Q} = \{(x, \xi) \in \mathbb{R}^2, x \geq 0, \xi \geq 0\},$$

we have $A_0 = \mathrm{Op_w}(\mathbf{1}_{\mathcal{Q}})$, and although

$$\mathbf{1}_{\mathcal{Q}}(x, \xi)^2 = \mathbf{1}_{\mathcal{Q}}(x, \xi), \tag{151}$$

it turns out that the quantization $\mathrm{Op_w}(\mathbf{1}_{\mathcal{Q}})$ is indeed bounded and self-adjoint but is quite far from a projection and in particular has a norm strictly greater than 1. It is not really surprising that the identity (151) could not survive the quantization process, which involves the Heisenberg Uncertainty Principle, by producing non-commuting operators from the quantization of a commutative algebra of Hamiltonians.

Remark 3.7 Moreover, we must point out here that we do not have at our disposal a semiclassical parameter, similar to the Planck constant h. For instance, if we had to deal with a semiclassical symbol of order 0, say $a(x, \xi, h)$, smooth function of x, ξ such that for all multi-indices α, β,

$$\sup_{(x,\xi,h)\in\mathbb{R}^n\times\mathbb{R}^n\times(0,1]} |(\partial_x^\alpha \partial_\xi^\beta a)(x, \xi, h)|h^{-|\beta|} < +\infty, \tag{152}$$

satisfying a hypothesis such as

$$a(x, \xi, h) \in [0, 1], \tag{153}$$

we would get from the Fefferman–Phong Inequality (see [14] or Theorem 18.6.8 in [23])

$$-Ch^2 \le \mathrm{Op_w}(a) \le 1 + Ch^2, \tag{154}$$

providing a sharp control of the spectrum of the operator $\mathrm{Op_w}(a)$.

The most important point for us is to take a look at the singular integrals involved in the expression of the kernel k_0 of A_0: We have indeed, according to (128) and (131),

$$\begin{aligned} k_0(x, y) = &\underbrace{\frac{1}{2}H(x)\Big(1 + \frac{1}{i\pi(y-x)}\Big)H(y)}_{\text{diagonal terms}} \\ &+ \underbrace{\frac{H(x+y)\big(H(x)\check{H}(y) + \check{H}(x)H(y)\big)}{2i\pi(y-x)}}_{=\frac{\mathfrak{h}_1(x,y)}{2i\pi},\ \text{off-diagonal terms}}. \end{aligned} \tag{155}$$

The diagonal terms are producing the operator

$$H\frac{1}{2}(1 + \operatorname{sign} D)H = HH(D)H,$$

which is a quite standard singular integral operator (here H, as above in the text, is standing for the operator of multiplication by $H(x)$, i.e., the orthogonal projection from $L^2(\mathbb{R})$ onto $L^2(\mathbb{R}_+)$). The operator $H(D)$ is the Fourier multiplier with symbol $H(\xi)$. The off-diagonal terms are more unusual and, in particular, the kernel $\mathfrak{h}_1$ is supported in the set

$$\{(x, y) \in \mathbb{R}^2, x \ge 0, y \le 0, y + x \ge 0\} \cup \{(x, y) \in \mathbb{R}^2, x \le 0, y \ge 0, y + x \ge 0\}.$$

Condition (177) is obviously satisfied, and the kernel $\mathfrak{h}_1$ has singularities at

$$\Sigma = \{(x, -x)\}_{x \in \mathbb{R}} \cup \{(0, y)\}_{y \ge 0} \cup \{(x, 0)\}_{x \ge 0}.$$

Although Σ intersects the diagonal only at the origin, the fact that the singular set is not located near the diagonal creates some difficulties that we check right now. Let us take a look at

$$\begin{aligned} &\mathfrak{h}_1(x, y) - \mathfrak{h}_1(x, z) \\ &= \frac{H(x+y)\big(H(x)\check{H}(y) + \check{H}(x)H(y)\big)}{y - x} - \frac{H(x+z)\big(H(x)\check{H}(z) + \check{H}(x)H(z)\big)}{z - x}. \end{aligned}$$

Let us show that the Hörmander integral condition (180) is violated: *reductio ad absurdum.* Let us assume that we have

$$\sup_{y\neq z}\int_{|x-y|>2|y-z|} |\mathfrak{h}_1(x,y)-\mathfrak{h}_1(x,z)|dx<+\infty.$$

Then we would have for $y=-1, z=1$ (note $(3,+\infty)\subset\{x,|x+1|>4\}$)

$$\int_3^{+\infty}|\mathfrak{h}_1(x,-1)-\mathfrak{h}_1(x,1)|dx\le\int_{|x+1|>4}|\mathfrak{h}_1(x,-1)-\mathfrak{h}_1(x,1)|dx<+\infty.$$

We note that when $x>0$, we have $\mathfrak{h}_1(x,1)=0$. Also, for $x>0$, we have

$$\mathfrak{h}_1(x,-1)=\frac{H(x-1)}{-1-x},$$

so that $|\mathfrak{h}_1(x,-1)|=\frac{1}{x+1}$ for $x>1$. Thus we would have

$$\int_3^{+\infty}\frac{dx}{x+1}<+\infty,\qquad\text{which is untrue,}$$

a contradiction proving that (180) is violated. Similarly, (181) is violated, and the off-diagonal terms do not correspond to a standard Calderón–Zygmund singular integral. Nevertheless it was not so difficult to obtain L^2-boundedness for the operators associated with these kernels, and it is even possible to provide a full diagonalization for A_0. It is interesting to see that the kernel of the Hardy operator (with kernel $\mathfrak{h}$ given in (93)) in Sect. 2.3 displays a quite similar behavior, although the simple factorization (94) in Sect. 2.3 allowed us to avoid any complication in the analysis of that operator. However, if we go back to the kernel

$$\mathfrak{h}(x,y)=\frac{H(x)H(y)}{\pi(x+y)},$$

we see that its singularities are located on

$$\Sigma_0=\{(0,y)\}_{y\ge0}\cup\{(x,0)\}_{x\ge0}\tag{156}$$

and that Hörmander's integral condition (180) is violated. Let us assume that we have

$$\sup_{y\neq z}\int_{|x-y|>2|y-z|}|\mathfrak{h}(x,y)-\mathfrak{h}(x,z)|dx<+\infty.$$

Then for $z=1, y=-1$, we have $(3,+\infty)\subset\{x,|x+1|>4\}$, and thus we would get

$$\int_3^{+\infty} |\underbrace{\mathfrak{h}(x,-1)}_{=0} - \mathfrak{h}(x,1)|dx < +\infty, \text{ i.e. } \int_3^{+\infty} \frac{1}{x+1} dx < +\infty, \text{ which is untrue.}$$

As a conclusion, we have seen that the kernels $\mathfrak{h}$ (given in (93)) in Sect. 2.3 and $\mathfrak{h}_1$ (given in (155)) both give rise to $L^2(\mathbb{R})$-bounded operators but are not standard kernels (cf. Definition A.3). Both of these kernels are playing an important role in two different counterexamples, $\mathfrak{h}$ for a counterexample to optimal solvability with the fact that, using the notation (192),

$$\|\mathrm{Op_w}(\mathsf{a}[\mathfrak{h}])\|_{\mathcal{B}(L^2(\mathbb{R}))} = 1, \quad \mathtt{spectrum}\big(\mathrm{Op_w}(\mathsf{a}[\mathfrak{h}])\big) = [0,1], \tag{157}$$

and $\mathfrak{h}_1$ for a counterexample to Flandrin's conjecture with the fact that

$$\exists \beta_1 > 0, \beta_2 > 0, \quad [-\beta_1, 1+\beta_2] \subset \mathtt{spectrum}\big(\mathrm{Op_w}(\mathsf{a}[\mathfrak{h}_1])\big). \tag{158}$$

It is then certainly interesting to study these types of kernels and, for instance, to take a look in one dimension to a relevant condition ensuring $L^2(\mathbb{R})$-boundedness of operators with kernels

$$\mathfrak{k}_\alpha(x,y) = \frac{\alpha(x,y)}{i\pi(x-y)}, \quad \alpha \in L^\infty(\mathbb{R}^2). \tag{159}$$

Of course some additional conditions on α must be required to ensure that $\mathrm{Op_w}(\mathsf{a}[\mathfrak{k}_\alpha])$ is $L^2(\mathbb{R})$-bounded. However, having some additional information on the support of α yields the following result.

Lemma 3.8 *Let $\mathfrak{k}_\alpha$ be the kernel given by* (159) *such that*

$$\operatorname{supp}\alpha \subset \{(x,y), xy \le 0\}. \tag{160}$$

Then the operator $\mathrm{Op_w}(\mathsf{a}[\mathfrak{k}_\alpha])$ *is bounded on* $L^2(\mathbb{R})$, *and we have*

$$\mathrm{Op_w}(\mathsf{a}[\mathfrak{k}_\alpha]) \le \|\alpha\|_{L^\infty(\mathbb{R}^2)}. \tag{161}$$

Proof Indeed we have then

$$\mathfrak{k}_\alpha(x,y) = \frac{H(x)H(-y)\alpha(x,y)}{i\pi(x-y)} + i\frac{H(-x)H(y)\alpha(x,y)}{\pi(-x+y)},$$

and thus, using (141),

$$\begin{aligned}\mathrm{Op_w}(\mathsf{a}[\mathfrak{k}_\alpha]) &= -i\mathscr{H}_\alpha C + iC\mathscr{H}_{\check{\alpha}} \\ &= -iH\mathscr{H}_\alpha C\check{H} + i\check{H}C\mathscr{H}_{\check{\alpha}}H = -i\mathscr{H}_\alpha C + iC\mathscr{H}_{\check{\alpha}},\end{aligned} \tag{162}$$

where $\mathcal{C}$ is given by (170) and

$$\mathtt{kernel}(\mathscr{H}_\alpha) = \frac{H(x)H(y)\alpha(x,-y)}{\pi(x+y)}. \tag{163}$$

From Lemma 2.5, we see that the operator $\mathscr{H}_\alpha$ is $L^2(\mathbb{R})$-bounded with

$$\|\mathscr{H}_\alpha\|_{\mathcal{B}(L^2(\mathbb{R}))} \le \|\alpha\|_{L^\infty(\mathbb{R}^2)}, \tag{164}$$

and this gives

$$\begin{aligned}
&|\langle \mathrm{Op_w}(\mathsf{a}[\Bbbk_\alpha])u, v\rangle_{L^2(\mathbb{R})}| \\
&\quad \le \|\alpha\|_{L^\infty(\mathbb{R}^2)}\big(\|\check{H}u\|_{L^2(\mathbb{R})}\|Hv\|_{L^2(\mathbb{R})} + \|Hu\|_{L^2(\mathbb{R})}\|\check{H}v\|_{L^2(\mathbb{R})}\big) \\
&\quad \le \|\alpha\|_{L^\infty(\mathbb{R}^2)}\|u\|_{L^2(\mathbb{R})}\|v\|_{L^2(\mathbb{R})},
\end{aligned}$$

proving the lemma.

We remark that, dropping Condition (160), even assuming formal self-adjointness which amounts to

$$\overline{\alpha(y,x)} = \alpha(x,y), \tag{165}$$

the estimate (161) does not hold. Indeed, we have seen in Sect. 2 that the choice

$$\alpha_0(x,y) = H(x+y)\big[H(x)H(y) + H(x)\check{H}(y) + \check{H}(x)H(y)\big] \tag{166}$$

yields (self-adjointness and) $L^2(\mathbb{R})$-boundedness for the operator $\mathrm{Op_w}(\mathsf{a}[\Bbbk_{\alpha_0}])$ (cf. (159)), but we have seen also that[12]

$$\mathtt{spectrum}\Big(\frac{1}{2}H + \frac{1}{2}\mathrm{Op_w}(\mathsf{a}[\Bbbk_{\alpha_0}])\Big) \supset [-\beta_1, 1+\beta_2], \quad \beta_1 > 0, \beta_2 > 0,$$

entailing that

$$\|\mathrm{Op_w}(\mathsf{a}[\Bbbk_{\alpha_0}])\|_{\mathcal{B}(L^2(\mathbb{R}))} \ge 1 + 2\beta_2 > 1, \tag{167}$$

since for any $\varepsilon > 0$, there exists v_ε normalized in $L^2(\mathbb{R})$ such that

[12] We have, according to (155),

$$k_0(x,y) = \frac{1}{2}H(x)\delta_0(y-x)H(y) + \frac{H(x+y)}{2i\pi(y-x)}\big(H(x)H(y) + H(x)\check{H}(y) + \check{H}(x)H(y)\big).$$

$$\frac{1}{2}\|Hv_\varepsilon\|^2_{L^2(\mathbb{R})} + \frac{1}{2}\langle \mathrm{Op_w}(\mathsf{a}[\mathfrak{k}_{\alpha_0}])v_\varepsilon, v_\varepsilon\rangle_{L^2(\mathbb{R})} \geq 1 + \beta_2 - \varepsilon,$$

yielding $\langle \mathrm{Op_w}(\mathsf{a}[\mathfrak{k}_{\alpha_0}])v_\varepsilon, v_\varepsilon\rangle_{L^2(\mathbb{R})} \geq 1 + 2\beta_2 - 2\varepsilon$. Since we have $\|\alpha_0\|_{L^\infty(\mathbb{R}^2)} = 1$, this proves that the estimate (161) cannot hold in general without the support assumption (160). As we have seen in this section, the kernel $\mathfrak{k}_{\alpha_0}$ is not a standard kernel and does not even satisfy Hörmander's integral conditions (180), (181), so that we cannot expect to use the many results known on the Calderón–Zygmund operators.

Appendix A: About Pseudo-Differential Operators

On the Fourier Transformation

The Fourier transformation is defined for $u \in \mathscr{S}(\mathbb{R}^n)$ by

$$\hat{u}(\xi) = \int_{\mathbb{R}^n} e^{-2i\pi x\cdot\xi} u(x)dx. \tag{168}$$

It is an elementary fact of Fourier analysis to prove that the Fourier transformation is an isomorphism of the Schwartz space and that we have for $u \in \mathscr{S}(\mathbb{R}^n)$

$$u(x) = \int_{\mathbb{R}^n} e^{2i\pi x\cdot\xi} \hat{u}(\xi)d\xi. \tag{169}$$

Introducing the notation

$$\check{u}(x) = u(-x), \quad Cu = \check{u}, \tag{170}$$

$$\text{we can write this inversion formula as} \quad \check{\hat{\hat{u}}} = u, \tag{171}$$

quite easily proven true for $u \in \mathscr{S}(\mathbb{R}^n)$ and extended trivially by abstract nonsense to temperate distributions u on $\mathbb{R}^n$. Defining

$$H = \mathbf{1}_{\mathbb{R}_+}, \quad \text{as the Heaviside function}, \tag{172}$$

we get from Section 4.1.4 of [29] that

$$\hat{H}(\xi) = \frac{1}{2}\left(\delta_0 + \frac{1}{i\pi}\mathrm{pv}\frac{1}{\xi}\right), \quad \text{where} \quad \mathrm{pv}\frac{1}{\xi} = \frac{d}{d\xi}\{\ln|\xi|\}, \tag{173}$$

$$\frac{1}{i\pi}\mathcal{F}\big(\mathrm{pv}\frac{1}{x}\big) = -\operatorname{sign}\xi, \quad (\mathcal{F}(\mathrm{sign}))(\xi) = \frac{1}{i\pi}\mathrm{pv}\frac{1}{\xi}. \tag{174}$$

We calculate the Fourier transform of the function sech with

$$\int e^{2i\pi x\xi}\operatorname{sech} x\,dx = \pi \operatorname{sech}(\pi^2\xi). \tag{175}$$

Indeed we have with $y = e^x$, $t = (1+y^2)^{-1}$,

$$\begin{aligned}\int e^{2i\pi x\xi}\operatorname{sech} x\,dx &= 2\int_{\mathbb{R}_+}\frac{e^{2i\pi\xi\ln y}}{1+y^2}dy = \int_0^1 (1-t)^{i\pi\xi-\frac12}t^{-i\pi\xi-\frac12}dt\\ &= B\Big(i\pi\xi+\frac12, -i\pi\xi+\frac12\Big) \quad \text{(and thanks to Euler's reflection formula)}\\ &= \frac{\pi}{\sin\pi\,(i\pi\xi+\frac12)} = \pi\operatorname{sech}(\pi^2\xi).\end{aligned}$$

The Hilbert Transform, Singular Integrals

Lemma A.1 *The Hilbert transform is defined as the Fourier multiplier* $\operatorname{sign}(D)$*: That operator can be extended to a bounded self-adjoint operator on* $L^2(\mathbb{R})$ *with distribution kernel*

$$\frac{1}{i\pi}\operatorname{pv}\frac{1}{y-x}. \tag{176}$$

The $\mathcal{B}(L^2(\mathbb{R}))$ *norm of the Hilbert transform is 1.*

Proof Since the Fourier transformation is a unitary operator on $L^2(\mathbb{R})$ and thus $\mathcal{F}^* = \mathcal{F}^{-1}$, we find that

$$\operatorname{sign}(D)u = \mathcal{F}^{-1}\big((\operatorname{sign}\xi)\hat{u}(\xi)\big),$$

that is, $(\operatorname{sign}(D)u)(x) = \lim_{\varepsilon\to 0_+}\int_{|y-x|\ge\varepsilon}\frac{1}{i\pi(y-x)}u(y)dy = -\frac{d}{i\pi dx}\big\{\ln|\cdot|*u\big\}$. The operator $\operatorname{sign}(D)$ is unitarily equivalent to the operator of multiplication by $\operatorname{sign}\xi$ and thus is a bounded and self-adjoint operator on $L^2(\mathbb{R})$ with norm 1.

Remark A.2 We note that the Hilbert transform is simply the Fourier multiplier $\operatorname{sign}(D)$, so has symbol $\operatorname{sign}\xi$; as a result, the Hilbert transform is *not* a pseudo-differential operator since its symbol displays a singularity at 0. The Hilbert transform nevertheless belongs to a very interesting class of operators, extensively studied in Harmonic Analysis, the so-called *Singular Integral Operators*.

Following the book [8], published in 1978 by R. Coifman and Y. Meyer, we recall the notion of *standard kernel* in order to obtain general Calderón–Zygmund operators.

Definition A.3 An L^1_{loc} function k defined on $(\mathbb{R}^n \times \mathbb{R}^n)\backslash\mathtt{diagonal}(:= \mathbb{R}^{2n}_{\mathrm{o}})$ is said to be a *standard kernel* when there exist positive constants δ, C, such that for all $(x, y), (x, z), (w, y) \in \mathbb{R}^{2n}_{\mathrm{o}}$,

$$|k(x,y)| \le \frac{C}{|x-y|^n}, \tag{177}$$

$$|k(x,y) - k(x,z)| \le C\frac{|y-z|^\delta}{|x-y|^{n+\delta}} \quad \text{if } |x-y| > 2|y-z|, \tag{178}$$

$$|k(x,y) - k(w,y)| \le C\frac{|x-w|^\delta}{|x-y|^{n+\delta}} \quad \text{if } |x-y| > 2|x-w|. \tag{179}$$

Remark A.4 We could replace (178), (179) by a weaker integral condition due to L. Hörmander in [18], namely

$$\sup_{(y,z)\in\mathbb{R}^{2n}_{\mathrm{o}}} \int_{|x-y|>2|y-z|} |k(x,y) - k(x,z)|dx < +\infty, \tag{180}$$

$$\sup_{(x,w)\in\mathbb{R}^{2n}_{\mathrm{o}}} \int_{|x-y|>2|x-w|} |k(x,y) - k(w,y)|dy < +\infty. \tag{181}$$

Definition A.5 A Calderón–Zygmund operator on $\mathbb{R}^n$ is a bounded operator on $L^2(\mathbb{R}^n)$ such that its distribution kernel is a standard kernel.

The main result of the theory of Calderón–Zygmund operators is the following result.

Theorem A.6 *Let A be a Calderón–Zygmund operator on $\mathbb{R}^n$. Then, for $p \in (1, +\infty)$, A and A^* are continuous from $L^p(\mathbb{R}^n)$ into itself and are also sending continuously $L^1(\mathbb{R}^n)$ into $L^1_w(\mathbb{R}^n)$.*

The Symplectic Group

On the phase space $\mathbb{R}^{2n} = \mathbb{R}^n_x \times \mathbb{R}^n_\xi$, we define the canonical symplectic form

$$[(x,\xi);(y,\eta)] = \xi \cdot y - \eta \cdot x. \tag{182}$$

With

$$\sigma = \begin{pmatrix} 0 & I_n \\ -I_n & 0 \end{pmatrix}, \quad \text{we have for } X, Y \in \mathbb{R}^{2n}, \quad [X,Y] = (\sigma X)\cdot Y. \tag{183}$$

The symplectic group $Sp(2n, \mathbb{R})$ is the subgroup of $S \in Sl(2n, \mathbb{R})$, such that

$$\forall X, Y \in \mathbb{R}^{2n}, \quad [SX, SY] = [X, Y], \quad \text{i.e., such that} \quad S^*\sigma S = \sigma. \tag{184}$$

For a positive-definite quadratic form Q on $\mathbb{R}^{2n}$ (identified with a positive-definite real symmetric $2n \times 2n$ matrix), we define

$$Q^\sigma = \sigma^* Q^{-1} \sigma. \tag{185}$$

In particular, if

$$Q(x,\xi) = \sum_{1\le j\le n} \left(\frac{x_j^2}{a_j^2} + \frac{\xi_j^2}{\alpha_j^2}\right), \quad \text{we find} \quad Q^\sigma(x,\xi) = \sum_{1\le j\le n} \left(\alpha_j^2 x_j^2 + a_j^2 \xi_j^2\right).$$

Let Q_1, Q_2 be two positive-definite quadratic forms on $\mathbb{R}^{2n}$. We define the Harmonic Mean of Q_1, Q_2 as

$$Q_1 \wedge Q_2 = 2\left(Q_1^{-1} + Q_2^{-1}\right)^{-1}. \tag{186}$$

The Weyl Quantization

Let u, v be in the Schwartz space $\mathscr{S}(\mathbb{R}^n)$. Then we define the *Wigner distribution* of u, v by

$$\mathcal{W}(u,v)(x,\xi) = \int_{\mathbb{R}^n} e^{-2i\pi z\cdot\xi} u\left(x+\frac{z}{2}\right)\bar{v}\left(x-\frac{z}{2}\right) dz, \tag{187}$$

and it is easy to show that $\mathcal{W}(u,v)$ belongs to $\mathscr{S}(\mathbb{R}^{2n})$. Let a be in $\mathscr{S}'(\mathbb{R}^{2n})$: We define the Weyl quantization $\operatorname{Op_w}(a)$ of a by the formula

$$\langle \operatorname{Op_w}(a)u, v\rangle_{\mathscr{S}'(\mathbb{R}^n),\mathscr{S}(\mathbb{R}^n)} = \langle a, \mathcal{W}(u,v)\rangle_{\mathscr{S}'(\mathbb{R}^{2n}),\mathscr{S}(\mathbb{R}^{2n})}, \tag{188}$$

and this provides a weak definition for

$$\left(\operatorname{Op_w}(a)u\right)(x) = \iint e^{2i\pi\langle x-y,\xi\rangle} a\left(\frac{x+y}{2}, \xi\right) u(y)\, dy\, d\xi. \tag{189}$$

We have in particular

$$\left(\operatorname{Op_w}(a)\right)^* = \operatorname{Op_w}(\overline{a}). \tag{190}$$

Claim A.7 Let a be a tempered distribution on $\mathbb{R}^{2n}$. Then the operator $\operatorname{Op_w}(a)$ has the distribution kernel

$$k(x, y) = \widehat{a}^{[2]}\left(\frac{x+y}{2}, y - x\right), \tag{191}$$

where $\widehat{a}^{[2]}$ stands for the Fourier transformation with respect to the second variable on $\mathbb{R}^n \times \mathbb{R}^n$. Conversely, if k is a tempered distribution on $\mathbb{R}^{2n}$, the operator with distribution kernel k has the Weyl symbol

$$a(x, \xi) = \int_{\mathbb{R}^n} k\left(x + \frac{z}{2}, x - \frac{z}{2}\right) e^{-2i\pi z\cdot\xi} dz := \mathsf{a}[k](x, \xi), \tag{192}$$

i.e., the Fourier transform with respect to the variable z of the distribution

$$k\left(x + \frac{z}{2}, x - \frac{z}{2}\right) = (k \circ S)(x, z), \quad S(x, z) = \begin{pmatrix} I_n & \frac{I_n}{2} \\ I_n & -\frac{I_n}{2} \end{pmatrix}\begin{pmatrix} x \\ z \end{pmatrix},$$

on $\mathbb{R}^n_x \times \mathbb{R}^n_z$.

Proof of the Claim Formula (191) follows readily from (189) and gives

$$k\left(x + \frac{z}{2}, x - \frac{z}{2}\right) = \widehat{a}^{[2]}(x, -z).$$

Applying the Fourier inversion formula yields $a(x, \xi) = \int e^{-2i\pi z\cdot\xi}\widehat{a}^{[2]}(x, -z)dz$, which is the sought result.

Symplectic Covariance of the Weyl Calculus

Another important property of the Weyl quantization is its *symplectic covariance.* For $S \in Sp(2n, \mathbb{R})$ (see the subsection on page 269), we have the following formula, say for $a \in L^1(\mathbb{R}^{2n})$:

$$\mathrm{Op_w}(a \circ S) = \mathcal{M}^*\mathrm{Op_w}(a)\mathcal{M}, \tag{193}$$

where $\mathcal{M}$ belongs to the metaplectic group, which is a group of unitary transformations of $L^2(\mathbb{R}^n)$. This translates for the Wigner distribution as

$$\mathcal{W}(\mathcal{M}u, \mathcal{M}v) = \mathcal{W}(u, v) \circ S^{-1}. \tag{194}$$

For instance, for $T \in Gl(n, \mathbb{R})$, we have with $A, C, n \times n$ real matrices

$$\mathrm{Op_w}\big(a(T^{-1}x, {}^tT\xi)\big) = \mathcal{M}_T^*\mathrm{Op_w}(a)\mathcal{M}_T, \quad \text{with } (\mathcal{M}_T u)(x) = (\det T)^{1/2}u(Tx),$$

$$\mathrm{Op_w}\big(a(\xi, -x)\big) = \mathcal{F}^*\mathrm{Op_w}(a)\mathcal{F}, \quad \text{where } \mathcal{F} \text{ is the Fourier transformation,}$$

$$\operatorname{Op_w}\bigl(a(x,\xi+Ax)\bigr)=\mathcal{L}_A^*\operatorname{Op_w}(a)\mathcal{L}_A,\ \text{ where } (\mathcal{L}_A v)(x)=e^{i\pi\langle Ax,x\rangle}v(x),\ {}^tA=A,$$

$$\operatorname{Op_w}\bigl(a(x-C\xi,\xi)\bigr)=\mathcal{R}_C^*\operatorname{Op_w}(a)\mathcal{R}_C,\ \text{ where } (\mathcal{R}_C v)(x)=\bigl(e^{i\pi\langle CD,D\rangle}v\bigr)(x),$$
$${}^tC=C.$$

Estimating an Integral

Lemma A.8 *For $\varepsilon>-1$, we have*

$$\frac{2}{\pi}\int_0^{\pi/2}(\cos\theta)^{\varepsilon}d\theta\ge 1-\varepsilon\operatorname{Log}2.$$

Proof Defining $J(\varepsilon)$ as the left-hand side of the above inequality, we have

$$J'(\varepsilon)=⨍_0^{\pi/2}(\cos\theta)^{\varepsilon}\operatorname{Log}(\cos\theta)d\theta,\quad J''(\varepsilon)=⨍_0^{\pi/2}(\cos\theta)^{\varepsilon}\bigl(\operatorname{Log}(\cos\theta)\bigr)^2d\theta,$$

so that the function J is convex on $\mathbb{R}_+$. Moreover, we have

$$J'(0)=⨍_0^{\pi/2}\operatorname{Log}(\cos\theta)d\theta=⨍_0^{\pi/2}\operatorname{Log}(\sin\theta)d\theta=⨍_0^{\pi/2}\operatorname{Log}(\sin 2\theta)d\theta,$$

and thus $2J'(0)=\frac{2}{\pi}\int_0^{\pi/2}\operatorname{Log}(\frac{1}{2}\sin 2\theta)d\theta=-\operatorname{Log}2+J'(0)$ so that $J'(0)=-\operatorname{Log}2$. We find then $J(\varepsilon)\ge J(0)+\varepsilon J'(0)$, which is the sought result.

Appendix B: Metrics on the Phase Space, Examples, and Properties

The Classes $S^m_{\rho,\delta}$

These standard classes are in fact

$$S\bigl(\langle\xi\rangle^m,g^{[\rho,\delta]}\bigr),\ \text{with } g^{[\rho,\delta]}_{x,\xi}(y,\eta)=\langle\xi\rangle^{2\delta}|y|^2+\langle\xi\rangle^{-2\rho}|\eta|^2,\quad \langle\xi\rangle=(1+|\xi|^2)^{1/2}.$$

When $0\le\delta\le\rho\le 1,\delta<1$, it is not difficult to prove that $g^{[\rho,\delta]}$ is admissible. Also the metric $g^{[1,1]}$ satisfies (56) and (57) but fails to satisfy (58) (references of Sect. 1.3).

Proper Classes

Let us consider a semiclassical symbol of order 2, i.e., a smooth function a defined on $\mathbb{R}^n \times \mathbb{R}^n \times [1, +\infty)$ such that

$$\sup_{\substack{(x,\xi,\Lambda)\in\\ \mathbb{R}^n\times\mathbb{R}^n\times[1,+\infty)}} |(\partial_x^\alpha \partial_\xi^\beta a)(x, \xi, \Lambda)|\Lambda^{-2+\frac{|\alpha|+|\beta|}{2}} < +\infty. \tag{195}$$

We want to find a class of symbols tailored to the symbol a, analogously to what we have done in Sect. 1.2. In fact we would like to find a suitable weight $\lambda(x, \xi) \geq 1$ such that for all $\alpha, \beta \in \mathbb{N}^n$,

$$\sup_{\substack{(x,\xi,\Lambda)\in\\ \mathbb{R}^n\times\mathbb{R}^n\times[1,+\infty)}} |(\partial_x^\alpha \partial_\xi^\beta a)(x, \xi, \Lambda)|\lambda(x, \xi)^{-2+\frac{|\alpha|+|\beta|}{2}} < +\infty, \tag{196}$$

which can be written as

$$\forall k \in \mathbb{N}, \qquad \sup_{\substack{(X,\Lambda)\in\\ \mathbb{R}^{2n}\times[1,+\infty)}} |(\partial_X^k a)(X, \Lambda)|\lambda(X)^{-2+\frac{k}{2}} < +\infty. \tag{197}$$

A dimensional analysis will provide the right—and unique—choice for this Calderón–Zygmund decomposition. The quantity $\lambda^{1/2}$ must have the dimension of a length (radius of the typical ball), so that writing $[\alpha] = \texttt{unit of}\ \alpha$, we get

$$[\lambda^{1/2}] = \mathtt{L},\ [a] = [\lambda^2] = \mathtt{L}^4,\ [a'] = [\lambda^{3/2}] = \mathtt{L}^3,\ [a''] = [\lambda] = \mathtt{L}^2,$$
$$[a'''] = [\lambda^{1/2}] = \mathtt{L},$$

and we find that

$$\lambda(X)^2 = \max\big(|a(X)|, |a'(X)|^{4/3}, |a''(X)|^2, |a'''(X)|^4\big) + 1. \tag{198}$$

The following lemma is easily proven.

Lemma B.1 (See, e.g., Lemma 3.2.12 in [29]) *Let g be the metric on $\mathbb{R}^{2n}$ defined by*

$$g_X(T) = \frac{|T|^2}{\lambda(X)}. \tag{199}$$

Then g is an admissible metric on $\mathbb{R}^{2n}$, $\lambda_g = \lambda$ as defined by (198) *and $a \in S(\lambda^2, g)$.*

Proof The admissibility of g is left to the reader (see the aforementioned reference); the fact that $a \in S(\lambda^2, g)$ is obvious since (197) holds true for $k \leq 3$. On the other hand, for $k \geq 4$, we may use (195) and the fact that $\lambda \leq C\Lambda$, which is a consequence of the very definition of λ and of (195).

There are further simplifications if a takes only nonnegative values: Using Lemma 4.3.11 in [29], we obtain the following result.

Lemma B.2 *Let g be the metric on $\mathbb{R}^{2n}$ defined by*

$$\tilde{g}_X(T) = \frac{|T|^2}{\widetilde{\lambda}(X)}, \quad \widetilde{\lambda}(X) = \max\big(a(X), |a''(X)|^2\big) + 1, \tag{200}$$

where a is a nonnegative function satisfying (195). *Then $\tilde{g}$ is an admissible metric on $\mathbb{R}^{2n}$, $\lambda_{\tilde{g}} = \widetilde{\lambda}$, and $a \in \mathcal{S}((\tilde{\lambda})^2, g)$.*

We found thus a new class of symbols, tailored to the symbol a. We note that the huge progress following this construction is that we get now two types of points X: First the "elliptic" points where $\widetilde{\lambda}(X)$ is "equivalent" to $a(X)$, which means that it will be possible to consider $a^{1/2}$ as a symbol in $\mathcal{S}(\tilde{\lambda}, g)$. Next the "non-degenerate" points where $\widetilde{\lambda}(X)$ is "equivalent" to $|a''(X)|$: In that case, we could use a partial Morse lemma and start an induction on the dimension. What we have shortly discussed above is the first important step in the proof of the Fefferman–Phong Inequality, proven in 1978 by C. Fefferman and D.H. Phong in [14] (see also Theorem 18.6.8 in L. Hörmander's [23], J.-M. Bony's [4], and Corollary 2.5.11 in the author's [29]).

N.B The original proof of [14] used explicitly a Calderón–Zygmund procedure, a very suggestive and intuitive dynamical process for cutting $\mathbb{R}^{2n}$ into relevant "boxes"; moreover C. Fefferman and D.H. Phong adjusted this procedure to the phase space by implementing a stopping time to take into account the uncertainty principle. In fact, using the metrics-on-the-phase-space terminology is certainly less intuitive, but it represents indeed a shortcut for the proof.

References

1. Richard Beals and Charles Fefferman, *Classes of spatially inhomogeneous pseudodifferential operators*, Proc. Nat. Acad. Sci. U.S.A. **70** (1973), 1500–1501. MR 338840
2. Richard Beals and Charles Fefferman, *On local solvability of linear partial differential equations*, Ann. of Math. (2) **97** (1973), 482–498. MR 352746
3. Jean-Michel Bony, *Principe du maximum, inégalite de Harnack et unicité du problème de Cauchy pour les opérateurs elliptiques dégénérés*, Ann. Inst. Fourier (Grenoble) **19** (1969), 277–304 xii. MR 262881
4. Jean-Michel Bony, *Sur l'inégalité de Fefferman-Phong*, Seminaire: Équations aux Dérivées Partielles, 1998–1999, Sémin. Équ. Dériv. Partielles, École Polytech., Palaiseau, 1999, pp. Exp. No. III, 16. MR 1721321
5. Jean-Michel Bony and Jean-Yves Chemin, *Espaces fonctionnels associés au calcul de Weyl-Hörmander*, Bull. Soc. Math. France **122** (1994), no. 1, 77–118. MR 1259109

6. Jean-Michel Bony and Nicolas Lerner, *Quantification asymptotique et microlocalisations d'ordre supérieur. I*, Ann. Sci. École Norm. Sup. (4) **22** (1989), no. 3, 377–433. MR 1011988
7. Haïm Brezis, *On a characterization of flow-invariant sets*, Comm. Pure Appl. Math. **23** (1970), 261–263. MR 257511
8. Ronald R. Coifman and Yves Meyer, *Au delà des opérateurs pseudo-différentiels*, Astérisque, vol. 57, Société Mathématique de France, Paris, 1978, With an English summary. MR 518170
9. Paulo D. Cordaro and Jorge Hounie, *Semi-global solvability with loss of one derivative of partial differential operators on surfaces*, Adv. Math. **301** (2016), 458–485. MR 3539381
10. Bérangère Delourme, Thomas Duyckaerts, and Nicolas Lerner, *On integrals over a convex set of the Wigner distribution*, J. Fourier Anal. Appl. **26** (2020), no. 1, Paper No. 6. MR 4054880
11. Nils Dencker, *The solvability of pseudo-differential operators*, Phase space analysis of partial differential equations. Vol. I, Pubbl. Cent. Ric. Mat. Ennio Giorgi, Scuola Norm. Sup., Pisa, 2004, pp. 175–200. MR 2144409
12. Nils Dencker, *The resolution of the Nirenberg-Treves conjecture*, Ann. of Math. (2) **163** (2006), no. 2, 405–444. MR 2199222
13. Nils Dencker, *Nirenberg's contributions to linear partial differential equations: pseudo-differential operators and solvability*, Bull. Amer. Math. Soc. (N.S.) **60** (2023), no. 2, 159–166. MR 4557378
14. Charles Fefferman and Duong H. Phong, *On positivity of pseudo-differential operators*, Proc. Nat. Acad. Sci. U.S.A. **75** (1978), no. 10, 4673–4674. MR 507931
15. Patrick Flandrin, *Maximum signal energy concentration in a time-frequency domain*, Proc. IEEE Int. Conf. Acoustics **4** (1988), no. 1, 2176–2179.
16. Lars Hörmander, *On the theory of general partial differential operators*, Acta Math. **94** (1955), 161–248. MR 76151
17. Lars Hörmander, *Differential equations without solutions*, Math. Ann. **140** (1960), 169–173. MR 147765
18. Lars Hörmander, *Estimates for translation invariant operators in L^p spaces*, Acta Math. **104** (1960), 93–140. MR 121655
19. Lars Hörmander, *Pseudo-differential operators and non-elliptic boundary problems*, Ann. of Math. (2) **83** (1966), 129–209. MR 233064
20. Lars Hörmander, *The Weyl calculus of pseudodifferential operators*, Comm. Pure Appl. Math. **32** (1979), no. 3, 360–444. MR 517939
21. Lars Hörmander, *Pseudodifferential operators of principal type*, Singularities in boundary value problems (Proc. NATO Adv. Study Inst., Maratea, 1980), NATO Adv. Study Inst. Ser. C: Math. Phys. Sci., vol. 65, Reidel, Dordrecht-Boston, Mass., 1981, pp. 69–96. MR 617227
22. Lars Hörmander, *On the solvability of pseudodifferential equations*, Structure of solutions of differential equations (Katata/Kyoto, 1995), World Sci. Publ., River Edge, NJ, 1996, pp. 183–213. MR 1445340
23. Lars Hörmander, *The analysis of linear partial differential operators. III*, Classics in Mathematics, Springer, Berlin, 2007, Pseudo-differential operators, Reprint of the 1994 edition. MR 2304165
24. Lars Hörmander, *The analysis of linear partial differential operators. IV*, Classics in Mathematics, Springer-Verlag, Berlin, 2009, Fourier integral operators, Reprint of the 1994 edition. MR 2512677
25. Lars Hörmander, *Unpublished manuscripts—from 1951 to 2007*, Springer, Cham, 2018. MR 3837517
26. Nicolas Lerner, *Sufficiency of condition (ψ) for local solvability in two dimensions*, Ann. of Math. (2) **128** (1988), no. 2, 243–258. MR 960946
27. Nicolas Lerner, *Nonsolvability in L^2 for a first order operator satisfying condition (ψ)*, Ann. of Math. (2) **139** (1994), no. 2, 363–393. MR 1274095
28. Nicolas Lerner, *Cutting the loss of derivatives for solvability under condition (Ψ)*, Bull. Soc. Math. France **134** (2006), no. 4, 559–631. MR 2364944

29. Nicolas Lerner, *Metrics on the phase space and non-selfadjoint pseudo-differential operators*, Pseudo-Differential Operators. Theory and Applications, vol. 3, Birkhäuser Verlag, Basel, 2010. MR 2599384
30. Hans Lewy, *An example of a smooth linear partial differential equation without solution*, Ann. of Math. (2) **66** (1957), 155–158. MR 88629
31. Elliott H. Lieb and Yaron Ostrover, *Localization of multidimensional Wigner distributions*, J. Math. Phys. **51** (2010), no. 10, 102101, 6. MR 2761287
32. Sigeru Mizohata, *Solutions nulles et solutions non analytiques*, J. Math. Kyoto Univ. **1** (1961/62), 271–302. MR 142873
33. Robert D. Moyer, *Local solvability in two dimensions: necessary conditions for the principal type case*, Mimeographed manuscript, University of Kansas, 1978.
34. Louis Nirenberg and François Trèves, *Solvability of a first order linear partial differential equation*, Comm. Pure Appl. Math. **16** (1963), 331–351. MR 163045
35. Louis Nirenberg and François Trèves, *On local solvability of linear partial differential equations. I. Necessary conditions*, Comm. Pure Appl. Math. **23** (1970), 1–38. MR 264470
36. Louis Nirenberg and François Trèves, *On local solvability of linear partial differential equations. II. Sufficient conditions*, Comm. Pure Appl. Math. **23** (1970), 459–509. MR 264471
37. Louis Nirenberg and François Trèves, *A correction to: "On local solvability of linear partial differential equations. II. Sufficient conditions" (Comm. Pure Appl. Math. 23 (1970), 459–509)*, Comm. Pure Appl. Math. **24** (1971), no. 2, 279–288. MR 435641
38. André Unterberger, *Oscillateur harmonique et opérateurs pseudo-différentiels*, Ann. Inst. Fourier (Grenoble) **29** (1979), no. 3, xi, 201–221. MR 552965
39. J.G. Wood and A.J. Bracken, *Bounds on integrals of the Wigner function: the hyperbolic case*, J. Math. Phys. **46** (2005), no. 4, 042103, 14. MR 2131219

On Spectral Properties of Certain Semiregular Systems of PDEs with Polynomial Coefficients

Alberto Parmeggiani

To Jorge Hounie

2010 Mathematics Subject Classification Primary 35P20; Secondary 35S05, 81Q10

1 Introduction

In this chapter, I will be giving a survey of some recent joint work [12, 13] with Marcello Malagutti, concerning two basic properties in the spectral theory of systems of PDEs with polynomial coefficients: the asymptotics of the Weyl counting function in a two-term form and in a refined three-term form and the quasi-clustering property (a weak form of clustering property, i.e., a concentration of the spectrum near an arithmetic sequence).

Recall that if $-\infty < \lambda_1 \leq \lambda_2 \leq \ldots \lambda_j \leq \ldots \to +\infty$ is the sequence of the repeated eigenvalues (according to multiplicity) of an elliptic (bounded from below) operator A, then the Weyl counting function is given by

$$\mathsf{N}_A(\lambda) = \sum_{\lambda_j \leq \lambda} 1.$$

It is well known that understanding the asymptotics as $\lambda \to +\infty$ of N_A is a problem of fundamental importance (also related to Number Theory). Several important

Member of the Research Group GNAMPA of INdAM

A. Parmeggiani (✉)
Department of Mathematics, University of Bologna, Bologna, Italy
e-mail: alberto.parmeggiani@unibo.it

S. Berhanu et al. (eds.), *Geometric Analysis of PDEs and Several Complex Variables*, Latin American Mathematics Series – UFSCar subseries,
https://doi.org/10.1007/978-3-031-69702-9_12

contributions were given by Weyl (1911), Hörmander (in his paper on sharp remainders by using wave equation methods), Seeley, Ivrii, Shubin, Duistermaat and Guillemin, Melrose, Safarov, and Vassiliev for *scalar* elliptic operators on a compact (boundaryless) manifold (see [3, 11, 20, 22] and the references in [10] and [7]). In the case of *scalar* globally elliptic operators (see, e.g., [7, 22], also in the semiclassical setting), important contributions were given by Hörmander, Shubin, Ivrii, Chazarain, Helffer, and Robert (see [22, 6, 7, 11]) and recently by Doll, Gannot, and Wunsch in [4] and Doll and Zelditch in [5] in refining, under suitable geometric conditions, the asymptotics of N_A for semiregular operators (see Definition 2.1 below) that was obtained by Helffer and Robert in [6]. (See also Coriasco and Doll [1] for an extension of [4] to the SG calculus.) Note that the important, and wonderful, point is that all the asymptotics of the Weyl function are described in terms of dynamical symplectic invariants of the symbol.

For *systems* very little is known. It was Ivrii (see [11]) who started studying the asymptotics for systems on compact manifolds and myself (see [14, 15, 16, 17]) in the case of globally elliptic systems with polynomial coefficients in the class of Non-Commutative Harmonic Oscillators (NCHOs), introduced by M. Wakayama and myself some time ago (see [18, 19]). Here, I will explain the approach in [12] that provides the Weyl asymptotics in case the system with polynomial coefficients is of the *semiregular* type, under further suitable conditions that include in the considered class also important models coming from Quantum Optics, namely the Jaynes-Cumming system (which describes the interaction of photons with an atom with a fixed number of energy levels in a cavity; see [21]). An important feature of the class considered here is that the system is semiregular with scalar principal part and smoothly diagonalizable semiprincipal part (the SMGES class; see Sect. 2). The Weyl asymptotics I obtain here are the analogues for semiregular systems of those of Helffer and Robert [6] in the scalar case and may be refined, under suitable conditions, to those obtained by Doll, Gannot, and Wunsch [4] (and by Doll and Zelditch [5]).

As for the "quasi-clustering," it was Weinstein [23] (in the case of a scalar Schrödinger operator on a compact boundaryless manifold) and Colin de Verdiere [2] (in the case of a scalar elliptic operator on a compact boundaryless manifold; see also Duistermaat and Guillemin [3]) who proved that when the bicharacteristics of the principal symbol are all periodic with period independent of the energy, then one has that the spectrum of the operator clusters around an arithmetic sequence, determined in terms of invariants of the symbol, in the sense that it is contained in the union of intervals centered in that sequence with vanishing diameters as the centers go to infinity. The arithmetic sequence is such that eventually the intervals become disjoint. The *quasi-clustering* property, weaker than the clustering, does not require this latter property, but that the number of intersections of the intervals is bounded by a universal constant. For systems, the situation is vastly unknown, apart from some results I obtained for regular NCHOs (see [14, 15, 16, 17]).

For systems of the SMGES class, the quasi-clustering property of the spectrum may be viewed as a more general property than the refined spectral asymptotics. In fact, the former holds for systems (as we shall see below) for which the eigenvalues of the semiprincipal part are functions of combinations with positive coefficients of the harmonic oscillators in each symplectic plane $\mathbb{R}^2_{(x_j,\xi_j)}$ inside $\mathbb{R}^{2n}$, which gives, for some choices of the coefficients, an obstruction to the refined Weyl law.

Here, I will describe the approach in [13] that yields quasi-clustering properties of the spectrum for operators in the SMGES class. A delicate point in the approach is to exploit the fact that these systems may be smoothly diagonalized (modulo smoothing operators) through almost unitary diagonalizers (i.e., unitary modulo smoothing operators). Having almost unitary diagonalizers for controlling the spectrum of the system in terms of that of the diagonalized one is not enough, and we have to deform the diagonalizer (or its adjoint, depending on the index of its principal symbol) into a genuine isometry in the global pseudodifferential calculus we are using (i.e., that of the quantum harmonic oscillator; see Sect. 2), but in certain cases this requires augmenting the system and hence throwing in new centers (explicitly determined) that give a little noise in the description of the spectrum that, however, does not effectively change the quasi-clustering property.

I end this Introduction by giving the plan of the chapter. In Sect. 2, I recall what a semiregular symbol is and introduce the class of systems of the kind SMGES, along with two basic examples. In Sect. 3, I give the statement of a fundamental step in the study of spectral properties of this system: the diagonalizability modulo smoothing in terms of ψdos that are unitary up to smoothing. Then, in Sect. 4, I give an idea of the proof of the Weyl asymptotics in both the case of the two-term and three-term asymptotics, showing also an example (the Jaynes-Cummings system) for which the refined asymptotics does not hold and one, a perturbation of a Laplacian of a connection, for which it does. In Sect. 5, I survey the proof of the quasi-clustering property, passing through the "unitarization" theorem and giving, in the end, the quasi-clustering property in the case of the Jaynes-Cummings system. In the final section, I will illustrate two open problems related to this class of systems.

2 The Class SMGES

I will be considering pseudodifferential operators that are Weyl quantizations of symbols in the Weyl-Hörmander classes: For $\mu \in \mathbb{R}$ and $X = (x, \xi) \in \mathbb{R}^n \times \mathbb{R}^n = T^*\mathbb{R}^n$, consider $\langle X\rangle = (1 + |X|^2)^{1/2}$, and write

$$S^\mu := S(\langle X\rangle^\mu, |dX|^2/\langle X\rangle^2; \mathsf{M}_N)$$

for the class of *global symbols* of order μ, where M_N denotes the algebra of $N \times N$ complex valued matrices. In other words, $a \in S^\mu$ iff

$$|\partial_X^\alpha a(X)|_{\mathsf{M}_N} \lesssim \langle X\rangle^{\mu-|\alpha|}, \quad \forall \alpha \in \mathbb{Z}_+^{2n}.$$

We denote by $a^{\mathrm{w}}(x, D) \in \Psi^{\mu}$ the corresponding pseudodifferential operator obtained by Weyl quantization

$$a^{\mathrm{w}}(x, D)u(x) = (2\pi)^{-n} \iint e^{i\langle x-y,\xi\rangle} a\Big(\frac{x+y}{2}, \xi\Big) u(y)dyd\xi, \quad u \in \mathscr{S}(\mathbb{R}^n; \mathbb{C}^N).$$

Throughout the chapter, I will write

$$p_2(X) = |X|^2/2, \quad \psi_j(X) = \frac{x_j + i\xi_j}{\sqrt{2}}, \quad 1 \le j \le n,$$

so that $p_2^{\mathrm{w}}(x, D) \in \Psi^2$ is the usual n-dimensional quantum harmonic oscillator, and $\psi_j^{\mathrm{w}}(x, D) \in \Psi^1$ is the annihilation operator in the jth variables, with $\psi^{\mathrm{w}}(x, D)^* = \bar{\psi}_j^{\mathrm{w}}(x, D)$ the corresponding creation operator, $j = 1, \ldots, n$.

Definition 2.1 We say that $a \in S^{\mu}$ is a regular symbol and write $a \in S^{\mu}_{\mathrm{reg}}$ when there is $(a_{\mu-2j})_{j\ge 0} \subset C^{\infty}(\dot{\mathbb{R}}^{2n}; \mathsf{M}_N)$, where $\dot{\mathbb{R}}^{2n} = \mathbb{R}^{2n} \setminus \{0\}$, such that $a \sim \sum_{j\ge 0} a_{\mu-2j}$, that is, ($\chi$ being an excision function)

$$a - \chi \sum_{j=0}^{N} a_{\mu-2j} \in S^{\mu-2(N+1)}, \quad \forall N \in \mathbb{Z}_+.$$

I say that $a \in S^{\mu}$ is **semiregular** and write $a \in S^{\mu}_{\mathrm{sreg}}$, when $a = a^0 + a^1$ with $a^k \in S^{\mu-k}_{\mathrm{reg}}$ for $k = 0, 1$. The corresponding classes of pseudodifferential operators are $\Psi^{\mu}_{\mathrm{reg}}$ and $\Psi^{\mu}_{\mathrm{sreg}}$, respectively.

The terms

$$a^0_{\mu}, \quad a^1_{\mu-1}, \quad a^0_{\mu-2}$$

are, respectively, called **principal**, **semiprincipal**, and **subprincipal** symbols, and they are symplectically invariant.

We finally say that $a^{\mathrm{w}} \in \Psi^{\mu}_{\mathrm{sreg}}$ is **globally elliptic** when

$$|\det a^0_{\mu}(X)| \approx |X|^{\mu N}, \quad \forall X \neq 0.$$

I may now introduce the class of $N \times N$ symbols we consider here, the SMGES class, which stands for *Semiregular Metric Globally Elliptic System*, "metric" coming from the differential geometric terminology for which a metric operator is one which has a scalar principal symbol, given by a (smooth) Riemannian metric on the manifold on which one is working.

Definition 2.2 I say that $a \in S^{\mu}_{\mathrm{sreg}}$ is SMGES when

$$a(X) = a(X)^* = p_{\mu}(X) I_N + a_{\mu-1}(X) + a_{\mu-2}(X) + S^{\mu-2}_{\mathrm{sreg}}, \quad X \neq 0,$$

where $p_{\mu}(X) \approx |X|^{\mu}$, and the semiprincipal part $a_{\mu-1}$ is such that there exists $e_0 \in C^{\infty}(\dot{\mathbb{R}}^{2n}; \mathsf{M}_N)$, unitary and positively homogeneous of degree 0 such that

$$e_0^* a_{\mu-1} e_0 = \operatorname{diag}\big(\lambda_{\mu-1,j} I_{N_j};\ 1 \le j \le r\big),$$

where $N = N_1 + N_2 + \ldots + N_r$ and $\lambda_{\mu-1,j} \in C^{\infty}(\dot{\mathbb{R}}^{2n}; \mathbb{R})$ are positively homogeneous of degree $\mu - 1$ and such that

$$j < k \Longrightarrow \lambda_{\mu-1,j}(X) < \lambda_{\mu-1,k}(X), \quad \forall X \neq 0.$$

In particular, the eigenvalues $\lambda_{\mu-1,j}$ of $a_{\mu-1}$ have constant multiplicity.

Given $a \in S^{\mu}$, $\mu > 0$, I will think of $a^{\mathrm{w}}(x, D)$ also as the unbounded operator $A\colon D(A) = B^{\mu} \subset L^2 \longrightarrow L^2$ represented by $a^{\mathrm{w}}(x, D)$ on $\mathscr{S}$. Hence, capital letters will be used to indicate abstract operators between Hilbert spaces.

2.1 Examples

I next give two examples of such a class of systems (see [12]). The first example is the Jaynes-Cummings model [21] which describes the interaction of a two-level atom with a photon. Consider the Pauli matrices

$$\boldsymbol{\sigma}_0 = I_2,\ \boldsymbol{\sigma}_1 = \begin{bmatrix} 0 & 1 \\ 1 & 0 \end{bmatrix},\ \boldsymbol{\sigma}_2 = \begin{bmatrix} 0 & -i \\ i & 0 \end{bmatrix},\ \boldsymbol{\sigma}_3 = \begin{bmatrix} 1 & 0 \\ 0 & -1 \end{bmatrix}.$$

Set $\boldsymbol{\sigma}_{\pm} := (\boldsymbol{\sigma}_1 \pm i\boldsymbol{\sigma}_2)/2$. For $n = 1$, the Jaynes-Cummings operator is

$$a^{\mathrm{w}}_{JC}(x, D) = p^{\mathrm{w}}_2(x, D) I_2 + \alpha\Big(\boldsymbol{\sigma}_- \psi^{\mathrm{w}}(x, D)^* + \boldsymbol{\sigma}_+ \psi^{\mathrm{w}}(x, D)\Big) + a_0 \in \Psi^2_{\mathrm{sreg}},$$

where $a_0 = \begin{bmatrix} \gamma_1 & 0 \\ 0 & \gamma_2 \end{bmatrix}$, with $\alpha, \gamma_1, \gamma_2 \in \mathbb{R}$. In this case

$$a_1(X) = \alpha \begin{bmatrix} 0 & \psi(X) \\ \overline{\psi(X)} & 0 \end{bmatrix},$$

with eigenvalues

$$\lambda_{1,\pm}(X) = \pm|\alpha||\psi(X)|/\sqrt{2}.$$

As we will see below, a^{w}_{JC} is actually a Laplacian of a connection on the trivial bundle $\mathbb{R}^n \times \mathbb{C}^N \to \mathbb{R}^n$ (in general possessing a nonzero curvature).

The second example is a Laplacian associated with a connection over $\mathbb{R}^n$. If $\star$ denotes the usual Hodge-star operator in $\mathbb{R}^n$, let $\Omega^k(\mathbb{R}^n)$ be the space of smooth k-differential forms on $\mathbb{R}^n$, and let d_k be the exterior differential on $\Omega^k(\mathbb{R}^n)$. Then $d_k^* = (-1)^{nk+1} \star d_k \star$ is the adjoint operator on $\Omega^{k+1}(\mathbb{R}^n)$ forms. I next take

$$D_k = \frac{1}{\sqrt{2}}\left(d_k + \sum_{j=1}^{n} x_j dx_j \wedge\right) : \Omega^k(\mathbb{R}^n) \longrightarrow \Omega^{k+1}(\mathbb{R}^n),$$

$$D_k^* = \frac{1}{\sqrt{2}}\left(d_k^* + \sum_{j=1}^{n} x_j \mathbf{i}_{\partial/\partial x_j}\right) : \Omega^{k+1}(\mathbb{R}^n) \longrightarrow \Omega^{k}(\mathbb{R}^n),$$

where $\mathbf{i}_{\partial/\partial x_j}$ denotes the contraction in the jth direction. Let $N = n+1$, and let E_{jk} be the $N \times N$ matrix that acts on the vectors $\{e_1, \dots, e_N\}$ of the canonical basis of $\mathbb{C}^N$ as $E_{jk} w = \langle w, e_k\rangle e_j$, $1 \le j, k \le N$. For $\alpha_1, \dots, \alpha_n \in \mathbb{R} \setminus \{0\}$, consider

$$\mathbf{D}_k = D_k \otimes I_N + \sum_{j=1}^{n} \alpha_j (dx_j \wedge) \otimes E_{j,j+1} \colon \Omega^k(\mathbb{R}^n; \mathbb{C}^N) \longrightarrow \Omega^{k+1}(\mathbb{R}^n; \mathbb{C}^N).$$

Hence

$$\mathbf{D}_k^* = D_k^* \otimes I_N + \sum_{j=1}^{n} \alpha_j \mathbf{i}_{\partial/\partial x_j} \otimes E_{j+1,j} \colon \Omega^{k+1}(\mathbb{R}^n; \mathbb{C}^N) \longrightarrow \Omega^{k}(\mathbb{R}^n; \mathbb{C}^N).$$

In this case the curvature of the connection is

$$\mathsf{F}_{\mathbf{D}} = \sum_{j=1}^{n-1} \alpha_j \alpha_{j+2} (dx_j \wedge dx_{j+1} \wedge) \otimes E_{j,j+2} \in \Omega^2(\mathbb{R}^n; \mathsf{M}_N).$$

We may consider the associated Laplacian

$$\Box_k^{(N)} = \mathbf{D}_k^* \mathbf{D}_k + \mathbf{D}_{k-1} \mathbf{D}_{k-1}^* \colon \Omega^k(\mathbb{R}^n; \mathbb{C}^N) \longrightarrow \Omega^k(\mathbb{R}^n; \mathbb{C}^N), \tag{1}$$

which is in Ψ^2_{sreg} with principal part $p_2^{\mathrm{w}}(x, D)\mathbf{1}_k \otimes I_N$, semiprincipal part

$$\sum_{j=1}^{n} \alpha_j \Big(\psi_j^{\mathrm{w}}(x, D)^* \mathbf{1}_k \otimes E_{j,j+1} + \psi_j^{\mathrm{w}}(x, D) \mathbf{1}_k \otimes E_{j+1,j}\Big),$$

and subprincipal part

$$\left(k-\frac{n}{2}\right)\mathbf{1}_k\otimes I_N+\sum_{j=1}^{n}\alpha_j^2\mathbf{1}_k\otimes E_{j+1,j+1}+\sum_{j=1}^{n}\alpha_j^2 dx_j\wedge\mathbf{i}_{\partial/\partial x_j}\mathbf{1}_k\otimes(E_{j,j}-E_{j+1,j+1}),$$

where $\mathbf{1}_k$ denotes the identity on $\Omega^k(\mathbb{R}^n)$.

Therefore, a^{w}_{JC} is related to $\square_0^{(2)}$ ($n=1$, $N=2$), and $\square_k^{(N)}$ is a generalization. Both operators $\left(\square_0^{(2)}\text{ and }\square_k^{(N)}\right)$ are SMGES, for they have *semiprincipal part which is smoothly diagonalizable* (see [12]).

3 A Fundamental Property: Diagonalizability

A fundamental step in the study of the Weyl asymptotics and the possible quasi-clustering is to pass from the given system to a system in diagonal form (in the quasi-clustering setting, we will also need to make quasi-isometries into true isometries in order to control the spectrum of the considered system in terms of that of its diagonalization). For the class SMGES introduced above, we have the following result [12].

Theorem 3.1 *Let $\mu>0$, $a=a^*\in S^{\mu}_{\mathsf{sreg}}$ with $a\sim\sum_{j\geq 0}a_{\mu-j}$, such that $a_\mu = p_\mu I_N$ with $p_\mu\in C^\infty(\dot{\mathbb{R}}^{2n};\mathbb{R})$ positively homogeneous of degree μ. Suppose there is $e_0\in C^\infty(\mathbb{R}^{2n};\mathsf{M}_N)$ unitary and positively homogeneous of degree 0 such that*

$$e_0^* a_{\mu-1} e_0 = \left[\begin{array}{c|c}\Lambda_{\mu-1,1} & 0\\ \hline 0 & \Lambda_{\mu-1,2}\end{array}\right],\quad X\neq 0,$$

where $N=N_1+N_2$ and $\Lambda_{\mu-1,j}\in C^\infty(\dot{\mathbb{R}}^{2n};\mathsf{M}_{N_j})$ is positively homogeneous of degree $\mu-1$ such that

$$\mathsf{Spec}(\Lambda_{\mu-1,1}(X))\cap\mathsf{Spec}(\Lambda_{\mu-1,2}(X))=\emptyset,\ \forall X\neq 0.$$

Then there exists $e\in S^0_{\mathsf{sreg}}$ with principal symbol e_0 such that

$$e^{\mathsf{w}}(x,D)e^{\mathsf{w}}(x,D)^*-I\in\Psi^{-\infty},\quad e^{\mathsf{w}}(x,D)^*e^{\mathsf{w}}(x,D)-I\in\Psi^{-\infty},$$

and

$$e^{\mathsf{w}}(x,D)^*a^{\mathsf{w}}(x,D)e^{\mathsf{w}}(x,D)=b^{\mathsf{w}}(x,D)+\Psi^{-\infty},$$

where

$$b \sim \sum_{j\geq 0} \left[\begin{array}{c|c} b_{\mu-j,1} & 0 \\ \hline 0 & b_{\mu-j,2} \end{array}\right] \in S^{\mu}_{\mathrm{sreg}},$$

and

$$b_\mu = a_\mu, \quad b_{\mu-1} = \left[\begin{array}{c|c} \Lambda_{\mu-1,1} & 0 \\ \hline 0 & \Lambda_{\mu-1,2} \end{array}\right].$$

Hence in the case of an SMGES, we have, by iterating the above result with $N = N_1 + \ldots + N_r$, that

$$b_\mu(X) = p_\mu(X) I_N, \quad b_{\mu-1}(X) = \mathrm{diag}\Big(\lambda_{\mu-1,j}(X) I_{N_j};\ 1 \leq j \leq r\Big).$$

As for the subprincipal part, we consider

$$\pi_j \colon \mathbb{C}^N \longrightarrow \mathbb{C}^N = \bigoplus_{j=1}^{r} \mathbb{C}^{N_j},$$

the jth orthogonal projection onto the jth $\mathbb{C}^{N_j} \subset \mathbb{C}^N$. Then the subprincipal symbol of the diagonalized operator is given by (see [12])

$$b_{\mu-2}(X) = \mathrm{diag}\Big(b^{(j)}_{\mu-2}(X);\ 1 \leq j \leq r\Big), \quad X \neq 0,$$

where

$$b^{(j)}_{\mu-2} = \pi_j e_0^* a_{\mu-2} e_0 \pi_j^* - i\pi_j e_0^* \{p_\mu, e_0\}\pi^*, \quad j = 1, \ldots, r.$$

Of course, one has (see [12]) transformation formulas for the subprincipal symbol in terms of the principal symbol e_0 of the diagonalizer.

There is also a formula for the semi-subprincipal term (the term of order $\mu - 3$) along with its transformation rules with respect to e_0, but we will not be needing it in the sequel.

4 Asymptotics for the Weyl Spectral Counting Function

We are given $a^{\mathrm{w}} \in \Psi^2_{\mathrm{sreg}}$ in the class SMGES of order $\mu = 2$ with principal symbol p_2 (*the standard harmonic oscillator*), realized as an unbounded operator A in L^2

with domain the Shubin Sobolev space B^2, which is therefore semibounded. It is thus a globally elliptic operator and has a discrete spectrum

$$-\infty < \lambda_1 \leq \lambda_2 \leq \ldots \leq \lambda_j \to +\infty,$$

where the eigenvalues are repeated according to multiplicity. As recalled in the Introduction, the Weyl function associated with A is defined by

$$\mathsf{N}_A(\lambda) = \sum_{\lambda_j \leq \lambda} 1.$$

Such function has a well-known crucial role in spectral theory and in many related problems. At first, we observe that

$$\mathscr{F}_{\lambda \to t}\left(\sum_{j \geq 1} \delta(\lambda - \lambda_j)\right) = \sum_{j \geq 1} e^{-it\lambda_j}. \tag{2}$$

The latter belongs to $\mathscr{S}'(\mathbb{R})$. In fact, for all $\phi \in \mathscr{S}(\mathbb{R})$, we may consider, using the Schrödinger evolution equation associated with a^{w} and integration by parts,

$$U_\phi = \int \phi(t) e^{-ita^{\mathrm{w}}} dt = \int (-i)^k (\partial_t^k \phi)(t) (a^{\mathrm{w}})^{-k} e^{-ita^{\mathrm{w}}} dt,$$

which is therefore, for sufficiently large k, a trace-class operator. Hence the operator-trace $\mathrm{Tr}_{\mathrm{op}}(e^{-ita^{\mathrm{w}}})\colon \mathscr{S}(\mathbb{R}) \ni \phi \longmapsto \mathrm{Tr}_\Delta \mathsf{Tr}(U_\phi) \in \mathbb{C}$ is a tempered distribution ($\mathrm{Tr}_{\mathrm{op}}$ is the restriction Tr_Δ on the diagonal Δ of $\mathbb{R}^n \times \mathbb{R}^n$, of the matrix-trace Tr of the Schwartz kernel of U_ϕ). Therefore the right-hand side of (2) is a temperate distribution and so is the argument of the Fourier transform in the left-hand side, which is therefore the derivative with respect to λ of N_A. We thus study the group $t \longmapsto e^{-ita^{\mathrm{w}}}$, and for that we use, following Chazarain and Helffer-Robert, an FIO (Fourier integral operator) parametrix of it. The point is therefore to obtain the parametrix (using the WKB method) in a sufficiently friendly class to allow a control on compositions, which are needed because we have to pass through the diagonalization procedure. The class of FIOs introduced by Helffer and Robert in [6] is not sufficiently friendly, and here the approach of Doll, Gannot, and Wunsch [4] is crucial. We consider therefore a parametrix of the Schrödinger group $t \longmapsto e^{-itb^{\mathrm{w}}}$ of the diagonalized system which is given by the composition of the evolution $t \mapsto e^{-itp_2^{\mathrm{w}}}$, which has a nice representation by Mehler's formula and is the Weyl quantization of an exponential of a quadratic form (see Hörmander [8]), with the parametrix of the first-order time-dependent operator

$$t \mapsto P(t) := e^{itp_2^{\mathrm{w}}}\left(B^{\mathrm{w}} - p_2^{\mathrm{w}} I_N\right)e^{-itp_2^{\mathrm{w}}},$$

which can be easily computed by Egorov's theorem and the fact that the characteristics in the WKB expansion exist for all times. We therefore get, recalling that the blocks of the principal symbol of b^{w} are given by $\Lambda_{1,j} = \lambda_{1,j} I_{N_j}$, $1 \leq j \leq r$, that the operator

$$t \longmapsto F(t) := \operatorname{diag}\left(\left(e^{i\phi_{j,1}(t)}\alpha_j(t)\right)^{\mathrm{w}};\ 1 \leq j \leq r\right),$$

where

$$\phi_{j,1}(t, X) = -\int_0^t (\lambda_{1,j} \circ \exp sH_{p_2})(X)ds,$$

H_{p_2} being the Hamilton vector-field associated with p_2, and where α_j is an $N_j \times N_j$ matrix with $\alpha_j \in C^\infty(\mathbb{R}_t; S^0_{\mathrm{sreg}})$, is a parametrix for $P(t)$,

$$(i\partial_t - P(t))F(t) \in C^\infty(\mathbb{R}_t; \mathscr{L}(\mathscr{S}', \mathscr{S}) \otimes \mathsf{M}_N), \quad F(0) = I_N + \text{smoothing}.$$

We therefore have the following result [12].

Theorem 4.1 *Let $\varepsilon > 0$ be sufficiently small so that $(2k \pm 1)\pi \notin (2k\pi - \varepsilon, 2k\pi + \varepsilon) := I_{k,\varepsilon}$, for all $k \in \mathbb{Z}$. Then, for $j = 1, \ldots, r$, there is $\tilde{\alpha}_j \in C^\infty(I_{k,\varepsilon}; S^0)$ such that, with*

$$\phi_2(t, X) = -2\tan(t/2)p_2(X), \quad \tilde{\phi}_{j,1}(t, X) = \phi_{j,1}(t, X - 2\tan(t/2)JX)$$

(where $J = \begin{bmatrix} 0 & I \\ -I & 0 \end{bmatrix}$) and $U_0(t) = \cos(t/2)^{-n}(e^{-\phi_2(t)})^{\mathrm{w}}$, we have that

$$U_B(t) = U_0(t)F(t) = \operatorname{diag}\left(\left(e^{i\phi_2(t)+i\tilde{\phi}_{j,1}(t)}\tilde{\alpha}_j(t)\right)^{\mathrm{w}};\ 1 \leq j \leq r\right)$$

is a parametrix of $e^{-itB^{\mathrm{w}}}$. Hence,

$$I_{k,\varepsilon} \ni t \longmapsto U_A(t) := E^{\mathrm{w}}U_B(t)(E^{\mathrm{w}})^*$$

is a parametrix for $t \mapsto e^{-itA^{\mathrm{w}}}$.

Once this is established, one can therefore follow the classical construction of Hörmander [10] to obtain the two-term asymptotics that extends that of Helffer and Robert [6] to the vector-valued case.

Theorem 4.2 *Let* $A = a^{\mathrm{w}} \in \Psi^2_{\mathrm{sreg}}$, *where* a *is an SMGES. Let* $\rho \in \mathscr{S}(\mathbb{R})$ *be such that* $\hat{\rho} \in C_c^\infty(-\varepsilon, \varepsilon)$ *and* $\hat{\rho} \equiv 1$ *near* 0. *Then (ds being the Riemannian hypersurface measure)*

$$(\mathsf{N}_A \star \rho)(\lambda)$$
$$= (2\pi)^{-n} \left(\sum_{j=1}^{r} N_j \int_{p_2+\lambda_{1,j} \le \lambda} dX - \int_{p_2=\lambda} \mathsf{Tr}(a_0) \frac{ds}{|\nabla_X p_2|} \right) + O(\lambda^{n-3/2}) \tag{3}$$

as $\lambda \to +\infty$, *where* Tr *denotes the matrix-trace. Hence,*

$$\mathsf{N}_A(\lambda)$$
$$= (2\pi)^{-n} \left(\left(N \int_{p_2 \le 1} dX \right) \lambda^n - \left(\int_{p_2=1} \mathsf{Tr}(a_1) \frac{ds}{|\nabla_X p_2|} \right) \lambda^{n-1/2} \right) + O(\lambda^{n-1}) \tag{4}$$

as $\lambda \to +\infty$.

We have (4) from (3) by using a Taylor expansion as $\lambda \to +\infty$ of the volume of the set $\{X;\ p_2 + \lambda_{1,j} \le \lambda\}$ and using the polynomial growth of $\mathsf{N}_A(\lambda)$ (which follows from the global ellipticity and the minimax principle).

Using the improvement that Doll, Gannot, and Wunsch have given in the scalar semiregular case, we may refine the above theorem. We need to impose a condition on the eigenvalues of the semiprincipal part: *Suppose that for* $j = 1, \dots, r$ *the sets*

$$Z_j := \left\{ \omega \in \mathbb{S}^{2n-1};\ \partial_\omega^\alpha \int_0^{2\pi} (\lambda_{1,j} \circ \exp t H_{p_2})(\omega) dt = 0,\ \forall |\alpha| \ge 1 \right\} \tag{5}$$

have measure zero. (I have denoted by ∂_ω^α the iterated tangential derivatives to $\mathbb{S}^{2n-1}$.) We refer to this condition as Condition **DGW**.

That condition allows one to use Stein's theorem on the control of the oscillatory integrals that enter the description of N_A. Since from Theorem 4.2 it follows in particular that $\mathsf{N}'_A \star \rho = O(\lambda^{n-1})$, $\lambda \to +\infty$, one is in a position to use the following Tauberian theorem [20].

Theorem 4.3 *With* ρ *as in Theorem 4.2, if there is* $\gamma \in \mathbb{R}$ *such that* $(\mathsf{N}'_A \star \rho)(\lambda) = O(\lambda^\gamma)$ *and* $(\mathsf{N}'_A \star \chi)(\lambda) = o(\lambda^\gamma)$ *for all* $\chi \in \mathscr{S}(\mathbb{R})$ *such that* $\hat{\chi} \in C_c^\infty(0, +\infty)$, *then*

$$\mathsf{N}_A(\lambda) = (\mathsf{N}_A \star \rho)(\lambda) + o(\lambda^\gamma), \quad \lambda \to +\infty.$$

One then has the following refined Weyl asymptotics.

Theorem 4.4 *If Condition* ***DGW*** *is satisfied, then*

$$\mathsf{N}_A(\lambda) = (2\pi)^{-n}\left(\sum_{j=1}^{r} N_j \int_{p_2+\lambda_{1,j}\leq\lambda} dX - \int_{p_2=\lambda} \mathsf{Tr}(a_0)\frac{ds}{|\nabla_X p_2|}\right) + o(\lambda^{n-1})$$

as $\lambda \to +\infty$. *In particular,*

$$\mathsf{N}_A(\lambda) = (2\pi)^{-n}\Bigg(N\lambda^n \int_{p_2\leq 1} dX - \lambda^{n-1/2}\int_{p_2=1}\mathsf{Tr}(a_0)\frac{ds}{|\nabla_X p_2|}$$

$$+\lambda^{n-1}\int_{p_2=1}\left(\frac{n}{2}\mathsf{Tr}(a_1^2) - \mathsf{Tr}(a_0)\right)\frac{ds}{|\nabla_X p_2|}\Bigg) + o(\lambda^{n-1})$$

as $\lambda \to +\infty$.

Notice that Condition **DGW** *cannot* hold when the semiprincipal part has the same symmetries of the principal part, for instance, when the eigenvalues of the semiprincipal part are functions of $|X|$. In fact, in such a case the average of the eigenvalues along the bicharacteristics of the principal symbol is a constant, and hence the angular derivatives are identically zero. And this is indeed the case of A_{JC}. In fact, the eigenvalues of the semiprincipal part are $\lambda_{1,\pm}(X) = \pm|\alpha||X|/\sqrt{2}$, whence the sets $Z_\pm$ have full measure. In this case we have only

$$\mathsf{N}_{A_{JC}}(\lambda) = (2\pi)^{-1}2\lambda\int_{p_2\leq 1} dX + O(1) = 2\lambda + O(1), \quad \lambda \to +\infty.$$

4.1 A Deformation of $\square_0^{(3)}$ Which Has a Refined Asymptotics

In the next example, I will break the symmetry of the eigenvalues of the semiprincipal part of $\square_0^{(3)}$ (see (1)), one of which is 0, and construct a perturbation for which Condition **DGW** is satisfied.

Recalling that $\alpha_1, \alpha_2 \neq 0$, consider $\square_0^{(3)}: \mathscr{S}(\mathbb{R}^2;\mathbb{C}^3) \longrightarrow \mathscr{S}(\mathbb{R}^2;\mathbb{C}^3)$. The eigenvalues of the semiprincipal part are

$$\lambda_0(X) = 0, \quad \lambda_{1,\pm}(X) = \pm\underbrace{(\alpha_1^2|\psi_1(X)|^2 + \alpha_2^2|\psi_2(X)|^2)^{1/2}}_{=:|\alpha\psi(X)|},$$

with constant multiplicity 1. Here the problem is generated by Z_0 which has full measure (the same occurs for $Z_\pm$ when $|\alpha_1| = |\alpha_2|$). Consider then, for the parameter $\mu \in (0,1)$, the new eigenvalue $\lambda_\mu(X) := \mu\lambda_+(X) + (1-\mu)\lambda_-(X)$. Notice that when $\mu = 1/2$, then $\lambda_\mu(X) \equiv 0$. So we suppose that $\mu \neq 1/2$ and that

$|\alpha_1| \neq |\alpha_2|$, whence (as one may easily check) Condition **DGW** holds. Then we consider the SMGES

$$a(X) = p_2(X)I_N + a_{1,\mu}(X) + a_0(X),$$

where, putting $f_j = \alpha_j \psi_j / |\alpha\psi|$,

$$a_{1,\mu}(X) = \begin{bmatrix} \lambda_\mu(X)|f_2(X)|^2 & \lambda_+(X)\overline{f_1(X)} & -\lambda_\mu(X)\overline{f_1(X)f_2(X)} \\ \lambda_+(X)f_1(X) & 0 & \lambda_+(X)\overline{f_2(X)} \\ -\lambda_\mu(X)f_1(X)f_2(X) & \lambda_+(X)f_2(X) & \lambda_\mu(X)|f_1(X)|^2 \end{bmatrix},$$

and

$$a_0(X) = -I_3 + \sum_{j=1}^{2} \alpha_j^2 E_{j+1,j+1} + \sum_{j=1}^{2} (E_{j,j} - E_{j+1,j+1}).$$

Then $\mathsf{Tr}(a_0) = \alpha_1^2 + \alpha_2^2 - 3$, and we have the following refined Weyl law:

$$\mathsf{N}_A(\lambda) = (2\pi)^{-2} \Big(3|\mathbb{S}^3|\lambda^2 - \frac{1}{2}\lambda^{3/2} \int_{|X|=2} \lambda_\mu ds$$

$$+ \frac{1}{2}\lambda \int_{|X|=2} (2\lambda_+^2 + \lambda_\mu^2 - \alpha_1^2 - \alpha_2^2 + 3) ds \Big) + o(\lambda)$$

as $\lambda \to +\infty$.

5 Quasi-Clustering

In this section we study *quasi-clustering properties* of SMGES, that is, the presence of the spectrum of the system located in the union of a sequence of intervals, whose diameters go to zero as the centers go to infinity, with universally bounded number of overlappings. Note, as already mentioned, that a *clustering property*, which is stronger, is the presence of the spectrum in the union of a sequence of intervals of the above kind that become disjoint for sufficiently large centers. Looking for quasi-clustering is quite natural to do for a system for two main reasons. On the one hand, the principal symbol of a second-order SMGES may have bicharacteristics that are all periodic with period independent of the energy. On the other, thinking of a diagonal system, the spectrum is contained in the union of that of the diagonal components, so that, even if each component clusters, we do have overlapping intervals, of course, with an a priori bounded number of overlappings. In the case of our class, the systems having as a principal symbol the harmonic oscillator

$p_2(X) = |X|/2$, we are exactly in the position of having all the bicharacteristics periodic of the same period. In this case it is crucial that the principal symbol be of order 2 (see, e.g., [14], Section 11.1).

As already mentioned in the Introduction, quasi-clustering here will be considered for SMGES for which the semiprincipal part has eigenvalues that are functions of

$$p_{2,\alpha}(X) = \sum_{j=1}^{n} \alpha_j \left(x_j^2 + \xi_j^2\right)/2, \tag{6}$$

with $\alpha_1, \ldots, \alpha_n > 0$. For certain values of the αs (e.g., when they are all equal, a case that occurs for A_{JC}), the condition for the three-term Weyl asymptotics (i.e., the sets Z_j in Condition **DGW** (5) should have measure zero), does not hold, and happens for instance when the level sets of p_2 are also level sets of $p_{2,\alpha}$. Hence, in this case, we may think of the quasi-clustering property as more general than having the refined Weyl formula.

Once more, a fundamental ingredient to approach the problem is the use of the diagonalization Theorem 3.1, but that turns out to be not enough, for the fact that $(e^{\mathrm{w}})^* a^{\mathrm{w}} e^{\mathrm{w}} = b^{\mathrm{w}} + \Psi^{-\infty}$ with e^{w} unitary up to smoothing operators is not sufficiently strong to control the spectrum of a^{w} in terms of that of b^{w} (using the minimax). The way out of this is to correct $(e^{\mathrm{w}})^*$ into a true L^2-isometry (i.e., $e^{\mathrm{w}}(e^{\mathrm{w}})^* = I$) and that depends on the nonnegativity of the index of e^{w}. When the index of e^{w} is negative, one has to "augment" the considered system into a new one which is still diagonalizable, with diagonalizer which now has the right index, and whose diagonalized operator has two diagonal blocks: One is given by b^{w} and the other is given by a system of which we know exactly the spectrum. Hence, we obtain a quasi-clustering in which the only addition is a sequence of new centers that, however, are exactly known.

I next explain the main steps in achieving all that.

Recall that we say that a linear, bounded operator $T\colon H \longrightarrow H$ (H a separable Hilbert space) is said to be a *quasi-isometry* if $T^*T = I + F_1$ and $TT^* = I + F_2$, where $F_1, F_2\colon H \longrightarrow H$ are compact.

We have the following "isometrization" theorem [13].

Theorem 5.1 *Suppose* $T \in \Psi^0_{\mathrm{sreg}}$ *is such that* T *is a quasi-isometry, with* $F_j \in \Psi^{-m_j}$, $m_j > 0$, $j = 1, 2$. *Then:*

(i) When $\operatorname{ind}(T) \le 0$, *there is* $R_1 \in \Psi^{-m_1}$ *such that* $T + R_1$ *is an isometry, i.e., one has* $(T + R_1)^*(T + R_1) = I$;
(ii) When $\operatorname{ind}(T) \ge 0$, *there is* $R_2 \in \Psi^{-m_2}$ *such that* $T^* + R_2$ *is an isometry, i.e., one has* $(T^* + R_2)^*(T^* + R_2) = I$.

When F_1 *(respectively,* F_2*) is smoothing, then so is* R_1 *(respectively,* R_2*).*

Once we have that theorem, we may use the minimax to relate the spectrum of A to that of B, where A and B are the realizations, as unbounded selfadjoint operators with domains the Shubin Sobolev space of order 2, of a^{w} and b^{w}, respectively. In fact, if E^* is the *isometrization* of $(e^{\mathrm{w}})^*$, which is therefore (by virtue of its pseudodifferential nature) continuous on any given Shubin Sobolev space into itself, supposing $A > 0$ (which we may assume without loss of generality by the Sharp-Gårding inequality) and that $EE^* = I$, one has

$$\mathrm{Spec}(A) = \mathrm{Spec}(E^*AE) \setminus \{0\}. \tag{7}$$

Recall that $E^*AE = B +$ compact operator. When the eigenvalues $\lambda_{1,h}$ of the semiprincipal symbol a_1 are functions of $p_{2,\alpha}$ (see (6)), i.e.,

$$\lambda_{1,h}(X) = p^{(h)}_{1/2}(p_{2,\alpha}(X)), \quad X \neq 0, \ h = 1, \ldots r \tag{8}$$

(with $p^{(h)}_{1/2}$ smooth and positively homogeneous of degree $1/2$ in $\dot{\mathbb{R}}^{2n}$), we have to look at the bicharacteristics of $p_{2,\alpha}$. Note that the smooth diagonalizability of a_1 yields then that a sufficient condition for having (8) is that the characteristic polynomial $\lambda \mapsto \det(\lambda - a_1)$ has coefficients that are smooth functions of $p_{2,\alpha}$.

When the constants $\alpha_1, \ldots, \alpha_n > 0$ are such that for positive coprime integers $m_1, \ldots, m_n$, one has

$$m_j/\alpha_j = q, \quad \forall j = 1, \ldots, n, \tag{9}$$

and for a diagonalizer e_0 of the semiprincipal symbol, we have that the resulting subprincipal symbol b_0 of the diagonalization B is constant on the Hamilton flows of both p_2 and $p_{2,\alpha}$, then one may find an orthonormal basis of $L^2(\mathbb{R}^n; \mathbb{C}^N)$, made of Schwartz functions such that the eigenspaces of P_2 are invariant for $P_{2,\alpha}$ and such that the eigenspaces of $P_2 + P_{2,\alpha}$ are left invariant by B_0. Hence one obtains the following theorem.

Theorem 5.2 *Let $0 < a^{\mathrm{w}} \in \Psi^2_{\mathrm{sreg}}$ be an SMGES. Suppose that the coefficients of the characteristic polynomial $\lambda \mapsto \det(\lambda - a_1(X))$ are smooth functions of $p_{2,\alpha}$ (see (8)) and that for a diagonalizer e_0 the subprincipal symbol b_0 is constant along both the bicharacteristic flows of p_2 and $p_{2,\alpha}$, where the α_js are such that (9) holds. Then there are $M > 0$ and $c > 0$ such that*

$$\mathrm{Spec}(A) \subset \bigcup_{h=1}^{r+1} S_h,$$

where

$$S_h = \bigcup_{k \geq 0} \bigcup_{\gamma \in \mathbb{Z}^n_+} \bigcup_{j=1}^{N_h} \left(c(k, \gamma, j) + \left[-\frac{M}{\sqrt{k+n/2}}, \frac{M}{\sqrt{k+n/2}} \right] \right),$$

and where for the centers $c(k,\gamma,j)$ *we have*

$$c(k,\gamma,j)=k+n/2+p_{1/2}^{(h)}\left(\sum_{j=1}^{n}\alpha_j(\gamma_j+1/2)\right)+\mu_{\gamma,j}^{(h)}.$$

One has that $N_{r+1}=N$, $p_{1/2}^{(r+1)}=0$, $\mu_{\gamma,j}^{(r+1)}=c$, *and for* $h=1,\dots,r$, *the* $\mu_{\gamma,j}^{(h)}$*s are the eigenvalues of the block* $B_0^{(h)}$ *of the subprincipal operator* B_0 *restricted to* $\mathrm{Ker}(P_2\otimes I_{N_h}-(k+n/2))$.

When $\mathrm{ind}(e_0)\geq 0$, *we may take* $S_{r+1}=\emptyset$.

For $h=1,\dots,r$, one has that $|\mu_{\gamma,j}^{(h)}|\leq\|B_0^{(h)}\|_{L^2\to L^2}$, for all $\gamma\in\mathbb{Z}_+^n$ and all $j=1,\dots N_h$. Hence they are uniformly bounded by $\|B_0\|_{L^2\to L^2}$.

Notice also that the set S_h is the contribution to the spectrum related to the block B_h of the diagonalization of A, $h=1,\dots,r$, and that the presence of S_{r+1} and $c\in\mathbb{R}$ is due to the case when $\mathrm{ind}(e_0)\geq 0$. In fact, in such a case one augments the system into a $2N\times 2N$ system of the form

$$\tilde{A}=\left[\begin{array}{c|c}A & 0_N\\\hline 0_N & P_2I_N+P_0\end{array}\right],$$

where the symbol p_0 of P_0 is chosen as

$$p_0=-e_0^*\Big(e_{-2}e_0^*p_2+p_2e_0e_{-2}^*-\frac{i}{2}\big(e_0\{p_2,e_0^*\}+\{e_0,p_2e_0^*\}\big)+e_{-1}p_2e_{-1}^*\Big)e_0+cI_N,$$

where $c>0$ is a constant such that $P_0>0$ (the expression of p_0 comes from the diagonalization). Hence

$$\tilde{E}^*\tilde{A}\tilde{E}=\left[\begin{array}{c|c}B+\Psi^{-\infty} & 0_N\\\hline 0_N & (P_2+c)I_N+R\end{array}\right],$$

where $\tilde{e}:=\left[\begin{array}{c|c}e & 0_N\\\hline 0_N & e^*\end{array}\right]$ (e and e^* being the total symbols of E and E^*, respectively), and $R\in\Psi_{\mathrm{sreg}}^{-1}$. In this case the index of $\tilde{E}$ is zero, and the subprincipal symbol of $E(P_2I_N+P_0)E^*$ is cI_N by the definition of p_0. One therefore may proceed by exploiting the hypotheses on the bicharacteristic flows and constancy of b_0 along the bicharacteristics of p_2 and of $p_{2,\alpha}$ that grant the eigenspaces invariance and simultaneous diagonalization that lead to the result.

As a matter of example, let us see what Theorem 5.2 says about the Jaynes-Cummings system A_{JC} (recalling that $\psi(X)=(x+i\xi)/\sqrt{2}$)

$$a_{JC}(X)=p_2(X)I_2+\alpha\begin{bmatrix}0 & \overline{\psi(X)}\\ \psi(X) & 0\end{bmatrix}+\begin{bmatrix}\gamma_1 & 0\\ 0 & \gamma_2\end{bmatrix},\quad X\in\mathbb{R}^2\setminus\{0\}.$$

In this case the eigenvalues of the semiprincipal symbol are given by

$$\lambda_{\pm}(X) = \pm|\alpha||X|/\sqrt{2},$$

the diagonalizer e_0 is given by

$$e_0(X) = \frac{1}{\sqrt{2}}\begin{bmatrix} 1 & 1 \\ \psi(X)/|\psi(X)| & -\psi(X)/|\psi(X)| \end{bmatrix}, \quad X \neq 0,$$

and subprincipal term b_0 is given by

$$b_0(X) = \frac{\gamma_1 + \gamma_2 - 1}{2} I_2.$$

Note that $p_{2,\alpha} = |\alpha| p_2$ and that b_0 is constant on the bicharacteristics of p_2 (and hence of $p_{2,\alpha}$). Using the index formula from [9], one computes $\mathrm{ind}(e_0) = -1$, so that we have to augment the system into a 4×4 one. Theorem 5.2 then gives that

$$\mathrm{Spec}(A_{JC}) \subset S_0 \cup \bigcup_{\pm} S_{\pm},$$

where the structure of $S_{\pm}$ and S_0 is as follows. Let, for $k \in \mathbb{Z}_+$,

$$c_{\pm}(k) = k + 1/2 \pm |\alpha|\sqrt{k+1/2} + (\gamma_1 + \gamma_2 - 1)/2, \quad \mu_k^{(\pm)} = (\gamma_1 + \gamma_2 - 1)/2,$$

and

$$c_0(k) = k + 1/2 + c,$$

where $c > 0$ is so large to ensure the positivity of the operator P_0. Then

$$S_{\pm} = \bigcup_{k \geq 0} \left(c_{\pm}(k) + \left[-\frac{M}{\sqrt{k+1/2}}, \frac{M}{\sqrt{k+1/2}} \right] \right),$$

$$S_0 = \bigcup_{k \geq 0} \left(c_0(k) + \left[-\frac{M}{\sqrt{k+1/2}}, \frac{M}{\sqrt{k+1/2}} \right] \right).$$

6 Problems

I end this survey by considering two open problems.

- As mentioned in Sect. 3, one has the expression of the semi-subprincipal term of the diagonalized operator, along with its transformation rules in respect of

the principal symbol of the diagonalizer. Is it possible to obtain a four-term asymptotics of N_A in terms of the principal, semiprincipal, subprincipal, and semi-subprincipal terms?
- In the quasi-clustering setting, what happens in case the bicharacteristic flows of the principal part p_2 and that of $p_{2,\alpha}$ do not commute? And how about the requirement on the subprincipal part of the *diagonalized* operator, having fixed the gauge e_0, to be constant on the bicharacteristics? Can it be relaxed? And is it possible to understand the constancy condition in terms of the initial symbol, without having to fix a diagonalizer? Note that the Jaynes-Cummings system satisfies the constancy of a_0 and b_0 on the bicharacteristics, which are commuting since p_2 and $p_{2,\alpha}$ are in this case proportional.

Acknowledgments I wish to thank the referee for her/his careful reading and important observations. I wish also to thank Shif Berhanu, Paulo Cordaro, Gustavo Hoepfner, and Nordine Mir for the kind invitation to contribute with a paper in honor of Jorge Hounie.

References

1. S. Coriasco, M. Doll. Weyl law on asymptotically Euclidean manifolds. Ann. Henri Poincaré **22** (2021), no. 2, 447–486. https://doi.org/10.1007/s00023-020-00995-1
2. Y. C. de Verdière. Sur le spectre des opŕateurs elliptiques a bicaractŕistiques toutes periodiques. Commentarii Mathematici Helvetici **54**, 508–522 (1979). https://doi.org/10.1007/BF02566290
3. J. J. Duistermaat, V. W. Guillemin. The spectrum of positive elliptic operators and periodic bicharacteristics. Invent Math **29**, 39–79 (1975). https://doi.org/10.1007/BF01405172
4. M. Doll,O. Gannot, J. Wunsch. Refined Weyl law for homogeneous perturbations of the harmonic oscillator. Comm. Math. Phys. **362** (2018), no. 1, 269–294. https://doi.org/10.1007/s00220-018-3100-5
5. M. Doll, S. Zelditch. Schrödinger trace invariants for homogeneous perturbations of the harmonic oscillator. J. Spectr. Theory **10** (2020), 1303–1332 https://doi.org/10.4171/JST/328
6. B. Helffer, D. Robert. Comportement asymptotique précisé du spectre d'opérateurs globalement elliptiques dans $\mathbb{R}^n$ Séminaire Équations aux dérivées partielles (Polytechnique) (1980-1981), exp. no. 2, p. 1–22.
7. B. Helffer. *Théorie spectrale pour des opérateurs globalement elliptiques*. Astérisque **112** (1984)
8. L. Hörmander. Symplectic classification of quadratic forms, and general Mehler formulas. Math. Z. **219** (1995), no. 3, 413–449.
9. L. Hörmander. The analysis of linear partial differential operators. III. Pseudo-differential operators. Reprint of the 1994 edition. Classics in Mathematics. Springer, Berlin, 2007. viii+525 pp.
10. L. Hörmander. The Analysis of Linear Partial Differential Operators IV. Classics in Mathematics. Springer, Berlin (2009).
11. V. Ivrii. *Microlocal analysis and precise spectral asymptotics*. Springer Monogr. Math. Springer-Verlag, Berlin, 1998, xvi+731 pp.
12. M. Malagutti and A. Parmeggiani. *Spectral asymptotic properties of semiregular Non-Commutative Harmonic Oscillators*. Comm. Math. Phys. **405** (2024), Paper No. 46. https://doi.org/10.1007/s00220-024-04934-7
13. M. Malagutti and A. Parmeggiani. *Spectral quasi-clustering estimates for certain semiregular systems*. Bull. Sci. Math. **193** (2024), Paper No. 103423, 21 pp. https://doi.org/10.1016/j.bulsci.2024.103423

14. A. Parmeggiani. *Spectral theory of Non-Commutative Harmonic Oscillators: An Introduction*. Lecture Notes in Mathematics, 1992. Springer-Verlag, Berlin, 2010. https://doi.org/10.1007/978-3-642-11922-4
15. A. Parmeggiani. Non-commutative harmonic oscillators and related problems. Milan J. Math. **82** (2014), 343–387. https://doi.org/10.1007/s00032-014-0220-z
16. A. Parmeggiani. On the spectrum of certain noncommutative harmonic oscillators. Ann. Univ. Ferrara Sez. VII Sci. Mat. **52** (2006), 431–456. https://doi.org/10.1007/s11565-006-0030-5
17. A. Parmeggiani. On the spectrum of certain non-commutative harmonic oscillators and semiclassical analysis. Comm. Math. Phys. **279** (2008), 285–308. https://doi.org/10.1007/s00220-008-0436-2
18. A. Parmeggiani, M. Wakayama. Oscillator representations and systems of ordinary differential equations. Proceedings of the National Academy of Sciences U.S.A. **98** (2001), 2630. https://doi.org/10.1073/pnas.98.1.26
19. A. Parmeggiani, M. Wakayama. Non-commutative harmonic oscillators-I, -II. Forum Mathematicum **14** (2002), 539–604 ibid. 669–690. https://doi.org/10.1515/form.2002.025, https://doi.org/10.1515/form.2002.029
20. Y. Safarov, D. Vassiliev. *The asymptotic distribution of eigenvalues of partial differential operators*. Translations of Mathematical Monographs, vol. 155. American Mathematical Society, Providence (1997).
21. B. W. Shore, P. L. Knight. The Jaynes-Cummings model. J. Modern Opt. **40** (1993), no. 7, 1195–1238. https://doi.org/10.1080/09500349314551321
22. M. A. Shubin. *Pseudodifferential operators and spectral theory* (second edition). Springer-Verlag, Berlin, 2001.
23. A. Weinstein. Asymptotics of eigenvalue clusters for the Laplacian plus a potential. Duke Math. J. **44**(4):883-892 (1977). https://doi.org/10.1215/S0012-7094-77-04442-8

Friedrichs' Lemma for Besov and Triebel-Lizorkin Spaces

Evandro Raimundo da Silva and José R. dos Santos Filho

1 Introduction

Suppose $\phi \in \mathcal{S}(\mathbb{R}^N)$ (=Schwartz space), and let J_ϵ be the regularizing operator of Friedrichs associated with ϕ, defined by

$$J_\epsilon f(x) = \phi_\epsilon * f(x), \quad \phi_\epsilon(x) = \epsilon^{-N}\phi(x/\epsilon), \quad \epsilon > 0.$$

Young's inequality says that if $1 \leq p \leq \infty$, then

$$\|J_\epsilon f\|_{L^p(\mathbb{R}^N)} \leq \|\phi\|_{L^1(\mathbb{R}^N)} \|f\|_{L^p(\mathbb{R}^N)} \tag{1}$$

for all $f \in L^p(\mathbb{R}^N)$ and $\epsilon > 0$. Furthermore, if $\int \phi\, dx = 1$, then

$$\lim_{\epsilon \to 0} \|J_\epsilon f - f\|_{L^p(\mathbb{R}^N)} = 0. \tag{2}$$

An extension of (1)–(2) was long time ago obtained, which is as follows: Let $a(x)$ be a Lipschitz function, for $j = 1, \cdots, N$, $D_j = -i\partial/\partial x_j$, and consider the commutators

$$[J_\epsilon, a\, D_j]f(x) = \phi_\epsilon * (a\, D_j f)(x) - a(x) D_j(\phi_\epsilon * f)(x).$$

E. R. da Silva
Instituto de Ciências Matemáticas e Computação, Universidade de São Paulo, São Carlos, São Paulo, Brazil
e-mail: evandro@icmc.usp.br

J. R. dos Santos Filho (✉)
Departamento de Matemática, Universidade Federal de São Carlos, São Carlos, São Paulo, Brazil
e-mail: ruidival@ufscar.br

S. Berhanu et al. (eds.), *Geometric Analysis of PDEs and Several Complex Variables*, Latin American Mathematics Series – UFSCar subseries,
https://doi.org/10.1007/978-3-031-69702-9_13

Following L. Hörmander, see [2], pg. 9, Friedrichs' Lemma says that there exists $C = C(N, p) > 0$ such that

$$\|[J_\epsilon, a\, D_j]f\|_{L^p(\mathbb{R}^N)} \leq C\|a\|_{Lip}\|f\|_{L^p(\mathbb{R}^N)}$$

for all $f \in L^p(\mathbb{R}^N)$, and $\epsilon > 0$. Furthermore,

$$\lim_{\epsilon\to 0} \|[J_\epsilon, a\, D_j]f\|_{L^p(\mathbb{R}^N)} = 0.$$

J. Hounie and dos Santos Filho, in [3], extended this result when $a(x)$ is a Hölder function, and the operator D_j is replaced by a pseudo-differential operator in Hörmander's class $L^r_{1,\delta}$. They showed that if $a \in \mathcal{C}^t(\mathbb{R}^N)$ (= Hölder space), $b(x, D) \in L^r_{1,\delta}$, $0 \leq \delta < 1$, $0 < s < r < t < 1$, there exists $C = C(s, r, t) > 0$ such that

$$\|\, [J_\epsilon, a]b(x, D)u \,\|_{\mathcal{C}^s(\mathbb{R}^N)} \leq C\|a\|_{\mathcal{C}^t(\mathbb{R}^N)}\|u\|_{\mathcal{C}^s(\mathbb{R}^N)}, \quad u \in \mathcal{S}(\mathbb{R}^N).$$

On the other hand, J. Hounie, in [4], showed that if $\frac{N}{N+1} < p \leq 1$ and $a \in \mathcal{C}^{\rho+1}(\mathbb{R}^N)$, $\rho > N(\frac{1}{p} - 1)$, there exists $C = C(p) > 0$, such that

$$\|\, [J_\epsilon, b]D_j f\|_{h^p(\mathbb{R}^N)} \leq C\|b\|_{\mathcal{C}^{\rho+1}(\mathbb{R}^N)}\|f\|_{h^p(\mathbb{R}^N)}, \quad f \in h^p(\mathbb{R}^N);$$

here $h^p(\mathbb{R}^N)$ is the local Hardy space. The spaces L^p, $\mathcal{C}^s$, and h^p are special cases of Triebel-Lizorkin $F^s_{p,q}$ and Besov spaces $B^s_{p,q}$. In fact, $F^0_{p,2} = L^p$, $1 < p < \infty$, $F^s_{\infty,\infty} = B^s_{\infty,\infty} = \mathcal{C}^s$, $s > 0$ not integer, and $F^0_{p,2} = h^p$, $0 < p \leq 1$. It is natural to ask if Friedrichs' Lemma can be extended to Besov and Triebel-Lizorkin spaces. In fact, E. da Silva, in [5] and [6], extended, without precise constants; in this article we will fill such gap. More precisely, our results show that

$$\|\, [J_\epsilon, a]b(x, D)f \,\|_{A^s_{p,q}(\mathbb{R}^N)} \leq C\|a\|_{\mathcal{C}^{\rho+1}(\mathbb{R}^N)}\|f\|_{A^s_{p,q}(\mathbb{R}^N)}, \quad f \in A^s_{p,q}(\mathbb{R}^N),$$

for $1 \leq p < \infty$, $1 \leq q \leq \infty$, $s \in \mathbb{R}$ e $a \in \mathcal{C}^{\rho+1}(\mathbb{R}^N)$, $\rho > |s|$, where $A^s_{p,q(\mathbb{R}^N)}$ denotes either $B^s_{p,q}(\mathbb{R}^N)$ or $F^s_{p,q}(\mathbb{R}^N)$. This article is organized in the following way: In Sect. 2 we recall some definitions and properties of the spaces considered here. In Sect. 3 we extend Young's Inequality. Finally, in Sect. 4 we extend Friedrichs' Lemma. In Sect. 5 we present applications to the regularity of solutions of elliptic systems of complex vectors fields with non-smooth coefficients. The second author is an associate researcher of the Thematic project no. 2018/14316-3 (FAPESP).

2 The Spaces of Besov and Triebel-Lizorkin

In this section we define and recall some properties of the spaces of Besov e Triebel-Lizorkin; for those spaces we use more flexible definitions, as given by H. Triebel, that is, instead of multipliers we use the convolution's versions, and the notations therein, see [7, 8], and [9].

Let $B(0,1) = \{x \in \mathbb{R}^N;\ \|x\| < 1\}$. Consider k_0, k^0 in $C_c^\infty(B(0,1))$, where $\int k_0(x)dx \neq 0$ and $\int k^0(x)dx \neq 0$. If $M \in \mathbb{N}$, consider $k_M(x) = \Delta^M k^0(x)$, where Δ is the Laplace operator in $\mathbb{R}^N$. If $f \in \mathcal{S}'(\mathbb{R}^N)$, we define

$$K_0(1, f)(x) = \int_{\mathbb{R}^N} k_0(y-x) f(y) dy\,,$$

and for $j \in \mathbb{N}$,

$$K_M(2^{-j}, f)(x) = 2^{jN} \int_{\mathbb{R}} k_M(2^j(y-x)) f(y) dy, \quad x \in \mathbb{R},$$

as usually is interpreted, for $f \in \mathcal{S}'(\mathbb{R}^N)$.

Definition 2.1 (Besov Space) Let $1 \leq p \leq \infty$, $1 \leq q \leq \infty$, and $s \in \mathbb{R}$. If $M \in \mathbb{N}$ with $2M > \max\{s, 0\}$, take

$$B^s_{p,q}(\mathbb{R}^N) = \{f \in S'(\mathbb{R}^N)\ ;\ \|f\|_{B^s_{p,q}(\mathbb{R}^N)} < \infty\},$$

with the respective norm

$$\|f\|_{B^s_{p,q}(\mathbb{R}^N)} = \|K_0(1, f)\|_{L^p(\mathbb{R}^N)} + \left(\sum_{j=1}^{\infty} 2^{jsq} \|K_M(2^{-j}, f)\|^q_{L^p(\mathbb{R}^N)}\right)^{\frac{1}{q}}.$$

For $q = \infty$, we take the usual modification.

Definition 2.2 (Triebel-Lizorkin Space) Let $1 \leq p < \infty$, $1 \leq q \leq \infty$, and $s \in \mathbb{R}$. If $M \in \mathbb{N}$, $2M > max\{s, 0\}$, take

$$F^s_{p,q}(\mathbb{R}^N) = \{f \in S'(\mathbb{R}^N)\ ;\ \|f\|_{F^s_{p,q}(\mathbb{R}^N)} < \infty\},$$

with the respective norm

$$\|f\|_{F^s_{p,q}(\mathbb{R}^N)} = \|K_0(1, f)\|_{L^p(\mathbb{R}^N)} + \left\| \left(\sum_{j=1}^{\infty} 2^{jsq} |K_M(2^{-j}, f)(\cdot)|^q\right)^{\frac{1}{q}} \right\|_{L^p(\mathbb{R}^N)}.$$

For $q = \infty$, we take the usual modification.

The spaces Besov and Triebel-Lizorkin are Banach spaces. For different choices of k_0 and k^0, the norms are equivalents. Furthermore, pseudo-differential operators of order 0 are bounded in $B^s_{p,q}(\mathbb{R}^N)$ and in $F^s_{p,q}(\mathbb{R}^N)$.

The theorems below give equivalent characterization of $B^s_{p,q}(\mathbb{R}^N)$ and $F^s_{p,q}(\mathbb{R}^N)$, see [7] on page 58.

Theorem 2.3 *Let $s \in \mathbb{R}$, $m \in \mathbb{N}$, $1 \leq q \leq \infty$.*
(A) If $1 \leq p \leq \infty$, then

$$\|f\|_{B^{s-m}_{p,q}(\mathbb{R}^N)} + \sum_{j=1}^{N} \|D^m_j f\|_{B^{s-m}_{p,q}(\mathbb{R}^N)}$$

is an equivalent norm of $B^s_{p,q}(\mathbb{R}^N)$.
(B) If $1 \leq p < \infty$, then

$$\|f\|_{F^{s-m}_{p,q}(\mathbb{R}^N)} + \sum_{j=1}^{N} \|D^m_j f\|_{F^{s-m}_{p,q}(\mathbb{R}^N)}$$

is an equivalent norm of $F^s_{p,q}(\mathbb{R}^N)$.

We will use the notation $A^s_{p,q}(\mathbb{R}^N)$ to denote either $B^s_{p,q}(\mathbb{R}^N)$ or $F^s_{p,q}(\mathbb{R}^N)$.

Theorem 2.4 *Let $f \in A^s_{p,q}(\mathbb{R}^N)$, given $\epsilon > 0$, then there exists $k_\epsilon \in \mathbb{N}$ such that*

$$\|f\|_{A^s_{p,q}(\mathbb{R}^N)} \leq \epsilon \|f\|_{A^{s+1}_{p,q}(\mathbb{R}^N)} + 2^{k_\epsilon} \|f\|_{A^{s-1}_{p,q}(\mathbb{R}^N)}.$$

Proof It is enough to observe that

$$\begin{aligned}\sum_{j=1}^{\infty} 2^{jsq} &= \sum_{j=1}^{k} 2^{jsq} + \sum_{j=k+1}^{\infty} 2^{jsq} \\ &= \sum_{j=1}^{k} 2^{j(s-1)q} 2^{jq} + \sum_{j=k+1}^{k} 2^{j(s+1)q} 2^{-jq}.\end{aligned}$$

Also that given $\epsilon > 0$, there exists $k_\epsilon \in \mathbb{N}$ such that $2^{-jq} < \epsilon$ for all $j > k_\epsilon$, then

$$\sum_{j=1}^{\infty} 2^{jsq} \leq 2^{k_\epsilon q} \sum_{j=1}^{k_\epsilon} 2^{j(s-1)q} + \epsilon \sum_{j=k_\epsilon+1}^{\infty} 2^{j(s+1)q}.$$

□

Theorem 2.5 *Let $0 < \rho < 1$ and $B(0,\rho) = \{x \in \mathbb{R}^N; \|x\| < \rho\}$. Suppose $s \in \mathbb{R}$, $1 < p, q < \infty$, and p' with $\frac{1}{p} + \frac{1}{p'} = 1$. There exists $C = C(s, p, q) > 0$ such that:*

(A) If $s < 0$, then

$$\|\varphi\|_{A^s_{p,q}(\mathbb{R}^N)} \leq C\|\varphi\|_{L^p(\mathbb{R}^N)} \max\{\rho^{\frac{N}{p'}}, \rho^{\frac{N}{p}}\}$$

for all $\varphi \in C_c^\infty(B(0,\rho))$.

(B) If $s \geq 0$ and $m \in \mathbb{N}$, with $m > s$, then

$$\|\varphi\|_{A^s_{p,q}(\mathbb{R}^N)} \leq C\left(\|\varphi\|_{L^p(\mathbb{R}^N)} + \sum_{j=1}^{N} \|D_j^m \varphi\|_{L^p(\mathbb{R}^N)}\right) \max\left\{\rho^{\frac{N}{p'}}, \rho^{\frac{N}{p}}\right\}$$

for all $\varphi \in C_c^\infty(B(0,\rho))$.

(C) If $s > 0$, then

$$\|\varphi\|_{A^{s-2}_{p,q}(\mathbb{R}^N)} \leq C\|\varphi\|_{A^s_{p,q}(\mathbb{R}^N)} \max\left\{\rho^{\frac{N}{p'}}, \rho^{\frac{N}{p}}\right\}$$

for all $\varphi \in C_c^\infty(B(0,\rho))$.

Proof

Case 1 $A^s_{p,q}(\mathbb{R}^N) = B^s_{p,q}(\mathbb{R}^N)$

Let us show (A). By Young's inequality and Hölder's inequality, we have

$$\begin{aligned}\|K_0(1,\varphi)\|_{L^p(\mathbb{R}^N)} &\leq \|k_0\|_{L^p(\mathbb{R}^N)}\|\varphi\|_{L^1(\mathbb{R}^N)}\\ &\leq C\|\varphi\|_{L^p(\mathbb{R}^N)}\rho^{\frac{N}{p'}}\end{aligned}$$

and

$$\begin{aligned}\|K_M(2^{-j},\varphi)\|_{L^p(\mathbb{R}^N)} &\leq \|k^0\|_{L^p(\mathbb{R}^N)}\|\varphi\|_{L^1(\mathbb{R}^N)}\\ &\leq C\|\varphi\|_{L^p(\mathbb{R}^N)}\rho^{\frac{N}{p'}}.\end{aligned}$$

Then

$$\begin{aligned}\|\varphi\|_{B^s_{p,q}(\mathbb{R}^N)} &\leq C\|\varphi\|_{L^p(\mathbb{R}^N)}\rho^{\frac{N}{p'}} + \left(\sum_{j=1}^{\infty} 2^{jsq}\right)^{\frac{1}{q}} \|\varphi\|_{L^p(\mathbb{R}^N)}\rho^{\frac{N}{p'}}\\ &= C\|\varphi\|_{L^p(\mathbb{R}^N)}\,\rho^{\frac{N}{p'}}.\end{aligned}$$

Let us show (B). By Theorem 2.3, if $m > s$, then

$$\|\varphi\|_{B^s_{p,q}(\mathbb{R}^N)} \leq C\left(\|\varphi\|_{B^{s-m}_{p,q}(\mathbb{R}^N)} + \sum_{j=1}^{N} \|D^m_j \varphi\|_{B^{s-m}_{p,q}(\mathbb{R}^N)}\right).$$

Since $s - m < 0$, by A) we conclude that

$$\|\varphi\|_{B^s_{p,q}(\mathbb{R}^N)} \leq C\left(\|\varphi\|_{L^p(\mathbb{R}^N)} + \sum_{j=1}^{N} \|D^m_j \varphi\|_{L^p(\mathbb{R}^N)}\right) \rho^{\frac{N}{p'}}.$$

Let us show (C).

First, suppose that $0 < s < 2$. There exists $C = C(s, p, q) > 0$ such that

$$\|\varphi\|_{L^p(\mathbb{R}^N)} \leq C\|\varphi\|_{B^s_{p,q}(\mathbb{R}^N)}. \tag{3}$$

Then, by A) and (3), we have

$$\begin{aligned} \|\varphi\|_{B^{s-2}_{p,q}} &\leq C\|\varphi\|_{L^p(\mathbb{R}^N)} \max\left\{\rho^{\frac{N}{p'}}, \rho^{\frac{N}{p}}\right\} \\ &\leq C\|\varphi\|_{B^s_{p,q}(\mathbb{R}^N)} \max\left\{\rho^{\frac{N}{p'}}, \rho^{\frac{N}{p}}\right\}. \end{aligned}$$

Suppose $s - 2 \geq 0$. Let $m \in \mathbb{R}$, such that, $m < s < m + 2$. Then $B^{s-m}_{p,q}(\mathbb{R}^N) \subset L^p(\mathbb{R}^N)$ (continuous inclusion); this implies that there exists $C = C(s, p, q) > 0$ such that

$$\|\varphi\|_{L^p(\mathbb{R}^N)} \leq C\|\varphi\|_{B^{s-m}_{p,q}(\mathbb{R}^N)} \tag{4}$$

and

$$\|D^m_j \varphi\|_{L^p(\mathbb{R}^N)} \leq C\|D^m_j \varphi\|_{B^{s-m}_{p,q}(\mathbb{R}^N)}, \tag{5}$$

$j = 1, \cdots, N$. Therefore, by (B), (4), and (5), we have

$$\begin{aligned} \|\varphi\|_{B^{s-2}_{p,q}(\mathbb{R}^N)} &\leq C\left(\|\varphi\|_{L^p(\mathbb{R}^N)} + \sum_{j=1}^{N} \|D^m_j \varphi\|_{L^p(\mathbb{R}^N)}\right) \max\left\{\rho^{\frac{N}{p'}}, \rho^{\frac{N}{p}}\right\} \qquad (6) \\ &\leq C\left(\|\varphi\|_{B^{s-m}_{p,q}(\mathbb{R}^N} + \sum_{j=1}^{N} \|D^m_j \varphi\|_{B^{s-m}_{p,q}(\mathbb{R}^N}\right) \max\left\{\rho^{\frac{N}{p'}}, \rho^{\frac{N}{p}}\right\}. \end{aligned}$$

Recall that $B^s_{p,q}(\mathbb{R}^N) \subset B^{s-m}_{p,q}(\mathbb{R}^N)$ and $D^m_j : B^s_{p,q}(\mathbb{R}^N) \to B^{s-m}_{p,q}(\mathbb{R}^N)$ is a bounded operator (pseudo-differential operator of order m), for $j = 1, \cdots, N$, we have of (6) that

$$\|\varphi\|_{B^{s-2}_{p,q}(\mathbb{R}^N)} \le C\|\varphi\|_{B^s_{p,q}(\mathbb{R}^N)} \max\left\{\rho^{\frac{N}{p'}}, \rho^{\frac{N}{p}}\right\}.$$

Case 2 $A^s_{p,q}(\mathbb{R}^N) = F^s_{p,q}(\mathbb{R}^N)$

Let us show (A). By Young's inequality and Hölder's inequality, we have

$$\begin{aligned}\|K_0(1,\varphi)\|_{L^p(\mathbb{R}^N)} &\le \|k_0\|_{L^p(\mathbb{R}^N)}\|\varphi\|_{L^1(\mathbb{R}^N)}\\ &\le C\|\varphi\|_{L^p(\mathbb{R}^N)}\rho^{\frac{N}{p'}},\end{aligned}$$

and by Hölder's inequality, we have

$$|K_M(2^{-j},\varphi)(x)| \le \|k^0\|_{L^{p'}(\mathbb{R}^N)}\|\varphi\|_{L^p(\mathbb{R}^N)}.$$

Therefore

$$\sum_{j=1}^{\infty} 2^{jsq}|K_M(2^{-j},\varphi)(x)|^q \le C\|\varphi\|^q_{L^p(\mathbb{R}^N)}.$$

Suppose $\operatorname{supp}(k^0) \subset B(0,\rho)$, then we have that $\operatorname{supp}((K_M(2^{-j},\varphi)) \subset B(0,2\rho)$, for all j, and then

$$\begin{aligned}\|\varphi\|_{F^s_{p,q}(\mathbb{R}^n)} &\le C\|\varphi\|_{L^p(\mathbb{R}^N)}\rho^{\frac{N}{p'}} + C\|\varphi\|_{L^p(\mathbb{R}^N)}\rho^{\frac{N}{p}}\\ &\le C\|\varphi\|_{L^\infty(\mathbb{R}^N)} \max\left\{\rho^{\frac{N}{p'}}, \rho^{\frac{N}{p}}\right\}.\end{aligned}$$

Let us show (B). By Theorem 2.3, if $m > s$, then

$$\|\varphi\|_{F^s_{p,q}(\mathbb{R}^N)} \le C\left(\|\varphi\|_{F^{s-m}_{p,q}(\mathbb{R}^N)} + \sum_{j=1}^{N}\|D^m_j\varphi\|_{F^{s-m}_{p,q}(\mathbb{R}^N)}\right).$$

With $s - m < 0$, by (A) we conclude that

$$\|\varphi\|_{F^s_{p,q}(\mathbb{R}^N)} \le C\left(\|\varphi\|_{L^\infty(\mathbb{R}^N)} + \sum_{j=1}^{N}\|D^m_j\varphi\|_{L^\infty(\mathbb{R}^N)}\right)\max\left\{\rho^{\frac{N}{p'}}, \rho^{\frac{N}{p}}\right\}.$$

Let us show (C). We use (A) and (B), and we proceed as in proof of (C) for $B^s_{p,q}(\mathbb{R}^N)$.

□

Theorem 2.6 *Suppose* $s > 0$, $1 \le p, q < \infty$, *and* $C = C(s, p, q) > 0$ *given in Theorem 2.3. Let* $0 < \epsilon < \frac{1}{2C}$, *then there exists* $0 < \rho_\epsilon < 1$ *such that*

$$\|\varphi\|_{A^s_{p,q}(\mathbb{R}^N)} \le \epsilon\, C\, \|\nabla\varphi\|_{A^s_{p,q}(\mathbb{R}^N)},$$

for all $\varphi \in C_c^\infty(B(0, \rho_\epsilon))$, *where*

$$\|\nabla\varphi\|_{A^s_{p,q}(\mathbb{R}^N)} = \sum_{j=1}^{N} \|D_j\varphi\|_{A^s_{p,q}(\mathbb{R}^N)}.$$

Proof By Theorem 2.3 there exists $C = C(s, p, q) > 0$ such that

$$\|f\|_{A^s_{p,q}(\mathbb{R}^N)} \le C\left(\|f\|_{A^{s-1}_{p,q}(\mathbb{R}^N)} + \|\nabla f\|_{A^{s-1}_{p,q}(\mathbb{R}^N)}\right) \tag{7}$$

for all $f \in A^s_{p,q}(\mathbb{R}^N)$. Let $0 < \epsilon < \frac{1}{2C}$, and if $f \in A^s_{p,q}(\mathbb{R}^N)$ is fixed, by Theorem 2.4, there exist $k_\epsilon > 0$ and $C_\epsilon > 0$ such that

$$\|f\|_{A^{s-1}_{p,q}(\mathbb{R}^N)} \le \epsilon\|f\|_{A^s_{p,q}(\mathbb{R}^N)} + 2^{k_\epsilon}\|f\|_{A^{s-2}_{p,q}(\mathbb{R}^N)} \tag{8}$$

and

$$\|\nabla f\|_{A^{s-1}_{p,q}(\mathbb{R}^N)} \le \epsilon\|\nabla f\|_{A^s_{p,q}(\mathbb{R}^N)} + C_\epsilon\|\nabla f\|_{A^{s-2}_{p,q}(\mathbb{R}^N)}. \tag{9}$$

Therefore, from (7), (8), and (9), we have

$$\begin{aligned}\|f\|_{A^s_{p,q}(\mathbb{R}^N)} \le{}& \epsilon\, C\|f\|_{A^s_{p,q}(\mathbb{R}^N)} + C2^{k_\epsilon}\|f\|_{A^{s-2}_{p,q}(\mathbb{R}^N)}\\ &+ \epsilon\, C\|\nabla f\|_{A^s_{p,q}(\mathbb{R}^N)} + CC_\epsilon\|\nabla f\|_{A^{s-2}_{p,q}(\mathbb{R}^N)}\end{aligned} \tag{10}$$

Now, by Theorem 2.5 and (10), we have

$$\begin{aligned}\|\varphi\|_{A^s_{p,q}(\mathbb{R}^N)} \le{}& \epsilon\, C\|\varphi\|_{A^s_{p,q}(\mathbb{R}^N)} + C2^{k_\epsilon}\|\varphi\|_{A^s_{p,q}(\mathbb{R}^N)}\left\{\rho^{\frac{N}{p}}, \rho^{\frac{N}{p'}}\right\}\\ &+ \epsilon\, C\|\nabla\varphi\|_{A^s_{p,q}(\mathbb{R}^N)} + CC_\epsilon\|\nabla\varphi\|_{A^s_{p,q}(\mathbb{R}^N)} \max\left\{\rho^{\frac{N}{p}}, \rho^{\frac{N}{p'}}\right\}\end{aligned}$$

for all $\varphi \in C_c^\infty(B(0,\rho))$. Finally, taking $\rho > 0$ such that

$$0 < \max\left\{\rho^{\frac{N}{p}}, \rho^{\frac{N}{p'}}\right\} < \min\left\{\frac{\epsilon}{2C2^{k_\epsilon}}, \frac{\epsilon}{C_\epsilon}\right\}$$

we obtain

$$\|\varphi\|_{A^s_{p,q}(\mathbb{R}^N)} \le \frac{1}{2}\|\varphi\|_{A^s_{p,q}(\mathbb{R}^N)} + 2\epsilon\, C\|\nabla\varphi\|_{A^s_{p,q}(\mathbb{R}^N)},$$

which implies

$$\|\varphi\|_{A^s_{p,q}(\mathbb{R}^N)} \le \epsilon\, C\|\nabla\varphi\|_{A^s_{p,q}(\mathbb{R}^N)}.$$

□

When $\rho > 0$ is not an integer, we denote $\rho = [\rho] + \{\rho\}$, where $[\rho]$ is an integer and $0 < \{\rho\} < 1$, and we take $\alpha = (\alpha_1, \cdots, \alpha_N) \in \mathbb{N}^N$ and $D^\alpha = D_1^{\alpha_1} \cdots D_N^{\alpha_N}$. It is known that $B^\rho_{\infty,\infty} = C^\rho$, where C^ρ is a Hölder space.

Definition 2.7 (Hölder Space)

$$C^\rho(\mathbb{R}^N) = \{f \in C^{[\rho]}(\mathbb{R}^N)) \;\text{ such that }\; \|f\|_{C^\rho(\mathbb{R}^N)} < \infty\},$$

where

$$\|f\|_{C^\rho(\mathbb{R}^N)} = \sum_{|\alpha|\le[\rho]} \|D^\alpha f\|_{L^\infty(\mathbb{R}^N)} + \sum_{|\alpha|=[\rho]} \sup_{z\neq w} \frac{|D^\alpha f(z) - D^\alpha f(w)|}{|z-w|^{\{\rho\}}}.$$

The following result is a consequence of the theorem proved in [8], see page 205:

Theorem 2.8 *Let $s \in \mathbb{R}$, $1 \le q \le \infty$, and $g \in C^\rho(\mathbb{R}^N)$ with $\rho > |s|$.*
(A) If $1 \le p \le \infty$, then there exists $C = C(s,p,q) > 0$ such that

$$\|gf\|_{B^s_{p,q}(\mathbb{R}^N)} \le C\|g\|_{C^\rho(\mathbb{R}^N)}\|f\|_{B^s_{p,q}(\mathbb{R}^N)},$$

for all $f \in B^s_{p,q}(\mathbb{R}^N)$.
(B) If $1 \le p < \infty$, then there exists $C = C(s,p,q) > 0$ such that

$$\|gf\|_{F^s_{p,q}(\mathbb{R}^N)} \le C\|g\|_{C^\rho(\mathbb{R}^N)}\|f\|_{F^s_{p,q}(\mathbb{R}^N)},$$

for all $f \in F^s_{p,q}(\mathbb{R}^N)$.

3 Young's Inequality

The following holds:

Theorem 3.1 *Let $\phi \in \mathcal{S}(\mathbb{R}^N)$, $1 \leq q \leq \infty$, $1 \leq p \leq \infty$, and $s \in \mathbb{R}$. Then*

$$\|J_\epsilon f\|_{B^s_{p,q}(\mathbb{R}^N)} \leq \|\phi\|_{L^1(\mathbb{R}^N)} \|f\|_{B^s_{p,q}(\mathbb{R}^N)}, \tag{11}$$

for all $f \in B^s_{p,q}(\mathbb{R}^N)$ and $\epsilon > 0$. Furthermore, if $\int \phi\, dx = 1$, then

$$\lim_{\epsilon \to 0} \|J_\epsilon f - f\|_{B^s_{p,q}(\mathbb{R}^N)} = 0. \tag{12}$$

Proof For (11) observe that

$$K_0(1, J_\epsilon f) = J_\epsilon K_0(1, f)$$

and

$$K_M(2^{-j}, J_\epsilon f) = J_\epsilon K_M(2^{-j}, f);$$

hence from (1), we have

$$\|K_0(1, J_\epsilon f)\|_{L^p(\mathbb{R}^N)} \leq \|\phi\|_{L^1(\mathbb{R}^N)} \|K_0(1, f)\|_{L^p(\mathbb{R}^N)}.$$

$$\|K_M(2^{-j}, J_\epsilon f)\|_{L^p(\mathbb{R}^N)} \leq \|\phi\|_{L^1(\mathbb{R}^N)} \|K_M(2^{-j}, f)\|_{L^p(\mathbb{R}^N)}.$$

Therefore

$$\|J_\epsilon f\|_{B^s_{p,q}(\mathbb{R}^N)} \leq \|\phi\|_{L^1(\mathbb{R}^N)} \|f\|_{B^s_{p,q}(\mathbb{R}^N)}.$$

For (12), suppose $s < 0$ and $f \in \mathcal{S}(\mathbb{R}^N)$. Since $L^p(\mathbb{R}^N) = F^0_{p,2}(\mathbb{R}^N) \subset B^0_{p,\max\{p,2\}}(\mathbb{R}^N) \subset B^s_{p,q}(\mathbb{R}^N)$ (continuous inclusion), we have

$$\|J_\epsilon f - f\|_{B^s_{p,q}(\mathbb{R}^N)} \leq C \|J_\epsilon f - f\|_{L^p(\mathbb{R}^N)}; \tag{13}$$

therefore from (2) and (13), we have (12).

Suppose $s \geq 0$ and $f \in S(\mathbb{R}^N)$. Let $m \in \mathbb{N}$ with $m > s$. From Theorem 2.3. A) we have

$$\begin{aligned} \|J_\epsilon f - f\|_{B^s_{p,q}(\mathbb{R}^N)} \leq C\Big(& \|J_\epsilon f - f\|_{B^{s-m}_{p,q}(\mathbb{R}^N)} \\ & + \sum_{j=1}^{N} \|J_\epsilon D^m_j f - D^m_j f\|_{B^{s-m}_{p,q}(\mathbb{R}^N)}\Big). \end{aligned} \tag{14}$$

On the other hand, since $s - m < 0$, we have $L^p(\mathbb{R}^N) \subset B^{s-m}_{p,q}(\mathbb{R}^N)$. Therefore

$$\|J_\epsilon f - f\|_{B^{s-m}_{p,q}(\mathbb{R}^N)} \leq C\|J_\epsilon f - f\|_{L^p(\mathbb{R}^N)}, \tag{15}$$

and for each $j = 1, \cdots, N$,

$$\|J_\epsilon D^m_j f - D^m_j f\|_{B^{s-m}_{p,q}(\mathbb{R}^N)} \leq C\|J_\epsilon D^m_j f - D^m_j f\|_{L^p(\mathbb{R}^N)}. \tag{16}$$

Hence, from (14), (15), (16), and (2), we obtain

$$\|J_\epsilon f - f\|_{B^s_{p,q}(\mathbb{R}^N)} \to 0, \quad \epsilon \to 0.$$

Now, suppose $f \in B^s_{p,q}(\mathbb{R}^N)$. From the triangular inequality of the norm and (11), we have for all f and g in $B^s_{p,q}(\mathbb{R}^N)$

$$\|J_\epsilon f - f\|_{B^s_{p,q}(\mathbb{R}^N)} \leq (\|\phi\|_{L^1(\mathbb{R}^N)} + 1)\|f - g\|_{B^s_{p,q}(\mathbb{R}^N)} + \|J_\epsilon g - g\|_{B^s_{p,q}(\mathbb{R}^N)}. \tag{17}$$

From (17) and the density $S(\mathbb{R}^N)$ in $B^s_{p,q}(\mathbb{R}^N)$, we obtain

$$\|J_\epsilon f - f\|_{B^s_{p,q}(\mathbb{R}^N)} \to 0, \quad \epsilon \to 0.$$

□

Also we have the following theorem:

Theorem 3.2 *Let $\phi \in \mathcal{S}(\mathbb{R}^N)$, $1 \leq q \leq \infty$, $1 \leq p < \infty$, and $s \in \mathbb{R}$. Then*

$$\|J_\epsilon f\|_{F^s_{p,q}(\mathbb{R}^N)} \leq \|\phi\|_{L^1(\mathbb{R}^N)}\|f\|_{F^s_{p,q}(\mathbb{R}^N)}, \tag{18}$$

for all $f \in F^s_{p,q}(\mathbb{R}^N)$ and $\epsilon > 0$. Furthermore, if $\int \phi \, dx = 1$, then

$$\lim_{\epsilon \to 0} \|J_\epsilon f - f\|_{F^s_{p,q}(\mathbb{R}^N)} = 0. \tag{19}$$

Proof For (18) observe that

$$|K_0(1, J_\epsilon f)(x)| \leq \int |\phi_\epsilon(y)| \, |K_0(1, f)(x - y)| dy,$$

and

$$|K_M(2^{-j}, J_\epsilon f)(x)| \leq \int |\phi_\epsilon(y)| \, |K_M(2^{-j}, f)(x - y)| dy.$$

Hence from Minkowski's inequality for integrals, we have

$$\|K_0(1, J_\epsilon f)\|_{L^p(\mathbb{R}^N)} \leq \int |\phi_\epsilon(y)| \|K_0(1, f_y)\|_{L^p(\mathbb{R}^N)} dy,$$

where $f_y(z) = f(z - y)$ and

$$\begin{aligned}
&\left(\sum_{j=1}^{\infty} 2^{jsq} |K_M(2^{-j}, J_\epsilon f)(x)|^q\right)^{\frac{1}{q}} \\
&\leq \left(\sum_{j=1}^{\infty} 2^{jsq} \left(\int |\phi_\epsilon(y)|\, |K_M(2^{-j}, f)(x-y)| dy\right)^q\right)^{\frac{1}{q}} \\
&\leq \int |\phi_\epsilon(y)| \left(\sum_{j=1}^{\infty} 2^{jsq} |K_M(2^{-j}, f)(x-y)|^q\right)^{\frac{1}{q}} dy.
\end{aligned}$$

Again, from Minkowski's inequality for integrals, we have

$$\begin{aligned}
&\left\| \left(\sum_{j=1}^{\infty} 2^{jsq} |K_M(2^{-j}, J_\epsilon f)(\cdot)|^q\right)^{\frac{1}{q}} \right\|_{L^p(\mathbb{R}^N)} \\
&\leq \int |\phi_\epsilon(y)| \left\| \left(\sum_{j=1}^{\infty} 2^{jsq} |K_M(2^{-j}, f)(\cdot - y)|^q\right)^{\frac{1}{q}} \right\|_{L^p(\mathbb{R}^N)} dy \\
&= \int |\phi_\epsilon(y)| \left\| \left(\sum_{j=1}^{\infty} 2^{jsq} |K_M(2^{-j}, f_y)(\cdot)|^q\right)^{\frac{1}{q}} \right\|_{L^p(\mathbb{R}^N)} dy.
\end{aligned}$$

Therefore, since $\|f_y\|_{F^s_{p,q}(\mathbb{R}^N)} = \|f\|_{F^s_{p,q}(\mathbb{R}^N)}$, we have

$$\begin{aligned}
&\|J_\epsilon f\|_{F^s_{p,q}(\mathbb{R}^N)} = \|K_0(1, J_\epsilon f)\|_{L^p(\mathbb{R}^N)} + \left\| \left(\sum_{j=1}^{\infty} 2^{jsq} |K_M(2^{-j}, J_\epsilon f)(\cdot)|^q\right)^{\frac{1}{q}} \right\|_{L^p(\mathbb{R}^N)} \\
&\leq \int |\phi_\epsilon(y)| \left(\|K_0(1, f_y)\|_{L^p(\mathbb{R}^N)} + \left\| \left(\sum_{j=1}^{\infty} 2^{jsq} |K_M(2^{-j}, f_y)(\cdot)|^q\right)^{\frac{1}{q}} \right\|_{L^p(\mathbb{R}^N)} \right) dy \\
&= \int |\phi_\epsilon(y)|\, \|f_y\|_{F^s_{p,q}(\mathbb{R}^N)} dy = \|\phi\|_{L^1(\mathbb{R}^N)}\, \|f\|_{F^s_{p,q}(\mathbb{R}^N)}.
\end{aligned}$$

For (19) suppose that $s < 0$ and $f \in \mathcal{S}(\mathbb{R}^N)$. Since $L^p(\mathbb{R}^N) = F^0_{p,2}(\mathbb{R}^N) \subset F^s_{p,q}(\mathbb{R}^N)$(continuous inclusion), we have

$$\|J_\epsilon f - f\|_{F^s_{p,q}(\mathbb{R}^N)} \leq C\|J_\epsilon f - f\|_{L^p(\mathbb{R}^N)}; \tag{20}$$

then from (2) and (20), we have (19).

Suppose that $s \geq 0$ and $f \in \mathcal{S}(\mathbb{R}^N)$. Let $m \in \mathbb{N}$, $m > s$. From Theorem 2.3 B), we have

$$\|J_\epsilon f - f\|_{F^s_{p,q}(\mathbb{R}^N)} \leq C\Big(\|J_\epsilon f - f\|_{F^{s-m}_{p,q}(\mathbb{R}^N)} + \sum_{j=1}^{N} \Big\| J_\epsilon D^m_j f - D^m_j f\|_{F^{s-m}_{p,q}(\mathbb{R}^N)}\Big). \tag{21}$$

Now, since $s - m < 0$, we have $L^p(\mathbb{R}^N) \subset F^{s-m}_{p,q}(\mathbb{R}^N)$, hence

$$\|J_\epsilon f - f\|_{F^{s-m}_{p,q}(\mathbb{R}^N)} \leq C\|J_\epsilon f - f\|_{L^p(\mathbb{R}^N)}, \tag{22}$$

and for each $j = 1, \cdots, N$,

$$\|J_\epsilon D^m_j f - D^m_j f\|_{F^{s-m}_{p,q}(\mathbb{R}^N)} \leq C\|J_\epsilon D^m_j f - D^m_j f\|_{L^p(\mathbb{R}^N)}. \tag{23}$$

Therefore, from (21), (22), (23), and (2), we have that

$$\|J_\epsilon f - f\|_{F^s_{p,q}(\mathbb{R}^N)} \to 0, \quad \epsilon \to 0.$$

Suppose that $f \in F^s_{p,q}(\mathbb{R}^N)$ and $s \in \mathbb{R}$. From the triangular inequality for the norm and (18), we have that for any f and g in $F^s_{p,q}(\mathbb{R}^N)$,

$$\|J_\epsilon f - f\|_{F^s_{p,q}(\mathbb{R}^N)} \leq (\|\phi\|_{L^1(\mathbb{R}^N)} + 1)\|f - g\|_{F^s_{p,q}(\mathbb{R}^N)} + \|J_\epsilon g - g\|_{F^s_{p,q}(\mathbb{R}^N)}. \tag{24}$$

From (24) and the density of $S(\mathbb{R}^N)$ in $F^s_{p,q}(\mathbb{R}^N)$, we obtain

$$\|J_\epsilon f - f\|_{F^s_{p,q}(\mathbb{R}^N)} \to 0, \quad \epsilon \to 0.$$

□

4 Friedrichs' Lemma

The main results of this chapter are as follows:

Theorem 4.1 *Let $1 \leq p \leq \infty$, $1 \leq q \leq \infty$, $s \in \mathbb{R}$, and $a \in \mathcal{C}^{\rho+1}(\mathbb{R}^N)$ with $\rho > |s|$. Then there exists $C = C(\phi, s, p, q) > 0$ such that*

$$\left\| [J_\epsilon, a\, D_j] f \right\|_{B^s_{p,q}(\mathbb{R}^N)} \leq C \|a\|_{C^{\rho+1}(\mathbb{R}^N)} \|f\|_{B^s_{p,q}(\mathbb{R}^N)} \tag{25}$$

for all $f \in B^s_{p,q}(\mathbb{R}^N)$ and $\epsilon > 0$. Furthermore,

$$\lim_{\epsilon \to 0} \left\| [J_\epsilon, a\, D_j] f \right\|_{B^s_{p,q}(\mathbb{R}^N)} = 0 \tag{26}$$

for all $f \in B^s_{p,q}(\mathbb{R}^N)$.

Also we have the following theorem:

Theorem 4.2 *Let $1 \leq p < \infty$, $1 \leq q \leq \infty$, $s \in \mathbb{R}$, and $a \in \mathcal{C}^{\rho+1}(\mathbb{R}^N)$ with $\rho > |s|$. Then there exists $C = C(\phi, s, p, q) > 0$ such that*

$$\left\| [J_\epsilon, a\, D_j] f \right\|_{F^s_{p,q}(\mathbb{R}^N)} \leq C \|a\|_{C^{\rho+1}(\mathbb{R}^N)} \|f\|_{F^s_{p,q}(\mathbb{R}^N)} \tag{27}$$

for all $f \in F^s_{p,q}(\mathbb{R}^N)$ and for all $\epsilon > 0$. Furthermore,

$$\lim_{\epsilon \to 0} \left\| [J_\epsilon, a\, D_j] f \right\|_{F^s_{p,q}(\mathbb{R}^N)} = 0 \tag{28}$$

for all $f \in F^s_{p,q}(\mathbb{R}^N)$.

As extensions we have the following corollaries.

Corollary 4.3 *Let $1 \leq p \leq \infty$, $1 \leq q \leq \infty$, $s \in \mathbb{R}$, and $a \in \mathcal{C}^{\rho+1}(\mathbb{R}^N)$ com $\rho > |s|$. Suppose that $b(x, D)$ is a pseudo-differential operator, $b(x, D) \in L^r_{1,\delta}$, and $0 < r \leq 1$. Then there exists constant $C = C(\phi, p, q, s, r) > 0$ such that*

$$\left\| [J_\epsilon, a] b(x, D) f \right\|_{B^s_{p,q}(\mathbb{R}^N)} \leq C \|a\|_{C^{\rho+1}(\mathbb{R}^N)} \|f\|_{B^s_{p,q}(\mathbb{R}^N)} \tag{29}$$

for all $f \in B^s_{p,q}(\mathbb{R}^N)$ and $\epsilon > 0$.

Corollary 4.4 *Let $1 \le p < \infty$, $1 \le q \le \infty$, $s \in \mathbb{R}$, and $a \in C^{\rho+1}(\mathbb{R}^N)$ with $\rho > |s|$. Suppose that $b(x, D)$ is a pseudo-differential, $b(x, D) \in L^r_{1,\delta}$, and $0 < r \le 1$. Then there exists a constant $C = C(\phi, p, q, s, r) > 0$ such that*

$$\left\| [J_\epsilon, a] b(x, D) f \right\|_{F^s_{p,q}(\mathbb{R}^N)} \le C \|a\|_{C^{\rho+1}(\mathbb{R}^N)} \|f\|_{F^s_{p,q}(\mathbb{R}^N)} \tag{30}$$

for all $f \in F^s_{p,q}(\mathbb{R}^N)$ and $\epsilon > 0$.

4.1 Preliminaries

Without loss of generality, we will show Friedrichs' Lemma for $N = 1$. We denote $Df = \frac{df}{dx} = f'$.

We have that

$$\begin{aligned} K_0(1, [J_\epsilon, a\, D] f) &= K_0(1, (\phi_\epsilon * af')) - K_0(1, a(\phi_\epsilon * f)') \\ &= \phi_\epsilon * \check{k}_0 * (af') - \check{k}_0 * (a(\phi_\epsilon * f)'), \end{aligned}$$

where $\check{k}_0(x) = k_0(-x)$. From integration by parts we have

$$\begin{aligned} \phi_\epsilon * \check{k}_0 * (af') &= -\phi_\epsilon * \check{k}_0 * (a' f) - \phi_\epsilon * \check{k}_0' * (af) \\ &= -k_0(1, \phi_\epsilon * (a' f)) - \phi_\epsilon * \check{k}_0' * (af), \end{aligned}$$

and

$$\begin{aligned} -\check{k}_0 * (a(\phi_\epsilon * f)') &= \check{k}_0 * (a'(\phi_\epsilon * f)) + \check{k}_0' * (a(\phi_\epsilon * f)) \\ &= k_0(1, a'(\phi_\epsilon * f)) + \check{k}_0' * (a(\phi_\epsilon * f)). \end{aligned}$$

Also,

$$\begin{aligned} K_0(1, [J_\epsilon, a\, D] f) \;=\; & -K_0(1, \phi_\epsilon * (a' f)) + K_0(1, a'(\phi_\epsilon * f)) \\ & - \phi_\epsilon * \check{k}_0' * (af) + (\check{k}_0') * (a(\phi_\epsilon * f)). \end{aligned} \tag{31}$$

On the other hand,

$$\begin{aligned} -\phi_\epsilon * \check{k}_0' * (af) + (\check{k}_0') * (a(\phi_\epsilon * f)) &= (-\phi_\epsilon * \check{k}_0' * (af) + \phi_\epsilon * ((\check{k}_0' * f)) \\ &\quad + (\check{k}_0' * (a(\phi_\epsilon * f) - a(\check{k}_0{}' * \phi_\epsilon * f)) \\ &\quad + (a(\check{k}_0{}' * \phi_\epsilon * f) - \phi_\epsilon * (a(\check{k}_0' * f)) \\ &= I_1 + I_2 + I_3. \end{aligned} \tag{32}$$

Let $\alpha(x, y) = \frac{a(x)-a(y)}{x-y}$, if $x \neq y$ e $\alpha(x, x) = a'(x)$. Then

$$\begin{aligned} a(x)(\check{k}_0' * f)(x) - \check{k}_0' * (af)(x) &= \int (x-y)\check{k}_0'(x-y)\alpha(x, y) f(y)dy \\ &= \tau_0(f)(x), \end{aligned}$$

which gives us

$$I_1 = \phi_\epsilon * (a(\check{k}_0' * f)) - \phi_\epsilon * \check{k}_0' * (af) = \phi_\epsilon * (\tau_0(f)). \tag{33}$$

Similarly,

$$I_2 = \check{k}_0' * (a(\phi_\epsilon * f)) - a(\check{k_0}' * (\phi_\epsilon * f)) = -\tau_0(\phi_\epsilon * f). \tag{34}$$

From integration by parts, we have

$$\begin{aligned} I_3 = a(\check{k_0}' * \phi_\epsilon * f) - \phi_\epsilon * (a((\check{k}_0' * f)) &= -a(\phi_\epsilon' * \check{k}_0 * f) + \phi_\epsilon * (a'(\check{k}_0 * f)) \\ &\quad + \phi_\epsilon' * (a\check{k}_0 * f). \end{aligned}$$

Taking $\psi(x) = x\phi'(x)$, then

$$\begin{aligned} \phi_\epsilon' * (a\check{k}_0 * f)(x) - a(\phi_\epsilon' * \check{k}_0 * f)(x) &= \int \psi_\epsilon(x-y)\alpha(x, y) K_0(1, f)(y)dy \\ &= T_\epsilon(K_0(1, f))(x). \end{aligned}$$

Therefore,

$$I_3 = \phi_\epsilon * (a' K_0(1, f)) + T_\epsilon(K_0(1, f)). \tag{35}$$

From (31), (32), (33), (34), and (35), we have

$$\begin{aligned} K_0(1, [J_\epsilon, a D]f) &= -K_0(1, \phi_\epsilon * (a' f)) + K_0(1, a'(\phi_\epsilon * f)) \\ &\quad + \phi_\epsilon * (\tau_0(f)) - \tau_0(\phi_\epsilon * f) + \phi_\epsilon * (a' k_0(1, f)) \\ &\quad + T_\epsilon(K_0(1, f)). \end{aligned} \tag{36}$$

Similarly,

$$\begin{aligned} K_M(2^{-j}, [J_\epsilon, a D]f) &= -K_M(2^{-j}, \phi_\epsilon * (a' f)) + K_M(2^{-j}, (a'(\phi_\epsilon * f)) \\ &\quad + \phi_\epsilon * (\tau_j(f))(x) - \tau_j(\phi_\epsilon * f) + \phi_\epsilon * (a' K_M(2^{-j}, f)) \\ &\quad + T_\epsilon(K_M(2^{-j}, f)), \end{aligned} \tag{37}$$

where

$$\tau_j(f)(x) = \int 2^j(x-y)\check{k}'_M(2^j(x-y))\alpha(x,y)f(y)dy,$$

and

$$T_\epsilon(K_M(2^{-j},f))(x) = \int \psi_\epsilon(x-y)\alpha(x,y)K_M(2^{-j},f)(y)dy.$$

The following theorems were proved in [5] and [6], respectively:

Theorem 4.5 (See Theorem 2.6 in [5]) *Let $s \in \mathbb{R}$, $1 \le q \le \infty$, $1 \le p < \infty$, and $a \in C^{\rho+1}(\mathbb{R}^N)$, with $\rho > |s|$. Then there exists a constant $C = C(s,p,q) > 0$ such that*

$$\begin{aligned}\tau(f) &= \|\tau_0(f)\|_{L^p(\mathbb{R}^N)} + \left(\sum_{j=1}^{\infty} 2^{jsq}\|\tau_j(f)\||^q_{L^p(\mathbb{R}^N)}\right)^{\frac{1}{q}} \\ &\le C\|a\|_{C^{\rho+1}(\mathbb{R}^N)}\|f\|_{B^s_{p,q}(\mathbb{R}^N)},\end{aligned}$$

for all $f \in B^s_{pq}(\mathbb{R}^N)$.

Theorem 4.6 (See Theorem 2.11 in [6]) *Let $s \in \mathbb{R}$, $1 \le q \le \infty$, $1 \le p < \infty$, $a \in C^{\rho+1}(\mathbb{R}^N)$, with $\rho > |s|$. Then there exists a constant $C = C(b,s,p,q) > 0$ such that*

$$\begin{aligned}\tau(f) &= \|\tau_0(f)\|_{L^p(\mathbb{R}^N)} + \left\| \left(\sum_{j=1}^{\infty} 2^{jsq}|\tau_j(f)(\cdot)|^q\right)^{\frac{1}{q}} \right\|_{L^p(\mathbb{R}^N)} \\ &\le C\|a\|_{C^{\rho+1}(\mathbb{R}^N)}\|f\|_{F^s_{p,q}(\mathbb{R}^N)},\end{aligned}$$

for all $f \in F^s_{p,q}(\mathbb{R}^N)$.

4.2 *Proof of Theorem 4.1*

Since $\|\psi_\epsilon\|_{L^1} = \|\psi\|_{L^1(\mathbb{R})}$ for all $\epsilon > 0$, by Theorem 6.18 of [1] we have that $\|T_\epsilon g\| \le \|\psi\|_{L^1}\|a'\|_{C^\rho}\|g\|_{L^p}$ for all $g \in L^p$. From this, (36), and Theorem of Young, we have

$$\begin{aligned}\left\|K_0(1,[J_\epsilon,aD]f)\right\|_{L^p} &\le \|K_0(1,a'f)\|_{L^p} + \|K_0(1,a'(\phi_\epsilon * f))\|_{L^p} \\ &+ \|\phi\|_{L^1}\|\tau_0(f)\|_{L^p} + \|\tau_0(\phi_\epsilon * f)\|_{L^p} \\ &+ (\|\phi\|_{L^1} + \|\psi\|_{L^1})\|a'\|_{C^\rho}\|K_0(1,f)\|_{L^p}. \qquad (38)\end{aligned}$$

In a similar way, we have

$$\begin{aligned}\left\|K_M(2^{-j},[J_\epsilon,a\,D]f)\right\|_{L^p} &\leq \|K_M(2^{-j},a'f)\|_{L^p}+\|K_M(2^{-j},a'(\phi_\epsilon*f))\|_{L^p}\\ &+\|\phi\|_{L^1}\|\tau_j(f)\|_{L^p}+\|\tau_j(\phi_\epsilon*f)\|_{L^p}\\ &+(\|\phi\|_{L^1}+\|\psi\|_{L^1})\|a'\|_{C^\rho}\|K_M(2^{-j},f)\|_{L^p}. \quad (39)\end{aligned}$$

Therefore, from (38) and (39) it follows that

$$\begin{aligned}\left\|[J_\epsilon,a\,D]f\right\|_{B^s_{p,q}} &\leq C(\|a'f\|_{B^s_{p,q}}+\|a'(\phi_\epsilon*f)\|_{B^s_{p,q}}+\tau(f)\\ &+\tau(\phi_\epsilon*f)+\|a'\|_{C^\rho}\|f\|_{B^s_{p,q}}).\end{aligned}$$

From (11) of Theorem 3.1, Theorem 2.8, and Theorem 4.5, we get

$$\begin{aligned}\left\|[J_\epsilon,a\,D]f\right\|_{B^s_{p,q}} &\leq C(\|a'\|_{C^\rho}\|f\|_{B^s_{p,q}}+\|a'\|_{C^\rho}\|\phi_\epsilon*f\|_{B^s_{p,q}})\\ &\leq C\|a\|_{C^{\rho+1}}\|f\|_{B^s_{p,q}}.\end{aligned}$$

To show (26), note that since h is bounded and uniformly continuous, then $\phi_\epsilon * h$ converges uniformly to h and therefore follows (26) when $f \in C_c^\infty$. The general case follows from density of C_c^∞ in $B^s_{p,q}$ and (25).

□

4.3 *Proof of Theorem 4.2*

Let us show (27). From the above we have

$$\begin{aligned}\left\|K_0(1,[J_\epsilon,a\,D]f)\right\|_{L^p} &\leq \|K_0(1,a'f)\|_{L^p}+\|K_0(1,a'(\phi_\epsilon*f))\|_{L^p}\\ &+\|\phi\|_{L^1}\|\tau_0(f)\|_{L^p}+\|\tau_0(\phi_\epsilon*f)\|_{L^p}\\ &+(\|\phi\|_{L^1}+\|\psi\|_{L^1})\|a'\|_{C^\rho}\|K_0(1,f)\|_{L^p}. \quad (40)\end{aligned}$$

From Minkowski's inequality for integrals, we have

$$\left(\sum_{j=1}^{\infty} 2^{jsq}|\phi_\epsilon * K_M(2^{-j},(a'f))(x)|^q\right)^{\frac{1}{q}}$$

$$\leq \int |\phi_\epsilon(x-y)| \left(\sum_{j=1}^{\infty} 2^{jsq}|K_M(2^{-j},(a'f))(y)|^q\right)^{\frac{1}{q}} dy;$$

hence from Young's inequality it follows

$$\left\| \left(\sum_{j=1}^{\infty} 2^{jsq}|\phi_\epsilon * K_M(2^{-j},(a'f))(\cdot)|^q\right)^{\frac{1}{q}} \right\|_{L^p}$$

$$\leq \|\phi\|_{L^1} \left\| \left(\sum_{j=1}^{\infty} 2^{jsq}|K_M(2^{-j},(a'f))(\cdot)|^q\right)^{\frac{1}{q}} \right\|_{L^p}. \tag{41}$$

Similarly,

$$\left\| \left(\sum_{j=1}^{\infty} 2^{jsq}|\phi_\epsilon * (\tau_j(f))(\cdot)|^q\right)^{\frac{1}{q}} \right\|_{L^p} \leq \|\phi\|_{L^1}\,\tau(f). \tag{42}$$

Again, from Minkowski's inequality,

$$\left(\sum_{j=1}^{\infty} 2^{jsq}|T_\epsilon(K_M(2^{-j},f))(x)|^q\right)^{\frac{1}{q}}$$

$$\leq \int |\psi_\epsilon(x-y)||\alpha(x,y)| \left(\sum_{j=1}^{\infty} 2^{jsq}|K_M(2^{-j},f)(y)|^q\right) dy$$

$$\leq \|a'\|_{C^\rho} \int |\psi_\epsilon(x-y)| \left(\sum_{j=1}^{\infty} 2^{jsq}|K_M(2^{-j},f)(y)|^q\right) dy,$$

and hence from Young's inequality,

$$\left\| \left(\sum_{j=1}^{\infty} 2^{jsq} |T_\epsilon(K_M(2^{-j}, f))(\cdot)|^q \right)^{\frac{1}{q}} \right\|_{L^p} \leq$$

$$|\psi\|_{L^1} \|a'\|_{\mathcal{C}^\rho} \left\| \left(\sum_{j=1}^{\infty} 2^{jsq} |K_M(2^{-j}, f)(\cdot)|^q \right)^{\frac{1}{q}} \right\|_{L^p}. \tag{43}$$

Therefore, from (40), (37), (41), (42), and (43), we have

$$\begin{aligned} \left\| [J_\epsilon, a\, D] f \right\|_{F^s_{p,q}} &\leq C(\|a' f\|_{F^s_{p,q}} + \|a'(\phi_\epsilon * f)\|_{F^s_{p,q}} + \tau(f) \\ &\quad + \tau(\phi_\epsilon * f) + \|a'\|_{\mathcal{C}^\rho} \|f\|_{F^s_{p,q}}). \end{aligned}$$

From Theorem 3.2 (18), Theorem 4.6, and Theorem 2.8, we get

$$\begin{aligned} \left\| [J_\epsilon, a\, D] f \right\|_{F^s_{p,q}} &\leq C(\|a'\|_{\mathcal{C}^\rho} \|f\|_{F^s_{p,q}} + \|a'\|_{\mathcal{C}^\rho} \|\phi_\epsilon * f\|_{F^s_{p,q}}) \\ &\leq C \|a\|_{\mathcal{C}^{\rho+1}} \|f\|_{F^s_{p,q}}. \end{aligned}$$

To show (28) we repeat the same argument as in Sect. 4.2.

□

4.4 Proofs of Corollaries 4.3 and 4.4

Let $b(x, D)$ be a pseudo-differential operator in Hörmander's class $L^r_{1,\delta}$, $0 \leq \delta < 1$, and $r \in \mathbb{R}$, that is,

$$b(x, D)u(x) = (2\pi)^{-n} \int e^{ix\xi} b(x, \xi)\, \widehat{u}(\xi)\, d\xi, \quad u \in \mathcal{S},$$

where the symbol $b(x, \xi)$ belongs to class $\mathcal{S}^r_{1,\delta}$, meaning that $b(x, \xi) \in C^\infty(\mathbb{R}^N \times \mathbb{R}^N)$ and satisfies the following estimates:

$$|D^\alpha_x D^\beta_\xi b(x, \xi)| \leq C_{\alpha,\beta}(1 + |\xi|)^{r+\delta|\alpha|-|\beta|}, \quad (x, \xi) \in \mathbb{R}^N \times \mathbb{R}^N.$$

Proof As before, for simplicity, we assume that $N = 1$. In which follows $A^s_{p,q}$ denotes either $B^s_{p,q}$ or $F^s_{p,q}$. From Taylor's expansion with remainder integral, we have

$$b(x,\xi) = b(x,0) + \xi \int_0^1 D_\xi b(x, t\,\xi)\,dt = b(x,0) + \xi\,\tilde{b}(x,\xi),$$

and then

$$\begin{aligned} b(x,D)f(x) &= (2\pi)^{-n} \int e^{ix\xi} b(x,0)\widehat{f}(\xi)d\xi + (2\pi)^{-n} \int e^{ix\xi}\xi\,\tilde{b}(x,\xi)\widehat{f}(\xi)\,d\xi \\ &= b(x,0)f(x) + iD_x(\tilde{b}(x,D)f)(x). \end{aligned} \tag{44}$$

From (44) we have

$$[J_\epsilon, a]\,b(x,D)f(x) = [J_\epsilon, a]b(x,0)f(x) + i[J_\epsilon, a\,D_x](\tilde{b}(x,D)f)(x). \tag{45}$$

From Theorems 3.1, 3.2, and 2.8 and that the multiplication operator by $b(x,0)$ is a pseudo-differential of order 0, therefore bounded in $A^s_{p,q}$, we obtain

$$\begin{aligned} \|[J_\epsilon, a]b(\cdot,0)f)\|_{A^s_{p,q}} &\le \|\phi_\epsilon * (a(b(\cdot,0)f))\|_{A^s_{p,q}} + \|a\phi_\epsilon * (b(\cdot,0)f)\|\|_{A^s_{p,q}} \\ &\le \|\phi\|_{L^1}\|a(b(\cdot,0)f)\|_{A^s_{p,q}} + \|a\|_{C^\rho}\|\phi_\epsilon * (b(\cdot,0)f)\|\|_{A^s_{p,q}} \\ &\le C\|\phi\|_{L^1}\|a\|_{C^\rho}\|b(\cdot,0)f)\|_{A^s_{p,q}} \\ &\le C\|\phi\|_{L^1}\|a\|_{C^\rho}\|f\|\|_{A^s_{p,q}}. \end{aligned} \tag{46}$$

Now, from Theorems 4.1 and 4.2 and that $\tilde{b}(x,D)$ is a pseudo-differential of order 0 and, hence bounded in $A^s_{p,q}$, it follows

$$\begin{aligned} \|i[J_\epsilon, a\,D_x](\tilde{b}(x,D)f)\|_{A^s_{p,q}} &\le C\|a'\|_{C^\rho}\|\tilde{b}(x,D)f\|_{A^s_{p,q}} \\ &\le C\|a'\|_{C^\rho}\|\|f\|_{A^s_{p,q}}. \end{aligned} \tag{47}$$

Finally, from (45), (46), and (47), we have

$$\begin{aligned} \left\| [J_\epsilon, a]\,b(x,D)f \right\|_{A^s_{p,q}} &\le C(\|a\|\|_{C^\rho} + \|a'\|_{C^{\rho+1}})\|f\|_{A^s_{p,q}} \\ &\le C\|a\|_{C^{\rho+1}}\|f\|_{A^s_{p,q}}. \end{aligned}$$

We observe that the constant on the right-hand side of our last inequality is a multiple of norm of the operator $b(x,D)$ in the appropriate space.

□

5 Elliptic Regularity in Besov-Sobolev and Triebel-Lizorkin-Sobolev Spaces

5.1 Besov-Sobolev and Triebel-Lizorkin-Sobolev Spaces

We start with the following definitions:

Definition 5.1 (Besov-Sobolev) Suppose $1 \leq p \leq \infty$, $1 \leq q \leq \infty$, and $s \in \mathbb{R}$. If $f \in C_c^\infty(\mathbb{R}^N)$, consider the norm

$$\|f\|_{B^{s,1}_{p,q}(\mathbb{R}^N)} = \sum_{j=1}^{N} \|\partial_j f\|_{B^s_{p,q}(\mathbb{R}^N)} = \|\nabla f\|_{B^s_{p,q}(\mathbb{R}^N)}.$$

The completion of $C_c^\infty(\mathbb{R}^N)$ in relation to the norm $\|\nabla f\|_{B^s_{p,q}(\mathbb{R}^N)}$ can be identified with a subspace of $S'(\mathbb{R}^N)$ denoted by $B^{s,1}_{p,q}(\mathbb{R}^N)$ and called Besov-Sobolev space of order 1. If $f \in B^{s,1}_{p,q}(\mathbb{R}^N)$, then

$$\|f\|_{B^{s,1}_{p,q}(\mathbb{R}^N)} = \|\nabla f\|_{B^s_{p,q}(\mathbb{R}^N)}.$$

In the same way, we have the following definition:

Definition 5.2 (Triebel-Lizorkin-Sobolev) Suppose $1 \leq p \leq \infty$, $1 \leq q \leq \infty$, and $s \in \mathbb{R}$. If $f \in C_c^\infty(\mathbb{R}^N)$, consider the norm

$$\|f\|_{F^{s,1}_{p,q}(\mathbb{R}^N)} = \sum_{j=1}^{N} \|\partial_j f\|_{F^s_{p,q}(\mathbb{R}^N)} = \|\nabla f\|_{F^s_{p,q}(\mathbb{R}^N)}.$$

The completion of $C_c^\infty(\mathbb{R}^N)$ in relation to the norm $\|\nabla f\|_{F^s_{p,q}(\mathbb{R}^N)}$ can be identified with a subspace of $S'(\mathbb{R}^N)$ denoted by $F^{s,1}_{p,q}(\mathbb{R}^N)$ and called Triebel-Lizorkin-Sobolev spaces of order 1. If $f \in F^{s,1}_{p,q}(\mathbb{R}^N)$, then

$$\|f\|_{F^{s,1}_{p,q}(\mathbb{R}^N)} = \|\nabla f\|_{F^s_{p,q}(\mathbb{R}^N)}.$$

In what follows $A^s_{p,q}$ can denote $B^s_{p,q}$ or $F^s_{p,q}$. Suppose Ω is an open subset of $\mathbb{R}^N$.

Definition 5.3 We define

$$(A^s_{p,q})_{\text{loc}}(\Omega) = \{\, f \in \mathcal{D}'(\Omega);\ \ \psi f \in A^s_{p,q}(\mathbb{R}^N), \forall\ \psi \in C_c^\infty(\Omega)\,\}.$$

Definition 5.4 We define

$$(A^{s,1}_{p,q})_{\text{loc}}(\Omega) = \{\, f \in \mathcal{D}'(\Omega);\ \ \psi f \in A^{s,1}_{p,q}(\mathbb{R}^N), \forall\ \psi \in C_c^\infty(\Omega)\,\}.$$

Definition 5.5 Given a sequence $f_j \in (A^s_{p,q})_{loc}(\Omega)$ and $f \in (A^s_{p,q})_{loc}(\Omega)$, we say that f_j converges for f in $(A^s_{p,q})_{loc}(\Omega)$ if $\psi f_j \to \psi f$ in $A^s_{p,q}(\mathbb{R}^N)$ for each $\psi \in C^\infty_c(\Omega)$. Similarly, given a sequence $f_j \in (A^{s,1}_{p,q})_{\mathrm{loc}}(\Omega)$ and $f \in (A^{s,1}_{p,q})_{\mathrm{loc}}(\Omega)$, we say that f_j converges for f in $(A^{s,1}_{p,q})_{\mathrm{loc}}(\Omega)$ if $\psi f_j \to \psi f$ in $A^{s,1}_{p,q}(\mathbb{R}^N)$ for each $\psi \in C^\infty_c(\Omega)$.

Definition 5.6 Suppose $\Omega \subset \mathbb{R}^N$ an open set. We define

$$(A^s_{p,q})_c = A^s_{p,q} \cap \mathcal{E}'(\Omega)$$

$$(A^{s,1}_{p,q})_c = A^{s,1}_{p,q} \cap \mathcal{E}'(\Omega),$$

where $\mathcal{E}'(\Omega)$ denotes the set of compact support distributions in Ω.

5.2 Elliptic Regularity

Suppose that $\mathbf{L} = \{L_1, \cdots, L_n\}$ is a system of linearly independent vector fields with complex coefficients defined on an open set $\Omega \subset \mathbb{R}^N$. Consider the operator $\nabla_{\mathbf{L}} u = (L_1 u, \cdots, L_n u)$ for $u \in C^\infty(\Omega)$, which corresponds to the gradient ∇ when $n = N$ and $L_j = \partial_{x_j}$. We will deal with the regularity for the equation

$$\nabla_{\mathbf{L}} u = \mathbf{f} \tag{48}$$

in Besov and Triebel-Lizorkin spaces. More precisely, if the coefficients of the vector fields, L_j, $j = 1, \cdots, n$, are in $C^\rho(\Omega)$ with $\rho > |s|$ and $u \in A^s_{p,q}(\mathbb{R}^N)$, $1 \leq p, q < \infty$, then $\nabla_{\mathbf{L}} u$ define a distribution in $D'(\Omega)$ due to the duality between $A^s_{p,q}(\mathbb{R}^N)$ and $A^{-s}_{p',q'}(\mathbb{R}^N)$ com $\frac{1}{p} + \frac{1}{p'} = \frac{1}{q} + \frac{1}{q'} = 1$ (where $A^s_{p,q}$ can denote $B^s_{p,q}$ or $F^s_{p,q}$). We want to conclude that if $\mathbf{f} = (f_1, \cdots, f_n)$ with $f_j \in (A^s_{p,q})_{\mathrm{loc}}$, $j = 1, \cdots, n$, on a neighborhood of a point $x_0 \in \Omega$, then $u \in (A^{s,1}_{p,q})_{\mathrm{loc}}$ on some neighborhood of x_0.
Suppose:

(a) $L_1, \cdots, L_n$ are linearly independent in Ω.
(b) The system $\mathbf{L} = \{L_1, \cdots, L_n\}$ is elliptic.

The condition **(b)** means that for any real 1-form ω (i,e., any section of $T^*(\Omega)$) such that $< \omega, L_j >= 0$, for all j, implies $\omega = 0$. Consequently, the number n of vector fields must satisfy $\frac{N}{2} \leq n \leq N$.

Let $L^*_j = \bar{L}'_j$, $j = 1, \cdots, n$, where $\bar{L}_j$ denotes the vector field obtained from L_j by conjugating its coefficients and L'_j is the formal transpose of L_j. If the coefficients of the vector fields are of class C^2, then **b)** is equivalent to say that the second-order operator

$$\Delta_{\mathbf{L}} = L_1^* L_1 + \cdots + L_n^* L_n$$

is elliptic. We define $\mathrm{div}_{\mathbf{L}}\mathbf{f} = \mathrm{div}_{\mathbf{L}}(f_1, \cdots, f_n) = L_1^* f_1 + \cdots + L_n^* f_n$; then we have that

$$\Delta_{\mathbf{L}} u = \mathrm{div}_{\mathbf{L}} \nabla_{\mathbf{L}} u.$$

In the following theorem we will closely follow what was done in Theorem 2, [2], page 13, but with adaptations to our case.

Theorem 5.7 *Suppose* $s > 0$, $\rho > s$, $1 < p, q < \infty$, *and the coefficients of the vector fields* $L_1, \cdots, L_n$ *are in* $C^{\rho+1}(\Omega)$, $u \in (A^s_{p,q})_{loc}$ *satisfying (48) with* $\mathbf{f} = (f_1, \cdots, f_n)$ *and* $f_j \in (A^s_{p,q})_{\mathrm{loc}}$, $j = 1, \cdots, n$. *Then* $u \in (A^{s,1}_{p,q})_{\mathrm{loc}}(\Omega)$.

Proof Since we are dealing with a local question, it suffices to prove that given a point $x_0 \in \Omega$ there is $\eta \in C_c^\infty(\Omega)$ such that $\eta(x_0) \neq 0$ and $\eta u \in A^{s,1}_{p,q}(\mathbb{R}^N)$, or equivalently, that $\eta \nabla u \in A^s_{p,q}(\mathbb{R}^N)$. Therefore, without loss of generality we can assume that $x_0 = 0$, and Ω is an open ball centered at the origin. We split the proof into two cases:

Case 1 Suppose that the coefficients of vector fields $L_1, \cdots, L_n$ are smooth.

By a standard result in the theory of pseudo-differential operators, we can find scalar operators $q(x, D)$ and $r(x, D)$ with symbols $q(x, \xi)$ of order -2 and $r(x, \xi)$ of order $-\infty$, respectively, such that

$$v = q(x, D)\Delta_{\mathbf{L}} v + r(x, D)v, \quad v \in \mathcal{E}'(\Omega). \tag{49}$$

Consider $\psi \in C_c^\infty(\Omega)$ equal to 1 in a neighborhood of the origin. By Leibniz's rule and as u satisfies (48), we have

$$\Delta_{\mathbf{L}}(\psi u) = \psi \Delta_{\mathbf{L}} u + u \Delta_{\mathbf{L}} \psi = \psi\, \mathrm{div}_{\mathbf{L}} f + w \tag{50}$$

with $w \equiv 0$ in a neighborhood at the origin. Writing (49) for $v = \psi u$ and taking the gradient in the resulting identity and using (50), we have

$$\nabla \psi u = \nabla q(x, D)(\psi \mathrm{div}_{\mathbf{L}} f) + \left[\nabla\left(q(x, D)w + r(x, D)\psi u\right)\right] = g + w_1, \tag{51}$$

where ω_1 is smooth in a neighborhood of the origin. Now, observing that $g = \nabla q(x, D)(\psi \mathrm{div}_{\mathbf{L}} f)$ can be written as a sum of zero-order pseudo-differential operators acting on $\psi f \in A^s_{p,q}(\mathbb{R}^N)$ or in $(\partial_j \psi) f \in A^s_{p,q}(\mathbb{R}^N)$, $j = 1, \cdots, N$, and zero-order pseudo-differential operators are bounded in $A^s_{p,q}(\mathbb{R}^N)$ since $g \in$

$A^s_{p,q}(\mathbb{R}^N)$. Thus, if $\eta \in C_c^\infty(\Omega)$ is supported in a neighborhood where $\psi \equiv 1$ and $\eta(0) \neq 0$, multiplying (51) by η we conclude that $\eta \nabla u \in A^s_{p,q}(\mathbb{R}^N)$.

Furthermore, we observe that in the case of smooth coefficients, we have (49), and taking gradient on both sides of the equation, we get

$$\nabla v = p_1(x, D)\nabla_{\mathbf{L}} v + p_2(x, D)v$$

with $p_1(x, D)$ and $p_2(x, D)$ zero-order pseudo-differential operators and thus bounded in $A^s_{p,q}(\mathbb{R}^N)$. This implies a priori estimative

$$\|\nabla v\|_{A^s_{p,q}} \leq C_1\|\nabla_{\mathbf{L}} v\|_{A^s_{p,q}} + C_2\|v\|_{A^s_{p,q}}, \quad v \in C_c^\infty(\Omega). \tag{52}$$

By Theorem 2.6, there exists $C > 0$, such that given $0 < \epsilon < \frac{1}{2C}$, shrinking Ω we have

$$\|v\|_{A^s_{p,q}} \leq \epsilon C\|\nabla v\|_{A^s_{p,q}}, \quad v \in C_c^\infty(\Omega). \tag{53}$$

Thus, taking $\epsilon > 0$ sufficiently small from (52) and (53), we obtain

$$\|\nabla v\|_{A^s_{p,q}} \leq C\|\nabla_{\mathbf{L}} v\|_{A^s_{p,q}}, \quad v \in C_c^\infty(\Omega). \tag{54}$$

Case 2 Suppose that the coefficients of the fields $L_1, \cdots, L_n$ are in $C^{\rho+1}$.

Denoting by $\mathbf{L}_0$ the system of fields obtained by freezing the coefficients of $\mathbf{L}$ at the origin, we have from (54) that

$$\|\nabla v\|_{A^s_{p,q}} \leq C\|\nabla_{\mathbf{L_0}} v\|_{A^s_{p,q}}, \quad v \in C_c^\infty(\Omega).$$

Considering $\mathbf{L}$ as a perturbation of $\mathbf{L_0}$, we have

$$\|\nabla v\|_{A^s_{p,q}} \leq C\|\nabla_{\mathbf{L}} v\|_{A^s_{p,q}} + C\|\nabla_{(\mathbf{L}-\mathbf{L_0})} v\|_{A^s_{p,q}}, \quad v \in C_c^\infty(\Omega). \tag{55}$$

The coefficients of $\mathbf{L} - \mathbf{L_0}$ are in $C^{\rho+1}$ and vanish at the origin, so we can assume that its $C^{\rho+1}$ norm is arbitrarily small. From Theorem 2.6 follows

$$\|\nabla_{(\mathbf{L}-\mathbf{L_0})} v\|_{A^s_{p,q}} \leq O(\operatorname{diam}(\Omega))\|\nabla v\|_{A^s_{p,q}}, \tag{56}$$

and from (55) and (56), shrinking Ω again, we obtain

$$\|\nabla v\|_{A^s_{p,q}} \leq C\|\nabla_{\mathbf{L}} v\|_{A^s_{p,q}}, \quad v \in C_c^\infty(\Omega). \tag{57}$$

Now, consider $u \in (A^s_{p,q})_{\rm loc}$ satisfying (48) with $\mathbf{f} = (f_1, \cdots, f_n)$, $f_j \in (A^s_{p,q})_{\rm loc}$, $j = 1, \cdots, n$. Let $\eta \in C_c^\infty(\Omega)$ satisfying $\eta(x_0) \neq 0$. From (56) for $v = J_\epsilon(\eta u)$ and $\epsilon > 0$, we have

$$\begin{aligned} \|\nabla J_\epsilon(\eta u)\|_{A^s_{p,q}} &\le C\|\nabla_{\mathbf{L}} J_\epsilon(\eta u)\|_{A^s_{p,q}} \\ &\le C\|J_\epsilon \nabla_{\mathbf{L}}(\eta u)\|_{A^s_{p,q}} + C\|[\nabla_{\mathbf{L}}, J_\epsilon](\eta u)\|_{A^s_{p,q}}. \end{aligned} \tag{58}$$

We have $g = \nabla_{\mathbf{L}}(\eta u) = u\nabla_{\mathbf{L}}\eta + \eta\mathbf{f} \in A^s_{p,q}(\mathbb{R}^N)$, and by Friedrichs' Lemma (Theorem (4.1) and Theorem (4.2)), we obtain that $\|[\nabla_{\mathbf{L}}, J_\epsilon](\eta u)\|_{A^s_{p,q}} \to 0$ if $\epsilon \to 0$. Therefore, if we choose a sequence $\epsilon_k \to 0$, taking $w_k = \nabla J_{\epsilon_k}(\eta u) = J_{\epsilon_k}\nabla(\eta u)$, $g_k = J_{\epsilon_k} g$, and applying (58) for $v = J_{\epsilon_k}(\eta u) - J_{\epsilon_j}(\eta u)$, we obtain

$$\|w_k - w_j\|_{A^s_{p,q}} \le C\|g_k - g_j\|_{A^s_{p,q}} + \rho(k, j),$$

and by Friedrichs' Lemma, we have $\rho(k, j) \to 0$, when $j, k \to \infty$. By Theorem 3.1 and Theorem 3.2, $\|g_k - g_j\|_{A^s_{p,q}} \to 0$ when $j, k \to \infty$. This implies that w_k is a sequence of Cauchy in $A^s_{p,q}$. This implies that $\nabla(\eta u) \in A^s_{p,q}$, and then $\eta\nabla u \in A^s_{p,q}$ as we want to proof. □

References

1. G. B. Folland, *Real Analysis, Modern Techniques and Their Applications*, 1984
2. L. Hörmander, *The Analysis of Linear Partial Differential Operators III*, Springer, 1983.
3. J. Hounie, J. R. dos Santos, *A fractional Friedrich's Lemma for Hölder norms and applications*, Nonlinear Analysis, Theory and Applications, **vol. 28** (1997), pp. 1063–1077.
4. J. Hounie, *On Friedrichs commutators lemma for Hardy spaces and applications*, Springer International Publishing, 2015, p. 1–14.
5. E. R. da Silva, *Local solvability for a class of linear operators in Besov and Hölder spaces*, J. Math. Anal. Appl. **465** (2018), pp. 175–195.
6. E. R. da Silva, *Local solvability for a class of linear operators in Triebel-Lizorkin spaces*, Journal of Differential Equations **267** (2019), pp. 3199–3231.
7. H. Triebel, *Theory of Function Spaces*, Birkhauser, 1983.
8. H. Triebel, *Theory of Function Spaces II*, Birkhauser, 1992.
9. H. Triebel, *Fractals and Spectra*, Birkhauser, 1997.

On the Solvability of Pseudodifferential Equations in Microdistributions

François Treves

To Jorge Hounie, in friendship and admiration for his mathematical achievements

AMS Classification Numbers (2020) 35A01, 35A18, 35S30

1 Introduction

This chapter presents a proof of the distribution version of Trépreau's Theorem [Trépreau, 1984] here stated:

Theorem 1.1 *Let* $\boldsymbol{P}$ *be an analytic classical microdifferential operator of principal type, defined in an open neighborhood of* $(x^\circ, \xi^\circ) \in \mathbb{R}^n \times \mathbb{S}^{n-1}$ $(n \geq 2)$*. For* $\boldsymbol{P}$ *to be solvable in microfunctions at* (x°, ξ°)*, it is necessary and sufficient that* $\boldsymbol{P}$ *satisfy Condition* $(\Psi_{(x^\circ, \xi^\circ)})$*.*

This statement involves important notions of microlocal analysis, possibly not familiar to potential readers. Microfunctions are the microlocal versions of *hyperfunctions*; we shall not make use of either of these concepts and refer the interested reader to [Sato-Kawai-Kashiwara, 1973] and also to [Treves, 2022]. We shall deal with *microdistributions*, the microlocal versions of distributions. But the reasoning will be limited to objects in the *Real-Analytic* (or $\mathcal{C}^\omega$) *category*, not the $\mathcal{C}^\infty$ category. In the $\mathcal{C}^\infty$ category much stronger, meaning *local* or *semiglobal*, results are known (see [Dencker, 2006]). In the $\mathcal{C}^\omega$ category microdistributions are special microfunctions just as distributions are special hyperfunctions. Microdifferential operators are the microlocal version of *pseudodifferential operators*, and we shall directly define the former and, on the way to do that, define *classical analytic*

F. Treves (✉)
Department of Mathematics, Rutgers University, NJ, USA

S. Berhanu et al. (eds.), *Geometric Analysis of PDEs and Several Complex Variables*, Latin American Mathematics Series – UFSCar subseries,
https://doi.org/10.1007/978-3-031-69702-9_14

pseudodifferential operators acting on compactly supported distributions and valued in $\mathcal{D}'/\mathcal{C}^\omega$ (the space of "*singularity distributions*").

We shall prove the following statement:

Theorem 1.2 *Let* $\boldsymbol{P}$ *be as in Theorem 1.1. For* $\boldsymbol{P}$ *to be solvable in microdistributions at* (x°, ξ°)*, it is necessary and sufficient that* $\boldsymbol{P}$ *satisfy Condition* $(\Psi_{(x^\circ,\xi^\circ)})$*.*

Theorem 1.2 does not follow directly from Theorem 1.1 since the existence of a microfunction $\mathbf{u}$ solution of the equation $\boldsymbol{P}\mathbf{u} = \mathbf{f}$ when $\mathbf{f}$ is a microdistribution is not necessarily a microdistribution. As a matter of fact, there are examples of linear partial differential equations with $\mathcal{C}^\omega$ coefficients, solvable in hyperfunctions but not in distributions (see [Cordaro-Hanges, 2009]). To contrast our proof of Theorem 1.2 to Trépreau's proof of Theorem 1.1, it might be worthwhile to recall the three "pillars" of the latter:

(1) A local symplectomorphism transforming a neighborhood of (x°, ξ°) in $\mathbb{R}^n \times (\mathbb{R}^n \setminus \{0\})$ into one of (x°, ζ^*), a point in the *conormal bundle* of the boundary Σ (assumed to be real-analytic) of a strictly pseudoconvex domain in $\mathbb{C}^n$ and the analytic pseudodifferential operator under study, $P(x, D_x)$, into $\partial/\partial z_1$ tangent to Σ at x°
(2) The characterization, in terms of the defining function ρ of Σ in a neighborhood of x°, of the local solvability of $\partial/\partial z_1$ acting on the functions that are holomorphic in the strictly pseudoconvex domain $\rho < 0$ and have arbitrary growth up to Σ
(3) The characterization of hyperfunctions on Σ as relative cohomology classes and the use of highly technical homological algebra to derive the sought surjectivity of the homomorphism, defined by $\partial/\partial z_1$, between these classes

Condition $(\Psi_{(x^\circ,\xi^\circ)})$ appears in Step 2, as the *quasiconvexity* of a part of the defining function ρ. (Step 2 is no simple matter: Chapter 25 in [Treves, 2022] is entirely devoted to it.)

The necessity of $(\Psi_{(x^\circ,\xi^\circ)})$ for microlocal solvability in Theorem 1.2 is an old result (see [Nirenberg-Treves, 1970] and also [Treves, 2022], Theorem 23.2.10). To prove the sufficiency, we construct a *parametrix* of $P(x, D_x)$ in the form of a Fourier Integral Operator (FIO) with a complex phase-function φ of the FBI type in the sense of [Sjöstrand, 1982]. Perhaps the main point of this chapter is that the proof of Theorem 1.2 relies on equating Condition $(\Psi_{(x^\circ,\xi^\circ)})$ to the *quasiconvexity* of a part of Imφ (which we call its *exponent weight*). The construction of the FIO can fairly be said to be routine:

(1) φ is the solution of the eikonal equation defined by the vanishing of the principal symbol of $P(x, D_x)$ with the phase-function of the standard FBI transform as Cauchy data.
(2) The amplitude is determined by the corresponding transport equations.

Actually, the proof in the present article is a simplification of the proof in Section 26.3 [Treves, 2022] and relies on the latter for the proofs of the (few) needed technical lemmas.

The proof of Theorem 1.2 leads naturally to a number of open questions, foremost among them the possibility of patching up the microlocal parametrices of the type introduced in this chapter, to construct an explicit *local right-inverse* of the analytic pseudodifferential operator $P(x, D_x)$ (regarded as an endomorphism of $\mathcal{D}'/\mathcal{C}^\omega$). In turn, if this is feasible and assuming order $P(x, D_x) = 1$, can the right-inverse be made to map (locally) L^2_c into $L^2/\mathcal{C}^\infty$ (which is not true in the $\mathcal{C}^\infty$ case, as shown in [Lerner, 1994])? A further question is the possible use of the parametrices to determine the propagation of analytic singularities (on this topic, cf. [Treves, 2022], Ch. 24) and thereby enable us to transition from microlocal to semiglobal solvability (this time, in hyperfunctions; cf. Theorem 24.8.12, *loc. cit.*).

A large portion of the article is devoted to the definitions of the technical terms in the statement of Theorem 1.2 (this part of the article may be regarded as a primer on microlocal analysis) and to the precise formulation of Condition $(\Psi_{(x^\circ,\xi^\circ)})$. But before this we give a list of the basic notations and say a few words about the unavoidable "germ terminology."

1.1 Basic Notation and Terminology

We use $\equiv$ to mean *"identically equal,"* not "equivalent mod something"; we use $\cong$ to mean the latter.

If f, g are two real-valued function in a set E, $f \lesssim g$ means that $f(t) \leq Cg(t)$ for some $C > 0$ and all $t \in E$.

If A and B are subsets of a topological space $\mathbb{X}$, $B \subset\subset A$ means that the closure of B in $\mathbb{X}$ is a compact subset contained in A.

$x = (x_1, ..., x_n)$: the variable in $\mathbb{R}^n$; $\xi = (\xi_1, ..., \xi_n)$: the variable in $\mathbb{R}^n \setminus \{0\}$

$z = (z_1, ..., z_n)$: the variable in $\mathbb{C}^n$; $\zeta = (\zeta_1, ..., \zeta_n)$: the variable in $\mathbb{C}^n \setminus \{0\}$

$u \cdot v = \sum_{j=1}^n u_j v_j$ if $u = (u_1, ..., u_n)$, $v = (v_1, ..., v_n)$, $u_j, v_j \in \mathbb{R}$ or $\mathbb{C}$.

$|z| = \left(|z_1|^2 + \cdots + |z_n|^2\right)^{1/2}$; likewise with x, y, $\zeta = \xi + i\eta$, ξ, η, in lieu of $z = x + iy$.

$\langle\zeta\rangle$: the principal branch of the square root of $\zeta \cdot \zeta = \sum_{j=1}^n \zeta_j^2$ if $|\eta| < \kappa |\xi|$, $\kappa \ll 1$

$\alpha = (\alpha_1, ..., \alpha_n) \in \mathbb{Z}_+^n$, $|\alpha| = \alpha_1 + \cdots + \alpha_n$, $\alpha! = \alpha_1! \cdots \alpha_n!$.

$D_{x_j} = \frac{1}{\sqrt{-1}} \frac{\partial}{\partial x_j}$, $D_x = \left(D_{x_1}, ..., D_{x_n}\right)$, $D_x^\alpha = D_{x_1}^{\alpha_1} \cdots D_{x_n}^{\alpha_n}$; likewise with ξ, z, y or ζ in lieu of x.

$\frac{\partial}{\partial z_j} = \frac{1}{2}\left(\frac{\partial}{\partial x_j} - i\frac{\partial}{\partial y_j}\right)$, $\frac{\partial}{\partial \zeta_j} = \frac{1}{2}\left(\frac{\partial}{\partial \xi_j} - i\frac{\partial}{\partial \eta_j}\right)$.

$\mathbb{S}^{n-1} = \{\xi \in \mathbb{R}^n;\ |\xi| = 1\}$.

Ω (resp. $\Omega^{\mathbb{C}}$) an open subset of $\mathbb{R}^n$ ($\mathbb{C}^n$)

$\mathcal{C}_c(\Omega)$ [respectively, $\mathcal{C}_c^\infty(\Omega)$]: the ring of complex-valued, compactly supported, continuous (respectively, $\mathcal{C}^\infty$) functions in Ω

$\mathcal{C}^\omega(\Omega)$: the ring of real-analytic, complex-valued functions in Ω

$\mathcal{D}'(\Omega)$: the space of distributions in Ω;

supp f: the support of the distribution (or the continuous function) f
$\mathcal{E}'(\Omega)$: the space of compactly supported distributions in Ω
$\mathcal{O}(\Omega^{\mathbb{C}})$: the ring of holomorphic functions in $\Omega^{\mathbb{C}}$

A **conic** subset of $\mathbb{C}^n \times (\mathbb{C}^n \setminus \{0\})$ is a set E with the property that $(z, \zeta) \in E \Longrightarrow (z, \lambda\zeta) \in E$ whatever $\lambda > 0$. All cones in $\mathbb{C}^n \setminus \{0\}$ or $\mathbb{R}^n \setminus \{0\}$ shall have their vertices at the origin.

A function f in $E \longrightarrow \mathbb{C}$ shall be said to be **homogeneous of degree** m if $f(z, \lambda\zeta) = \lambda^m f(z, \zeta)$ whatever $(z, \zeta) \in E$, $\lambda > 0$.

Germs Suppose that to each open set $\mathcal{U}$ in a topological space $\mathbb{X}$ we attach a set $\mathbf{E}(\mathcal{U})$ belonging to a certain class (a *category*) $\mathbf{E}$, say, of functions defined in $\mathcal{U}$—or of subsets of $\mathcal{U}$, or equipped with an algebraic structure (vector space, ring,...) based on $\mathcal{U}$—or of "whatever." At this stage we only need a notion of *restriction* from every open $\mathcal{U}$ to every open $\mathcal{V} \subset \mathcal{U}$, i.e., a map $\rho^{\mathcal{U}}_{\mathcal{V}} : \mathbf{E}(\mathcal{U}) \longrightarrow \mathbf{E}(\mathcal{V})$ such that $\mathcal{W} \subset \mathcal{V} \Longrightarrow \rho^{\mathcal{U}}_{\mathcal{W}} = \rho^{\mathcal{V}}_{\mathcal{W}} \circ \rho^{\mathcal{U}}_{\mathcal{V}}$. Let then S be an arbitrary subset of $\mathbb{X}$. We posit as *equivalent at* S any two elements $\mathbf{e}_{\mathcal{U}} \in \mathbf{E}(\mathcal{U})$ and $\mathbf{e}_{\mathcal{V}} \in \mathbf{E}(\mathcal{V})$ with $\mathcal{U} \cap \mathcal{V} \supset S$ such that $\exists \mathcal{W} \subset \mathcal{U} \cap \mathcal{V}$, $\mathcal{W} \supset S$, $\rho^{\mathcal{U}}_{\mathcal{W}}(\mathbf{e}_{\mathcal{U}}) = \rho^{\mathcal{V}}_{\mathcal{W}}(\mathbf{e}_{\mathcal{V}})$. An equivalence class $(\mathbf{e}_{\mathcal{U}})_{\mathcal{U} \supset S}$ is referred to as a **germ** at S in the $\mathbf{E}$ -class. Most often, $S = \{x\}$, x a point in $\mathbb{X}$. But it may be important in more general cases, e.g., a closed subet S of $\mathbb{R}^n = \mathbb{X}$ and $\mathbf{E}(\mathcal{U}) = \mathcal{C}^\infty(\mathcal{U}), \mathcal{U} \subset \mathbb{R}^n$ *open:* $(\mathbf{e}_{\mathcal{U}})_{\mathcal{U} \supset S}$ *is the germ of a* $\mathcal{C}^\infty$ *function at* S.

2 FBI Transforms and Microlocalization

2.1 The FBI Transform and Microlocalization

We shall refer to the following integral:

$$\mathcal{F} f(x, \xi) = \int_{\mathbb{R}^n} e^{i\xi\cdot(x-y) - |\xi||x-y|^2} f(y) \left(1 + i(x-y) \cdot \frac{\xi}{|\xi|}\right) dy \tag{1}$$

as the **FBI transform** of $f \in L^1(\mathbb{R}^n)$.[1] By interpreting the integral as a duality bracket, (1) is extended to sums of (distribution) derivatives of L^1 functions. Most of the time in this chapter, these will be *compactly supported* distributions.

Throughout the chapter we will reason in a neighborhood of a fixed (but arbitrary) point (x°, ξ°) in $\mathbb{R}^n \times (\mathbb{R}^n \setminus \{0\})$; it will be often convenient to assume $|\xi^\circ| = 1$.

Definition 2.1 We shall say that $f \in \mathcal{E}'(\mathbb{R}^n)$ is **microanalytic at** $(x^\circ, \xi^\circ) \in \mathbb{R}^n \times (\mathbb{R}^n \setminus \{0\})$ if there exist a neighborhood U of x° in $\mathbb{R}^n$, an open cone Γ in $\mathbb{R}^n \setminus \{0\}$ containing ξ°, and positive constants M, c, such that $|\mathcal{F} f(x, \xi)| \le M \exp(-c|\xi|)$

[1] The factor $1 + i(x-y)\cdot\xi/|\xi|$ is often omitted, but its presence will simplify notation in the final part of the chapter.

for all $(x, \xi) \in U \times \Gamma$. We shall say that $f \in \mathcal{D}'(\mathbb{R}^n)$ is microanalytic at (x°, ξ°) if there is $g \in \mathcal{E}'(\mathbb{R}^n)$ such that $f = g$ in a neighborhood of x° and g is microanalytic at (x°, ξ°).

Let Ω be an open subset of $\mathbb{R}^n$. The set of points $(x, \xi) \in \Omega \times (\mathbb{R}^n \setminus \{0\})$ where $f \in \mathcal{D}'(\Omega)$ is *not* microanalytic is called the **analytic wave front set** of f; we denote it by $WF_{\mathrm{a}}(f)$, and it is a closed conic subset of $\Omega \times (\mathbb{R}^n \setminus \{0\})$. The base projection $WF_{\mathrm{a}}(u) \ni (x, \xi) \longmapsto x \in \Omega$ is the **analytic singular support** of u, $\mathrm{singsupp}_{\mathrm{a}} u$, the smallest closed subset E of Ω such that $u \in \mathcal{C}^\omega(\Omega \setminus E)$. To say that $WF_{\mathrm{a}}(u) = \varnothing$ is the same as saying that $u \in \mathcal{C}^\omega(\Omega)$.

We denote by $\mathbf{D}'_{x^\circ}$ the vector space of germs at x° of distributions in $\mathbb{R}^n$ and by $\mathbf{A}_{(x^\circ, \xi^\circ)}$ the subspace of $\mathbf{D}'_{x^\circ}$ consisting of the germs at x° of distributions that are microanalytic at (x°, ξ°) [hence in a conic neighborhood of (x°, ξ°)]. We shall systematically disregard the elements of $\mathbf{A}_{(x^\circ, \xi^\circ)}$, thus the following:

Definition 2.2 By a **microdistribution** at (x°, ξ°) we shall mean an element of the quotient vector space $\mathbf{D}'_{x^\circ} / \mathbf{A}_{(x^\circ, \xi^\circ)}$, henceforth denoted by $\mathbf{D}'^{\mathrm{micro}}_{(x^\circ, \xi^\circ)}$.

For the reader familiar with sheaves, as (x, ξ) ranges over $\mathbb{R}^n \times (\mathbb{R}^n \setminus \{0\})$, the vector spaces $\mathbf{D}'^{\mathrm{micro}}_{(x, \xi)}$ are the stalks of a sheaf $\mathbf{D}'^{\mathrm{micro}}$ over $\mathbb{R}^n \times (\mathbb{R}^n \setminus \{0\})$. An arbitrary distribution u in an open subset Ω in $\mathbb{R}^n$ defines a continuous section $\boldsymbol{u}$ of $\mathbf{D}'^{\mathrm{micro}}$ (equipped with its sheaf topology) over $\Omega \times (\mathbb{R}^n \setminus \{0\})$ whose support is equal to $WF_{\mathrm{a}}(u)$. Because the concepts just introduced are invariant under dilations $(x, \xi) \mapsto (x, \lambda\xi)$, $\lambda > 0$, it is convenient to regard $\mathbf{D}'^{\mathrm{micro}}$ as a sheaf over $\mathbb{R}^n \times \mathbb{S}^{n-1}$; the support of the continuous section $\boldsymbol{u}$, i.e., $WF_{\mathrm{a}}(u) \cap \left(\Omega \times \mathbb{S}^{n-1}\right)$, is often referred to as the ***microsupport*** of u.

We introduce some microlocal concepts about continuous linear operators on $\mathcal{D}'(\Omega)$ [meaning continuous linear maps $\mathcal{D}'(\Omega) \hookleftarrow$, i.e., of $\mathcal{D}'(\Omega)$ into itself].

Definition 2.3 We shall say that a continuous linear operator $\boldsymbol{T}$ on $\mathcal{D}'(\Omega)$ is **analytic microlocal** if $WF_{\mathrm{a}}(\boldsymbol{T}u) \subset WF_{\mathrm{a}}(u)$ for every $u \in \mathcal{D}'(\Omega)$.

If $\boldsymbol{T}$ is analytic microlocal, it defines, for each $(x, \xi) \in \Omega \times \mathbb{S}^{n-1}$, a linear operator $\boldsymbol{T}_{(x,\xi)}$ on $\mathbf{D}'_{(x,\xi)}$ such that $\boldsymbol{T}_{(x,\xi)} \mathbf{A}_{(x,\xi)} \subset \mathbf{A}_{(x,\xi)}$; it follows that $\boldsymbol{T}$ defines a linear operator $\boldsymbol{T}^{\mathrm{micro}}_{(x,\xi)}$ on $\mathbf{D}'^{\mathrm{micro}}_{(x,\xi)}$. As (x, ξ) varies in $\Omega \times \mathbb{S}^{n-1}$, the $\boldsymbol{T}^{\mathrm{micro}}_{(x,\xi)}$ are the values of a continuous operator (stalkwise linear) on the sheaf $\mathbf{D}'^{\mathrm{micro}}|_{\Omega}$.

A continuous linear operator $\boldsymbol{T}$ on $\mathcal{D}'(\Omega)$ is said to be **analytic regularizing** in an open subset U of Ω if $U \cap \mathrm{singsupp}_{\mathrm{a}} \boldsymbol{T}f = \varnothing$ whatever $f \in \mathcal{D}'(\Omega)$. [If $\boldsymbol{T}$ is analytic regularizing in Ω, then $\boldsymbol{T}\mathcal{D}'(\Omega) \subset \mathcal{C}^\omega(\Omega)$.]

Definition 2.4 We shall say that $\boldsymbol{T}$ is **analytic regularizing at** $(x^\circ, \xi^\circ) \in \mathbb{R}^n \times \mathbb{S}^{n-1}$ if $(x^\circ, \xi^\circ) \notin WF_{\mathrm{a}}(\boldsymbol{T}f)$ whatever $f \in \mathcal{D}'(\Omega)$.

Definition 2.4 can be restated as follows: $\boldsymbol{T} : \mathcal{D}'(\Omega) \hookleftarrow$ is analytic regularizing at (x°, ξ°) if $\boldsymbol{T}_{(x^\circ, \xi^\circ)} \mathbf{D}'_{(x^\circ, \xi^\circ)} \subset \mathbf{A}_{(x^\circ, \xi^\circ)}$, i.e., $\boldsymbol{T}^{\mathrm{micro}}_{(x^\circ, \xi^\circ)} = \mathbf{0}$. The operators of interest to us will all be analytic microlocal, and we shall mod off those that are analytic regularizing at (x°, ξ°).

The statement of Theorem 1.2 is clarified by the next definition.

Definition 2.5 We shall say that an analytic microlocal linear operator $\boldsymbol{T}$ on $\mathcal{D}'(\Omega)$ is **solvable at** $(x^\circ, \xi^\circ) \in \mathbb{R}^n \times \mathbb{S}^{n-1}$ **in microdistributions** if to every $f \in \mathcal{D}'(\Omega)$ there is $u \in \mathcal{D}'(\Omega)$ such that $(x^\circ, \xi^\circ) \notin WF_{\mathrm{a}}(\boldsymbol{T}u - f)$.

If $\boldsymbol{T}$ is solvable at (x°, ξ°) and if the continuous linear operator $\boldsymbol{R}$ on $\mathcal{D}'(\Omega)$ is analytically regularizing at (x°, ξ°), then $\boldsymbol{T} + \boldsymbol{R}$ is also solvable at (x°, ξ°). Clearly, $\boldsymbol{T}$ and $\boldsymbol{R}$ define linear operators $\boldsymbol{T}_{(x^\circ,\xi^\circ)}$ and $\boldsymbol{R}_{(x^\circ,\xi^\circ)}$ on $\mathbf{D}'_{(x^\circ,\xi^\circ)}$ and $\boldsymbol{R}_{(x^\circ,\xi^\circ)}\mathbf{D}'_{(x^\circ,\xi^\circ)} \subset \mathbf{A}_{(x,\xi)}$. It follows that the following properties are equivalent:

(1) $\boldsymbol{T}$ *is microlocally solvable at* (x°, ξ°).
(2) $\boldsymbol{T}^{\mathrm{micro}}_{(x^\circ,\xi^\circ)}$ *is surjective.*

2.2 Classical Analytic Symbolic Calculus

We recall some basic concepts of analytic symbolic calculus. We essentially follow [Boutet-Krée, 1967].

Definition 2.6 Let Ω be an open subset of $\mathbb{R}^n$ and Γ an open cone in $\mathbb{R}^n \setminus \{0\}$. By a **classical analytic symbol of order** m ($\in \mathbb{R}$) in $\Omega \times \Gamma$ we shall mean a series $a(x, \xi) = \sum_{j=0}^{\infty} a_j(x, \xi)$ having the following properties:

(1) $a_j(x, \lambda\xi) = \lambda^{m-j} a_j(x, \xi)$ for all $(x, \xi) \in \Omega \times \Gamma$, $\lambda > 0$.
(2) To every compact set $K \subset \Omega$, there is $C_K > 0$, such that, for all $(\alpha, \beta) \in \mathbb{Z}_+^n \times \mathbb{Z}_+^n$,

$$\forall \xi \in \Gamma, \max_{x \in K} \left| D_x^\alpha D_\xi^\beta a_j(x, \xi) \right| \leq C_K^{j+|\alpha+\beta|+1} j!\alpha!\beta! \, |\xi|^{m-j-|\beta|}. \tag{2}$$

We denote by $S^m_{\mathrm{class}}(\Omega \times \Gamma)$ the vector space of classical analytic symbols of order m in $\Omega \times \Gamma$, and we define

$$S_{\mathrm{class}}(\Omega \times \Gamma) = \bigcup_{m \in \mathbb{R}} S^m_{\mathrm{class}}(\Omega \times \Gamma).$$

Note that $\bigcap_{m \in \mathbb{R}} S^m_{\mathrm{class}}(\Omega \times \Gamma) = \{0\}$.

Remark 2.1 We underline the fact that the growth of $a_j(x, \xi)$ at the boundary of the cone Γ and that of each one of its derivatives is bounded. This is needed in the forthcoming definitions of operators through integrals with respect to ξ over Γ in which the $a_j(x, \xi)$ (suitably truncated about $\xi = 0$) are factors of the integrand. It is also of no importance since, as we are dealing with germs of microdistributions and microdifferential operators, $\Gamma \cap \mathbb{S}^{n-1}$ is going to be contracted about ξ° as much as needed.

Remark 2.2 The Cauchy inequalities imply directly the following restatement of Condition (2) in Definition 2.6: There are an open set $\Omega^{\mathbb{C}}$ in $\mathbb{C}^n$, an open cone $\Gamma^{\mathbb{C}}$ in $\mathbb{C}^n \backslash \{0\}$ such that $\Omega \subset \Omega^{\mathbb{C}} \cap \mathbb{R}^n$, $\Gamma \cap \mathbb{S}^{n-1} \subset\subset \Gamma^{\mathbb{C}} \cap \mathbb{R}^n$, and a holomorphic extension $a_j(z, \zeta)$ of $a_j(x, \xi)$ to $\Omega^{\mathbb{C}} \times \Gamma^{\mathbb{C}}$ such that $a_j(z, \lambda\zeta) = \lambda^{m-j} a_j(z, \zeta)$ for all $\lambda > 0$ and satisfies the following condition:

To every compact subset $\mathcal{K}^{\mathbb{C}}$ of $\Omega^{\mathbb{C}} \times \Gamma^{\mathbb{C}}$, there is $C > 0$, independent of $j \in \mathbb{Z}_+$, such that

$$\max_{(z,\zeta)\in\mathcal{K}^{\mathbb{C}}} \left|a_j(z, \zeta)\right| \leq C^{j+1} j!. \tag{3}$$

Naturally, $\Omega^{\mathbb{C}}$ and $\Gamma^{\mathbb{C}}$ can vary with $a(x, \xi)$.

Definition 2.7 By a **finite realization** of $a(x, \xi) = \sum_{j=0}^{\infty} a_j(x, \xi) \in S^m_{\text{class}}(\Omega \times \Gamma)$ we shall mean a finite sum

$$a_{(R,g)}(x, \xi) = g(\xi) \sum_{j<R/|\xi|} a_j(x, \xi), \ (x, \xi) \in \Omega \times \Gamma,$$

where $R > 0$ is large and $g \in \mathcal{C}^\infty\left(\mathbb{R}^N\right)$, $g(\xi) = 0$ if $|\xi| < \rho_1$, $g(\xi) = 1$ if $|\xi| > \rho_2$ $(0 < \rho_1 < \rho_2)$.

We also need the following:

Definition 2.8 By an **exponentially decaying symbol** in $\Omega \times \Gamma$ we shall mean a function $h \in \mathcal{C}^\infty(\Omega \times \Gamma)$ having the following two properties:

(1) Given arbitrarily a compact subset K of Ω and a pair of multi-indices $(\alpha, \beta) \in \mathbb{Z}^n_+ \times \mathbb{Z}^n_+$, the function $\left|\partial^\alpha_x \partial^\beta_\xi h(x, \xi)\right|$ is bounded in $K \times \Gamma$.
(2) There exist an open subset $\Omega^{\mathbb{C}}$ of $\mathbb{C}^n$ and an open cone $\Gamma^{\mathbb{C}}$ in $\mathbb{C}^n \backslash \{0\}$, $\Omega \subset \Omega^{\mathbb{C}} \cap \mathbb{R}^n$, $\Gamma \cap \mathbb{S}^{n-1} \subset \Gamma^{\mathbb{C}} \cap \mathbb{R}^n$, and a holomorphic function $\widetilde{h}$ in $\Omega^{\mathbb{C}} \times \left\{\zeta \in \Gamma^{\mathbb{C}};\ |\zeta| > \rho_\circ\right\}$, $\rho_\circ > 0$, such that $\widetilde{h}(x, \xi) = h(x, \xi)$ if $(x, \xi) \in \Omega \times \Gamma$, $|\xi| > \rho_\circ$, and the following condition is satisfied:
To every compact subset $K^{\mathbb{C}}$ of $\Omega^{\mathbb{C}}$, there are positive constants ε, $\rho > \rho_\circ$, C, such that

$$\forall \zeta \in \Gamma^{\mathbb{C}},\ |\zeta| > \rho,\ \max_{z\in K^{\mathbb{C}}} \left|\widetilde{h}(z, \zeta)\right| \leq C \exp(-\varepsilon |\zeta|).$$

We denote by $S^{-\omega}_{\text{a}}(\Omega \times \Gamma)$ the algebra (for ordinary addition and multiplication) of exponentially decaying analytic symbols in $\Omega \times \Gamma$.

Note that the requirements on h in Definition 2.8 allow $h(x, \xi)$ to vanish identically for $|\xi|$ small.

Lemma 2.1 *Let $a(x, \xi) \in S_{class}(\Omega \times \Gamma)$. To every domain $\Omega' \subset\subset \Omega$, there is $R' > 0$ such that the following is true: If $a_{(R,g)}(x, \xi)$ and $a_{(R^\flat,g^\flat)}(x, \xi)$ are* *two finite realizations of* $a(x,\xi)$ *and if* $\min\left(R, R^{\flat}\right) \geq R'$, *then* $a_{(R,g)}(x,\xi) - a_{(R^{\flat},g^{\flat})}(x,\xi) \in S_a^{-\omega}\left(\Omega' \times \Gamma\right)$.

The proof is an easy exercise (cf. [Treves, 2022], Lemma 19.1.22). The next definition is standard.

Definition 2.9 The symbol $a(x,\xi) = \sum_{j=0}^{\infty} a_j(x,\xi) \in S_{\mathrm{class}}^m(\Omega \times \Gamma)$ is said to be **elliptic of order** m if the following condition is satisfied:

(ELL) To every compact subset K of Ω, there is $c_K > 0$ such that $|a_0(x,\xi)| \geq c_K|\xi|^m$ for every $(x,\xi) \in K \times \Gamma$.

We point again to the uniformity with respect to $\xi \in \Gamma$ of the requirement in **(ELL)**, usually not a feature of ellipticity.

We introduce the fundamental *composition law* in $S_{\mathrm{class}}(\Omega \times \Gamma) \cup S_{\mathrm{a}}^{-\omega}(\Omega \times \Gamma)$:

$$a(x,\xi)\#b(x,\xi) = \sum_{\alpha \in \mathbb{Z}_+^n} \frac{1}{\alpha!}\left(D_\xi^\alpha a(x,\xi)\right)\partial_x^\alpha b(x,\xi). \tag{4}$$

With addition and the composition law #, $S_{\mathrm{class}}(\Omega \times \Gamma) \cup S_{\mathrm{a}}^{-\omega}(\Omega \times \Gamma)$ becomes a noncommutative ring in which the symbol $\Omega \times \Gamma \ni (x,\xi) \mapsto 1$ is the unit. It is not difficult to prove that $S_{\mathrm{a}}^{-\omega}(\Omega \times \Gamma)$ is an *ideal* in this ring. If $a \in S_{\mathrm{class}}^m(\Omega \times \Gamma)$ and $b \in S_{\mathrm{class}}^{m'}(\Omega \times \Gamma)$, then $a\#b \in S_{\mathrm{class}}^{m+m'}(\Omega \times \Gamma)$ and

$$[a,b]_\# = a\#b - b\#a \in S_{\mathrm{class}}^{m+m'-1}(\Omega \times \Gamma). \tag{5}$$

We recall the "classical" result ([Treves, 2022], Theorem 19.1.17):

Proposition 2.1 *If* $a \in S_{class}(\Omega \times \Gamma)$ *is elliptic of order* $m \in \mathbb{R}$, *then there is an elliptic symbol* $b \in S_{class}(\Omega \times \Gamma)$ *of order* $-m$ *such that* $a\#b = b\#a = 1$.

Thus the elliptic symbols in $S_{\mathrm{class}}(\Omega \times \Gamma)$ form a (noncommutative) group for the product #.

2.3 Classical Analytic Microdifferential Operators

First suppose $h(x,\xi) \in S_{\mathrm{a}}^{-\omega}(\Omega \times \Gamma)$, and consider the linear operator

$$\mathcal{E}'\left(\mathbb{R}^n\right) \ni u \mapsto (2\pi)^{-n}\int_\Gamma e^{ix\cdot\xi}h(x,\xi)\widehat{u}(\xi)\,d\xi. \tag{6}$$

Recall the Paley-Wiener-Schwartz Theorem: The Fourier transform of $u \in \mathcal{E}'(\mathbb{R}^n)$

$$\widehat{u}(\xi) = \int_{\mathbb{R}^n} e^{-iy\cdot\xi}u(y)\,dy$$

is the restriction to $\mathbb{R}^n$ of an entire function of exponential type in $\mathbb{C}^n$, and there is $r \in \mathbb{R}$ such that $(1+|\xi|)^{-r}|\widehat{u}(\xi)| \in L^1(\mathbb{R}^n)$. [The integrals with respect to y are, generally speaking, duality brackets between $\mathcal{C}^\infty$ and $\mathcal{E}'$.] The Paley-Wiener-Schwartz Theorem and Definition 2.8 imply the following:

Proposition 2.2 *If $h \in S_a^{-\omega}(\Omega \times \Gamma)$, the operator (6) maps $\mathcal{E}'(\mathbb{R}^n)$ into $\mathcal{C}^\omega(\Omega)$.*

Thus, if $h \in S_{\mathrm{a}}^{-\omega}(\Omega \times \Gamma)$ and $\chi \in \mathcal{C}_{\mathrm{c}}^\infty(\Omega)$, then the operator (6), where u is replaced by χu, is analytic regularizing in Ω (Definition 2.4).

Now suppose $P(x,\xi) \in S_{\mathrm{a}}^m(\Omega \times \Gamma)$, $m \in \mathbb{R}$, and let $P_{R,g}(x,\xi)$ be a finite realization of P (Definition 2.7). We define, for $u \in \mathcal{E}'(\mathbb{R}^n)$,

$$\begin{aligned}\boldsymbol{P}_{R,g}u(x) &= (2\pi)^{-n}\lim_{\varepsilon\searrow 0}\int\int_\Gamma e^{i(x-y)\cdot\xi-\varepsilon|\xi|^2}P_{R,g}(x,\xi)\,u(y)\,dyd\xi \qquad (7)\\ &= (2\pi)^{-n}\lim_{\varepsilon\searrow 0}\int_\Gamma e^{ix\cdot\xi-\varepsilon|\xi|^2}P_{R,g}(x,\xi)\,\widehat{u}(\xi)\,d\xi.\end{aligned}$$

What counts here is that $P_{R,g}(x,\xi) \in \mathcal{C}^\infty(\Omega \times \Gamma)$ and $|\xi|^{-m+|\beta|}\left|\partial_x^\alpha\partial_\xi^\beta P_{R,g}(x,\xi)\right| \in L^\infty(K \times \Gamma)$ for arbitrary $K \subset \Omega$ compact, $\alpha,\beta \in \mathbb{Z}_+^n$. The limit in (7) is uniform if $(1+|\xi|)^s\,\widehat{u}(\xi) \in L^2(\mathbb{R}^n)$, i.e., u belongs to the *Sobolev space* $H^s(\mathbb{R}^n)$, $s > m + n/2$. Since every $u \in \mathcal{E}'(\mathbb{R}^n)$ is a finite sum of distribution derivatives of functions $f \in H^s(\mathbb{R}^n)$, it is not difficult to prove that the limit in (7) is valid in the sense of distributions in Ω: (7) defines a continuous linear map $\boldsymbol{P}_{R,g} : \mathcal{E}'(\Omega) \longrightarrow \mathcal{D}'(\Omega)$. The most important property of the operators $\boldsymbol{P}_{R,g}$ is that they are analytic microlocal (Definition 2.3):

$$\forall u \in \mathcal{E}'(\Omega),\ WF_{\mathrm{a}}\left(\boldsymbol{P}_{R,g}u\right) \subset WF_{\mathrm{a}}(u). \qquad (8)$$

For a proof, see [Treves, 2022], Theorem 17.3.10.

Let Ω' be an open subset of $\mathbb{R}^n$, $\Omega' \subset\subset \Omega$, and $u \in \mathcal{E}'(\Omega)$, supp $u \subset \Omega'$. If $\left(R^\flat, g^\flat\right)$ is another pair like (R,g), then Lemma 2.1 asserts that there is $R' > 0$ such that

$$\min\left(R, R^\flat\right) \geq R' \Longrightarrow P_{R,g} - P_{R^\flat,g^\flat} \in S_{\mathrm{a}}^{-\omega}\left(\Omega' \times \Gamma\right).$$

It follows that $\boldsymbol{P}_{R,g}$ defines a linear operator $\boldsymbol{P}^{\Omega'} : \mathcal{E}'\left(\Omega'\right) \longrightarrow \mathcal{D}'(\Omega)/\mathcal{C}^\omega(\Omega)$ independent of R (sufficiently large) and of g. Given another open subset of $\mathbb{R}^n$, $\Omega'' \supset \Omega'$, the operator $\boldsymbol{P}^{\Omega''} : \mathcal{E}'\left(\Omega''\right) \longrightarrow \mathcal{D}'(\Omega)/\mathcal{C}^\omega(\Omega)$ coincides with $\boldsymbol{P}^{\Omega'}$ when both act on $\mathcal{E}'\left(\Omega'\right)$ and we mod off $\mathcal{C}^\omega(\Omega)$. Thus, by letting $\Omega' \nearrow \Omega$, we end up with a linear operator $\mathcal{E}'(\Omega) \longrightarrow \mathcal{D}'(\Omega)/\mathcal{C}^\omega(\Omega)$ which we shall denote by $P(x,D)$ and refer to as a **classical analytic pseudodifferential operator of order** m in Ω. We can obviously transfer the concept of WF_{a} to $\mathcal{D}'(\Omega)/\mathcal{C}^\omega(\Omega)$ and derive from (8)

$$\forall u \in \mathcal{E}'(\Omega),\ WF_{\mathrm{a}}(P(x,D)\,u) \subset WF_{\mathrm{a}}(u). \qquad (9)$$

The set

$$\operatorname{Char} P = \{(x,\xi) \in \Omega \times \Gamma;\ P_0(x,\xi) = 0\}$$

is the **characteristic set** of $P(x, D)$.

Now let $(x^\circ, \xi^\circ) \in \Omega \times \Gamma$ and $\chi \in \mathcal{C}_c^\infty(\Omega)$, $\chi \equiv 1$ in an open neighborhood U of x°; then $u \mapsto P(x, D)(\chi u)$ is a continuous linear operator $\mathcal{D}'(\Omega) \longrightarrow \mathcal{D}'(\Omega)/\mathcal{C}^\omega(\Omega)$. If $\chi^\flat \in \mathcal{C}_c^\infty(\Omega)$, $\chi^\flat \equiv \chi$ in U, then, by (9),

$$P(x, D)\left(\chi^\flat u\right) - P(x, D)(\chi u) = P(x, D)\left(\left(\chi^\flat - \chi\right) u\right) \in \mathcal{C}^\omega(U).$$

This and Proposition 2.2 allow us to follow up the map

$$u \mapsto P(x, D)(\chi u) \in \mathcal{D}'(\Omega)/\mathcal{C}^\omega(\Omega)$$

with the restriction maps from Ω to U to obtain a linear operator $\boldsymbol{P} : \mathcal{D}'(\Omega) \longrightarrow \mathcal{D}'(U)/\mathcal{C}^\omega(U)$. By contracting U about x° and $\Gamma \cap \mathbb{S}^{n-1}$ about ξ°, we see that $\boldsymbol{P}$ defines a linear map of $\mathbf{D}'_{(x^\circ,\xi^\circ)}$ into itself which we (provisionally) denote by $\boldsymbol{Op}_{(x^\circ,\xi^\circ)}P$.

Definition 2.10 Let $P(x,\xi) \in S_{\mathrm{a}}^m(\Omega \times \Gamma)$, $m \in \mathbb{R}$. We shall refer to the operator $\boldsymbol{Op}_{(x^\circ,\xi^\circ)}P : \mathbf{D}'_{(x^\circ,\xi^\circ)} \hookleftarrow$ as a **classical microdifferential operator** in Ω **of order** m **at** (x°, ξ°). We shall denote by $\boldsymbol{\Psi}^{(m)\mathrm{micro}}_{(x^\circ,\xi^\circ)}$ the set of classical microdifferential operators of order m at (x°, ξ°) and define

$$\boldsymbol{\Psi}^{\mathrm{micro}}_{(x^\circ,\xi^\circ)} = \bigcup_{m\in\mathbb{R}} \boldsymbol{\Psi}^{(m)\mathrm{micro}}_{(x^\circ,\xi^\circ)}. \tag{10}$$

The set $\boldsymbol{\Psi}^{\mathrm{micro}}_{(x^\circ,\xi^\circ)}$ is a subalgebra of the algebra (for standard addition and composition) of all the linear endomorphisms of $\mathbf{D}'^{\mathrm{micro}}_{(x^\circ,\xi^\circ)}$. There is a one-to-one correspondence between classical microdifferential operators at (x°, ξ°) and *germs at* (x°, ξ°) of symbols belonging to $S_{\mathrm{class}}(\Omega \times \Gamma)$ [with Ω and $\Gamma \cap \mathbb{S}^{n-1}$ ranging over the family of open neighborhoods of x° and ξ° in $\mathbb{R}^n$ and $\mathbb{S}^{n-1}$, respectively]. This correspondence becomes an *algebra isomorphism* if we equip $S_{\mathrm{class}}(\Omega \times \Gamma)$ with the composition law #: If $P, Q \in S_{\mathrm{class}}(\Omega \times \Gamma)$, then

$$\left(\boldsymbol{Op}_{(x^\circ,\xi^\circ)}P\right) \circ \boldsymbol{Op}_{(x^\circ,\xi^\circ)}Q = \boldsymbol{Op}_{(x^\circ,\xi^\circ)}(P\#Q). \tag{11}$$

Definition 2.11 We shall refer to $P(x,\xi) = \sum_{j=0}^\infty P_j(x,\xi) \in S^m_{\mathrm{class}}(\Omega \times \Gamma)$ as the **total symbol** of the microdifferential operator $\boldsymbol{Op}_{(x^\circ,\xi^\circ)}P$ and to $P_0(x,\xi)$ as its **principal symbol**, provided P_0 does not vanish identically.

As (x,ξ) ranges over $\mathbb{R}^n \times \mathbb{S}^{n-1}$, the algebras $\boldsymbol{\Psi}^{\mathrm{micro}}_{(x,\xi)}$ are the stalks of the sheaf of classical analytic microdifferential operators; if $P(x,\xi) \in S_{\mathrm{class}}(\Omega \times \Gamma)$, then

$(x, \xi) \mapsto \boldsymbol{Op}_{(x,\xi)} P$ is a continuous section of this sheaf over $\Omega \times \left(\Gamma \cap \mathbb{S}^{n-1}\right)$. To lighten notation, from now on we write $\boldsymbol{P}_{(x,\xi)}$ instead of $\boldsymbol{Op}_{(x,\xi)} P$.

Definition 2.12 The microdifferential operator $\boldsymbol{P}_{(x^\circ,\xi^\circ)}$ is said to be **elliptic** if $P_0(x^\circ, \xi^\circ) \neq 0$.

The following statement is a direct consequence of Proposition 2.1 and (11).

Proposition 2.3 *If* $\boldsymbol{P}_{(x^\circ,\xi^\circ)}$ *is elliptic, it is a bijection of* $\mathbf{D}'^{micro}_{(x^\circ,\xi^\circ)}$ *onto itself.*

The next class of microdifferential operators is the subject of our study; we will abide by the next definition (there is a symplectic invariant extension of it).

Definition 2.13 The microdifferential operator $\boldsymbol{P}_{(x^\circ,\xi^\circ)}$ is said to be of **principal type** if $|P_0(x^\circ, \xi^\circ)| + \sum_{j=1}^n \left|\frac{\partial P_0}{\partial \xi_j}(x^\circ, \xi^\circ)\right| \neq 0$.

Obviously, if $\boldsymbol{P}_{(x^\circ,\xi^\circ)}$ is of principal type, the same is true of $\boldsymbol{P}_{(x,\xi)}$ provided $(x, \xi) \in \Omega \times \mathbb{S}^{n-1}$ is sufficiently close to (x°, ξ°). If $\boldsymbol{E}_{(x^\circ,\xi^\circ)}$ is elliptic and $\boldsymbol{P}_{(x^\circ,\xi^\circ)}$ is of principal type, then $\boldsymbol{P}_{(x^\circ,\xi^\circ)}\boldsymbol{E}_{(x^\circ,\xi^\circ)}$ and $\boldsymbol{E}_{(x^\circ,\xi^\circ)}\boldsymbol{P}_{(x^\circ,\xi^\circ)}$ are of principal type.

Remark 2.3 If $\boldsymbol{P}_{(x^\circ,\xi^\circ)}$ is of principal type, but P_0 is not real-valued, it does not ensure that Char P is a smooth hypersurface in a neighborhood of (x°, ξ°): At least one of the analytic *varieties* defined by $\mathrm{Re} P_0\,(z, \zeta) = 0$ or $\mathrm{Im} P_0\,(z, \zeta) = 0$ must be an analytic *manifold*, but, generally speaking, their intersection will not be.

With the terminology introduced in this and the preceding subsection, we can state Theorem 1.2 as follows:

Theorem 2.1 *If* $\boldsymbol{P} \in \boldsymbol{\Psi}^{micro}_{(x^\circ,\xi^\circ)}$ *is of principal type and satisfies Condition* $(\Psi_{(x^\circ,\xi^\circ)})$*, then* $\boldsymbol{P} : \mathbf{D}'_{(x^\circ,\xi^\circ)} \longrightarrow \mathbf{D}'_{(x^\circ,\xi^\circ)}$ *is surjective.*

Remark 2.4 If $\boldsymbol{P} \in \boldsymbol{\Psi}^{\mathrm{micro}}_{(x^\circ,\xi^\circ)}$ is surjective, the same is true of $\boldsymbol{PE}$ and $\boldsymbol{EP}$ whatever the *elliptic* microdifferential operator $\boldsymbol{E}$ at (x°, ξ°) *(cf. Proposition 2.3)*. Since every distribution with compact support contained in an open neighborhood U of x° is of the form $(1 - \Delta_x)^N f$, $f \in \mathcal{C}_c(U)$ (Δ_x: the Laplacian in $\mathbb{R}^n$, $N \in \mathbb{Z}_+$), it follows that it suffices to prove the solvability of $\boldsymbol{Pu} = \boldsymbol{f}$ when the microdistribution $\boldsymbol{f}$ is represented by $f \in \mathcal{C}_c(U)$.

2.4 Condition $(\Psi_{(x^\circ,\xi^\circ)})$

The hypotheses in Theorem 2.1 only pertain to the principal symbol $P_0\,(x, \xi)$ of $\boldsymbol{P}$ in an open neighborhood of (x°, ξ°) in $\mathbb{R}^n \times (\mathbb{R}^n \setminus \{0\})$. If $\boldsymbol{P}$ is elliptic, Proposition 2.3 implies the conclusion. Henceforth we assume $P_0\,(x^\circ, \xi^\circ) = 0$, and therefore, by Definition 2.13, $\nabla_\xi P_0(x^\circ, \xi^\circ) = \left(\frac{\partial P_0}{\partial \xi_1}, \ldots, \frac{\partial P_0}{\partial \xi_n}\right)(x^\circ, \xi^\circ) \neq 0$. For the sake of simplicity, let us (provisionally) assume that

$$\nabla_\xi \mathrm{Re} P_0(x^\circ, \xi^\circ) \neq 0. \tag{12}$$

The **Hamiltonian** of $\mathrm{Re} P_0$ is the following analytic vector field, in a conic open neighborhood $\mathcal{N}$ of (x°, ξ°) in $\mathbb{R}^n \times (\mathbb{R}^n \setminus \{0\})$ in which P_0 is defined

$$H_{\mathrm{Re} P_0} = \sum_{j=1}^{n} \frac{\partial \mathrm{Re} P_0}{\partial \xi_j} \frac{\partial}{\partial x_j} - \frac{\partial \mathrm{Re} P_0}{\partial x_j} \frac{\partial}{\partial \xi_j}.$$

It is evident that $H_{\mathrm{Re} P_0}(\mathrm{Re} P_0) \equiv 0$; (12) implies that the pushdown of $H_{\mathrm{Re} P_0}$ under the base projection $(x, \xi) \mapsto x$ is a nowhere vanishing analytic vector field in an open neighborhood of x°.

Definition 2.14 By a **null bicharacteristic** of $\mathrm{Re} P_0$ we shall mean an integral curve of $H_{\mathrm{Re} P_0}$ on which $\mathrm{Re} P_0 \equiv 0$.

This allows us to state Condition $(\Psi_{(x^\circ,\xi^\circ)})$ in $\mathcal{N}$ under Hypothesis (12):

$(\boldsymbol{\Psi}_{(x^\circ,\xi^\circ)})$ $\mathrm{Im} P_0$ does not change sign from $-$ to $+$ on any null bicharacteristic of $\mathrm{Re} P_0$ in $\mathcal{N}$.

To obviate the patent asymmetry between real and imaginary parts in this statement of $(\Psi_{(x^\circ,\xi^\circ)})$, we apply the following result in [Nirenberg-Treves, 1970] (see Theorem 26.2.8 in [Treves, 2022]):

Theorem 2.2 *If $(\Psi_{(x^\circ,\xi^\circ)})$ holds as stated in Definition 2.14, it also holds if P_0 is replaced by cP_0 whatever $c \in \mathbb{C}$, $c \neq 0$.*

2.5 Sjöstrand's FBI Transforms

The transform (1) can be extended to complex values (z, ζ) of (x, ξ) with z varying in an open neighborhood of x° in $\mathbb{C}^n$ and $\zeta = \xi + i\eta$ in an open cone $\Gamma^{\mathbb{C}} \subset \{\zeta \in \mathbb{C}^n;\ |\eta| < \kappa\,|\xi|\}, 0 < \kappa \ll 1, \xi^\circ \in \Gamma^{\mathbb{C}} \cap \mathbb{S}^{n-1}$; $|\xi|$ is then replaced by $\langle \zeta \rangle$ and $|x - y|^2$ by $\langle z - y \rangle^2$. Actually, the proof of Theorem 1.2 will make use of integral transforms in *complex phase-space* generalizing this complex extension of (1). This generalization, which we now describe, was introduced in [Sjöstrand, 1982] (see Ch. **22**, [Treves, 2022]; here we modify a little the formal definition in [Sjöstrand, 1982] to suit our special purposes). Let $\Omega^{\mathbb{C}}$ be an open subset of $\mathbb{C}^n$, $x^\circ \in \Omega \subset \Omega^{\mathbb{C}} \cap \mathbb{R}^n$; $\xi^\circ \in \Gamma \cap \mathbb{S}^{n-1} \subset \Gamma^{\mathbb{C}}$, Γ an open cone in $\mathbb{R}^n \setminus \{0\}$ as before.

Definition 2.15 By a **Sjöstrand FBI phase-function** in $\Omega^{\mathbb{C}} \times \Omega^{\mathbb{C}} \times \Gamma^{\mathbb{C}}$ at the point (x°, ξ°) we shall mean a holomorphic function $\varphi(z, w, \zeta)$ in $\Omega^{\mathbb{C}} \times \Omega^{\mathbb{C}} \times \Gamma^{\mathbb{C}}$ such that $\varphi(x^\circ, x^\circ, \xi^\circ) = 0$, and the following properties hold:

(1) $\varphi(z, w, \lambda\zeta) = \lambda\varphi(z, w, \zeta)$ for all $(z, w, \zeta) \in \Omega^{\mathbb{C}} \times \Omega^{\mathbb{C}} \times \Gamma^{\mathbb{C}}, \lambda > 0$.

(2) $\sum_{j=1}^{n} \left| \frac{\partial^2 \varphi}{\partial \zeta_j}(x^\circ, x^\circ, \xi^\circ) \right| \neq 0$.

(3) $\det\left(\frac{\partial^2\varphi}{\partial z_k \partial \zeta_j}(x^\circ, x^\circ, \xi^\circ)\right)_{1\leq j,k\leq n} \neq 0.$

(4) $\mathrm{Im}\left(\frac{\partial^2\varphi}{\partial z_j \partial z_k}(x^\circ, x^\circ, \xi^\circ)\right)_{1\leq j,k\leq n}$ is positive definite.

To define a **Sjöstrand FBI operator**, we also need, beside a Sjöstrand FBI phase-function φ, a **classical analytic amplitude** of order $m \in \mathbb{R}$, $a(x, y, \xi) = \sum_{j=0}^{\infty} a_j(x, y, \xi)$, which is defined like a classical analytic symbol (cf. Definition 2.6) except for the extra variable y: $a_j(x, y, \lambda\xi) = \lambda^{m-j} a_j(x, y, \xi)$ for all $(x, y, \xi) \in \Omega \times \Omega \times \Gamma$ and all $\lambda > 0$, satisfying the following condition:

To every compact set $K \subset \Omega \times \Omega$, there is $C_K > 0$, independent of $j \in \mathbb{Z}_+$, such that, for all $(\alpha, \beta) \in \mathbb{Z}_+^n \times \mathbb{Z}_+^n$, $\xi \in \Gamma$,

$$\max_{(x,y)\in K} \left| D_x^\alpha D_y^\beta D_\xi^\gamma a_j(x, y, \xi) \right| \leq C_K^{j+|\alpha+\beta+\gamma|+1} j!\alpha!\beta!\gamma! \, |\xi|^{m-j-|\gamma|} . \tag{13}$$

We shall denote by $S^m_{\mathrm{class}}(\Omega \times \Omega \times \Gamma)$ the vector space of classical analytic amplitudes of order m and define $S_{\mathrm{class}}(\Omega \times \Omega \times \Gamma) = \bigcup_{m\in\mathbb{R}} S^m_{\mathrm{class}}(\Omega \times \Omega \times \Gamma)$.

Many of the definitions related to symbols extend to amplitudes by simply inserting the extra variable $y \in \mathbb{R}^n$ (or $w \in \mathbb{C}^n$). This is true of **exponentially decaying analytic amplitudes** (cf. Definition 2.8); they form the vector space $S_{\mathrm{a}}^{-\omega}(\Omega \times \Omega \times \Gamma)$. Likewise, for ellipticity (cf. Definition 2.9), $a \in S^m_{\mathrm{class}}(\Omega \times \Omega \times \Gamma)$ is **elliptic of order** $m \in \mathbb{R}$ if to every compact subset K of $\Omega \times \Omega$ there is $c_K > 0$ such that

$$\forall \xi \in \Gamma, \min_{(x,y)\in K} |a_0(x, y, \xi)| \geq c_K |\xi|^m .$$

It also applies to ***finite realizations*** *of* $a(x, y, \xi) = \sum_{j=0}^{\infty} a_j(x, y, \xi)$ (cf. Definition 7):

$$a_{(R,g)}(x, y, \xi) = g(\xi) \sum_{j<R/|\xi|} a_j(x, y, \xi) . \tag{14}$$

Remark 2.5 Note, however, that we are not defining the composition law # in $S_{\mathrm{class}}(\Omega \times \Omega \times \Gamma)$, not needed in the sequel. What is straightforward is the extension of (4) to the case where one of the factors a or b is a classical symbol and the other is a classical amplitude.

Given (14) and $u \in \mathcal{E}'(\Omega)$ arbitrary, we introduce the duality bracket

$$\boldsymbol{A}_{R,g} u(x, \xi) = \int_\Omega e^{i\varphi(x,y,\xi)} a_{R,g}(x, y, \xi)\, u(y)\, dy, \ (x, \xi) \in \Omega \times \Gamma. \tag{15}$$

The following result in [Sjöstrand, 1982] (cf. Theorem 22.3.13 in [Treves, 2022]) is crucial for our purposes:

Theorem 2.3 *Let $\varphi \in \mathcal{O}\left(\Omega^{\mathbb{C}} \times \Omega^{\mathbb{C}} \times \Gamma^{\mathbb{C}}\right)$ be a Sjöstrand FBI phase-function at (x°, ξ°) and $a \in S_{class}(\Omega \times \Omega \times \Gamma)$ be elliptic. For a distribution $u \in \mathcal{E}'(\Omega)$ to be microanalytic at (x°, ξ°) (Definition 2.1), it is necessary and sufficient that there are an open neighborhood $U \subset \Omega$ of x° and an open cone Γ' in $\mathbb{R}^n$, $\xi^\circ \in \Gamma' \subset \Gamma$, such that $\boldsymbol{A}_{R,g}u$ is an exponentially decaying symbol in $U \times \Gamma'$ (Definition 2.8).*

The analog of Lemma 2.1 is valid for amplitudes, and, as a consequence, if $a_{R^\flat,g^\flat}$ is another finite realization *of* a, then $\boldsymbol{A}_{R,g}u - \boldsymbol{A}_{R^\flat,g^\flat}u$ *is an* exponentially decaying symbol in $U \times \Gamma'$ [possibly contracted about (x°, ξ°)]. This implies that the condition in Theorem 2.3 is independent of the finite realization $a_{R,g}$. The following notation

$$\boldsymbol{A}u(x,\xi) = \int_\Omega e^{i\varphi(x,y,\xi)} a(x,y,\xi)\, u(y)\, dy \tag{16}$$

will stand for the equivalence class (mod exponentially decaying symbols) of the integrals (15).

When $a(x,y,\xi)$ is elliptic of order 0, we shall refer to the (16) as a **Sjöstrand FBI transform** in $\Omega \times \Gamma$. Needless to say, the holomorphic extension of the standard FBI transform (1), alluded to at the beginning of this subsection, is a special Sjöstrand FBI transform in $\mathbb{R}^n \times (\mathbb{R}^n \setminus \{0\})$. Theorem 2.3 states that to determine $WF_a(u)$ we need not rely solely on (1); any of the Sjöstrand FBI transforms would do.

3 Microdifferential Operators of Principal Type: Reductions

3.1 The Operator Under Study: First Reduction

In accordance with the statements of Theorems 2.1 and 2.2, we denote by $P(x,\xi) = \sum_{j=0}^{\infty} P_j(x,\xi) \in S_a^m(\Omega \times \Gamma)$ the total symbol of the microdifferential operator $\boldsymbol{P}_{(x^\circ,\xi^\circ)}$ we propose to study. We will deal with the classical analytic pseudodifferential operator $P(x,D) : \mathcal{E}'(\Omega) \longrightarrow \mathcal{D}'(\Omega)/\mathcal{C}^\omega(\Omega)$ defined (see Sect. 2.3) by the (distribution) limit, as $\varepsilon \searrow 0$, of the integrals

$$\frac{1}{(2\pi)^n} \int_\Gamma e^{i(x-y)\cdot\xi - \varepsilon|\xi|^2} P_{R,g}(x,\xi)\, \widehat{u}(\xi)\, d\xi,\ u \in \mathcal{E}'(\Omega).$$

If $P_0(x^\circ,\xi^\circ) \neq 0$, Condition $(\Psi_{(x^\circ,\xi^\circ)})$ is trivially satisfied, and Theorem 2.1 is an immediate consequence of Proposition 2.3. Henceforth we assume $P_0(x^\circ,\xi^\circ) = 0$ and $\nabla_\xi P_0(x^\circ,\xi^\circ) \neq 0$.

Recall that $P_0(x^\circ,\lambda\xi^\circ) = \lambda^m P_0(x^\circ,\xi^\circ)$, $\lambda > 0$. Theorem 2.2 allows us to assume $\nabla_\xi \mathrm{Re} P_0(x,\xi) \neq 0$ in a conic neighborhood of (x°,ξ°) which we take to be $\Omega \times \Gamma$ [contracted about (x°,ξ°)]. Possibly after relabeling the coordinates, we can assume $\frac{\partial P_0}{\partial \xi_n}(x,\xi) \neq 0$ in $\Omega \times \Gamma$. The analytic Implicit Function Theorem yields the factorization

$$P_0(x,\xi) = E(x,\xi)\left(\xi_n + q\left(x,\xi'\right)\right) \text{ in } \Omega \times \Gamma,$$

where $\xi' = \left(\xi_1, ..., \xi_{n-1}\right)$, $q\left(x,\xi'\right)$ and $E(x,\xi)$ are homogeneous symbols in $\Omega \times \Gamma$, of degree 1 and $m-1$, respectively, $E(x,\xi)$ elliptic in $\Omega \times \Gamma$ (Definition 2.12), thus defining a linear automorphism $\boldsymbol{E}_{(x^\circ,\xi^\circ)}$ of $\mathbf{D}'_{(x^\circ,\xi^\circ)}$ (Proposition 2.1); *the surjectivity of* $\boldsymbol{P}_{(x^\circ,\xi^\circ)}$ *is equivalent to that of* $\boldsymbol{E}^{-1}_{(x^\circ,\xi^\circ)}\boldsymbol{P}_{(x^\circ,\xi^\circ)}$. This allows us to replace $P(x,\xi)$ by $E^{\#-1}(x,\xi)\#P(x,\xi)$; in other words, to assume $m=1$ and $P_0(x,\xi) = \xi_n + q\left(x,\xi'\right)$.

We carry out a first, rather routine, reduction of $P(x,\xi)$. We extend $q\left(x,\xi'\right)$ and $P_j(x,\xi)$, $j \in \mathbb{Z}_+$, as holomorphic functions in $\Omega^{\mathbb{C}} \times \Gamma^{\mathbb{C}}$, $q\left(z,\zeta'\right)$ and $P_j(z,\zeta)$, respectively, as indicated in Remark 2.2, and we apply the Weierstrass Division Theorem to obtain

$$P_1(z,\zeta) = F_1(z,\zeta)\left(\zeta_n + q\left(z,\zeta'\right)\right) + Q_1\left(z,\zeta'\right),$$

where $F_1, Q_1 \in \mathcal{O}\left(\Omega^{\mathbb{C}} \times \Gamma^{\mathbb{C}}\right)$, $F_1(z,\lambda\zeta) = \lambda^{-1}F_1(z,\zeta)$, $Q_1\left(z,\lambda\zeta'\right) = Q_1\left(z,\zeta'\right)$, for all $\lambda > 0$, $\left(z,\left(\zeta',0\right)\right) \in \Omega^{\mathbb{C}} \times \Gamma^{\mathbb{C}}$. The total symbol of $\boldsymbol{P}_{(z^\circ,\zeta^\circ)}$ becomes

$$P(z,\zeta) = \left(\zeta_n + q\left(z,\zeta'\right)\right)(1 + F_1(z,\zeta)) + Q_1\left(z,\zeta'\right) + \sum_{j=2}^{\infty} P_j(z,\zeta)$$

with $P_j(z,\lambda\zeta) = \lambda^{1-j}P_j(z,\zeta)$. We have

$$Q_1\left(z,\zeta'\right)\#(1+F_1(z,\zeta))^{\#-1} = Q_1\left(z,\zeta'\right) + \sum_{k=1}^{\infty} R_{1,k}(z,\zeta),$$

$$P_j(z,\zeta)\#(1+F_1(z,\zeta))^{\#-1} = P_j(z,\zeta) + \sum_{k=1}^{\infty} R_{j,k}(z,\zeta),\ j=2,3,...,$$

with $R_{j,k}(z,\lambda\zeta) = \lambda^{1-j-k}R_{j,k}(z,\zeta)$. We derive

$$P(z,\zeta)\#(1+F_1(z,\zeta))^{\#-1} = \zeta_n + q\left(z,\zeta'\right) + Q_1\left(z,\zeta'\right) + \sum_{j=2}^{\infty} P_j^{(1)}(z,\zeta)$$

with $P_j^{(1)}(z,\lambda\zeta) = \lambda^{1-j}P_j^{(1)}(z,\zeta)$. We repeat the preceding operation with $P_2^{(1)}$ substituted for P_1:

$$P_2^{(1)}(z,\zeta) = F_2(z,\zeta)\left(\zeta_n + q\left(z,\zeta'\right)\right) + Q_2\left(z,\zeta'\right),$$

where $F_2, Q_2 \in \mathcal{O}\left(\Omega^{\mathbb{C}} \times \Gamma^{\mathbb{C}}\right)$, $F_2(x, \lambda\zeta) = \lambda^{-2} F_1(z, \zeta)$, $Q_2\left(z, \lambda\zeta'\right) = \lambda^{-1} Q_2\left(z, \zeta'\right)$. We note that the order of

$$Q_1\left(z, \zeta'\right) - Q_1\left(z, \zeta'\right) \# \left(1 + F_2(z, \zeta)\right)^{\#-1}$$

is equal to order $(Q_1 F_2) = -2$; we obtain

$$P(z, \zeta) \# \left(1 + F_1(z, \zeta)\right)^{\#-1} \# \left(1 + F_2(z, \zeta)\right)^{\#-1} = \zeta_n + q\left(z, \zeta'\right)$$
$$+ Q_1\left(z, \zeta'\right) + Q_2\left(z, \zeta'\right) + \sum_{j=3}^{\infty} P_j^{(2)}(z, \zeta)$$

with $P_j^{(2)}(z, \lambda\zeta) = \lambda^{1-j} P_j^{(2)}(z, \zeta)$. Repeated *ad infinitum*, this procedure (and concomitant use of the Cauchy inequalities) leads to an elliptic symbol $G(z, \zeta) = F_1(z, \zeta)^{\#-1} \# F_2(z, \zeta)^{\#-1} \# \cdots$ and a symbol of order 0, $\sum_{j=1}^{\infty} Q_j^{(j)}(z, \zeta)$, both in $\Omega \times \Gamma$ [possibly contracted about the ray $(x^\circ, \lambda\xi^\circ)_{\lambda > 0}$], which allows us to write

$$P(z, \zeta) \# G(z, \zeta) = \zeta_n + q\left(z, \zeta'\right) + \sum_{j=1}^{\infty} Q_j\left(z, \zeta'\right).$$

Since elliptic factors can be disregarded, we shall now assume that

$$P(z, \zeta) = \zeta_n + q\left(z, \zeta'\right) + \sum_{j=1}^{\infty} Q_j\left(z, \zeta'\right).$$

3.2 Final Reduction

At this juncture it is convenient to carry out a change of notation. A translation in x-space allows us to take $x^\circ = 0$. In the remainder of this section, we replace x_n by t, ξ_n by τ, $n - 1$ by n, and $\left(x', \xi'\right)$ by (x, ξ) which varies in $\Omega \times \Gamma$ as before. We require

$$|t| < \delta, \; |\tau| < (\delta + C_\circ)\,|\xi|,$$
$$\forall (x, t) \in \Omega \times (-\delta, \delta), \forall \xi \in \Gamma, \; |\mathrm{Re}\, q(x, t, \xi)| < C_\circ\,|\xi|,$$

with $C_\circ > 0$, $\delta > 0$. Notice that (ξ, τ) varies in an open cone in $\mathbb{R}^{n+1} \setminus \{0\}$. We now have $P_0(x, t, \xi, \tau) = \tau + q(x, y, \xi)$, and if $j \geq 1$, $Q_j = Q_j(x, t, \xi)$, $Q_j(x, t, \lambda\xi) = \lambda^{1-j} Q_j(x, t, \xi)$ for all $\lambda > 0$. The "central point" becomes $((0, 0), (\xi^\circ, \tau^\circ))$, $\tau^\circ = \mathrm{Re}\, q(0, 0, \xi^\circ)$, and the total symbol of $\boldsymbol{P}$ becomes

$$P(x,t,\xi,\tau) = \tau + q(x,t,\xi) + \sum_{j=1}^{\infty} Q_j(x,t,\xi). \tag{17}$$

We denote by $\varphi_\circ(x, y, t, \xi)$ the $\mathcal{C}^\omega$ solution of the following *eikonal* Cauchy problem:

$$\frac{\partial \varphi_\circ}{\partial t} + \mathrm{Re}q(x,t,\partial_x\varphi_\circ) = 0, \ \varphi_\circ|_{t=0} = (x-y)\cdot\xi. \tag{18}$$

We can assume that $\varphi_\circ \in \mathcal{C}^\omega(\Omega\times\Omega\times(-\delta,\delta)\times\mathbb{R}^n)$ (with diam Ω and δ sufficiently small); since $\varphi_\circ$ is real-valued, (18) is a standard Cauchy problem. We then consider the Fourier integral operator

$$\boldsymbol{A}(t) : \mathcal{E}'(\Omega) \ni u \longmapsto (2\pi)^{-n} \int_\Gamma \int e^{i\varphi_\circ(x,y,t,\xi)} a(x,y,t,\xi) u(y)\, dy d\xi, \tag{19}$$

where $u \in \mathcal{E}'(\Omega)$ and the integral with respect to y stands for a duality bracket; $a(x,y,t,\xi) = \sum_{j=0}^{\infty} a_j(x,y,t,\xi)$ is a classical analytic amplitude (cf. Sect. 2.5) depending analytically on $t \in (-\delta,\delta)$:

(1) $a_j(x,y,t,\lambda\xi) = \lambda^{m-j} a_j(x,y,t,\xi)$ for all $\lambda > 0$.
(2) To every compact set $K \subset \Omega\times\Omega\times(-\delta,\delta)$, there is $C_K > 0$, such that, for all $(\alpha,\beta,r,\gamma) \in \mathbb{Z}_+^n \times \mathbb{Z}_+^n \times \mathbb{Z}_+ \times \mathbb{Z}_+^n$ [cf. (13)]

$$\begin{aligned}\forall \xi \in \Gamma, \quad & \sup_{(x,y,t)\in K} \left| D_x^\alpha D_y^\beta D_t^r D_\xi^\gamma a_j(x,y,t,\xi) \right| \\ & \le C_K^{j+|\alpha+\beta+\gamma|+r+1} j!\alpha!\beta!\gamma!r!\, |\xi|^{m-j-|\gamma|}. \end{aligned} \tag{20}$$

Actually we select $m = 0$. We require the following (formal) equation to be satisfied:

$$D_t \boldsymbol{A}(t) + (\mathrm{Re}q(x,t,D_x))\boldsymbol{A}(t) + \sum_{j=1}^{\infty} Q_j(x,t,D_x)\boldsymbol{A}(t) = 0 \tag{21}$$

with the “initial” condition

$$\boldsymbol{A}(0) u(x) = (2\pi)^{-n} \int_\Gamma \int e^{i(x-y)\cdot\xi} u(y)\, dy d\xi. \tag{22}$$

Explicitly, (21) and (22) are equivalent to the following system of equations:

$$a\frac{\partial\varphi_\circ}{\partial t}+D_t a-\operatorname{Re}q\left(x,t,D_x+\partial_x\varphi_\circ\right)+\sum_{j=1}^{\infty}Q_j\left(x,t,D_x+\partial_x\varphi_\circ\right)a=0, \tag{23}$$

$$a\left(x,y,0,\xi\right)=1. \tag{24}$$

Remark 3.1 Concerning (22), Lemma 17.3.9 in [Treves, 2022] implies

$$\forall u\in\mathcal{E}'\left(\Omega\right),\, WF_{\mathrm{a}}\left(u-\boldsymbol{A}\left(0\right)u\right)\subset\mathbb{R}^n\backslash\Gamma.$$

As a consequence, *whatever* $u\in\mathcal{E}'\left(\Omega\right)$, *the microdistribution represented by* $\boldsymbol{A}\left(0\right)u$ *in* $\Omega\times\left(\Gamma\cap\mathbb{S}^{n-1}\right)$ *is equal to that represented by* u.

We equate the terms of the same homogeneity degree in both sides of (23) and (24); the latter becomes

$$a_0\left(x,y,0,\xi\right)=1;\, a_j\left(x,y,0,\xi\right)=0\ \text{ if } j\geq 1.$$

By (18) the terms of homogeneity degree 1 in (23) are

$$a_0\left(\frac{\partial\varphi_\circ}{\partial t}+\operatorname{Re}q\left(x,t,\partial_x\varphi_\circ\right)\right)\equiv 0.$$

The terms $a_j\left(x,y,t,\xi\right)$, $j\geq 1$, are recursively determined by the standard *transport equations* (see, e.g., [Treves, 2022], p. 666). Condition (24) ensures that $a\left(x,y,t,\xi\right)$ is elliptic in $\Omega\times\Gamma$ for all $t\in(-\delta,\delta)$ [provided, as usual, that diam Ω, diam $\Gamma\cap\mathbb{S}^{n-1}$, and δ are sufficiently small].

In phase-space (x,t,ξ,τ) the (microlocal) solvability of $\boldsymbol{P}$ is equivalent to that of

$$\boldsymbol{A}\left(t\right)^{-1}\boldsymbol{P}\boldsymbol{A}\left(t\right)=D_t+i\boldsymbol{A}\left(t\right)^{-1}\operatorname{Im}\boldsymbol{q}\left(x,t,D_x\right)\boldsymbol{A}\left(t\right).$$

We now avail ourselves of the celebrated Egorov Theorem (see [Egorov, 1969], [Treves, 2022], Theorem 18.5.31): $\boldsymbol{A}\left(t\right)^{-1}\,\mathbf{Im}\,\boldsymbol{q}\left(x,t,D_x\right)\boldsymbol{A}\left(t\right)$ defines a microdifferential operator of the same type as $\boldsymbol{q}$ with principal symbol $\tilde{q}\left(x,t,\xi\right)$ where $\tilde{q}\left(x,t,\xi\right)$ is a symplectic transform of $\operatorname{Im}q\left(x,t,\xi\right)$ $\left[\text{in } (x,\xi)\text{-space, depending analytically on } t\right]$. In the sequel we omit the tildas and assume that the (principal and total) symbol of $\boldsymbol{P}$ is $\tau+iq\left(x,t,\xi\right)$ with q *real-valued*: The microdifferential operator under study has become

$$\boldsymbol{P}=\boldsymbol{D}_t+i\boldsymbol{q}\left(x,t,D_x\right) \tag{25}$$

in an open neighborhood of $\left((0,0),(\xi^\circ,0)\right)$ in (x,t)-phase-space $\Omega\times\mathbb{R}\times\Gamma\times\mathbb{R}$; Char $\boldsymbol{P}$ is defined by the equations $\tau=q\left(x,t,\xi\right)=0$. The null bicharacteristics of the real part τ of the symbol of $\boldsymbol{P}$ are the t straight-line intervals $(x,\xi)\in\Omega\times\Gamma$

fixed, $-\delta < t < \delta$. Assuming that diam Ω, diam $\left(\Gamma \cap \mathbb{S}^{n-1}\right)$, and δ are sufficiently small Condition ($\boldsymbol{\Psi}_{(0,\xi^\circ)}$) reads

($\boldsymbol{\Psi}_{(0,\xi^\circ)}$) The function $(-\delta, \delta) \ni t \mapsto q(x, t, \xi)$ does not change sign from $-$ to $+$ whatever $(x, \xi) \in \Omega \times \Gamma$.

In the sequel we preclude the trivial case $q \equiv 0$.

4 Proof of Theorem 2.1

4.1 Introducing the Phase-Function φ and the Associated Exponent Weight

We assume, throughout, that the *real-valued* function $q(x, t, \xi)$ has a holomorphic extension $q(z, t, \zeta)$ of $q(x, t, \xi)$ to $\Omega^{\mathbb{C}} \times \Gamma^{\mathbb{C}}$ ($\Omega^{\mathbb{C}}$ and $\Gamma^{\mathbb{C}}$ as in Remark 2.2); we have $q(z, t, \lambda\zeta) = \lambda q(z, t, \zeta)$ for all $(z, t, \zeta) \in \Omega^{\mathbb{C}} \times (-\delta, \delta) \times \Gamma^{\mathbb{C}}$ and $\lambda > 0$; $q(z, t, \zeta)$ depends analytically on $t \in (-\delta, \delta)$.

We seek a microlocal right-inverse of $P(x, t, D_x, D_t) = D_t + iq(x, t, D_x)$ acting on $\mathcal{E}'(\Omega \times (-\delta, \delta))$ (as usual, with diam Ω and $\delta > 0$ decreased as needed) in the form of an FIO with a *complex* phase-function φ, specifically the solution, holomorphic with respect to (z, w, ζ) in a conic neighborhood of $(0, 0, \xi^\circ)$ in $\mathbb{C}^{2n} \times (\mathbb{C}^n \setminus \{0\})$, of the eikonal equation

$$\frac{\partial \varphi}{\partial t}(z, t, w, s, \zeta) + iq(z, t, \partial_z \varphi(z, t, w, s, \zeta)) = 0 \tag{26}$$

$\left[(s, t) \in \mathbb{R}^2, |s| < \delta, |t| < \delta\right]$ satisfying the Cauchy condition

$$\varphi(z, t, w, s, \zeta)|_{t=s} = (z - w) \cdot \zeta + i \langle\zeta\rangle \langle z - w\rangle^2 \tag{27}$$

($\langle\zeta\rangle = \sqrt{\zeta \cdot \zeta}$); (27) ensures that $\varphi(z, s, \operatorname{Re} w, s, \zeta)$ is a Sjöstrand FBI phase-function at $(\operatorname{Re} w, \xi) \in \Omega \times \Gamma$ (Definition 2.15), depending analytically on s.

The crux of the proof of Theorem 2.1 is the interpretation of Condition ($\Psi_{(x^\circ,\xi^\circ)}$) in terms of the real part of the function

$$\pi_\varphi(w, t, s, \zeta) = i\varphi(w, t, w, s, \zeta). \tag{28}$$

The following equations are direct consequences of (26)–(27):

$$\frac{\partial \pi_\varphi}{\partial t}(w, t, s, \zeta) = q(w, t, \partial_z \varphi(w, t, w, s, \zeta)), \tag{29}$$

$$\pi_\varphi(w, s, s, \zeta) = 0.$$

The following consequence of (27)-(29) is most important

$$\frac{\partial \pi_{\varphi}}{\partial t}\left(w, s, s, \zeta\right) = q\left(w, s, \zeta\right). \tag{30}$$

We shall refer to $\mathrm{Re}\pi_{\varphi}\left(y, t, s, \xi\right)$ as the **exponent weight** (associated with φ).[2]

We also define

$$\begin{aligned} \varphi^{\natural}\left(z, t, w, s, \zeta\right) &= \varphi\left(z, t, w, s, \zeta\right) + i\pi_{\varphi}\left(w, t, s, \zeta\right) \\ &= \varphi\left(z, t, w, s, \zeta\right) - \varphi\left(w, t, w, s, \zeta\right). \end{aligned} \tag{31}$$

Note that $\varphi^{\natural}\left(w, t, w, s, \zeta\right) = 0$ and

$$\varphi^{\natural}\left(z, t, w, s, \zeta\right)\Big|_{t=s} = \left(z - w\right) \cdot \zeta + i \left\langle \zeta \right\rangle \left\langle z - w \right\rangle^{2}.$$

We shall use the notation

$$Q\left(y, t, s, \xi\right) = \int_{s}^{t} q\left(y, t', \xi\right) dt', \tag{32}$$

$$\nabla^{\mathrm{h}}_{y,\xi} Q\left(y, t, s, \xi\right) = \left(\partial_y Q\left(y, t, s, \xi\right), \left|\xi\right| \partial_{\xi} Q\left(y, t, s, \xi\right)\right);$$

$\nabla^{\mathrm{h}}_{y,\xi}$ preserves the homogeneity degree. The statements that follow presuppose that diam U, diam $\left(\Gamma^{\flat} \cap \mathbb{S}^{n-1}\right)$, and δ are suitably small; they are consequences of Eqs. (26)–(32).

Lemma 4.1 *We have, for all* $(y, t, s, \xi) \in U \times (-\delta, \delta)^2 \times \Gamma^{\flat}$,

$$\left|\partial_y \varphi\left(y, t, y, s, \xi\right) - \xi\right| \lesssim \left|\nabla^{h}_{y,\xi} Q\left(y, t, s, \xi\right)\right| + \int_{s}^{t} \left|\nabla^{h}_{y,\xi} Q\left(y, t', s, \xi\right)\right| dt'.$$

Lemma 4.2 *We have, for all* $(y, t, s, \xi) \in U \times (-\delta, \delta)^2 \times \Gamma^{\flat}$,

$$\begin{aligned} \left|\mathrm{Re}\pi_{\varphi}\left(y, t, s, \xi\right) - Q\left(y, t, s, \xi\right)\right| \lesssim{} & \left|\nabla^{h}_{y,\xi} Q\left(y, t, s, \xi\right)\right|^{2} \\ & + \int_{s}^{t} \left|\nabla^{h}_{y,\xi} Q\left(y, t', s, \xi\right)\right|^{2} dt'. \end{aligned} \tag{33}$$

The detailed (fairly elementary) proofs of these two lemmas can be found in [Treves, 2022], pp. 1145-1147.

---

[2] We are correcting the terminology in [Treves, 2022], p. 1144, as it is preferable to have a real-valued function as a *weight*.

4.2 Condition ($\Psi_{(x^\circ,\xi^\circ)}$) in Terms of the Exponent Weight

The link (33) between π_φ and Q can be made much stronger under Hypothesis ($\Psi_{(0,\xi^\circ)}$).

The analyticity of $t \mapsto q(y,t,\xi)$ implies that ($\Psi_{(0,\xi^\circ)}$) is equivalent to the partition of $\Omega \times \Gamma$ (appropriately contracted) into the following four sets:

(1) $\mathfrak{N} = \{(y,\xi) \in \Omega \times \Gamma;\ \forall t \in (-\delta,\delta),\ q(y,t,\xi) = 0\}$
(2) *The set* $\mathfrak{A}^+$ *of points* (y,ξ) *such that* $q(y,t,\xi) > 0$ *in the whole interval* $(-\delta,\delta)$ *except, possibly, on a finite set*
(3) *The set* $\mathfrak{A}^-$ *of points* (y,ξ) *such that* $q(y,t,\xi) < 0$ *in the whole interval* $(-\delta,\delta)$ *except, possibly, on a finite set*
(4) *The set* $\mathfrak{A}$ *of points* (y,ξ) *such that* $t \mapsto q(y,t,\xi)$ *changes sign from* $+$ *to* $-$ *at a single point* $t(y,\xi) \in (-\delta,\delta)$

The following statement is a direct consequence of the partition $\Omega \times \Gamma = \mathfrak{N} \cup \mathfrak{A}^+ \cup \mathfrak{A}^- \cup \mathfrak{A}$ and the equation $\frac{\partial Q}{\partial s}(y,t,s,\xi) = -q(y,s,\xi)$ [cf. (32)]. We continue to denote by $t(y,\xi)$ the point in $(-\delta,\delta)$ at which $q(y,t,\xi)$ changes sign when $(y,\xi) \in \mathfrak{A}$.

Lemma 4.3 *The validity of* ($\Psi_{(0,\xi^\circ)}$) *is equivalent to the following property:*

If $(y,\xi) \notin \mathfrak{N}$, *then* $Q(y,t,s,\xi) < 0$ *in every one of the following subsets of* $\Omega \times (-\delta,\delta)^2 \times \Gamma$:

(a) $(y,\xi) \in \mathfrak{A}^+$, $-\delta < t < s < \delta$.
(b) $(y,\xi) \in \mathfrak{A}^-$, $-\delta < s < t < \delta$.
(c) $(y,\xi) \in \mathfrak{A}$, $t(y,\xi) \le s < t < \delta$.
(d) $(y,\xi) \in \mathfrak{A}$, $-\delta < t < s \le t(y,\xi)$.

Moreover, we have $\frac{\partial Q}{\partial s}(y,t,s,\xi) \le 0$ *in* ***(a)*** *and* ***(d)***, $\frac{\partial Q}{\partial s}(y,t,s,\xi) \ge 0$ *in* ***(b)*** *and* ***(c)***.

It is obvious that $Q(y,t,s,\xi) \equiv 0$ if $(y,\xi) \in \mathfrak{N}$. Lemmas 4.1 and 4.3 imply the following:

Lemma 4.4 *If* ($\Psi_{(0,\xi^\circ)}$) *holds, then*

$$\left|\nabla^h_{y,\xi} Q(y,t,s,\xi)\right|^2 \lesssim -|t-s|\, Q(y,t,s,\xi)$$

in each of the sets ***(a)***, ***(b)***, ***(c)***, ***(d)***, *of Lemma 4.3.*

Lemma 4.4 is stated as Corollary 26.3.8 in [Treves, 2022]; its proof is presented in p. 1149, *ibid*. It is readily seen that Lemmas 4.2 and 4.4 imply

$$\left|\mathrm{Re}\pi_\varphi(y,t,s,\xi) - Q(y,t,s,\xi)\right| \lesssim -|t-s|\, Q(y,t,s,\xi) - \int_s^t \left|t'-s\right| Q\left(y,t',s,\xi\right) dt'$$

for every (y, t, s, ξ) in each of the sets **(a)**, **(b)**, **(c)**, **(d)**, of Lemma 4.3. From this it is easy to deduce (cf. Corollary 26.3.10, *loc. cit.)*

Corollary 4.1 *If* $(\Psi_{(0,\xi^\circ)})$ *holds, then*

$$\left|\mathrm{Re}\pi_\varphi(y, t, s, \xi) - Q(y, t, s, \xi)\right| \lesssim -\delta Q(y, t, s, \xi)$$

in each of the sets ***(a)***, ***(b)***, ***(c)***, ***(d)***, *of Lemma 4.3.*

The preceding results imply the following:

Proposition 4.1 *If* diam U, diam $\left(\Gamma \cap \mathbb{S}^{n-1}\right)$, *and* δ *are sufficiently small, then* $(\Psi_{(0,\xi^\circ)})$ *is equivalent to the following property:*

(R) $\mathrm{Re}\pi_\varphi(y, t, s, \xi) \leq 0$ *for every* (y, t, s, ξ) *in each of the sets* ***(a)***, ***(b)***, ***(c)***, ***(d)***, *of Lemma 4.3 or such that* $(y, \xi) \in \mathfrak{N}$.

If we combine Lemma 4.3, Corollary 4.1, and Proposition 4.1, we reach the following conclusion:

If $(y, \xi) \in \mathfrak{A}$, the graph of $t \mapsto -\mathrm{Re}\pi_\varphi(y, t, s, \xi)$ resembles a parabola in the upper half-plane, with its minimum at $t(x, \xi)$; more precisely, it is monotone decreasing as $t \nearrow t(x, \xi)$ and monotone increasing as $t(x, \xi) < t \nearrow \delta$. If $(y, \xi) \in \mathfrak{A}^{\pm}$, then $t \mapsto -\mathrm{Re}\pi_\varphi(y, t, s, \xi)$ is monotone, either increasing or decreasing.

In traditional terminology this is the same as saying that $t \mapsto -\mathrm{Re}\pi_\varphi(y, t, s, \xi)$ is quasiconvex in $(-\delta, \delta)$.

Definition 4.1 A continuous real-valued function f in an open subset $\mathcal{A}$ of $\mathbb{R}$ is said to be **quasiconvex** if given any compact segment $[t_1, t_2] \subset \mathcal{A}$ we have $f(t) \leq \max(f(t_1), f(t_2))$ for all $t \in [t_1, t_2]$.

4.3 Construction of a Microlocal Parametrix

The parametrix, acting on an arbitrary $f \in \mathcal{C}_c(\Omega \times (-\delta, \delta))$ (cf. Remark 2.4), will be the limit, as $\varepsilon \searrow 0$, of the integrals

$$A^{(\varepsilon)}_{(R,g)} f(x, t) = \tag{34}$$

$$\frac{1}{(2\pi)^n} i \int_{\mathfrak{A}^+} \int_{-\delta}^{t} e^{i\varphi(x,t,y,s,\xi) - \varepsilon|\xi|^2} a_{(R,g)}(x, t, y, s, \xi) f(y, s)\, ds d\xi dy$$

$$-\frac{1}{(2\pi)^n} i \int_{\mathfrak{A}^-} \int_{t}^{\delta} e^{i\varphi(x,t,y,s,\xi) - \varepsilon|\xi|^2} a_{(R,g)}(x, t, y, s, \xi) f(y, s)\, ds d\xi dy$$

$$+\frac{1}{(2\pi)^n} i \int_{\mathfrak{A}} \int_{t(y,\xi)}^{t} e^{i\varphi(x,t,y,s,\xi) - \varepsilon|\xi|^2} a_{(R,g)}(x, t, y, s, \xi) f(y, s)\, ds d\xi dy,$$

where φ is the solution of (26)–(27), $a_{(R,g)}$ is a finite realization (Definition 2.7) of a classical analytic symbol of order 0 $a(x,t,y,s,\xi) = \sum_{j=0}^{\infty} a_j(x,t,y,s,\xi)$; $t(y,\xi)$ is the point in $(-\delta,\delta)$, where $t \mapsto q(x,t,\xi)$ changes sign from $+$ to $-$ [in **(c)** and **(d)**, Lemma 4.3]; thus, in the third integral in (34), we can have $-\delta < t < t(y,\xi)$ and $t(y,\xi) < t < \delta$.

We can determine $a(z,t,w,s,\xi) = \sum_{j=0}^{\infty} a_j(z,t,w,s,\xi)$ in a conic neighborhood $\mathcal{U}^{\mathbb{C}}$ of $U \times U \times \Gamma$ in $\mathbb{C}^{n+1} \times \mathbb{C}^{n+1} \times (\mathbb{C}^n \setminus \{0\})$, with $-\delta < t, s < \delta$, as the holomorphic solution of the Cauchy problem

$$(D_t + iq(z,t,D_z))\left(e^{i\varphi(z,t,w,s,\zeta)} a(z,t,w,s,\zeta)\right) = 0, \tag{35}$$

$$a(z,s,w,s,\zeta) = 1 + i\langle\zeta\rangle^{-1}(z-w)\cdot\zeta. \tag{36}$$

The homogeneous parts $a_j \in \mathcal{O}\left(\mathcal{U}^{\mathbb{C}}\right)$ are recursively determined by the transport equations under the initial conditions

$$a_0(z,s,w,s,\zeta) = 1 + i\langle\zeta\rangle^{-1}(z-w)\cdot\zeta,$$

$$a_j(z,s,w,s,\zeta) = 0 \text{ if } j \geq 1.$$

With this choice of $a(x,t,y,s,\xi)$, the convergence of (34) to a distribution $A_{(R,g)}f$ in $\Omega \times (-\delta,\delta)$ is a consequence of the following facts:

(i) $\varphi(x,t,y,s,\xi) = \varphi^{\natural}(x,t,y,s,\xi) - i\pi_{\varphi}(y,t,s,\xi)$.
(ii) By Proposition 4.1, $\mathrm{Re}\pi_{\varphi}(y,t,s,\xi) \leq 0$ in each domain of integration in (34).
(iii) $\varphi(z,s,y,s,\zeta)$ is an FBI phase-function at (y,ξ) sufficiently near $(0,\xi^{\circ})$.

The details of the proof of the convergence of (34) are provided in [Treves, 2022], pp. 1152-1155. We can state the following proposition:

Proposition 4.2 *Let $\chi \in C_c^{\infty}(U)$, $\chi \equiv 1$ in a neighborhood of 0. It follows from (35) that replacing $e^{i\varphi(x,t,y,s,\xi)}\chi(x)a_{(R,g)}(x,t,y,s,\xi)$ by*

$$(D_t + iq(x,t,D_x))\left(e^{i\varphi(x,t,y,s,\xi)}\chi(x)a_{(R,g)}(x,t,y,s,\xi)\right)$$

in (34) yields an analytic regularizing operator $\mathfrak{T}$ at $(0,\xi^{\circ})$ in $\Omega \times \Gamma$ depending analytically on $t \in (-\delta,\delta)$.

Combining Proposition 4.2 with (27) and (35)-(36) yields, for every $f \in \mathcal{C}_c(\Omega)$ depending continuously on $t \in (-\delta,\delta)$,

$$(D_t + iq(x,t,D_x))\left(\chi(x)A_{(R,g)}f(x,t)\right) - \mathfrak{T}f(x,t) =$$

$$\frac{1}{(2\pi)^n}\chi(x)\int_{\mathbb{R}^n}\int_{\Gamma} e^{i\xi\cdot(x-y)-|\xi||x-y|^2} f(y,t)\left(1 + i(x-y)\cdot\frac{\xi}{|\xi|}\right)d\xi dy.$$

The inversion formula for the FBI transform (1) $\big[$cf. [Treves, 2022], Formula (3.4.10)$\big]$ yields

$$f(x,t) = \frac{1}{(2\pi)^n} \int_{\mathbb{R}^{2n}} \mathcal{F} f(x,t,\xi)\, d\xi, \tag{37}$$

implying

$$(D_t + iq(x,t,D_x))\left(\chi(x) A_{(R,g)} f(x,t)\right) - \chi(x) f(x,t) = \tag{38}$$

$$\frac{1}{(2\pi)^n}\chi(x) \int_{\mathbb{R}^n} \int_{\mathbb{R}^n \backslash \Gamma} e^{i\xi\cdot(x-y)-|\xi||x-y|^2} f(y,t) \left(1 + i(x-y)\cdot\frac{\xi}{|\xi|}\right) d\xi dy$$

$$+\mathfrak{T} f(x,t).$$

4.4 Completing the Proof of Theorem 2.1

The right-hand side of (38), which we denote by $f_1(x,t)$, is microanalytic with respect to (x,ξ) at $(0,\xi^\circ)$ (cf. Remark 3.1) but only continuous with respect to $t \in (-\delta,\delta)$. We introduce the open cone $\widehat{\Gamma}_\varepsilon = \left\{(\xi,\tau) \in \mathbb{R}^{n+1};\ \xi \in \Gamma,\ |\tau| < \varepsilon|\xi|\right\}$; if $(x,t) \in U \times (-\delta,\delta)$, we derive from (37) applied to f_1 and the inversion of the Fourier transform in (t,τ) space:

$$\begin{aligned}(2\pi)^{n+1} f_1(x,t) &= \int_{-\delta}^{\delta}\int_{\mathbb{R}^{n+1}} e^{i(t-s)\tau} \mathcal{F} f_1(x,s,\xi)\, d\xi d\tau ds \\ &= \int_{-\delta}^{\delta}\int_{\widehat{\Gamma}_\varepsilon} e^{i(t-s)\tau} \mathcal{F} f_1(x,s,\xi)\, d\xi d\tau ds \\ &\quad + \int_{-\delta}^{\delta}\int_{\mathbb{R}^{n+1}\backslash\widehat{\Gamma}_\varepsilon} e^{i(t-s)\tau} \mathcal{F} f_1(x,s,\xi)\, d\xi d\tau ds.\end{aligned}$$

We apply once again Lemma 17.3.9, [Treves, 2022]: The integral

$$\int_{-\delta}^{\delta}\int_{\mathbb{R}^{n+1}\backslash\widehat{\Gamma}_\varepsilon} e^{i(t-s)\tau} \mathcal{F} f_1(x,s,\xi)\, d\xi d\tau ds$$

is microlocally analytic at $((0,0),(\xi^\circ,0)) \in U \times (-\delta,\delta) \times \widehat{\Gamma}_\varepsilon$. In the integral

$$\int_{-\delta}^{\delta}\int_{\widehat{\Gamma}_\varepsilon} e^{i(t-s)\tau} \mathcal{F} f_1(x,s,\xi)\, d\xi d\tau ds, \tag{39}$$

we have $|\mathcal{F} f_1(x,s,\xi)| \leq M \exp(-c|\xi|)$. Since $|\tau| < \varepsilon|\xi|$, selecting $\varepsilon \ll c$ allows t to vary in a neighborhood V of 0 in $\mathbb{C}$: (39) is a holomorphic function of t in V valued in the space of distributions in U that are microlocally analytic at $(0, \xi^\circ)$ implying, as can be readily checked, that (39) is microlocally analytic at $((0,0), (\xi^\circ, 0))$. This completes the proof of Theorem 2.1.

References

Boutet De Monvel, L. and Krée, P., *Pseudodifferential operators and Gevrey classes,* Ann. Inst. Fourier, Grenoble **17** (1967), 295–323.

Cordaro, P. D. and Hanges, N., *Hyperfunctions and (analytic) hypoellipticity,* Math. Ann. **344** (2009), 329–339.

Dencker, N., *The resolution of the Nirenberg-Treves conjecture,* Ann. of Math. **163** (2006), 405–444.

Egorov, Yu. V., *On canonical transformations of pseudodifferential operators,* Uspehi Mat. Nauk. 5 (1969), 235–236.

Lerner, N., *Nonsolvability in L^2 for a first-order operator satisfying condition* (ψ), Ann. of Math. **139** (1994), 363–393.

Nirenberg, L. and Treves, F., *On local solvability of linear partial differential equations. I. Necessary conditions. II. Sufficient conditions. Correction,* Comm. Pure Appl. Math. **XXIII** (1970), 1-38 and 459–509, **XXIV** (1971), 279–288.

Sjöstrand, J., *Singularités analytiques microlocales,* Astérisque **95** (1982), Soc. Math. France.

Sato, M., Kawai, T. and Kashiwara, M., *Microfunctions and pseudo-differential equations,* in Hyperfunctions and pseudo-differential equations, Springer Lecture Notes in Math. **287** (1973), 265–529.

Trépreau, J.-M., *Sur la résolubilité analytique microlocale des opérateurs pseudodifférentiels de type principal,* Thèse, Université de Rennes 1984.

Treves, F., Analytic Partial Differential Equations, Grund. Math. Wiss. **359**, Springer 2022.

Contribution to the Study of Closed 1-Forms and Its Applications

Giuliano Angelo Zugliani

1 Pseudoperiodic Functions

Suppose we are given a closed connected manifold M where a real smooth closed and non-exact 1-form b is defined.

Recall that the group of deck transformations of the universal covering space of M is isomorphic to the first homotopy group $\pi_1(M, p_0)$. We will construct a covering space $\widetilde{M}$ of M with special properties, one of them being that the pullback of b to $\widetilde{M}$ is exact. More precisely, take the normal subgroup G of $\pi_1(M, p_0)$ equal to the kernel of the homomorphism $T : \pi_1(M, p_0) \to \mathbb{R}$ given by

$$\gamma \mapsto \int_\gamma b.$$

There exists a covering space $(\widetilde{M}, \widetilde{p_0}) \to (M, p_0)$ such that $\pi_1(\widetilde{M}, \widetilde{p_0})$ is isomorphic to G, and for each pair of liftings of p_0 in $(\widetilde{M}, \widetilde{p_0})$, there is a deck transformation (or translation) mapping one to the other. Moreover, the group D of deck transformations of $(\widetilde{M}, \widetilde{p_0})$ is $\pi_1(M, p_0)/G$, which is finitely generated. We will denote a minimal set of generators of D by $\{\sigma_k\}$, $k = 1, \ldots, \ell$.

We call $(\widetilde{M}, \widetilde{p_0})$ the *minimal covering* of M (with respect to b). It is minimal in the sense that G must contain a copy of the fundamental group of any covering space on which a primitive of b is defined. This covering space is unique up to homeomorphisms, and the basepoints can be omitted since G is normal.

G. A. Zugliani (✉)
Universidade Estadual de Campinas (UNICAMP) - Departamento de Matemática, IMECC, Campinas, São Paulo, Brazil
e-mail: giuzu@unicamp.br

S. Berhanu et al. (eds.), *Geometric Analysis of PDEs and Several Complex Variables*, Latin American Mathematics Series – UFSCar subseries,
https://doi.org/10.1007/978-3-031-69702-9_15

Let us fix the projection $\Pi : \widetilde{M} \to M$ and define a primitive $\widetilde{B}$ of the pullback $\Pi^*(b)$ in $\widetilde{M}$ by integration from $t_0 \in \widetilde{M}$. We also pick a complete Riemannian metric on M and consider the induced Riemannian metric on $\widetilde{M}$.

Note that, for each $\sigma \in \mathsf{D}$, we have

$$\widetilde{B}(\sigma(t)) = \int_{t_0}^{\sigma(t)} \Pi^*(b) = \int_{t_0}^{\sigma(t_0)} \Pi^*(b) + \int_{\sigma(t_0)}^{\sigma(t)} \Pi^*(b) = \int_{t_0}^{\sigma(t_0)} \Pi^*(b) + B(t),$$

since $\Pi = \Pi \circ \sigma$ and, thus, $\Pi^*(b) = (\Pi \circ \sigma)^*(b) = \sigma^*(\Pi^*(b))$. The numbers $\int_{t_0}^{\sigma(t_0)} \Pi^*(b)$ do not depend on t_0. Indeed, if $t_0' \in \widetilde{M}$,

$$\int_{t_0'}^{\sigma(t_0')} \Pi^*(b) = \int_{t_0'}^{t_0} \Pi^*(b) + \int_{t_0}^{\sigma(t_0)} \Pi^*(b) + \int_{\sigma(t_0)}^{\sigma(t_0')} \Pi^*(b) =$$

$$= -\int_{t_0}^{t_0'} \Pi^*(b) + \int_{t_0}^{\sigma(t_0)} \Pi^*(b) + \int_{t_0}^{t_0'} \sigma^*(\Pi^*(b)) = \int_{t_0}^{\sigma(t_0)} \Pi^*(b).$$

Thus, we may write

$$\widetilde{B}(\sigma(t)) = \widetilde{B}(t) + b_\sigma, \quad t \in \widetilde{M}, \tag{1}$$

where the constant b_σ is not 0 unless σ is the identity transformation. This gives rise to the definition of pseudoperiodic functions that appeared in [BPZZ, HZ], where a plenty of examples are furnished.

2 The Definition Given by Arnol'd

The first definition of a pseudoperiodic function given in [Ar] is the following. We say that $\widetilde{B} : \mathbb{R}^r \to \mathbb{R}$ is pseudoperiodic if

$$\widetilde{B}(\sigma(t)) = \widetilde{B}(t + (m_1, \ldots, m_r)) = \widetilde{B}(t) + w(m_1, \ldots, m_r),$$

where $w : \mathbb{Z}^r \to \mathbb{R}$ is a monomorphism. It is clear that $(m_1, \ldots, m_r)$ is associated with a translation $\sigma : \mathbb{R}^r \to \mathbb{R}^r$, and then $w(m_1 \ldots, m_r)$ is b_σ in the above notation.

Also, if we write β_j for $w(e_j)$, where $\{e_j\}_{j=1}^r$ is the canonical basis of $\mathbb{R}^r$, then

$$\widetilde{B}(t) - (\beta_1 t_1 + \cdots + \beta_r t_r) \tag{2}$$

is periodic, that is, can be projected, thus defined, on the torus $\mathbb{T}^r$.

Moreover, $\widetilde{B}$ is a primitive on $\mathbb{R}^r$ of a closed 1-form defined on $\mathbb{T}^r$.

On the other side, given a closed 1-form b on an r-dimensional manifold M, we can embed $\widetilde{M}$ in a Euclidean space and see $\Pi^* b$ as the pullback of a linear form.

Indeed, consider an embedding m of M in some $\mathbb{R}^N$, and consider a basis $b_1, \ldots, b_r$ of the de Rham cohomology group $H^1(M, \mathbb{R})$, in such a way that

$$\int_\tau b_j \in \mathbb{Z}$$

for each $j = 1, \ldots r$ and each cycle $\tau \in H_1(M, \mathbb{R})$.

We then define the following map from $\widetilde{M}$ to $\mathbb{R}^{r+N}$:

$$e(t) = \left(\int_{t_0}^{t} \Pi^* b_1, \ldots, \int_{t_0}^{t} \Pi^* b_r, m \circ \Pi(t) \right).$$

Notice that e is injective, since $\Pi(t) = \Pi(s)$ implies $s = \sigma(t)$ for some $\sigma \in \mathsf{D}$ and, hence,

$$\int_{t_0}^{t} \Pi^* b_j = \int_{t_0}^{\sigma(t)} \Pi^* b_j \implies \int_{t}^{\sigma(t)} \Pi^* b_j = 0;$$

thus

$$\int_{t}^{\sigma(t)} \Pi^* b = 0,$$

which means by (1) that $\sigma = 0$.

As $D_{\Pi(t)} m$ and $D_t \Pi$ are all injective maps, it remains to show the following fact:

Fact 1 *If* $e(t_m) \to e(t)$, *then* $t_m \to t$.

By hypothesis,

$$\int_{t}^{t_m} \Pi^* b_j \to 0;$$

moreover, there is $\sigma_m \in \mathsf{D}$ such that $\sigma_m(t_m) \to t$, and this gives

$$\int_{t}^{t_m} \Pi^* b_j + \int_{t_m}^{\sigma_m(t_m)} \Pi^* b_j \to 0.$$

We then get

$$\int_{t_m}^{\sigma_m(t)} \Pi^* b_j \to 0.$$

As these numbers are integers, we conclude that σ_m is the identity for a large m.

If $(x_1, \dots, x_{r+N})$ are the coordinates of $\mathbb{R}^{r+N}$, by construction, $e^*(x_j) = \int_{t_0}^{t} \Pi^* b_j$. Therefore, $\Pi^* b = e^*(dL)$, where L is a linear combination of $dx_1, \dots, dx_r$.

3 Uniqueness of the Unbounded Component of Semilevel Sets

As previously, consider the minimal covering $\Pi : \widetilde{M} \to M$ of M, and for any $r \in \mathbb{R}$ denote by Ω^r the superlevel set $\{t \in \widetilde{M} : \widetilde{B}(t) > r\}$ and by Ω_r the sublevel set $\{t \in \widetilde{M} : \widetilde{B}(t) < r\}$. (They are called semilevel sets, and we will focus on superlevel sets for the sake of simplicity; the corresponding claims for sublevel sets are true as well.)

The uniqueness of an unbounded component for each semilevel set is conjectured by Arnol'd in [Ar]—and proved for the real analytic setup and for the case when $M = \mathbb{T}^2$.

In the next section, we will prove it for the case when M is a closed orientable surface.

One key tool in [HZ] (Proposition 10) was the following result in this direction:

Theorem 3.1 *For every $r \in \mathbb{R}$, there is exactly one component of Ω^r on which $\widetilde{B}$ is unbounded.*

Remark 3.2 Let $\mathscr{M}$ be an arbitrary covering space of M where a primitive B' of b is defined. Notice that there exist $r \in \mathbb{R}$ and a component $\mathcal{O}$ of $\Omega^r = \{t \in \mathscr{M} : \widetilde{B}(t) > r\}$ such that $\widetilde{B}$ is bounded $\mathcal{O}$ if and only if B' is bounded on a component $\mathcal{O}'$ of $\{t \in \widetilde{M} : B'(t) > r'\}$ for some $r' \in \mathbb{R}$. Therefore, the previous theorem shows that, when b is non-exact, the non-connectedness of a semilevel set in the minimal covering space is equivalent to the existence of a component $\mathcal{O}'$ of a semilevel set such that B' is bounded on $\mathcal{O}'$.

In other words, the conjecture by Arnol'd can be rewritten as follows (Open Problem 3 in [HZ]):

Theorem 3.3 *The components of semilevel sets on which $\widetilde{B}$ is bounded are relatively compact.*

A smooth closed 1-form whose local primitives have only nondegenerate critical points will be called a *Morse* form (the primitives defined on a covering space own the same property). The set Σ of (Morse) critical points of such a primitive B is finite, since there is a local chart φ in a neighborhood of $p \in \Sigma$ such that $\varphi(p) = 0$ and $B \circ \varphi^{-1}(t_1, \dots, t_n) = B(p) \pm t_1^2 \pm \dots \pm t_n^2$.

A useful fact about components of semilevel sets on which $\widetilde{B}$ is bounded is the following (for a proof, see [HZ] (Lemma 12) or a version in [Ar] (Lemma 1)):

Proposition 3.4 *If $\widetilde{B}$ is bounded on a component $\mathcal{O}$ of Ω^r, a translate $\sigma(\mathcal{O})$, with $\sigma \neq 1$, is disjoint from $\mathcal{O}$.*

Now, a weaker version of the conjecture goes as follows for superlevel sets.

Theorem 3.5 *Let $\mathcal{O} \subset \widetilde{M}$ be a component of the superlevel set Ω^r, and suppose that $\widetilde{B}$ is bounded on $\mathcal{O}$. Then there exist $s > r$ such that Ω^s possesses a relatively compact component contained in $\mathcal{O}$.*

Proof We first prove the statement when b is a Morse form.

By Proposition 3.4, the component $\mathcal{O}$ in the assumption is disjoint from its nontrivial translates.

Take a sequence $\{\widetilde{p_n}\}_{n\in\mathbb{N}}$ of points in $\mathcal{O}$ such that $\widetilde{B}(p_1) = r + \varepsilon_0$, with $\varepsilon_0 > 0$ small enough, and $\widetilde{B}(p_n) \nearrow K \doteq \sup_{\mathcal{O}} \widetilde{B}(t)$. Set $p_n \doteq \Pi(\widetilde{p_n})$; we may suppose that $p_n \to q \in M$. Choose a sufficiently small connected neighborhood N of q evenly covered by Π such that if $\widetilde{x}$ and $\widetilde{y}$ belong to a component of $\Pi^{-1}(N)$ (thus isometric to N), then $|\widetilde{B}(\widetilde{x}) - \widetilde{B}(\widetilde{y})| < \varepsilon_0$. This shows that the components of $\Pi^{-1}(N)$ which intercept $\{\widetilde{p_n}\}_{n\in\mathbb{N}}$ are in $\mathcal{O}$.

Note that for any $\widetilde{q}_1, \widetilde{q}_2 \in \Pi^{-1}\{q\}$, there is some $\sigma \in \mathsf{D}$ such that $\sigma(\widetilde{q}_1) = \widetilde{q}_2$. Hence, only one component of $\Pi^{-1}(N)$ intercepts $\{\widetilde{p_n}\}_{n\in\mathbb{N}}$. Denote by q^* the point in this component such that $\Pi(q^*) = q$. Therefore, $\widetilde{B}(q^*) = K$.

This means that q^* is a Morse critical point on which the maximum value K of $\widetilde{B}$ on $\mathcal{O}$ is attained. It is plain that for values $s < K$, sufficiently close to K, Ω^s possesses a relatively compact component contained in $\mathcal{O}$.

Now we consider that b is a smooth closed form on M. We have a sequence of closed Morse forms $\{b_k\}_{k\in\mathbb{N}}$ on M such that

$$\|\widetilde{B}_k(t) - \widetilde{B}(t)\|_\infty < 1/k\,,$$

where $d\widetilde{B}_k = b_k$.

In fact, b_k can be chosen in a way that $b_k - b$ is exact (see, for instance, [BPZZ]).

Therefore, if $\mathcal{O} \subset \widetilde{M}$ is a component of the superlevel set Ω^r and $K \doteq \sup_{\mathcal{O}} \widetilde{B}(t)$, we choose $k > \frac{2}{K-r}$, and then inside $\mathcal{O}$ there is a nonempty component of

$$\left\{t \in \widetilde{M} : \widetilde{B}_k(t) > s\right\}$$

for $r + 1/k < s < K - 1/k$, which in turn contains a relatively compact component $\mathcal{O}(k)$ of a superlevel set of the Morse function $\widetilde{B}_k$, say $\Omega^{s'}$, with $s < s'$, by the previous argument. As $\mathcal{O}(k) \subset \mathcal{O}$ and the points $t \in \mathcal{O}(k)$ are such that

$$\widetilde{B}(t) > s' + 1/k\,,$$

and $s' - 1/k > r$, we have a relatively compact component of a semilevel set of $\widetilde{B}$. □

4 Construction of Transversal Curves

The topological results in [Ar] have played an important role, and one crucial point is a lemma that shows the existence of a nontrivial loop in the torus that does not meet the singular set of b (see also [BK]). We extended in [HZ2] these results to orientable surfaces (see Lemma 4.1 below).

A decisive tool in the original proof of such Arnol'd loop lemma for the torus is a result ([Ar], Lemma 4) that was stated without a complete proof. For the sake of completeness, in [HZ2] there is also a more detailed proof of that result for surfaces.

Lemma 4.1 *Suppose that a* 1*-form* $b^{\dagger}$ *is defined on a surface* $M^{\dagger}$ *(possibly with boundary components). Assume that the boundary consists of piecewise smooth curves not in the singular set of* $b^{\dagger}$ *and that the integral of* $b^{\dagger}$ *along them is zero. If* $b^{\dagger}$ *is not exact, then there exists a nontrivial loop* γ *in* $M^{\dagger} \setminus \Sigma(b^{\dagger})$.

We then obtained, from M and b, a surface $M^{\sharp}$ with boundary consisting of piecewise smooth curves such that:

- An exact 1-form $b^{\sharp}$ is defined on $M^{\sharp}$.
- $\partial M^{\sharp}$ does not intersect the critical set of $b^{\sharp}$.

It is then plain to assume that the restriction $B^{\sharp}\restriction_{\partial M^{\sharp}}$ has finitely many critical points J.

It should be interesting to notice that there is a diffeomorphism between $M^{\sharp}\setminus\partial M^{\sharp}$ and the open subset $M \setminus X$ of M, where X is a finite collection of loops in M (they are curves transversal to the foliation defined by b, except on J).

We now move on the proof of Theorem 3.3 when M is a closed orientable surface of genus $g > 1$.

Suppose that a certain semilevel set, say Ω^{r}, has two distinct unbounded components $\mathcal{O}$ and $\mathcal{O}'$. We know from Theorem 3.1 that $\widetilde{B}$ is bounded on one of them, say $\mathcal{O}$.

It is known that there is a compact polygon of $4g$ sides that can be pairwise identified in a way to yield M. Such a polygon P is called a *fundamental domain*. Moreover, there exist $2g$ transformations generating a group (isomorphic to $\pi_1(M, p_0)$) that, by acting on P, originate distinct and non-overlapping polygons for distinct elements, exhausting the universal covering space.

We then have a sequence of points $\widetilde{t}_n \in \widetilde{M}$, each of which living in a different fundamental domain $\sigma_n(\mathcal{K})$, where $\mathcal{K}$ is the projection of P in $\widetilde{M}$.

Also, there is a path in $\mathcal{O}$ connecting $\widetilde{t}_n$ to $\widetilde{t}_{n+1}$. The projections under Π of these paths can be then lifted to paths γ_n in $M^{\sharp}$.

The proof then follows at once after the next lemma.

Lemma 4.2 *Each path* γ_n *intercepts* $\partial M^{\sharp}$.

Proof In fact, t_n and t_{n+1} are distinct points in M lifted to distinct points in $M^{\sharp}$. Consider $\widetilde{s}_n$ a translate of $\widetilde{t}_n$ in $\sigma_{n+1}(\mathcal{K})$ and a path ς in $\sigma_{n+1}(\mathcal{K})$ connecting them.

The path in $\mathcal{O}$ connecting $\widetilde{t}_n$ to $\widetilde{t}_{n+1}$ followed by ς is projected on a closed path τ in M. If γ_n does not cross $\partial M^\sharp$, the lift of τ will be a closed path in $M^\sharp$, which would imply that $\widetilde{B}(\widetilde{t}_n) - \widetilde{B}(\widetilde{s}_n) = 0$, an absurd. □

Now, each path γ_n will furnish us distinct critical points of $B^\sharp \restriction \partial M^\sharp$. This is a consequence of Proposition 3.4. This means that the intersection occurs a finite number of times and finishes the proof.

A question if similar arguments hold for higher dimensional manifolds is raised by Arnol'd.

5 An Application of a Novikov Type Inequality

The goal of this section will be to make use of an important result due by M. Farber, which generalizes a Novikov inequality. This study arised in order to extend the Morse theory but to the framework of forms. All the details involving such study can be found in [Fa].

Theorem 5.1 *Suppose that b is a closed Morse* 1*-form on M with indexes* $c_k(b)$. *Then, for every* $j \geqslant 0$, *the following inequality holds:*

$$\sum_{i=0}^{j} c_{j-i}(b) \geqslant q_j([b]) + \sum_{i=0}^{j} \beta_{j-i}([b]) .$$

Let us take a moment to set the notation used above.

The numbers $\beta_k([b])$ and $q_k([b])$ are known as *Novikov-Betti numbers* and *Novikov torsion numbers*, respectively. They are positive as are rank of certain groups. Here $[b]$ denotes a class of de Rham cohomology in $H^1(M, \mathbb{R})$.

If $\xi = 0$, then $\beta_k(\xi)$ is indeed the *Betti number* of M. An Euler-Poincaré type theorem also holds for *Novikov-Betti numbers*. Here we emphasize the following:

Proposition 5.2 ([Fa], Corollary 1.33) *If* $\xi \neq 0$, *then* $\beta_0(\xi) = 0$.

5.1 *Local Solvability Versus Global Solvability*

In the paper [HZ], we were interested in the global solvability of the differential operator $\mathbb{L} : C^\infty(M \times \mathbb{S}^1) \to \bigwedge^1 C^\infty(M \times \mathbb{S}^1)$ defined by

$$\mathbb{L}u = d_t u + ib(t) \wedge \partial_x u, \tag{3}$$

where x belongs to the unit circle $\mathbb{S}^1$ and $d_t : C^\infty(M) \to \bigwedge^1 C^\infty(M)$ is the exterior derivative on M.

We say that $\mathbb{L}$ is (smoothly) locally solvable at $p = (t, x) \in M \times \mathbb{S}^1$ if every neighborhood U of p contains a neighborhood V of p such that for every $f \in \mathfrak{U}^1(U)$ satisfying $\mathbb{L}^1 f = 0$ there exists $u \in C^\infty(V)$ satisfying $\mathbb{L}u = f$ in V, where the space $\mathfrak{U}^1(U)$ is obtained by taking an appropriate quotient on the space of 1-forms on $M \times \mathbb{S}^1$.

Note that if $\mathbb{L}$ is elliptic at p, it is fairly easy to prove that it is locally solvable at p.

The relation between local and global solvabilities is not immediate. Each concept involves a different type of compatibility conditions (see the corresponding references for more details).

Nevertheless, for the case when b is a non-exact Morse form, one can compare local and global solvabilities for $\mathbb{L} = d_t + ib(t) \wedge \partial_x$.

Suppose then that the closed non-exact 1-form b has only nondegenerate critical points, that is, we have a finite and non-empty set $\mathcal{C} = \{p_k\}$.

The local solvability at each non-elliptic point (p_k, x) is characterized by the results of Treves in [Tr]:

Theorem 5.3 ([Tr]) *The operator $\mathbb{L}$ is locally solvable at (p_k, x) if and only if the index of the critical point p_k is neither* 1 *nor* $n - 1$.

The global solvability, in turn, is characterized by the results in [HZ]:

Theorem 5.4 ([HZ]) *The operator $\mathbb{L}$ is globally solvable if and only if, for each k, the index of the critical point p_k is neither* 0 *nor* n.

And here is the mentioned application:

Theorem 5.5 *If b is not exact and the operator $\mathbb{L}$ is locally solvable at each (p_k, x), then $\mathbb{L}$ is globally solvable.*

Proof We are going to apply Theorem 5.1 for $j = 1$. Therefore,

$$c_1(b) - c_0(b) \geqslant q_1([b]) + \beta_1([b]) - \beta_0([b]) .$$

The right-hand side is positive as, recall, $q_1([b]), \beta_1([b]) \geqslant 0$, and, by Proposition 5.2, $\beta_0([b]) = 0$ since b is not exact.

This implies that $c_1(b) \geqslant c_0(b)$. Now if we suppose that $\mathbb{L}$ is not globally solvable, then $c_0(b) > 0$. Hence

$$c_1(b) > 0 ,$$

and by Theorem 5.3, we have that $\mathbb{L}$ will not be locally solvable at some (p_j, x). □

The converse is not true, since [HZ] also shows an example of a globally solvable system on $\mathbb{T}^4$ but locally solvable at (p_j, x), for some j, and not locally solvable at (p_k, x), for some k.

As a final remark, we know that the global and local solvability notions can be carried to the degrees of the differential complex defined by $\mathbb{L}$. In particular, there is an intrinsic relation between overdetermined systems of gradient type as $\mathbb{L}$ and underdetermined systems of divergent type as $\mathbb{L}^{n-1}$.

It turns out that $\mathbb{L}$ is locally (respectively, globally) solvable if and only if $\mathbb{L}^{n-1}$ is (again with corresponding compatibility conditions). Therefore, we gain the following:

Theorem 5.6 *If b is not exact and the operator $\mathbb{L}^{n-1}$ is locally solvable at each (p_k, x), then $\mathbb{L}^{n-1}$ is globally solvable.*

This leads to another question: Is there any analog relation between the local and global solvabilities for the operators in the intermediate degrees of the complex?

Acknowledgments The author thanks FAPESP for the financial support (grant 2021/00693-2).

References

[Ar] ARNOL'D, V. I., *Topological and ergodic properties of closed 1-forms with incommensurable periods*, FUNKTSIONAL. ANAL. I PRILOZHEN. 2 (1991), 1–12.
[BK] BERGAMASCO, A. AND KIRILOV, A., *Global solvability for a class of overdetermined systems*, J. FUNCT. ANAL. 2 (2007), 603–629.
[BPZZ] BERGAMASCO, A. ET AL., *Classes of globally solvable involutive systems*, J. PSEUDO-DIFFER. OPER. APPL. 8 (2017), 551–583.
[FA] FARBER, M., *Topology of closed one-forms*, AMERICAN MATHEMATICAL SOCIETY (2004).
[HZ] HOUNIE, J. AND ZUGLIANI, G., *Global solvability of real analytic involutive systems on compact manifolds*, MATH. ANN. 369 (2017), 1177–1209.
[HZ2] HOUNIE, J. AND ZUGLIANI, G., *Tube structures of co-rank 1 with forms defined on compact surfaces*, J. GEOM. ANAL. 31 (2021), 2540–2567.
[TR] TREVES, F., *Study of a model in the theory of complexes of pseudodifferential operators*, ANN. OF MATH. 104 (1976), 269–324.

SPRINGER NATURE

GPSR Compliance

The European Union's (EU) General Product Safety Regulation (GPSR) is a set of rules that requires consumer products to be safe and our obligations to ensure this.

If you have any concerns about our products, you can contact us on ProductSafety@springernature.com

In case Publisher is established outside the EU, the EU authorized representative is:

Springer Nature Customer Service Center GmbH
Europaplatz 3
69115 Heidelberg, Germany

The manufacturer's authorised representative in the EU is Springer Nature Customer Service Centre GmbH, Europaplatz 3, 69115 Heidelberg, Germany. If you have any concerns regarding our products, please contact ProductSafety@springernature.com

Printed and bound by CPI Group (UK) Ltd, Croydon, CR0 4YY
30/09/2026
02225584-0011